Analysis of Messy Data

VOLUME I: DESIGNED EXPERIMENTS

Analysis of Messy Data

VOLUME I: DESIGNED EXPERIMENTS

George A. Milliken
Dallas E. Johnson

CHAPMAN & HALL
London · Weinheim · New York · Tokyo · Melbourne · Madras

Published by Chapman & Hall, 2–6 Boundary Row, London SE1 8HN, UK

Chapman & Hall, 2–6 Boundary Row, London SE1 8HN, UK

Chapman & Hall GmbH, Pappelallee 3, 69469 Weinheim, Germany

Chapman & Hall USA, 115 Fifth Avenue, New York, NY 10003, USA

Chapman & Hall Japan, ITP-Japan, Kyowa Building, 3F, 2-2-1 Hirakawacho, Chiyoda-ku, Tokyo 102, Japan

Chapman & Hall Australia, 102 Dodds Street, South Melbourne, Victoria 3205, Australia

Chapman & Hall India, R. Seshadri, 32 Second Main Road, CIT East, Madras 600 035, India

First edition 1992
Reprinted 1996 (twice), 1997

Printed in Great Britain by St Edmundsbury Press Ltd, Bury St Edmunds, Suffolk

ISBN 0 412 99081 4

Library of Congress Catalog Card Number 84-839

Contents

Preface

PURPOSE

Many People who analyze data generally use procedures suitable for analyzing "nice" data sets but are often unaware of methods for analyzing "messy" data sets. Typically, experimental data involve missing observations, outliers, or failures of the usual assumptions that render standard statistical methodology useless. Analyzing messy data with techniques developed for nice data can produce results that are often meaningless or uninterpretable, and many existing computing packages do not always yield correct results. Unfortunately, experimenters are frequently ignorant of these consequences.

The basic purpose of this book is to present several techniques and methods for analyzing nonstandard, or messy, data sets effectively and correctly. It is our aim to show experimenters which techniques are suitable and to help experimenters interpret and analyze the information contained in messy data sets.

Many of the methods and techniques discussed in this book are not currently available in other books. The topics presented are among those that we, as statistical consultants, have found most helpful in making sense of messy data. Many of these topics also seem to be among the least understood by many people who analyze data.

PRINCIPAL METHODS AND TECHNIQUES

This book primarily discusses the analysis of messy data that arise from *experimental design* situations. Many of the techniques and methods described can also be used to analyze data collected in other ways, such as through sample surveys and clinical trials. Users of this book will learn the following:

1. The advantages of a means model over an effects model
2. How to determine when the experimental error variance is not constant

3. How to use multiple comparisons effectively and how to choose an appropriate multiple comparison method
4. How to recognize the experimental design or sampling design used to collect the data
5. How to analyze experiments having interaction (particularly those with missing data), how to determine whether there is interaction, how to determine whether all treatment combinations interact or just a few, and how to use this information to obtain the best analysis of the data
6. The differences among the Type I through Type IV analyses for unbalanced experiments, when each type of analysis is appropriate, and the types of analyses given by SAS®*, SPSS, and BMD routines
7. How to analyze very large data sets effectively and economically using existing statistical computing packages
8. How to recognize and correctly analyze random effects models using existing statistical computing packages and how to make these packages compute expected mean squares
9. How to recognize and correctly analyze mixed models, with emphasis on split-plot designs, strip-plot designs, and repeated measures designs
10. When to use and how to analyze crossover designs
11. How to analyze split-plot and repeated measures experiments with unbalanced and missing data
12. How to analyze repeated measures experiments using multivariate methods.

APPROACH

Analysis of Messy Data strives to cover each topic from a practical viewpoint, emphasizing the implementation of the methods much more than the theory behind those methods. Many real-world examples with data are discussed so that the reader can appreciate the techniques more completely. Several of the techniques are presented by examples rather than by detailed technical descriptions. Our basic method is to teach by example and to appeal to the reader's intuition, but at the same time we want the reader to have faith in and believe in the results and techniques discussed. Consequently, references are included for those readers wishing to study the theoretical aspects of the results in more detail. Formulas are also included for those wanting to program mini- or microcomputers to do the necessary computations.

INTENDED AUDIENCE

This book is intended for everyone who analyzes data. We assume that the reader has had the equivalent of a course in analysis of variance as well as some experience in analyzing data. Two statistical methods courses should provide an adequate background for using most of this work. A few chapters of the book (6, 18, 22, 29, and 30) are fairly

*SAS® is the registered trademark of SAS Institute Inc., Cary, NC.

technical and can be skipped by most readers; however, readers who want a deeper understanding of the complexities involved or want the formulas to use on mini- and microcomputers should read those chapters.

BACKGROUND MATERIAL

While the book primarily describes techniques for analyzing messy data, some sections provide elementary descriptions of some of the basic theories underlying many of the techniques. Specifically, Chapters 1, 3, 6, and 7 are intended to provide a basic foundation for the material covered in the remainder of the book. These chapters cover the necessary computations for the basic one-way analysis of variance model, a discussion of multiple comparison methods for the one-way model, and the use of matrices to represent models. Readers well versed in statistics may want to skip these chapters, although they may wish to scan them in order to acquaint themselves with the notation used in this book. Also, the three sections written primarily for SAS® users, Sections 10.3, 10.5, and 28.2.5, can be skipped by other readers.

SPECIAL FEATURES

Analysis of Messy Data contains many special features, including the following:

1. Several tables that are not generally available elsewhere
2. A unique, simplified approach to experimental design involving treatment structure and design structure
3. A discussion of estimable functions for effects models with interpretations corresponding to each of the Type I through Type IV analyses
4. Formulas given in matrix form that can easily be used by those wanting to write their own programs for small computers
5. Lots of examples with data that will help readers recognize the need for a particular method and show how the method can be applied to their own situation
6. Worked-out examples that provide benchmark computations for readers who have their own software.

USE OF STANDARD STATISTICAL COMPUTING PACKAGES

The analysis of messy data is not always easy, and techniques used in the analysis of messy data generally involve large numbers of complicated computations. Many of the techniques presented in this book can easily be implemented via several existing statistical computing packages, such as SAS®, BMD, and SPSS. We have often used SAS® for our computations and sometimes include the control information necessary to analyze data sets with SAS®. However, this book is written for all analyzers of data, not just SAS® users. Where there are differences among available packages, comparisons are made.

ACKNOWLEDGMENTS

We want to express appreciation to all who helped us with the development of this book. We are particularly grateful to the graduate students in the department of statistics at Kansas State University who have taken our course, Analysis of Messy Data, and to those professional researchers who have taken the same course through the Institute of Professional Education. These students have provided numerous valuable suggestions which have greatly improved the content and readability of the book. Thanks are due to Mr. Alexander Kugushev and several anonymous reviewers for many helpful suggestions. We also extend thanks to Mrs. Betty Skidmore and Mrs. Sharon Luce for many hours spent typing the manuscript. Though we have had much assistance in writing and editing this work, any errors that remain are solely our responsibility.

We are both thankful for having had Dr. Franklin A. Graybill as our major professor, as his enthusiasm for statistics and the application of statistical techniques to difficult problems was passed on to us. Finally, we would like to thank our wives, Janet and Erma, and our children, Scott, April, Kelly, and Mark, for all the help and support they provided during this endeavor.

George A. Milliken

Dallas E. Johnson

Analysis of Messy Data

VOLUME I:
DESIGNED EXPERIMENTS

1

The Simplest Case—One-Way Treatment Structure in a Completely Randomized Design Structure with Homogeneous Errors

CHAPTER OUTLINE

Suppose an experimenter wants to compare the effects of several different treatments, such as the effects of different drugs on people's heart rates or the yields of several different varieties of wheat. Often the first step in analyzing the data from such experiments is to use a statistical model, known as a one-way analysis of variance model, to describe the data.

The one-way analysis of variance model is one of the most useful models in the field of statistics. Many experimental situations are simply special cases of this model. Other models that appear to be much more complicated can often be considered as one-way models.

This chapter is divided into several sections. In the first two sections, the one-way model is defined and the estimation of its parameters is discussed. In Sections 1.3 and 1.4, inference procedures for specified linear combinations of the treatment effects are provided. In Sections 1.5 and 1.6, we introduce two basic methods for developing test statistics. These two methods are used extensively throughout the remainder of the book. Finally, in Section 1.7, we discuss readily available computer analyses that use the above techniques.

1.1 MODEL DEFINITIONS AND ASSUMPTIONS

Suppose that a sample of N experimental units is selected completely at random from a population of possible experimental units. An experimental unit is defined as the basic unit to which a treatment will be applied. A more complete description of experimental units can be found in Chapters 4 and 5.

In order to compare the effects of t different treatments, the sample of N units is randomly divided into t groups so that there are n_i experimental units in the ith group, $i = 1, 2, \ldots, t$, and $N = \Sigma_{i=1}^{t} n_i$. Grouping the experimental units at random into t groups should remove any systematic biases. That is, randomness should ensure that the t groups of experimental units are similar in nature before the treatments are applied. Finally, one of the t treatments is randomly assigned to each group.

Let y_{ij} denote a response from the (i, j)th experimental unit to the ith treatment. We can think of $y_{11}, y_{12}, \ldots, y_{1n_1}$ as being a random sample of size n_1 from a population with mean μ_1 and variance σ_1^2, and $y_{21}, y_{22}, \ldots, y_{2n_2}$ as being a random sample of size n_2 from another population with mean μ_2 and variance σ_2^2, and similarly for $i = 3, 4, \ldots, t$. The parameter μ_i represents the population mean if one applied treatment i to the population of all experimental units.

We first consider the simplest case where $\sigma_1^2 = \sigma_2^2 = \cdots = \sigma_t^2 = \sigma^2$ (say). That is, the application of the treatments to the experimental units may affect the mean of the response from the ith treatment but not the variance.

The basic objectives of a good statistical analysis are to estimate the parameters and to make inferences about them. The methods of inference usually include testing hypotheses and constructing confidence intervals.

There are several ways to write a model for the above situation. The one that we use first is called the μ_i model, or a means model. The model is:

$$y_{ij} = \mu_i + \varepsilon_{ij} \qquad i = 1, 2, \ldots, t, \quad j = 1, 2, \ldots, n_i,$$

where it is assumed that

$$\varepsilon_{ij} \sim \text{i.i.d. } N(0, \sigma^2) \qquad i = 1, 2, \ldots, t, \quad j = 1, 2, \ldots, n_i. \quad (1.1.1)$$

The notation $\varepsilon_{ij} \sim$ i.i.d. $N(0, \sigma^2)$ is used extensively throughout this book. It means that the ε_{ij}'s, $i = 1, 2, \ldots, t$; $j = 1, 2, \ldots, n_i$, are independently and identically distributed and that the sampling distribution of each ε_{ij} is the normal distribution with mean equal to zero and variance equal to σ^2.

1.2 PARAMETER ESTIMATION

The most important aspect of a statistical analysis is to get a good estimate of the error variance per experimental unit, namely σ^2. The error variance measures the accuracy of an experiment—the smaller σ^2, the more accurate the experiment. One cannot make any statistically valid inferences in any experiment without some knowledge of the experimental error variance.

In the above situation, the ith sample, $i = 1, 2, \ldots, t$, provides an estimate of σ^2 when $n_i > 1$. That is,

$$\hat{\sigma}_i^2 = \sum_{j=1}^{n_i} \frac{(y_{ij} - \bar{y}_{i\cdot})^2}{n_i - 1}$$

is an unbiased estimate of σ^2 where

$$\bar{y}_{i\cdot} = \frac{\Sigma_j y_{ij}}{n_i}.$$

We say that the estimate $\hat{\sigma}_i^2$ is based on $n_i - 1$ degrees of freedom, since the sampling distribution of $(n_i - 1)\hat{\sigma}_i^2/\sigma^2$ is a chi-square distribution with $n_i - 1$ degrees of freedom.

A weighted average of these t independent estimates of σ^2 gives the best estimate for σ^2 possible for this situation; each estimate is weighted by its corresponding degrees of freedom. The best estimate of σ^2 is

$$\hat{\sigma}^2 = \sum_{i=1}^{t} (n_i - 1)\hat{\sigma}_i^2 \Big/ \sum_{i=1}^{t} (n_i - 1).$$

For computational purposes, we note that

$$\sum_{j=1}^{n_i} (y_{ij} - \bar{y}_{i\cdot})^2 = \sum_j y_{ij}^2 - n_i \bar{y}_{i\cdot}^2 = \sum_j y_{ij}^2 - \frac{(y_{i\cdot})^2}{n_i} = \text{SS}_i \text{ (say)}$$

where $y_{i\cdot} = \Sigma_j y_{ij}$, and that

$$\hat{\sigma}^2 = \frac{SS_1 + SS_2 + \cdots + SS_t}{n_1 + n_2 + \cdots + n_t - t} = \frac{\Sigma_i SS_i}{N - t}.$$

It is also true that $\hat{\sigma}^2$ is an estimate of σ^2 that is based on $N - t$ degrees of freedom and that the sampling distribution of $(N - t)\hat{\sigma}^2/\sigma^2$ is a chi-square distribution with $N - t$ degrees of freedom; that is,

$$(N - t)\hat{\sigma}^2/\sigma^2 \sim \chi^2(N - t).$$

The best estimate of each μ_i is

$$\hat{\mu}_i = \bar{y}_{i\cdot} \qquad i = 1, 2, \ldots, t.$$

Under the assumption given in (1.1.1), the sampling distribution of $\hat{\mu}_i$ is normal with mean μ_i and variance σ^2/n_i. That is,

$$\hat{\mu}_i \sim N\left(\mu_i, \frac{\sigma^2}{n_i}\right) \qquad i = 1, 2, \ldots, t. \tag{1.2.1}$$

Also,

$$T_i = \frac{\hat{\mu}_i - \mu_i}{\sqrt{\hat{\sigma}^2/n_i}} \sim t(N - t) \qquad i = 1, 2, \ldots, t. \tag{1.2.2}$$

That is, the sampling distribution of T_i is the t-distribution with $N - t$ degrees of freedom. In addition, $\hat{\mu}_1, \hat{\mu}_2, \ldots, \hat{\mu}_t$ and $\hat{\sigma}^2$ are statistically independent.

1.3 INFERENCES ON LINEAR COMBINATIONS—TESTS AND CONFIDENCE INTERVALS

This section provides tests of hypotheses and confidence intervals for functions of the parameters in the means model. The results in the previous section can be used to test hypotheses about the individual μ_i's. Those results can also be used to test hypotheses about linear combinations of the μ_i's or to construct confidence intervals for linear combinations of the μ_i's.

For this kind of an experiment, the investigator will often want to compare the effects of the different treatments or, equivalently, the means of the different treatment populations. For example, the experimenter may want to test the following kinds of hypotheses:

$$H_{01}: \sum c_i \mu_i = a$$

for some given set of coefficients $c_1, c_2, \ldots, c_t$ and for some given constant a,

$$H_{02}: \mu_1 = \mu_2 = \cdots = \mu_t, \qquad \text{and}$$

$$H_{03}: \mu_i = \mu_{i'} \qquad \text{for some } i \neq i'.$$

For H_{01} it is true that

$$\frac{\sum c_i\hat{\mu}_i - \sum c_i\mu_i}{\sqrt{\hat{\sigma}^2\sum(c_i^2/n_i)}} \sim t(N-t). \tag{1.3.1}$$

This result can be used to make inferences about $\Sigma c_i\mu_i$. Since the hypothesis in H_{03} can be written as H_{03}: $\mu_i - \mu_i' = 0$, it is a special case of H_{01} with $c_i = 1$, $c_i' = -1$, and $c_k = 0$ if $k \neq i$ or i'. A test for H_{02} is given in Section 1.5.

The estimated standard error of $\Sigma c_i\hat{\mu}_i$ is given by

$$\widehat{\text{s.e.}}\left(\sum c_i\hat{\mu}_i\right) = \sqrt{\hat{\sigma}^2\sum\frac{c_i^2}{n_i}}\,. \tag{1.3.2}$$

To test $\Sigma c_i\mu_i = a$, one must calculate

$$t_c = \frac{\Sigma c_i\hat{\mu}_i - a}{\widehat{\text{s.e.}}\left(\sum c_i\hat{\mu}_i\right)}. \tag{1.3.3}$$

If $|t_c| > t_{\alpha/2,\nu}$, where $\nu = N - t$, then H_0 is rejected at the $\alpha \cdot 100\%$ significance level, where $t_{\alpha/2,\nu}$ is the upper $\alpha/2$ critical point of a t-distribution with ν degrees of freedom. A $(1-\alpha)\,100\%$ confidence interval for $\Sigma c_i\mu_i$ is provided by

$$\sum c_i\hat{\mu}_i \pm t_{\alpha/2,\nu}\cdot\widehat{\text{s.e.}}\left(\sum c_i\hat{\mu}_i\right). \tag{1.3.4}$$

EXAMPLE 1.1

The data in Table 1.1 came from an experiment that was conducted to determine how six different kinds of work tasks affect a worker's pulse rate. In this experiment, 78 male workers were assigned at random to six different groups so that there were 13 workers in each group. Each group of workers was trained to perform its assigned task. On a selected day after training, the pulse rates of the workers were measured after the workers had performed their assigned tasks for one hour. Unfortunately, some individuals withdrew from the experiment during the training process so that some groups contained fewer than 13 individuals. The recorded data represent the number of pulsations in 20 seconds.
Also note that $Y_{..} = 2{,}197$ and $N = 68$.

For this data, the best estimate of σ^2 is

$$\hat{\sigma}^2 = \frac{\Sigma \text{SS}_i}{N-t} = \frac{1{,}916.0761}{62} = 30.9045$$

Table 1.1 Pulsation Data

	TASK					
	1	*2*	*3*	*4*	*5*	*6*
	27	29	34	34	28	28
	31	28	36	34	28	26
	26	37	34	43	26	29
	32	24	41	44	35	25
	39	35	30	40	31	35
	37	40	44	47	30	34
	38	40	44	34	34	37
	39	31	32	31	34	28
	30	30	32	45	26	21
	28	25	31	28	20	28
	27	29			41	26
	27	25			21	
	34					
$Y_{i.}$	415	373	358	380	354	317
n_i	13	12	10	10	12	11
$\bar{Y}_{i.}$	31.923	31.083	35.8	38.0	29.5	28.818
SS_i	294.9231	352.9167	253.6	392.0	397.0	225.6364

and is based on 62 degrees of freedom. The best estimates of the μ_i's are

$$\hat{\mu}_1 = 31.923, \quad \hat{\mu}_2 = 31.083, \quad \hat{\mu}_3 = 35.8, \quad \hat{\mu}_4 = 38.0,$$
$$\hat{\mu}_5 = 29.5, \quad \text{and} \quad \hat{\mu}_6 = 28.818.$$

Exercise 1.1

To further illustrate the results obtained, we shall address each of the following:

a. Test H_0: $\mu_3 = 30$ versus H_a: $\mu_3 \neq 30$.

b. Find a 95% confidence interval for μ_1.

c. Test H_0: $\mu_4 = \mu_5$ versus H_a: $\mu_4 \neq \mu_5$.

d. Test H_0: $\mu_1 = (\mu_2 + \mu_3 + \mu_4)/3$ versus H_a: $\mu_1 \neq (\mu_2 + \mu_3 + \mu_4)/3$.

e. Obtain a 90% confidence interval for $4\mu_1 - \mu_3 - \mu_4 - \mu_5 - \mu_6$.

Answers

a. A t-statistic for testing H_0: $\mu_3 = 30$ is

$$t_c = \frac{\hat{\mu}_3 - 30}{\widehat{\text{s.e.}}(\hat{\mu}_3)}.$$

Substituting into equation (1.3.3), we find

$$t_c = \frac{35.8 - 30}{\sqrt{30.9045/10}} = 3.30$$

The significance probability of this calculated value of t is

$$\hat{\alpha} = \Pr\{|t| > 3.30\} = .0016$$

where $\Pr\{|t| > 3.30\}$ is the area to the right of 3.30 plus the area to the left of -3.30 in a t-distribution with 62 degrees of freedom. The above value of $\hat{\alpha}$ was obtained from a desk calculator. Users of this book who lack such a calculator should compare $t_c = 3.30$ to $t_{\alpha/2,62}$ for their choice of α.

b. A 95% confidence interval for μ_1 is given by

$$\hat{\mu}_1 \pm t_{.025,62}\widehat{\text{s.e.}}(\hat{\mu}_1) = 31.923 \mp 2.00 \cdot \sqrt{\frac{30.9045}{13}}$$
$$= 31.923 \mp (2.00)(1.542).$$

Thus the 95% confidence interval is

$$28.839 \le \mu_1 \le 35.007.$$

c. To test H_0: $\mu_4 = \mu_5$, we let $l_1 = \mu_4 - \mu_5$. Then $\hat{l}_1 = \hat{\mu}_4 - \hat{\mu}_5 = 38.0 - 29.5 = 8.5$ and

$$\widehat{\text{s.e.}}(\hat{l}_1) = \sqrt{\hat{\sigma}^2 \sum \frac{c_i^2}{n_i}} = \sqrt{30.9045\left(\frac{1}{10} + \frac{1}{12}\right)} = 2.380$$

since $c_1 = c_2 = c_3 = c_6 = 0$, $c_4 = 1$, and $c_5 = -1$.

The t-statistic for testing H_0 is

$$t_c = \frac{8.5}{2.380} = 3.57.$$

The significance probability is $\hat{\alpha} = .0007$.

d. A test of H_0: $\mu_1 = (\mu_2 + \mu_3 + \mu_4)/3$ is equivalent to testing

$$\mu_1 - \frac{1}{3}\mu_2 - \frac{1}{3}\mu_3 - \frac{1}{3}\mu_4 = 0$$

or testing

$$3\mu_1 - \mu_2 - \mu_3 - \mu_4 = 0.$$

We choose the latter form since the calculations are somewhat easier; also, the value of t_c is invariant with respect to a constant multiplier. Thus, we take $l_2 = 3\mu_1 - \mu_2 - \mu_3 - \mu_4$ and hence,

$$\begin{aligned}\hat{l}_2 &= 3\hat{\mu}_1 - \hat{\mu}_2 - \hat{\mu}_3 - \hat{\mu}_4\\ &= 3(31.923) - 31.083 - 35.8 - 38.0 = -9.114.\end{aligned}$$

The estimate of the standard error of $\hat{l}_2$ is

$$\widehat{\text{s.e.}}(\hat{l}_2) = \sqrt{30.9045\left(\tfrac{9}{13} + \tfrac{1}{12} + \tfrac{1}{10} + \tfrac{1}{10}\right)} = 5.491.$$

A t-statistic for testing H_0 is

$$t_c = -\frac{9.114}{5.491} = -1.66.$$

The significance probability is $\hat{\alpha} = .1020$.

e. Let $l_3 = 4\mu_1 - \mu_3 - \mu_4 - \mu_5 - \mu_6$. Then $\hat{l}_3 = -4.426$ and $\widehat{\text{s.e.}}(\hat{l}_3) = 7.0429$. A 90% confidence interval for l_3 is

$$\begin{aligned}\hat{l}_3 \pm t_{.05,62} \cdot \widehat{\text{s.e.}}(\hat{l}_3) &= -4.426 \pm (1.671)(7.043)\\ &= -4.426 \pm 11.769.\end{aligned}$$

Thus, a 90% confidence interval is

$$-16.195 \le 4\mu_1 - \mu_3 - \mu_4 - \mu_5 - \mu_6 \le 7.343.$$

1.4 SIMULTANEOUS TESTS ON SEVERAL LINEAR COMBINATIONS

Suppose we want to test several linear combinations of the treatment effects simultaneously. For example, suppose we wish to test

$$\begin{aligned} H_0\colon c_{11}\mu_1 + c_{12}\mu_2 + \cdots + c_{1t}\mu_t &= a_1,\\ c_{21}\mu_1 + c_{22}\mu_2 + \cdots + c_{2t}\mu_t &= a_2,\\ \vdots \qquad\quad \vdots \qquad\qquad\quad \vdots \quad &\quad \vdots\\ c_{k1}\mu_1 + c_{k2}\mu_2 + \cdots + c_{kt}\mu_t &= a_k. \end{aligned} \tag{1.4.1}$$

The results presented in this section are given by utilizing vectors and matrices. However, a knowledge of vectors and matrices is not really necessary for readers having access to computers, since most computers allow even novice users to compute with matrices very easily.

The hypothesis in (1.4.1) can be written in matrix notation as

$$H_0\colon \mathbf{C\mu} = \mathbf{a} \quad \text{versus} \quad H_a\colon \mathbf{C\mu} \ne \mathbf{a} \tag{1.4.2}$$

where

$$\mathbf{C} = \begin{bmatrix} c_{11} & c_{12} & \cdots & c_{1t}\\ c_{21} & c_{22} & \cdots & c_{2t}\\ \vdots & \vdots & & \vdots\\ c_{k1} & c_{k2} & \cdots & c_{kt}\end{bmatrix}, \quad \boldsymbol{\mu} = \begin{bmatrix}\mu_1\\ \mu_2\\ \vdots\\ \mu_t\end{bmatrix}, \quad \text{and} \quad \mathbf{a} = \begin{bmatrix} a_1\\ a_2\\ \vdots\\ a_t\end{bmatrix} \tag{1.4.3}$$

We assume that the k rows in $\mathbf{C}$ are linearly independent, which means that none of the rows can be expressed as a linear combination of the remaining rows. If the k rows are not linearly independent, they can always be replaced by a subset of rows that are linearly independent and contain all the necessary information for the required calculations.

Denote the vector of sample means by $\hat{\boldsymbol{\mu}}$ and note that in matrix notation,

$$\hat{\boldsymbol{\mu}} \sim N_t(\boldsymbol{\mu}, \sigma^2\mathbf{D}) \qquad \text{where} \quad \mathbf{D} = \begin{bmatrix} 1/n_1 & 0 & \cdots & 0 \\ 0 & 1/n_2 & \cdots & 0 \\ \vdots & \vdots & & \vdots \\ 0 & 0 & \cdots & 1/n_t \end{bmatrix}.$$

This is read as follows: The elements of the vector $\hat{\boldsymbol{\mu}}$ have a sampling distribution that is the t-variate normal distribution with means given by the vector $\boldsymbol{\mu}$ and with variances and covariances given by the matrix $\sigma^2\mathbf{D}$. The ith diagonal element of $\sigma^2\mathbf{D}$ is the variance of $\hat{\mu}_i$, and the (i, j)th $i \neq j$ off-diagonal element gives the covariance between $\hat{\mu}_i$ and $\hat{\mu}_j$.

The sampling distribution of $\mathbf{C}\hat{\boldsymbol{\mu}}$ is

$$\mathbf{C}\hat{\boldsymbol{\mu}} \sim N_k(\mathbf{C}\boldsymbol{\mu}, \sigma^2\mathbf{CDC}').$$

The sum of squares for testing H_0: $\mathbf{C}\boldsymbol{\mu} = \mathbf{a}$ is given by

$$SS_{H_0} = (\mathbf{C}\hat{\boldsymbol{\mu}} - \mathbf{a})'(\mathbf{CDC}')^{-1}(\mathbf{C}\hat{\boldsymbol{\mu}} - \mathbf{a}) \tag{1.4.4}$$

and has k degrees of freedom. That is, $SS_{H_0}/\sigma^2 \sim \chi^2(k)$ if H_0 is true. The test statistic for testing H_0 is

$$F_c = \frac{SS_{H_0}/k}{\hat{\sigma}^2}.$$

The hypothesis H_0: $\mathbf{C}\boldsymbol{\mu} = \mathbf{a}$ is rejected at the α significance level if $F_c > F_{\alpha,k,N-t}$ where $F_{\alpha,k,N-t}$ is the upper α critical point of the F-distribution with k numerator degrees of freedom and $N - t$ denominator degrees of freedom. The result given here is a special case of Theorem 6.3.1 in Graybill (1976).

We note that if H_0 is true, then SS_{H_0}/k is an unbiased estimate of σ^2, which is then compared with $\hat{\sigma}^2$, which in turn is an unbiased estimate of σ^2 regardless of whether H_0 is true or not. Thus the F-statistic given above should be close to 1 if H_0 is true. If H_0 is false, the statistic SS_{H_0}/k is an unbiased estimate of

$$\sigma^2 + \frac{1}{k}(\mathbf{C}\boldsymbol{\mu} - \mathbf{a})'(\mathbf{CDC}')^{-1}(\mathbf{C}\boldsymbol{\mu} - \mathbf{a}).$$

Thus, if H_0 is false, the F-statistic should be larger than 1. The hypothesis H_0 is rejected if the calculated F-statistic is significantly larger than 1.

EXAMPLE 1.2

Consider the following summary information from Example 1.1:

i	1	2	3	4	5	6
$\hat{\mu}_i$	31.923	31.083	35.8	38.0	29.5	28.818
n_i	13	12	10	10	12	11

with $\hat{\sigma}^2 = 30.9045$ with 62 degrees of freedom.

We consider testing

$$H_0\colon \mu_4 - \mu_5 = 4 \quad \text{and} \quad 3\mu_1 - \mu_2 - \mu_3 - \mu_4 = 0.$$

We get

$$\mathbf{C} = \begin{bmatrix} 0 & 0 & 0 & 1 & -1 & 0 \\ 3 & -1 & -1 & -1 & 0 & 0 \end{bmatrix} \quad \text{and}$$

$$\mathbf{D} = \begin{bmatrix} \frac{1}{13} & 0 & 0 & 0 & 0 & 0 \\ 0 & \frac{1}{12} & 0 & 0 & 0 & 0 \\ 0 & 0 & \frac{1}{10} & 0 & 0 & 0 \\ 0 & 0 & 0 & \frac{1}{10} & 0 & 0 \\ 0 & 0 & 0 & 0 & \frac{1}{12} & 0 \\ 0 & 0 & 0 & 0 & 0 & \frac{1}{11} \end{bmatrix}.$$

Thus

$$\mathbf{C}\hat{\boldsymbol{\mu}} - \mathbf{a} = \begin{bmatrix} 8.5 - 4 \\ -9.114 - 0 \end{bmatrix} = \begin{bmatrix} 4.5 \\ -9.114 \end{bmatrix},$$

$$\mathbf{CDC}' = \begin{bmatrix} \frac{1}{10} + \frac{1}{12} & -\frac{1}{10} \\ -\frac{1}{10} & \frac{9}{13} + \frac{1}{12} + \frac{1}{10} + \frac{1}{10} \end{bmatrix} = \begin{bmatrix} .1833 & -.1 \\ -.1 & .9756 \end{bmatrix},$$

$$(\mathbf{CDC}')^{-1} = \begin{bmatrix} 5.778 & .592 \\ .592 & 1.086 \end{bmatrix}, \quad \text{and}$$

$$SS_{H_0} = (\mathbf{C}\hat{\boldsymbol{\mu}} - \mathbf{a})'(\mathbf{CDC}')^{-1}(\mathbf{C}\hat{\boldsymbol{\mu}} - \mathbf{a}) = 158.603$$

with 2 degrees of freedom. The test statistic is

$$F_c = \frac{158.603/2}{30.9045} = 2.57.$$

The significance probability of this F-statistic is $\hat{\alpha} = \Pr\{F > 2.57\} = .0858$.

1.5 COMPARING ALL MEANS

Next we consider the hypothesis H_0: $\mu_1 = \mu_2 = \cdots = \mu_t$ versus H_a: $\mu_i \neq \mu_j$ for some $i \neq j$. We shall examine two basic procedures for developing statistics to test this hypothesis. For the particular situation discussed in this chapter, the two procedures give rise to the same statistical test. However, for most messy data situations, the two procedures give rise to different tests. The first procedure is covered in this section, while the second is introduced in Section 1.6.

If

$$\mathbf{C} = \begin{bmatrix} 1 & -1 & 0 & \cdots & 0 \\ 1 & 0 & -1 & \cdots & 0 \\ \vdots & \vdots & \vdots & & \vdots \\ \vdots & \vdots & \vdots & & \vdots \\ 1 & 0 & 0 & \cdots & -1 \end{bmatrix} \quad \text{and} \quad \mathbf{a} = \begin{bmatrix} 0 \\ 0 \\ \vdots \\ \vdots \\ 0 \end{bmatrix},$$

equation (1.4.4) gives a test statistic for testing H_0: $\mu_1 = \mu_2 = \cdots = \mu_t$. This is true since $\mu_1 = \mu_2 = \cdots = \mu_t$ if and only if $\mu_1 - \mu_2 = 0$, $\mu_1 - \mu_3 = 0, \ldots, \mu_1 - \mu_t = 0$. Many other matrices exist, so that $\mathbf{C}\boldsymbol{\mu} = \mathbf{0}$ if and only if $\mu_1 = \mu_2 = \cdots = \mu_t$; however, all such matrices will produce the same sum of squares for H_0 and the same degrees of freedom, $t - 1$, and hence the same F-statistic. For this special case, equation (1.4.4) always reduces to

$$SS_{H_0:\mu_1=\mu_2=\cdots=\mu_t} = \sum_i n_i(\bar{y}_{i\cdot} - \bar{y}_{\cdot\cdot})^2 = \sum\left(\frac{y_{i\cdot}^2}{n_i}\right) - \frac{y_{\cdot\cdot}^2}{N} \quad (1.5.1)$$

EXAMPLE 1.3

Consider the data in Example 1.1. We shall compute $SS_{H_0:\mu_1=\cdots=\mu_6}$ by two methods. First, we shall use the simplified formula in equation (1.5.1) and then the matrix formula given in equation (1.4.4).

Using the formula in equation (1.5.1), we get

$$SS_{H_0} = \frac{415^2}{13} + \frac{373^2}{12} + \frac{358^2}{10} + \frac{380^2}{10} + \frac{354^2}{12} + \frac{317^2}{11} - \frac{2{,}197^2}{68}$$

$$= 694.4386$$

with $t - 1 = 5$ degrees of freedom.

Next, we calculate this same sum of squares using the formula in equation (1.4.4). First we need $\mathbf{C}\hat{\boldsymbol{\mu}} - \mathbf{a}$ and $\mathbf{CDC}'$ where

$$\mathbf{C} = \begin{bmatrix} 1 & -1 & 0 & 0 & 0 & 0 \\ 1 & 0 & -1 & 0 & 0 & 0 \\ 1 & 0 & 0 & -1 & 0 & 0 \\ 1 & 0 & 0 & 0 & -1 & 0 \\ 1 & 0 & 0 & 0 & 0 & -1 \end{bmatrix}, \quad \mathbf{a} = \begin{bmatrix} 0 \\ 0 \\ 0 \\ 0 \\ 0 \end{bmatrix}, \quad \text{and}$$

$$\mathbf{D} = \begin{bmatrix} \frac{1}{13} & 0 & 0 & 0 & 0 & 0 \\ 0 & \frac{1}{12} & 0 & 0 & 0 & 0 \\ 0 & 0 & \frac{1}{10} & 0 & 0 & 0 \\ 0 & 0 & 0 & \frac{1}{10} & 0 & 0 \\ 0 & 0 & 0 & 0 & \frac{1}{12} & 0 \\ 0 & 0 & 0 & 0 & 0 & \frac{1}{11} \end{bmatrix}.$$

We get

$$\mathbf{C}\hat{\boldsymbol{\mu}} - \mathbf{a} = \begin{bmatrix} .844 \\ -3.877 \\ -6.077 \\ 2.423 \\ 3.105 \end{bmatrix} \quad \text{and} \quad \mathbf{CDC}' = \begin{bmatrix} \frac{25}{156} & \frac{1}{13} & \frac{1}{13} & \frac{1}{13} & \frac{1}{13} \\ \frac{1}{13} & \frac{23}{130} & \frac{1}{13} & \frac{1}{13} & \frac{1}{13} \\ \frac{1}{13} & \frac{1}{13} & \frac{23}{130} & \frac{1}{13} & \frac{1}{13} \\ \frac{1}{13} & \frac{1}{13} & \frac{1}{13} & \frac{25}{156} & \frac{1}{13} \\ \frac{1}{13} & \frac{1}{13} & \frac{1}{13} & \frac{1}{13} & \frac{24}{143} \end{bmatrix}.$$

Next,

$$(\mathbf{CDC}')^{-1} = \begin{bmatrix} 9.882 & -1.765 & -1.765 & -2.118 & -1.941 \\ -1.765 & 8.529 & -1.471 & -1.765 & -1.618 \\ -1.765 & -1.471 & 8.529 & -1.765 & -1.618 \\ -2.118 & -1.765 & -1.765 & 9.882 & -1.941 \\ -1.941 & -1.618 & -1.618 & -1.941 & 9.221 \end{bmatrix}.$$

Finally we get

$$(\mathbf{C}\hat{\boldsymbol{\mu}} - \mathbf{a})'(\mathbf{CDC}')^{-1}(\mathbf{C}\hat{\boldsymbol{\mu}} - \mathbf{a}) = 694.4386.$$

Clearly, this formula is not easy to use if one must do the calculations by hand. However, we shall see that in many messy data situations, formulas such as this one are necessary in order to test meaningful hypotheses. Fortunately, by utilizing computers we can generate our own **C** matrices and hence our own hypotheses and then allow the computer to do the tedious calculations.

1.6 GENERAL METHOD FOR COMPARING TWO MODELS THE PRINCIPLE OF CONDITIONAL ERROR

A second procedure for determining a test statistic compares the fit of two models. In this section, the two models compared are $y_{ij} = \mu_i + \varepsilon_{ij}$, which is the general model, and $y_{ij} = \mu + \varepsilon_{ij}$, the model one would have if H_0: $\mu_1 = \mu_2 = \cdots = \mu_t = \mu$ (say) were true. The first model is called the full model or the unrestricted model, while the second model is called the reduced model or the restricted model.

When comparing two models where one model is obtained by placing restrictions upon the parameters of another model, we shall use a principle known as the *principle of conditional error*. The principle is very simple, requiring that one obtain the residual or error sums of squares for both the full model and the reduced model. We let ESS_F denote the error sum of squares after fitting the full model and ESS_R the error sum of squares after fitting the reduced model. Then the sum of squares due to the restrictions given by the hypothesis is $\text{SS}_{\text{H}_0} = \text{ESS}_\text{R} - \text{ESS}_\text{F}$. The degrees of freedom for both ESS_R and ESS_F are given by the difference between the number of observations made and the number of (essential) parameters estimated (essential parameters will be discussed in Chapter 6). Denote the degrees of freedom corresponding to ESS_R and ESS_F by df_R and df_F, respectively. The degrees of freedom corresponding to SS_{H_0} is $df_{\text{H}_0} = df_\text{R} - df_\text{F}$. An F-statistic for testing H_0 is given by

$$F_c = \frac{\text{SS}_{\text{H}_0}/df_{\text{H}_0}}{\text{ESS}_\text{F}/df_\text{F}}.$$

One rejects H_0 at the α significance level if $F_c > F_{\alpha, df_{\text{H}_0}, df_\text{F}}$.

For the situation being discussed in this chapter,

$$\text{ESS}_\text{F} = \sum_{i=1}^{t} \sum_{j=1}^{n_i} (y_{ij} - \bar{y}_{i\cdot})^2 = (N - t)\hat{\sigma}^2$$

with $df_\text{F} = N - t$, and

$$\text{ESS}_\text{R} = \sum_{i=1}^{t} \sum_{j=1}^{n_i} (y_{ij} - \bar{y}_{\cdot\cdot})^2$$

with $df_\text{R} = N - 1$. Thus,

$$\begin{aligned} \text{SS}_{\text{H}_0: \mu_1 = \mu_2 = \cdots = \mu_t} &= \text{ESS}_\text{R} - \text{ESS}_\text{F} \\ &= \sum n_i (\bar{y}_{i\cdot} - \bar{y}_{\cdot\cdot})^2 \end{aligned}$$

with $t - 1$ degrees of freedom. The sum of squares is the same as that obtained in (1.5.1).

The sum of squares that are of interest in testing situations are often put in a table called an analysis of variance table. Such a table often has a form like that in the table below. The entries under the column "Source of Variation" are grouped into sets. Usually only one of the labels in each set is used, the choice being determined entirely by the experimenter.

Source of Variation	*df*	SS	MS	*F*
Hypothesis ($\mu_1 = \cdots = \mu_t$) Between Samples Treatments	$t-1$	SS_{H_0}	$\frac{SS_{H_0}}{(t-1)}$	$\frac{SS_{H_0}/(t-1)}{\hat{\sigma}^2}$
Error Within Samples	$N-t$	ESS_F	$\hat{\sigma}^2$	

Note: *df* = degrees of freedom; SS = sum of squares; MS = mean squares. These standard abbreviations are used throughout the book.

EXAMPLE 1.4

For the previous data set on pulse rates (Example 1.1), we test

$$H_0: \mu_1 = \mu_2 = \cdots = \mu_6 \text{ versus } H_a: H_0 \text{ is false}$$

by using the principle of conditional error:

$$ESS_F = 1{,}916.076 \quad \text{with 62 degrees of freedom and}$$

$$ESS_R = 73{,}593 - \frac{2{,}197^2}{68} = 2{,}610.545 \quad \text{with 67 degrees of freedom.}$$

Hence,

$$SS_{H_0} = 2{,}610.545 - 1{,}916.076 = 694.439$$

with 67 − 62 = 5 degrees of freedom. An analysis of variance table is given below.

Source of Variation	*df*	SS	MS	*F*	$\hat{\alpha}$
Due to H_0	5	694.439	138.888	4.49	.0015
Error	62	1,916.076	30.9045		

1.7 COMPUTER ANALYSES

We conclude this chapter with some remarks about utilizing computers and statistical computing packages such as SAS® (1982), BMDP (1981), and SPSS (1975, 1981). All of the methods and formulas provided in the preceding sections can easily be used on most minicomputers and many home computers. If the computer utilizes a programming language such as BASIC (with MAT operations) or APL, the required matrix calculations are simple to do by following the matrix formulas we have provided.

SAS®, BMDP, and SPSS each contain procedures that enable users to generate their own linear combinations of treatment means about which to test hypotheses. In addition, these packages all provide an analysis of variance table, treatment means, and their standard errors. See, for example, SAS Institute, Inc. (1982, p. 146), BMDP (1981, pp. 406–408), and SPSS (1975, pp. 425–426) or SPSS (1981, pp. 72–75).

CONCLUDING REMARKS

In this chapter, the one-way analysis of variance model was analyzed. General procedures for making statistical inferences about the effects of different treatments were provided and illustrated for the case of homogeneous errors. Two basic procedures for obtaining statistical analyses of experimental design models were introduced. These procedures are used extensively throughout the remainder of the book for more complex experimental design models and for messier data.

A test for comparing all treatment effect means simultaneously was also given. Such a test may be considered an initial step in a statistical analysis. Which procedures should be used to complete the analysis of a data set depends on whether the hypothesis of equal treatment means is rejected.

2

One-Way Treatment Structure in a Completely Randomized Design Structure with Heterogeneous Errors

CHAPTER OUTLINE

In this chapter we consider the case where the treatments assigned to the experimental units may affect the variance of the responses as well as the mean. Suppose that $Y_{ij} = \mu_i + \varepsilon_{ij}$, as in Chapter 1. In Chapter 1 it was assumed that the experimental errors all had the same variance. That is, the different treatments were expected to change the mean of the population being sampled but not the variance. In this chapter we want to introduce some methods for analyzing data when the treatment affects the variance as well as the mean. The types of questions that the experimenter should want to answer are similar to those in Chapter 1. That is: (1) Are all means equal? (2) Can we compare the means pairwise? (3) Can we test $\Sigma c_i\mu_i = a$ and obtain confidence intervals for $\Sigma c_i\mu_i$?

In this chapter we explore answers to all of these questions. In addition, we consider tests of whether the variances are homogeneous. Before continuing, we note that if $t = 2$, the problem of unequal variances is usually known as the Behrens–Fisher problem. We also note that heterogeneous error variances pose a much more serious problem than nonnormality of the error variances. The procedures in Chapter 1 are robust with respect to nonnormality, but not quite so robust with respect to heterogeneous error variances.

In the analyses previously considered, it was assumed that the population variances were all equal, which is a reasonable assumption in many cases. One method for analyzing data when variances are unequal is simply to ignore the fact that they are unequal and calculate the same F-statistics or t-tests that are calculated in the case of equal variances. Surprisingly perhaps, studies have shown that these usual tests are quite good, particularly if the sample sizes are all equal or almost equal. Also, if the larger sample sizes correspond to the populations with the larger variances, then the usual tests are also quite good. The usual tests are so good, in fact, that many statisticians do not even recommend testing for equal variances.

In the remainder of this chapter, we look at several ways to analyze data when the variances of the populations are unequal and the usual techniques are suspect.

2.1 MODEL DEFINITION AND ASSUMPTIONS

The model being considered is

$$Y_{ij} = \mu_i + \varepsilon_{ij} \qquad i = 1, 2, \ldots, t, \quad j = 1, 2, \ldots, n_i,$$

where $\varepsilon_{ij} \sim$ independent $N(0, \sigma_i^2)$, $i = 1, 2, \ldots, t$, and $j = 1, 2, \ldots, n_i$. The notation "$\varepsilon_{ij} \sim$ independent $N(0, \sigma_i^2)$" means that the errors, ε_{ij}, are all independent, distributed according to the normal distribution, and that the variance of the normal distribution depends on i and hence may be different for each population or treatment applied.

2.2 PARAMETER ESTIMATION

The best estimates of the parameters in the model are:

$$\hat{\mu}_i = \bar{Y}_{i\cdot} \tag{2.2.1}$$

and

$$\hat{\sigma}_i^2 = \sum_j \frac{(Y_{ij} - \bar{Y}_{i\cdot})^2}{n_i - 1} \qquad i = 1, 2, \ldots, t. \tag{2.2.2}$$

We note that

$$\hat{\mu}_i \sim \text{independent } N(\mu_i, \sigma_i^2/n_i) \qquad i = 1, 2, \ldots, t, \tag{2.2.3}$$

and that

$$\frac{(n_i - 1)\hat{\sigma}_i^2}{\sigma_i^2} \sim \text{independent } \chi^2(n_i - 1) \qquad i = 1, 2, \ldots, t. \tag{2.2.4}$$

2.3 TESTS FOR HOMOGENEITY OF VARIANCES

In this section we give four tests for testing H_0: $\sigma_1^2 = \sigma_2^2 = \cdots = \sigma_t^2$. If an experimenter is doubtful about whether the variances are equal, one of the following tests can be performed. If one fails to reject H_0, then one proceeds with an analysis of the data using the techniques given in Chapter 1. If one rejects H_0, then the techniques given in Sections 2.4 and 2.5 of this chapter can be used.

2.3.1 Hartley's *F*-Max Test

The first test that we shall discuss is known as Hartley's *F*-max test. This test requires that all samples be of the same size. That is, $n_1 = n_2 = \cdots = n_t$. The test is based on the statistic

$$F_{\max} = \frac{\max_i\{\hat{\sigma}_i^2\}}{\min_i\{\hat{\sigma}_i^2\}}.$$

Percentage points are given in the Appendix in Table A.1. One rejects H_0 in favor of H_a if $F_{\max} > F_{\max,\alpha,\nu}$ where $\nu = n - 1$. If the n_i's are not all equal, a "liberal" test of H_0 versus H_a can be obtained by taking $\nu = \max_i\{n_i\} - 1$. This test is liberal in the sense that it protects one from doing the usual analysis of variance when there is even a remote chance of it being inappropriate. An example illustrating the use of this test is found in Section 2.3.5.

2.3.2 Bartlett's Test

A second test for testing for homogeneity of variances is Bartlett's test. This test has the advantage of not requiring the n_i's to be equal.

Bartlett's test statistic is

$$U = \frac{1}{C}\left[\nu \log_e(\hat{\sigma}^2) - \sum_i \nu_i \log_e \hat{\sigma}_i^2\right],$$

where

$$\nu_i = n_i - 1, \quad \nu = \sum \nu_i, \quad \hat{\sigma}^2 = \sum \frac{\nu_i \hat{\sigma}_i^2}{\nu}, \quad \text{and}$$

$$C = 1 + \frac{1}{3(t-1)}\left(\sum \frac{1}{\nu_i} - \frac{1}{\nu}\right).$$

The hypothesis of equal variances is rejected if $U > \chi^2_{\alpha,t-1}$.

2.3.3 Box's Test

One of the disadvantages of the preceding two tests for homogeneity of variance is that they are quite sensitive to departures from normality as well as to unequal variances. Box's test and Levene's test (presented in the next subsection) are much more robust in that they are less sensitive to departures from normality but still sensitive to heterogeneous variances.

To conduct Box's test, the sample corresponding to each treatment group is partitioned into subsamples of approximately equal size in a random manner. The variance of each subsample is determined and logs of the subsample variances are obtained. Finally, a one-way analysis of variance on the logs of the variances is conducted; if the F-statistic is significant, then variance homogeneity is rejected. An example illustrating Box's test is given in Section 2.3.5.

2.3.4 Levene's Test

Levene proposed doing a one-way analysis of variance on the variables $Z_{ij} = |y_{ij} - \bar{y}_{i\cdot}|$. If the $F-$ statistic is significant, homogeneity of the variances is rejected. The test is illustrated in the next section.

2.3.5 Examples

We shall now illustrate applications of the four test procedures described above.

EXAMPLE 2.1

The data in Table 2.1 are from a paired-association learning task experiment performed on subjects under the influence of two drugs. Group 1 is a control group (no drug), Group 2 was given drug 1, Group 3 was given drug 2, and Group 4 was given both drugs.

Hartley's F-max statistic is $F_{\max} = 16.286/1.867 = 8.723$. The liberal 5% critical point is obtained from Table A.1 with $k = t = 4$ and $\nu = 7$.

Table 2.1 Data from Paired-Association Learning Task Experiment

	Group 1 (No Drug)	*Group 2 (Drug 1)*	*Group 3 (Drug 2)*	*Group 4 (Both Drugs)*
	1	12	12	13
	8	10	4	14
	9	13	11	14
	9	13	7	17
	4	12	8	11
	0	10	10	14
	1		12	13
			5	14
n_i	7	6	8	8
$\sum_j Y_{ij}$	32	70	69	110
$\bar{Y}_{i\cdot}$	4.571	11.667	8.625	13.750
$\hat{\sigma}_i^2$	16.286	1.867	9.696	2.786

The critical point is 8.44. Since 8.723 > 8.44, H_0: $\sigma_1^2 = \sigma_2^2 = \sigma_3^2 = \sigma_4^2$ is rejected.

For Bartlett's test we get

$$C = 1 + \frac{1}{3 \cdot 3}\left(\frac{1}{6} + \frac{1}{5} + \frac{1}{7} + \frac{1}{7} - \frac{1}{25}\right) = 1.068 \quad \text{and}$$

$$\hat{\sigma}^2 = \frac{6(16.286) + 5(1.867) + 7(9.696) + 7(2.786)}{25} = 7.777.$$

Thus

$$U = \frac{1}{C}\left(\nu \log_e \hat{\sigma}^2 - \sum_i \nu_i \log_e \hat{\sigma}_i^2\right)$$

$$= \frac{1}{1.068}[25 \log_e(7.777) - 6 \log_e(16.286) - 5 \log_e(1.867)$$

$$- 7 \log_e(9.696) - 7 \log_e(2.786)]$$

$$= \frac{1}{1.068}[25(2.051) - 6(2.790) - 5(.624) - 7(2.272) - 7(1.025)]$$

$$= \frac{1}{1.068}(51.275 - 42.939) = 7.81.$$

The significance probability is obtained from a chi-square distribution with 3 degrees of freedom. We get $\hat{\alpha} = .0501$.

To illustrate the use of Box's test, we first partition the data from each treatment group into subgroups of approximately equal size. We choose (somewhat arbitrarily) to partition the first group of seven observations into two subgroups of four and three observations each. The second group of six is partitioned into two subgroups of three each, and the remaining two groups are partitioned into subgroups of four each. These partitions must be randomly selected. Our random partitioning produced the following subgroups:

Group 1	*Group 2*	*Group 3*	*Group 4*
1	12	11	14
8	10	10	11
9	12	12	13
4		5	14
9	13	12	13
0	13	4	14
1	10	7	17
		8	14

Next calculate the variance of each subgroup and take its logarithm:

	Group 1	*Group 2*	*Group 3*	*Group 4*
	2.615	.288	2.269	.693
	3.192	1.097	2.390	1.097
Total	5.807	1.385	4.659	1.790

An analysis of variance on this data produces $F_c = 2.338/.146 = 16.05$ with 3 and 4 degrees of freedom, which has a significance probability of $\hat{\alpha} = .0107$.

The Z_{ij}'s needed for Levene's test are given in Table 2.2. An analysis of variance of the data in Table 2.2 produces $F_c =$

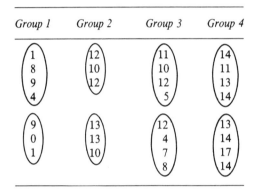

Table 2.2 Values of $Z_{ij} = |Y_{ij} - \bar{Y}_{i\cdot}|$ where Y_{ij} Is Taken from Table 2.1

	Group 1	*Group 2*	*Group 3*	*Group 4*
	3.571	.333	3.375	.75
	3.429	1.667	4.625	.25
	4.429	1.333	2.375	.25
	4.429	1.333	1.625	3.25
	.571	.333	.625	2.75
	4.571	1.667	1.375	.25
	3.571		3.375	.75
			3.625	.25
Total	24.571	6.667	21.000	8.5
n_i	7	6	8	8

10.2005/1.4644 = 6.966 with 3 and 25 degrees of freedom, which has a significance probability of $\hat{\alpha} = .0015$.

2.3.6 Comparisons of Tests and Recommendations

Conover, Johnson, and Johnson (1981) conducted a study of homogeneity of variance tests that included the ones above as well as numerous other ones. Based on their results, we make the following recommendations:

1. If the analyst is confident that the data are nearly normal, then use Bartlett's test or Hartley's test. If the sample sizes are very unequal, use Bartlett's test; otherwise, Hartley's test should be accurate enough to determine whether the usual F-tests and t-tests for comparing population means are appropriate.
2. For very large data sets, use Box's test. Box's test is very robust, but not very powerful for small sample sizes.
3. In all other instances, use Levene's test. Levene's test was shown to be nearly as good as Bartlett's test and Hartley's test for normally distributed data and superior to them for nonnormally distributed data. If the data tend to be very skewed, Levene's test can be improved by replacing $\bar{Y}_{i\cdot}$ by $\tilde{Y}_i$ where $\tilde{Y}_i$ is the median of the ith group. Thus, $Z_{ij} = |Y_{ij} - \tilde{Y}_i|$, and an analysis of variance is conducted on the Z_{ij}'s.

Conover, Johnson, and Johnson (1981) also recommend two other tests. However, because the test statistics required are somewhat more difficult to calculate than the test statistic for Levene's test, we do not recommend them.

2.4 INFERENCES ON LINEAR COMBINATIONS—TESTS AND CONFIDENCE INTERVALS

In this section we consider testing hypotheses about and constructing confidence intervals for $\Sigma c_i \mu_i$ when the $\hat{\sigma}_i^2$'s are too unequal to apply the tests discussed in Chapter 1. To decide whether the $\hat{\sigma}_i^2$'s are too unequal to use Chapter 1 methods, we recommend testing for homogeneity of variances. Use the techniques in this section and the next one if homogeneity is rejected at the 1% level. Otherwise, use the results in Chapter 1.

The best estimate of $\Sigma c_i \mu_i$ is $\Sigma c_i \hat{\mu}_i$ and

$$\sum_i c_i \hat{\mu}_i \sim N\left(\sum_i c_i \mu_i, \sum_i \frac{c_i^2 \sigma_i^2}{n_i} \right)$$

and hence,

$$z = \frac{\sum c_i \hat{\mu}_i - \sum c_i \mu_i}{\sqrt{\sum c_i^2 \sigma_i^2 / n_i}} \sim N(0, 1).$$

An obvious statistic to use for making inferences about $\Sigma c_i \mu_i$ when the variances are not known and are unequal is

$$Z = \frac{\sum c_i \hat{\mu}_i - \sum c_i \mu_i}{\sqrt{\sum_{i=1}^{t} c_i^2 \hat{\sigma}_i^2 / n_i}}.$$

If the n_i's corresponding to nonzero c_i's are all very large, one can reasonably assume that Z has an approximate $N(0, 1)$ distribution, and hence, Z can be used to make inferences about $\Sigma c_i \mu_i$. In this case, an approximate $(1 - \alpha)$ 100% confidence interval for $\Sigma c_i \mu_i$ is provided by

$$\sum c_i \hat{\mu}_i \pm z_{\alpha/2} \cdot \sqrt{\frac{\Sigma c_i^2 \hat{\sigma}_i^2}{n_i}}$$

where $z_{\alpha/2}$ is the upper $\alpha/2$ critical point of the standard normal probability distribution. To test H_0: $\Sigma c_i \mu_i = a$, we could calculate

$$Z_c = \frac{\sum c_i \hat{\mu}_i - a}{\sqrt{\sum c_i^2 \hat{\sigma}_i^2 / n_i}}$$

and reject H_0 at the α significance level if $|Z_c| > z_{\alpha/2}$.

In other instances, we note that Z can be written as

$$Z = \frac{\left(\sum c_i \hat{\mu}_i - \sum c_i \mu_i \right) \Big/ \sqrt{\sum c_i^2 \sigma_i^2 / n_i}}{\sqrt{\sum c_i^2 \hat{\sigma}_i^2 / n_i \Big/ \sum c_i^2 \sigma_i^2 / n_i}}. \tag{2.4.1}$$

Then, since the numerator and denominator in (2.4.1) are independent and the numerator has a standard normal distribution, Z would be approximately distributed as the $t(\nu)$ distribution if we could find ν so that

$$V = \nu \cdot \frac{\sum c_i^2 \hat{\sigma}_i^2 / n_i}{\sum c_i^2 \sigma_i^2 / n_i}$$

would be approximately distributed as $\chi^2(\nu)$. In order to get a good chi-square approximation to the distribution of V when the variances are unequal, we select that chi-square distribution that has the same first two moments as V. That is, to find ν for the case of unequal variances, we find ν so that the moments of V are equal to the first two moments of a $\chi^2(\nu)$ distribution. (This is known as Satterthwaite's method.) This results in determining that

$$\nu = \frac{\left(\sum_i c_i^2 \sigma_i^2 / n_i\right)^2}{\sum_i \left[c_i^4 \sigma_i^4 / n_i^2 (n_i - 1)\right]}$$

Unfortunately, since ν depends on $\sigma_1^2, \sigma_2^2, \ldots, \sigma_t^2$, it cannot be determined exactly. The usual procedure is to estimate ν by

$$\hat{\nu} = \frac{\left(\sum_i c_i^2 \hat{\sigma}_i^2 / n_i\right)^2}{\sum_i \left[c_i^4 \hat{\sigma}_i^4 / n_i^2 (n_i - 1)\right]}. \tag{2.4.2}$$

Summarizing, one rejects H_0: $\sum c_i \mu_i = a$ if

$$|t_c| = \frac{\left|\sum c_i \hat{\mu}_i - a\right|}{\sqrt{\sum c_i^2 \hat{\sigma}_i^2 / n_i}} > t_{\alpha/2, \hat{\nu}}$$

where $\hat{\nu}$ is given by (2.4.2). An approximate $(1 - \alpha)$ 100% confidence interval for $\sum c_i \mu_i$ is given by

$$\sum c_i \hat{\mu}_i \pm t_{\alpha/2, \hat{\nu}} \cdot \sqrt{\frac{\sum c_i^2 \hat{\sigma}_i^2}{n_i}}.$$

Unfortunately, every time we want to test a new hypothesis or construct another confidence interval, we must reestimate the degrees of freedom $\hat{\nu}$. It can be shown that $n_* - 1 \le \hat{\nu} \le t(n^* - 1)$ where $n_* = \min\{n_1, n_2, \ldots, n_t\}$ and $n^* = \max\{n_1, n_2, \ldots, n_t\}$. Thus, if $|t_c| > t_{\alpha/2, n_* - 1}$, one can be assured that $|t_c| > t_{\alpha/2, \hat{\nu}}$, and if $|t_c| < t_{\alpha/2, t(n^* - 1)}$, one can be assured that $|t_c| < t_{\alpha/2, \hat{\nu}}$. In these cases, one can avoid calculating $\hat{\nu}$. If

$$t_{\alpha/2,t(n^*-1)} < |t_c| < t_{\alpha/2,n_*-1},$$

then we must calculate $\hat{\nu}$ in order to be sure whether to accept or reject the hypothesis being tested. For confidence intervals, $\hat{\nu}$ should always be calculated. We now illustrate the preceding results with an example.

EXAMPLE 2.2

Consider the data in Table 2.1, and suppose the experimenter is interested in answering the following questions:

1. On the average, do drugs have any effect on learning at all?
2. Do subjects make more errors when given both drugs than when given only one?
3. Do the two drugs differ in their effects on the number of errors made?

To answer the first question, we test the hypothesis that the mean of the three drug groups is equal to the control mean. Thus, the hypothesis to be tested is

$$H_{01}: l_1 = \mu_1 - \tfrac{1}{3}\mu_2 - \tfrac{1}{3}\mu_3 - \tfrac{1}{3}\mu_4 = 0.$$

We first obtain

$$\hat{l}_1 = \hat{\mu}_1 - \tfrac{1}{3}(\hat{\mu}_2 + \hat{\mu}_3 + \hat{\mu}_4) = 4.571 - \tfrac{1}{3}(34.042) = -6.776 \quad \text{and}$$

$$\widehat{\text{s.e.}}(\hat{l}_1) = \sqrt{\sum\left(\frac{c_i^2\hat{\sigma}_i^2}{n_i}\right)}$$

$$= \sqrt{\frac{\hat{\sigma}_1^2}{7} + \frac{1}{9}\left(\frac{\hat{\sigma}_2^2}{6}\right) + \frac{1}{9}\left(\frac{\hat{\sigma}_3^2}{8}\right) + \frac{1}{9}\left(\frac{\hat{\sigma}_4^2}{8}\right)} = \sqrt{2.535} = 1.592.$$

Also we get

$$\sum \frac{c_i^4\hat{\sigma}_i^4}{n_i^2(n_i - 1)} = .9052 \quad \text{so that} \quad \hat{\nu} = \frac{(2.535)^2}{.9052} = 7.10.$$

Then $t_c = -6.776/1.592 = -4.256$, which is significant at the $\hat{\alpha} = .0038$ level.

A 95% confidence interval for l_1 is

$$\hat{l}_1 \pm t_{\alpha/2,\hat{\nu}} \cdot \widehat{\text{s.e.}}(\hat{l}_1) = -6.776 \pm (2.365)(1.592)$$

which gives

$$-10.54 < \mu_1 - \frac{\mu_2 + \mu_3 + \mu_4}{3} < -3.01.$$

Next we test to see if the mean of the group given both drugs is equal to the mean of the groups given a single drug. That is, we test H_{02}: $l_2 = \mu_4 - \frac{1}{2}\mu_2 - \frac{1}{2}\mu_3 = 0$. We obtain

$$\hat{l}_2 = \hat{\mu}_4 - \tfrac{1}{2}(\hat{\mu}_2 + \hat{\mu}_3) = 3.604 \quad \text{and}$$

$$\widehat{\text{s.e.}}(\hat{l}_2) = \sqrt{\sum\left(\frac{c_i^2\hat{\sigma}_i^2}{n_i}\right)}$$

$$= \left[\frac{\hat{\sigma}_4^2}{8} + \frac{1}{4}\left(\frac{\hat{\sigma}_2^2}{6}\right) + \frac{1}{4}\left(\frac{\hat{\sigma}_3^2}{8}\right)\right]^{1/2} = \sqrt{.7290} = .8538.$$

Thus, $t_c = 3.604/.8538 = 4.221$, which is significant at a level less than $\alpha = .01$ since $|t_c| > t_{.005,5}$. In this case, we did not calculate $\hat{\nu}$ but used $n_* - 1$ instead. Actually, $\hat{\nu} = 25.9$. If we were going to give a confidence interval for l_2, we would want to use $\hat{\nu}$ rather than $n_* - 1$.

Finally, we test to see if the two drug means differ; that is, we test H_{03}: $l_3 = \mu_2 - \mu_3 = 0$. We get

$$\hat{l}_3 = \hat{\mu}_2 - \hat{\mu}_3 = 3.042 \quad \text{and}$$

$$\widehat{\text{s.e.}}(\hat{l}_3) = \sqrt{\sum\left(\frac{c_i^2\hat{\sigma}_i^2}{n_i}\right)} = \left(\frac{\hat{\sigma}_2^2}{6} + \frac{\hat{\sigma}_3^2}{8}\right)^{1/2} = \sqrt{1.523} = 1.234.$$

Also,

$$\sum \frac{c_i^4\hat{\sigma}_i^4}{n_i^2(n_i - 1)} = .229 \quad \text{so that} \quad \hat{\nu} = \frac{(1.523)^2}{.229} = 10.1.$$

Thus, $t_c = 3.042/1.234 = 2.465$, which is significant at the $\hat{\alpha} = .0334$ level.

2.5 COMPARING ALL MEANS

As previously stated, the usual F-test is very robust for the unequal variances case provided that the sample sizes are nearly equal or, if not nearly equal, provided that the larger sample sizes correspond to the populations with the larger variances. In this section, two additional tests of the hypothesis of equal means are provided. The first was suggested by Box (1954) and requires equal sample sizes.

Box suggested that one calculate the usual F-statistic but compare it to the critical point of the F-distribution with degrees of freedom

$$\nu_1 = \frac{t - 1}{1 + c^2(t - 2)/(t - 1)} \quad \text{and} \quad \nu_2 = \frac{t(n - 1)}{1 + c^2}$$

where $c^2 = \Sigma(\sigma_i^2 - \bar{\sigma}^2)^2/t(\bar{\sigma}^2)^2$ and $\bar{\sigma}^2 = (\Sigma\sigma_i^2)/t$ rather than degrees of freedom $t - 1$ and $N - t$. Since the σ_i^2's are usually unknown, ν_1 and ν_2 must be estimated by replacing σ_i^2 with $\hat{\sigma}_i^2$ in the above formulas. Note that when all the σ_i^2's are equal, then $c^2 = 0$, and $\nu_1 = t - 1$ and $\nu_2 = N - t$ where $N = nt$ and n is the common sample size. Also note that the maximum value of c^2 is $t - 1$, and in this case $\nu_1 = 1$ and $\nu_2 = n - 1$. Hence, for an extremely conservative test, one would reject H_0: $\mu_1 = \mu_2 = \cdots = \mu_t$ if $F_c > F_{\alpha,1,n-1}$.

A second test for H_0: $\mu_1 = \mu_2 = \cdots = \mu_t$ is given by Welch (1951), known as Welch's test. Define $W_i = n_i/\hat{\sigma}_i^2$ and define $\bar{Y}^*$ by $\bar{Y}^* = \Sigma_i W_i \bar{Y}_{i\cdot}/\Sigma_i W_i$. Let

$$\Lambda = \sum_i \frac{(1 - W_i/W_\cdot)^2}{n_i - 1}$$

where $W_\cdot = \Sigma_i W_i$. Then

$$F_c = \frac{\sum_i W_i \dfrac{(\bar{Y}_{i\cdot} - \bar{Y}^*)^2}{(t-1)}}{1 + 2(t-2)\Lambda/(t^2 - 1)} \tag{2.5.1}$$

has an approximate F-distribution with $\nu_1 = t - 1$ and $\nu_2 = (t^2 - 1)/3\Lambda$. Thus H_0: $\mu_1 = \mu_2 = \cdots = \mu_t$ is rejected at the α level if $F_c > F_{\alpha,\nu_1,\nu_2}$.

The numerator of (2.5.1) may also be obtained from $(\Sigma W_i \bar{Y}_{i\cdot}^2 - W_\cdot \bar{Y}^{*2})/(t-1)$. We illustrate this latter procedure with an example.

EXAMPLE 2.3

Consider the drug data given in Example 5.1. We need the following information:

i	1	2	3	4
n_i	7	6	8	8
$\bar{Y}_{i\cdot}$	4.571	11.667	8.625	13.750
$\hat{\sigma}_i^2$	16.286	1.867	9.696	2.786
W_i	.430	3.214	.825	2.872

From the above information we get $W_{.} = 7.340$, $\bar{Y}^{*} = 11.725$,

$$\Lambda = \frac{(1 - .430/7.340)^2}{6} + \frac{(1 - 3.214/7.340)^2}{5} + \frac{(1 - .825/7.340)^2}{7} + \frac{(1 - 2.872/7.340)^2}{7} = .376,$$

and

$$\sum W_i \bar{Y}_i^2 - W_{.}\bar{Y}^{*2} = 1{,}050.6989 - 1{,}009.0711 = 41.6278.$$

Thus,

$$F_c = \frac{41.6278/3}{1 + 2(2)(.376)/15} = \frac{13.8759}{1.1003} = 12.61$$

with $\nu_1 = 3$ and $\nu_2 = 15/(3 \cdot .376) = 13.3$ degrees of freedom. The significance probability of F_c is approximately $\hat{\alpha} = .0004$. For comparison purposes, the usual F-statistic is $F_c = 14.91$ with 3 and 25 degrees of freedom.

In summary, for comparing all means, we recommend the following:

1. If the homogeneity of variance test is not significant at the 1% level, do the usual analysis of variance test.
2. If the homogeneity of variance test is significant at the 1% level and the group sample sizes are all equal, use Box's (mean) test.
3. If the homogeneity of variance test is significant at the 1% level and the group sample sizes are not equal, use Welch's test.

CONCLUDING REMARKS

This chapter discussed the statistical analysis of a one-way analysis of variance model with heterogeneous errors. The discussion included several statistical tests for determining homogeneity of the error variances and recommendations on when to use each test. Also reviewed were procedures appropriate for making statistical inferences about the effects of different treatments upon discovering heterogeneous error variances as well as examples illustrating the use of these procedures.

3

Simultaneous Inference Procedures and Multiple Comparisons

CHAPTER OUTLINE

Often an experimenter wants to compare several functions of the μ_i's in the same experiment. Experimenters should consider all functions of the μ_i's that are of interest; that is, they should attempt to answer all questions of interest. Often the treatments suggest comparisons of interest (see, for example, the drug experiment in Example 2.2). At other times, the experimenter may be interested in comparing each treatment to all other treatments. This would be the case, for example, when one is comparing the yields of several varieties of wheat.

One concern when making many comparisons in a single experiment is whether significant differences obtained are due to real differences in the functions being compared or simply due to the very large number of comparisons being made, which increases the chance of finding differences that appear to be significant. For example, if an experimenter conducts 25 independent tests in an experiment and finds one significant difference at the .05 level, he should not put too much faith in the result because he should expect to find (.05)(25) = 1.25 significant differences just by chance alone. Thus, if an experimenter is answering a large number of questions with one experiment (which we believe one should do), it is desirable to have a procedure that indicates whether the differences might be the result of chance alone.

In this chapter, several well-known and commonly used procedures for making multiple inferences are discussed and compared. Some of the procedures are primarily used for testing hypotheses, while others can also be used to obtain simultaneous confidence intervals—that is, a set of confidence intervals for a set of functions of the μ_i's can be derived for which we can be 95% confident that all the confidence intervals simultaneously contain their respective functions of the μ_i's.

3.1 ERROR RATES

One of the main ways to evaluate and compare multiple comparison procedures is to calculate *error rates*. If a given confidence interval does not contain the true value of the quantity being estimated, then an error occurs. Similarly, if a hypothesis test is used, an error is made whenever a true hypothesis is rejected or a false hypothesis is not. Next we define two kinds of error rates.

Definition 3.1 The *comparisonwise error rate* is equal to the ratio of the number of incorrect inferences made to the total number of inferences made in all experiments analyzed.

Definition 3.2 The *experimentwise error rate* is equal to the ratio of the number of experiments in which at least one error is made to the total number of experiments analyzed. It is the probability of making at least one error in an experiment when there are no differences between the treatments.

In order to prevent finding too many comparisons significant by chance alone in a single experiment, one quite often attempts to fix the experimentwise error rate at some prescribed level, such as .05. Whenever an experimenter is trying to answer many questions with a single experiment, it is a good idea to control the experimentwise error rate.

3.2 RECOMMENDATIONS

Carmer and Swanson (1973), conducted a Monte Carlo study to evaluate ten pairwise multiple comparison methods, including all those discussed later in this chapter. They simulated data from populations with unequal means as well as from populations with equal means. As a criterion for choosing among the methods, they observed "correct decision rates" for the various methods rather than Type I or Type II error rates. Based on their simulations, they recommend using either Fisher's LSD, which requires that the F-test for equal means from an ANOVA be significant at the 5% level, or the Waller–Duncan method. Duncan's method (to be distinguished from the Waller–Duncan method) came in a close third and could be used with a preliminary F-test by experimenters especially concerned about Type I errors.

Carmer and Swanson's (1973) simulations were done for cases with equal sample sizes only. For unequal sample sizes, our choice is between Fisher's LSD method and Duncan's method, since the Waller–Duncan method has not yet been generalized to the unequal-sample-size case. Those primarily concerned with Type II errors should choose Fisher's LSD, while those primarily concerned with Type I errors should use Duncan's procedure preceded by an F-test for equal means. However, Duncan's procedure should not be used if the sample sizes are extremely unequal for reasons described in Section 3.13.

Our recommendations for multiple comparisons in testing hypotheses are as follows:

1. Conduct an F-test for equal means.
2. If the F-statistic is significant at the 5% level, make any planned comparisons you wish to make by using the LSD method. This includes not only comparisons between pairs of means but comparisons based on any selected contrasts of the μ_i's. If one has equal sample sizes, the Waller–Duncan method can also be used. For data snooping and unplanned comparisons, use Scheffé's method.
3. If the F-statistic for equal means is not significant, the experimenter should still consider any individual comparisons that he or she had planned, but should do so using either the multivariate t-method or Bonferroni's method. The experimenter should not do any data snooping in this case. Indeed, since the F-test for equal means is nonsignificant, Scheffé's procedure would not yield any significant differences anyway.

We next examine each of the multiple comparison procedures recommended as well as a few other popular procedures available for the one-way treatment structure discussed in Chapter 1. Each of the procedures can also be used in much more complex situations, as will be illustrated throughout the remainder of this book. The parameter ν used during the remainder of this book represents the degrees of freedom corresponding to the estimator of σ^2. For the one-way case, $\nu = N - t$.

3.3 LEAST SIGNIFICANT DIFFERENCES

The LSD (least significant difference) multiple comparison method is possibly used more than any other method perhaps because it is one of the easiest to apply. It is usually used to compare each treatment mean to every other treatment mean, but it can be used for other comparisons as well. The LSD at the $\alpha \cdot 100\%$ significance level for comparing μ_i to μ_j is

$$\text{LSD}_\alpha = t_{\alpha/2,\nu}\hat{\sigma}\sqrt{\frac{1}{n_i} + \frac{1}{n_j}}. \tag{3.3.1}$$

One concludes that $\mu_i \neq \mu_j$ if $|\hat{\mu}_i - \hat{\mu}_j| > \text{LSD}_\alpha$. This procedure has a comparisonwise error rate equal to α. A corresponding $(1 - \alpha)100\%$ confidence interval for $\mu_i - \mu_j$ is

$$\hat{\mu}_i - \hat{\mu}_j \pm t_{\alpha/2,\nu}\hat{\sigma}\sqrt{\frac{1}{n_i} + \frac{1}{n_i}}. \tag{3.3.2}$$

If all sample sizes are equal (to n, say), then a single LSD value can be used for all pairwise comparisons. It is given by

$$\text{LSD}_\alpha = t_{\alpha/2,\nu}\hat{\sigma}\sqrt{\frac{2}{n}}. \tag{3.3.3}$$

Suppose we have t treatment means and that we are going to make all possible pairwise comparisons at the 5% significance level. The table below shows how experimentwise error rates vary for different values of t. The information in the table applies to cases where all treatment means are equal.

Table 3.1 shows that in an experiment involving six treatments, one would find at least one significant difference between the means 36.6% of

Table 3.1 Error Rates for the LSD Procedure

t	*2*	*3*	*4*	*5*	*6*	*10*	*20*
Comparisonwise error rate	.05	.05	.05	.05	.05	.05	.05
Experimentwise error rate	.05	.122	.203	.286	.366	.627	.918

the time, even when all the treatment means were equal. Obviously, using the LSD procedure could be very risky without some additional protection.

3.4 FISHER'S LSD PROCEDURE

Fisher's recommendation offers some protection for the LSD procedure discussed in the preceding section. In his procedure, LSD tests are made at the $\alpha \cdot 100\%$ significance level by utilizing equation (3.3.1) but only if $H_0\colon \mu_1 = \mu_2 = \cdots = \mu_t$ is first rejected by the F-test discussed in Chapter 1. This gives a rather large improvement over the straight LSD procedure since the experimentwise error rate is now approximately equal to α. However, it is possible to reject $H_0\colon \mu_1 = \mu_2 = \cdots = \mu_t$ and not reject any of $H_{0_{ij}}\colon \mu_i = \mu_j$ for $i \neq j$. It is also true that this procedure may not detect some differences between pairs of treatments when they really exist. In other words, differences between a few pairs of treatments may exist, but equality of the remaining treatments may cause the F-test to be nonsignificant, and this procedure does not allow the experimenter to make individual comparisons without first obtaining a significant F-statistic. We do not recommend using these two LSD procedures for constructing simultaneous confidence intervals on specified contrasts of the μ_i's, since the confidence intervals obtained will generally be too narrow.

Each of the above LSD procedures can be generalized to include any contrasts of the treatment means. The generalization is: Conclude that $\Sigma c_i \mu_i \neq 0$, if

$$\left|\sum c_i \hat{\mu}_i\right| > t_{\alpha/2,\nu} \hat{\sigma} \sqrt{\sum \frac{c_i^2}{n_i}}\,. \tag{3.4.1}$$

Examples are given in Section 3.9.

3.5 BONFERRONI'S METHOD

Although this procedure may be the least used, it is often the best. It is particularly good when the experimenter wants to make only a small number of comparisons. We recommend using this procedure on planned comparisons whenever the F-test for equal means is not significant.

Suppose the experimenter wants to make p such comparisons. He would conclude that the qth comparison $\Sigma c_{iq} \mu_i \neq 0$, $q = 1, 2, \ldots, p$, if

$$\left|\sum c_{iq} \hat{\mu}_i\right| > t_{\alpha/2p,\nu} \hat{\sigma} \sqrt{\sum_{i=1}^{t} \frac{c_{iq}^2}{n_i}}\,. \tag{3.5.1}$$

These p-tests will give an experimentwise error rate less than or equal to α and a comparisonwise error rate equal to α/p. Usually the experimentwise error rate is much less than α. Unfortunately, it is not possible to determine how much less. Values of $t_{\alpha/2p,\nu}$ for selected values of α, p, and ν are given in the Appendix in Table A.2. For example, if $\alpha = .05$,

$p = 5$, and $\nu = 24$, then from Table A.2 we get $t_{\alpha/2p,\nu} = 2.80$; see Section 3.9 for an example.

Confidence intervals obtained from the Bonferroni method, which can be recommended, have the form:

$$\sum_i c_{iq}\hat{\mu}_i \pm t_{\alpha/2p,\nu}\hat{\sigma}\sqrt{\sum_i \frac{c_{iq}^2}{n_i}}\,. \tag{3.5.2}$$

3.6 MULTIVARIATE *t*-METHOD

The multivariate t-method is a good method to use when the experimenter has to compare a linearly independent set of linear combinations of the μ_i's. It is important that this procedure not be used for a set of linear combinations that are linearly dependent. This restriction prevents one from using this method for making all possible comparisons between pairs of means, since the set of all possible comparisons between means is not a linearly independent set.

If the experimenter wants to make p linearly independent comparisons, then she would conclude that the qth comparison $\Sigma c_{iq}\mu_i \neq 0$, $q = 1, 2, \ldots, p$, if

$$\left|\sum c_{iq}\hat{\mu}_i\right| > t_{\alpha/2,p,\nu}\hat{\sigma}\sqrt{\sum_i \frac{c_{iq}^2}{n_i}} \tag{3.6.1}$$

where $t_{\alpha/2,p,\nu}$ is the upper $\alpha/2$ percentile of a p-variate multivariate t-distribution with ν degrees of freedom and correlation matrix I_p. Values of $t_{\alpha/2,p,\nu}$ are given in Appendix Table A.3 for selected values of α, p, and ν. Simultaneous confidence intervals based on the multivariate t-method are appropriate and can be recommended.

The multivariate t-method has an experimentwise error rate less than or equal to α. If the linear combinations $\Sigma_i c_{iq}\hat{\mu}_i$ are also statistically independent for $q = 1, 2, \ldots, p$, then the experimentwise error rate is exactly equal to α. For a specified set of linearly independent comparisons, the multivariate t-method will often give the best results. If the set of comparisons of interest to the experimenter is linearly independent, then the multivariate t-method will always be better than Bonferroni's method.

The following result enables one to extend the application of the multivariate t-procedure to a linearly dependent set of comparisons (however, the resulting tests or confidence intervals are not entirely satisfactory): Let $l_1, l_2, \ldots, l_p$ be a linearly independent set of linear combinations of the μ_i's. If $|l_q| \le c_q$ for $q = 1, 2, \ldots, p$, then

$$\left|\sum_q \lambda_q l_q\right| \le \sum_q |\lambda_q| \cdot c_q.$$

To make use of this result, use the following procedure:

1. Let $l_1, l_2, \ldots, l_p$ be a linearly independent set of linear combinations of the μ_i's. This set should be one of primary interest to the experi-

menter. For this set of comparisons, one concludes that $l_q = \Sigma_i c_{iq}\mu_i$ is significantly different from zero if

$$|\hat{l}_q| > t_{\alpha/2,p,\nu}\hat{\sigma}\sqrt{\sum_i \frac{c_{iq}^2}{n_i}}\,. \tag{3.6.2}$$

2. Let l^* be any comparison that is of secondary importance which is a linear combination of the l_q's, $q = 1, 2, \ldots, p$. That is, $l^* = \Sigma_{q=1}^{p}\lambda_q l_q$ for some set of λ_q's. One declares that l^* is significantly different from zero if

$$\hat{l}^* > t_{\alpha/2,p,\nu}\hat{\sigma}\sum_{q=1}^{p}\left(|\lambda_q|\sqrt{\sum_i \frac{c_{iq}^2}{n_i}}\right). \tag{3.6.3}$$

An experimenter can make as many l^*-type comparisons as he wants without increasing the experimentwise error rate. This extension of the multivariate t-method gives very powerful tests for those comparisons of primary importance but is extremely weak on those comparisons of secondary importance. This is illustrated in an example in Section 3.9.

3.7 SCHEFFÉ'S PROCEDURE

This procedure is recommended whenever the experimenter wants to make a large number of "unplanned" comparisons. Unplanned comparisons are comparisons that the experimenter had not thought of making when planning the experiment. These arise frequently, since the results of an experiment frequently suggest certain comparisons to the experimenter.

Consider testing $\Sigma c_i\mu_i = 0$ for a given contrast vector $\mathbf{c}$. It is true that

$$\Pr\left\{\frac{\left(\sum c_i\hat{\mu}_i - \sum c_i\mu_i\right)^2}{\sum\left(c_i^2/n_i\right)} \le (t-1)F_{\alpha,t-1,\nu}\hat{\sigma}^2 \quad \text{for all contrast vectors } \mathbf{c}\right\} = 1 - \alpha.$$

Thus a procedure with an experimentwise error rate equal to α for comparing all possible contrasts of the μ_i's to zero is as follows: Reject H_0: $\Sigma c_i\mu_i = 0$ if

$$\left|\sum c_i\hat{\mu}_i\right| > \sqrt{(t-1)F_{\alpha,t-1,\nu}}\cdot\hat{\sigma}\sqrt{\sum\frac{c_i^2}{n_i}}\,. \tag{3.7.1}$$

This procedure allows one to compare an infinite number of contrasts to zero while maintaining an experimentwise error rate equal to α. However, most experimenters will usually not be interested in an infinite number of comparisons—usually only a finite number of comparisons are of interest. Scheffé's procedure can still be used; in this case, the experimentwise error rate will generally be much smaller than α. Bonferroni's method or the multivariate t-method when appropriate will often be better than Scheffé's procedure for a finite number of comparisons. That is, a smaller value of $\Sigma c_i \hat{\mu}_i$ can often be declared significant by using Bonferroni's method or the multivariate t-method than can be declared significant by Scheffé's method. However, if one is going to "muck around" in the data to see if anything significant turns up, then one should use Scheffé's method, since the comparisons in this case are really unplanned comparisons rather than planned ones. It should be noted that Scheffé's method will not reveal any contrasts significantly different from zero unless the F-test discussed in Chapter 1 rejects H_0: $\mu_1 = \mu_2 = \cdots = \mu_t$.

Scheffé's procedure can also be used to obtain simultaneous confidence intervals for contrasts of the μ_i's. The result required is that for any set of contrasts $\mathbf{c}_1, \mathbf{c}_2, \ldots,$ one can be at least $(1 - \alpha)100\%$ confident that $\Sigma c_{iq}\mu_i$ will be contained within

$$\sum_i c_{iq}\hat{\mu}_i \pm \sqrt{(t-1)F_{\alpha,t-1,\nu}} \cdot \hat{\sigma}\sqrt{\sum_i \frac{c_{iq}^2}{n_i}} \quad \text{for all} \quad q = 1, 2, \ldots. \tag{3.7.2}$$

If one wants to consider all linear combinations of the μ_i's rather than just all contrasts, then $\sqrt{(t-1)F_{\alpha,t-1,\nu}}$ must be replaced by $\sqrt{tF_{\alpha,t,\nu}}$ in (3.7.1) and (3.7.2).

Examples can be found in Section 3.9.

3.8 TUKEY'S HONEST SIGNIFICANT DIFFERENCE

The preceding procedures can be used regardless of the values of the n_i's. Tukey's honest significant difference (HSD) procedure, however, requires equal n_i's, although an approximate procedure can be used if the n_i's are not too unequal. Tukey's method says to reject H_0: $\Sigma c_i\mu_i = 0$ if

$$\left|\sum c_i\hat{\mu}_i\right| > q(\alpha, t, \nu) \cdot \frac{\hat{\sigma}}{\sqrt{n}}\left(\frac{1}{2}\sum |c_i|\right) \tag{3.8.1}$$

for any contrast $\Sigma c_i\mu_i$. The critical point $q(\alpha, t, \nu)$ is the upper α percentile of the distribution of the Studentized range statistic. Values of $q(\alpha, t, \nu)$ are given in Appendix Table A.4 for selected values of α, t, and ν.

This procedure also gives an experimentwise error rate less than or equal to α. In some instances, the right-hand side of (3.8.1) will be smaller than the corresponding value given by Scheffé's method; in other instances, Scheffé's method will give the smaller value.

For comparisons of the form $\mu_i = \mu_j$, $\frac{1}{2}\Sigma|c_i| = 1$; in this case the procedure simplifies to the following: Declare $\mu_i \neq \mu_j$ if

$$|\hat{\mu}_i - \hat{\mu}_j| > q(\alpha, t, \nu) \cdot \frac{\hat{\sigma}}{\sqrt{n}}.$$

The right-hand side of this expression is usually known as Tukey's honest significant difference. For comparisons involving pairs of means only, Tukey's HSD will always be smaller than the corresponding point given by Scheffé's method, as well as that required by Bonferroni's method. While we do not recommend Tukey's method as a testing procedure for comparing all pairs of means, we do recommend it for constructing simultaneous confidence intervals for differences between all pairs of means when such intervals are desired.

A generalization of Tukey's HSD method to the case of unequal sample sizes was given by Spjøtvoll and Stoline (1973, p. 975). Their result is that one should declare $\mu_i \neq \mu_j$ if

$$|\hat{\mu}_i - \hat{\mu}_j| > q(\alpha, t, \nu) \cdot \frac{\hat{\sigma}}{\min\left(\sqrt{n_i}, \sqrt{n_j}\right)}.$$

This should be quite satisfactory if the sample sizes are not too unequal; when they are very unequal, the procedure is usually much less sensitive than Scheffé's method.

3.9 EXAMPLES

In this section, examples are given to illustrate the methods discussed thus far in this chapter. Unlike these methods, which are applicable both as multiple-hypothesis testing procedures and as simultaneous confidence interval procedures, the methods discussed in the following three sections are applicable only for multiple-hypothesis testing. Also, the remaining procedures can only be used to compare pairs of means, while the preceding ones have been generalized to include any contrasts of the μ_i's.

EXAMPLE 3.1

Consider the data given in Example 1.1, from which the following pieces of information are recalled:

	TASK					
i	*1*	*2*	*3*	*4*	*5*	*6*
n_i	13	12	10	10	12	11
$\hat{\mu}_i$	31.923	31.083	35.8	38.0	29.5	28.818

Also, $\hat{\sigma}^2 = 30.9045$ with 62 degrees of freedom.

We first consider comparisons between pairs of means. In this case, the critical values required in order to claim significance depend on the

Table 3.2 5% Critical Values for Comparing Pairs of Means

Sample Size	*Fisher's LSD*	*Bonferroni's*	*Scheffé's*	*Tukey's HSD (approximated)*
13, 12	4.450	6.809	7.658	6.677
13, 10	4.676	7.154	8.047	7.313
13, 11	4.556	6.971	7.841	6.972
12, 10	4.760	7.283	8.192	7.313
12, 12	4.540	6.946	7.813	6.677
12, 11	4.642	7.102	7.989	6.972
10, 10	4.972	7.607	8.557	7.313
10, 11	4.858	7.433	8.361	7.313

sample sizes of the pair being compared. Table 3.2 gives the required critical values for each procedure. Simultaneous confidence intervals are constructed by adding and subtracting the entries in Table 3.2 to the respective differences between the sample means. The multivariate t-method is not included, since the differences between all pairs of means are not linearly independent. We note that:

1. Bonferroni's method always beats Scheffé's method. In fact, one can make about 47 different comparisons using Bonferroni's method and still beat Scheffé's method; however, if one wants to make 48 or more comparisons, then Scheffé's will beat Bonferroni's.
2. Bonferroni's method is better than Tukey's method in some cases and worse in others. For nearly equal sample sizes, Tukey's HSD is usually smaller than Bonferroni's, but for more unequal sample sizes Bonferroni's is usually smaller.

Table 3.3 compares every mean to every other mean. The letters indicate which of the multiple comparison procedures would declare that the difference between the sample means is significant.

In order to illustrate the multivariate t-method, suppose that we are primarily interested in comparing μ_1 to each of the other means. Comparisons of this type are often of interest when one treatment is a control. Thus, the primary set of interest is $\{\mu_1 - \mu_2, \mu_1 - \mu_3, \mu_1 - \mu_4, \mu_1 - \mu_5, \mu_1 - \mu_6\}$. All other comparisons among pairs of means can be generated from this set. The point to which $\hat{\mu}_1 - \hat{\mu}_2$ will be compared is

$$t_{\alpha/2,5,62}\sigma\sqrt{\sum \frac{c_{iq}^2}{n_i}} = (2.649)(5.559)\sqrt{\frac{1}{13} + \frac{1}{12}} = 5.895.$$

Table 3.3 Comparisons between Means at .05 Level

Comparison	$\hat{\mu}_i - \hat{\mu}_j$	*Significance*
1 vs. 2	.840	None
1 vs. 3	−3.877	None
1 vs. 4	−6.077	F
1 vs. 5	2.423	None
1 vs. 6	3.105	None
2 vs. 3	−4.717	None
2 vs. 4	−6.917	F
2 vs. 5	1.583	None
2 vs. 6	2.265	None
3 vs. 4	−2.2	None
3 vs. 5	6.3	F
3 vs. 6	6.982	F
4 vs. 5	8.5	F, B, S, T
4 vs. 6	9.182	F, B, S, T
5 vs. 6	.682	None

Note: F = Fisher's LSD method; B = Bonferroni's method; S = Scheffé's method; T = Tukey's HSD method.

Similarly, the points to which $\hat{\mu}_1 - \hat{\mu}_3$, $\hat{\mu}_1 - \hat{\mu}_4$, $\hat{\mu}_1 - \hat{\mu}_5$, and $\hat{\mu}_1 - \hat{\mu}_6$ must be compared are 6.194, 6.194, 5.895, and 6.033, respectively. Comparing these LSD's with the results in the first three rows of Table 3.2, one sees that the multivariate t-method beats all but Fisher's LSD method.

However, the point to which $\hat{\mu}_2 - \hat{\mu}_3$ would be compared when using the extension to the multivariate t-method is

$$t_{\alpha/2,p,\nu}\hat{\sigma} \sum_{q=1}^{p} |\lambda_q| \sqrt{\sum_i \frac{c_{iq}^2}{n_i}} = (2.649)(5.559)\left(|1|\sqrt{\frac{1}{13} + \frac{1}{12}} + |-1|\sqrt{\frac{1}{13} + \frac{1}{10}} \right)$$

$$= 5.895 + 6.194 = 12.089.$$

Comparing this to the results in the fourth row of Table 3.2, we see that

12.089 is much bigger than any of the others. Thus, comparisons that are of secondary importance when using the multivariate t-method are much less sensitive than those using any of the other methods. Similar results hold for all other pairs of means.

3.10 NEWMAN–KEULS METHOD

This method requires that the n_i's be equal; however, a good approximation can be obtained provided that the sample sizes are not too unequal. In this case, the variable n in the following formulas is replaced by

$$\tilde{n} = t\left(\frac{1}{n_1} + \frac{1}{n_2} + \cdots + \frac{1}{n_t}\right)^{-1},$$

the harmonic mean of the n_i's. The exact experimentwise error rate for this procedure is not known.

To apply this method, rank the t means in ascending order. Let $\bar{y}_{(1)}$ denote the smallest mean, $\bar{y}_{(2)}$ denote the next smallest mean, and so on. The largest mean is denoted by $\bar{y}_{(t)}$.

Next, the Studentized range of the t means,

$$\frac{\bar{y}_{(t)} - \bar{y}_{(1)}}{\hat{\sigma}/\sqrt{n}}$$

is compared with the critical point $q(\alpha, t, \nu)$. If

$$\bar{y}_{(t)} - \bar{y}_{(1)} > \frac{\hat{\sigma}}{\sqrt{n}} \cdot q(\alpha, t, \nu),$$

one then compares the ranges of the two sets of $t - 1$ means $\{\bar{y}_{(1)}, \bar{y}_{(2)}, \ldots, \bar{y}_{(t-1)}\}$ and $\{\bar{y}_{(2)}, \bar{y}_{(3)}, \ldots, \bar{y}_{(t)}\}$ with $(\hat{\sigma}/\sqrt{n}) \cdot q(\alpha, t - 1, \nu)$. One continues to examine smaller subsets of means as long as the previous subset has a significant range. Each time a range proves nonsignificant, the means involved are included in a single group. No subset of means grouped in a nonsignificant group can later be deemed significant; that is, no further tests should be carried out on means that have previously been grouped in a common group. When all range tests prove nonsignificant, the procedure is complete. Any two means grouped in the same group are not significantly different; otherwise, they are significantly different.

We now illustrate this method with an example. Once again we use the task data from Example 1.1. First we get

$$\tilde{n} = 6\left(\tfrac{1}{13} + \tfrac{1}{12} + \tfrac{1}{10} + \tfrac{1}{10} + \tfrac{1}{12} + \tfrac{1}{11}\right)^{-1} = 11.23.$$

Ranking the means in ascending order, we get

Task	6	5	2	1	3	4
Mean	28.181	29.5	31.083	31.923	35.8	38.0

If we select $\alpha = .05$, we first must compare $38.0 - 28.181 = 9.819$ to

$$q(.05, 6, 62) \cdot \frac{\hat{\sigma}}{\sqrt{\tilde{n}}} = (4.16)\sqrt{\frac{30.9045}{11.23}} = 6.90.$$

Since $9.819 > 6.90$, we next examine the two subsets of $t - 1 = 5$ means.

In this case, we compare both $35.8 - 28.181 = 7.619$ and $38.0 - 29.5 = 8.5$ with

$$q(.05, 5, 62) \cdot \frac{\hat{\sigma}}{\sqrt{\tilde{n}}} = (3.98)(1.659) = 6.60.$$

Since $7.619 > 6.60$ and $8.5 > 6.60$, we must next look at subsets of $t - 2 = 4$ means, of which there are three.

Here we must compare $31.923 - 28.181 = 3.742$, $35.8 - 29.5 = 6.3$, and $38.0 - 31.083 = 6.917$ to

$$q(.05, 4, 62) \cdot \frac{\hat{\sigma}}{\sqrt{\tilde{n}}} = (3.74)(1.659) = 6.20.$$

Since $3.742 < 6.20$, the first four means are grouped in a single group, and since $6.3 > 6.20$ and $6.917 > 6.20$, both of the remaining groups of four means must be further subdivided into groups of three means. Before proceeding with this next step, consider the following schematic diagram, which illustrates our present position:

28.181 29.5 31.083 31.923 35.8 38.0

The second subset of four means, $\{29.5, 31.083, 31.923, 35.8\}$, contains two groups of three means that have already been grouped together, namely, 29.5 to 31.923 and 28.181 to 31.083. Hence, the ranges $31.923 - 29.5 = 2.423$ and $31.083 - 28.181 = 2.902$ are not compared with a critical point. The ranges that still must be compared are $35.8 - 31.083 = 4.717$ and $38.0 - 31.923 = 6.077$; these must be compared to

$$q(.05, 3, 62) \cdot \frac{\hat{\sigma}}{\sqrt{\tilde{n}}} = (3.40)(1.659) = 5.64.$$

Since $4.717 < 5.64$, these three means are now grouped in a single group.

whereas since 6.077 > 5.64, the second group must be further subdivided into groups of two means each. The following diagram illustrates our present position:

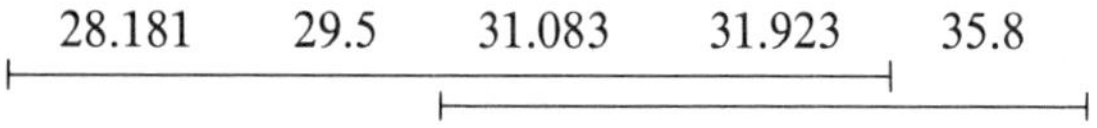

There is only one subset of size two that has not already been combined into a common group, that being {35.8, 38.0}. The range 38.0 − 35.8 = 2.2 is compared with

$$q(.05, 2, 62)\frac{\hat{\sigma}}{\sqrt{\tilde{n}}} = (2.83)(1.659) = 4.69.$$

Since 2.2 < 4.69, these final two means are grouped together.

Our final diagram is as follows:

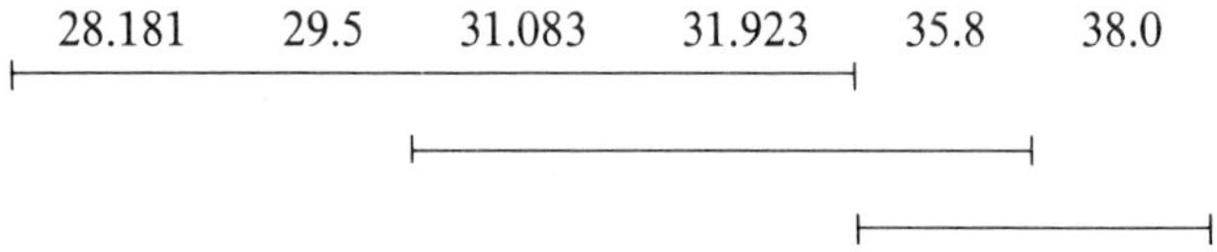

Another way to illustrate this information is to label means with the same letter if they occur in the same group and with different letters if they occur in different groups. Thus, for the data above we obtain:

Task	*Mean*
1	31.923ab
2	31.083ab
3	35.8bc
4	38.0c
5	29.5a
6	28.181a

3.11 DUNCAN'S NEW MULTIPLE RANGE METHOD

At present, this procedure is generally referred to as Duncan's method. It is one of the more popular methods, partly because it is often easier to find significant differences by using this method than by any other, except perhaps Fisher's LSD method. As one would guess, the method has a higher experimentwise error rate than Scheffé's, Bonferroni's, and Tukey's methods and the multivariate t-method, although the exact experimentwise error rate of this method is not known. This procedure also requires equal n_i's, but as in the preceding section, the variable n in the following formulas can be replaced by $\tilde{n}$ if the sample sizes are not too unequal for an approximate procedure.

Application of this procedure is similar to the application of the Newman-Keul method except that the studentized range critical point for comparing a group of p means, $q(\alpha, p, \nu)$, is replaced by $q(\alpha_p, p, \nu)$ where $\alpha_p = 1 - (1 - \alpha)^{p-1}$. Values of $q(\alpha_p, p, \nu)$ are given in the Appendix Table A.5. For the data in the preceding section, this procedure is applied as follows:

1. Compare 38 − 28.181 = 9.819 to

$$q(\alpha_6, 6, 62) \cdot \frac{\hat{\sigma}}{\sqrt{\tilde{n}}} = (3.198)(1.659) = 5.305.$$

   The range of six means is significant.
2. Compare 35.8 − 28.181 = 7.619 and 38.0 − 29.5 = 8.5 to (3.143)(1.659) = 5.214. Both ranges are significant.
3. Compare 31.923 − 28.181 = 3.742, 35.8 − 29.5 = 6.3, and 38.0 − 31.083 = 6.917 to (3.073)(1.659) = 5.098. The latter two are significant, while the first is not. We have:

   28.181   29.5   31.083   31.923   35.8   38.0

4. Compare 35.8 − 31.083 = 4.717 and 38 − 31.923 = 6.077 to (2.976)(1.659) = 4.937. The second is significant, while the first is not. We now have:

   28.181   29.5   31.083   31.923   35.8   38.0

5. Compare 38.0 − 35.8 = 2.2 to (2.829)(1.659) = 4.693. The range is not significant. Thus our final diagram is:

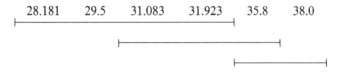

   28.181   29.5   31.083   31.923   35.8   38.0

In this case the Newman–Keuls method and Duncan's method give the same diagram; however, often they do not.

3.12 WALLER–DUNCAN PROCEDURE

This procedure is not applicable to most messy data situations because it also requires equal n_i's. As yet, no one has given a way to modify the procedure for unequal n_i's. We have decided to include the procedure here for several reasons, but primarily because we believe that the procedure is not well-known and because it seems to have some desirable properties, which are discussed later.

The Waller–Duncan procedure uses the sample data to help determine whether a conservative rule (like Tukey's HSD) or a nonconservative rule (like Fisher's LSD) is needed. The procedure makes use of the computed value of the F-test for testing H_0: $\mu_1 = \mu_2 = \cdots = \mu_t$. If the F-value is small, then the sample data tend to indicate that the means are homogeneous. In this case, the Waller–Duncan procedure requires a large absolute difference in sample means in order to declare significance, so as to prevent declaring too many differences as being significant. If the F-value is large, the sample data would tend to indicate that the means are heterogeneous. In this case, the procedure requires a smaller absolute difference in the sample means in order to declare significance, so as to prevent declaring too few differences significant.

The Waller–Duncan procedure requires choosing a constant K called the *error rate ratio*, which designates the seriousness of a Type I error relative to a Type II error.

	Typical Value		
α	.10	.05	.01
K	50	100	500

Thus, the critical points for this procedure depend on K, ν, t, and the computed value of the F-statistic for testing H_0: $\mu_1 = \mu_2 = \cdots = \mu_t$. Tables have not been included here; they are given in Ott (1977). To use the procedure, one calculates a Waller–Duncan LSD and compares all pairs of means to this single LSD value, just as one would do with Fisher's LSD procedure or Tukey's HSD procedure for the equal-sample-size problem. When one requests the MEANS/WALLER option in SAS® ANOVA, the Waller–Duncan LSD value is printed and the means are grouped into diagrams similar to those given in the last two sections.

3.13 A CAUTION

Before concluding this chapter, we point out that the underlining procedure can give rise to inconsistencies when the sample sizes are unequal. We illustrate this fact with an example.

EXAMPLE 3.2

Suppose $\hat{\sigma} = 2$ with 50 degrees of freedom. Now consider the means given in the table below:

i	*1*	*2*	*3*	*4*
$\hat{\mu}_i$	39.3	40.1	42.0	43.0
n_i	2	25	25	2

The F-statistic needed to test the hypothesis of equal means is

$$F_c = \frac{\sum n_i(\bar{y}_{i\cdot} - \bar{y}_{\cdot\cdot})^2/(t-1)}{\hat{\sigma}^2} = 4.90,$$

which is significant at the 5% level. The 5% LSD value for comparing μ_1 to μ_2, μ_1 to μ_3, μ_2 to μ_4, and μ_3 to μ_4 is $(2.008)(2)\sqrt{\frac{1}{2} + \frac{1}{25}} = 2.95$. The 5% LSD value for comparing μ_1 to μ_4 is $(2.008)(2)\sqrt{\frac{1}{2} + \frac{1}{2}} = 4.016$, and the 5% LSD value for comparing μ_2 to μ_3 is $(2.008)(2)\sqrt{\frac{1}{25} + \frac{1}{25}} = 1.135$. Thus, for these data we see that the difference between the largest and smallest mean, $\hat{\mu}_4 - \hat{\mu}_1 = 3.7$, is not significant, while the smaller difference between the two middle means, $\hat{\mu}_3 - \hat{\mu}_2 = 1.9$, is significant. One can explain this apparent inconsistency by noting that there is enough information to claim a statistically significant difference between μ_2 and μ_3, but not enough to claim a statistically significant difference between any other pairs of means.

CONCLUDING REMARKS

In this chapter, procedures for making many inferences from a single experimental data set were discussed. Such procedures are necessary to insure that differences between treatments that are observed are due to real differences in the parameter functions being compared and not to chance alone. Some procedures are more appropriate for planned comparisons, while other are more appropriate for data snooping.

Recommendations about which procedure to use in a given circumstance were also given.

4

Basics of Experimental Design

CHAPTER OUTLINE

Properly designed and analyzed experiments provide the maximum amount of information about the conditions investigated for the resources used. This chapter presents concepts and methods for the experimenter to use in designing and analyzing experiments. The basic concepts discussed in this chapter are *treatment structure* and *design structure*. Chapter 5 describes the concept of the *size of the experimental units* used at a particular phase of an experiment. The design structures presented include the completely randomized, randomized complete block, Latin square, and incomplete block designs. The treatment structures considered include the one-way, two-way with controls, and n-way structures. In this chapter, the models and analysis of variance tables with necessary sources and degrees of freedom are presented. The computation of sums of squares are discussed in later chapters. The basic approach in this chapter is to demonstrate the concepts with examples. The split-plot, repeated measures, and crossover designs use the concept of different sizes of experimental units and are described in Chapter 5. Designs involving nesting are also discussed in Chapter 5.

4.1 INTRODUCING BASIC IDEAS

Experimental design is concerned with planning experiments in order to obtain the maximum amount of information from the available resources. Often when designing an experiment, the experimenter has control over certain effects called treatments, populations, or treatment combinations. The experimenter generally controls the choice of the experimental units and whether those experimental units need to be put into groups, called blocks. A typical experiment involves t treatments (or treatment combinations) that are to be compared or whose effects are to be studied.

Before an experiment can be carried out, several questions must be answered:

1. How many treatments are to be studied? (This may already be specified, but most often it must be determined.)
2. How many times does each treatment need to be observed?
3. What are the experimental units?
4. How does the experimenter apply the treatments to the available experimental units and then observe the responses?
5. Can the resulting design be analyzed or can the desired comparisons be made?

The answers to these questions are not necessarily straightforward and cannot be answered in a general way. Hopefully, the ideas and concepts discussed here will help the experimenter put together enough information to provide answers.

To continue, consider an experiment involving t treatments in which each treatment is applied to r different experimental units. A mathematical model that can be used to describe y_{ij}, the response observed from the jth experimental unit of the ith treatment, is

$$y_{ij} = \mu_i + \varepsilon_{ij} \qquad i = 1, 2, \ldots, t, \quad j = 1, 2, \ldots, r, \qquad (4.1.1)$$

where μ_i is the true but unknown mean of the responses of the ith treatment and ε_{ij} is a random variable representing the noise resulting from natural variation and other possible sources of random and nonrandom error.

In order to conduct this experiment, the researcher must select rt experimental units and then randomly assign each treatment to r of the experimental units. The randomization part, which is very important, is used to develop the theory for a correct analysis. The very least that can be said about the use of randomization is that it prevents the introduction of systematic bias into the experiment. If the experimenter does not use randomization, then she cannot tell whether an observed difference is due to differences in treatments or due to the systematic method used to assign the treatments to the experimental units.

Since the objective of an experiment is to compare the observed response of treatments on experimental units, the more alike the experimental units are, the better the comparisons between treatments. In most experiments, it is impossible to select rt identical experimental units. The nonidentical experimental units contribute to the noise, ε_{ij}. Thus, experiments are improved if the experimenter can group the experimental units into groups of very nearly alike experimental units. Experimental units that are nearly alike are called *homogeneous*. When this is the case, the treatments can be compared on the similar experimental units where the group variation can be accounted for in the analysis. Those groups of similar experimental units are called *blocks*.

Let there be r blocks with t experimental units, each treatment occurring in each block. A model that represents the observed response of the ith treatment in the jth block is

$$y_{ij} = \mu_i + b_j + \varepsilon_{ij}^* \qquad i = 1, 2, \ldots, t, \quad j = 1, 2, \ldots, r. \qquad (4.1.2)$$

For model (4.1.2), the ε_{ij}'s in model (4.1.1) have been replaced by $\varepsilon_{ij} = b_j + \varepsilon_{ij}^*$; that is, the variation between groups or blocks of experimental units has been identified and isolated from ε_{ij}^*.

Two treatments can be compared, free of block effects, by taking within-block differences of the responses of the two treatments as

$$y_{ij} - y_{i'j} = \mu_i - \mu_{i'} + \varepsilon_{ij}^* - \varepsilon_{i'j}^*,$$

which does not depend on the block effect b_j.

An objective of experimental design is to select and group the experimental material so that the noise or experimental error in the experiment is reduced. Thus, the experimental units on which the treatments are to be compared should be as much alike as possible so that a smaller significant difference between two treatments can be detected.

If there are t treatments and t experimental units, an experiment can be conducted and the mean of each treatment can be estimated from the data. But an estimate of the error variance cannot be obtained unless some or all of the treatments are replicated. A *replication* of a treatment is an independent observation of the treatment, and thus two replications of a treatment must involve two experimental units. It is very important that this definition be observed during an experiment. Too often researchers use duplicate or split samples to generate two observations and call them replicates, when, in reality, they are actually subsamples or repeated measures. They certainly are not replicates. For example, two independent measurements of the height of one person does not provide a measure of the true variation in the heights of the population of people. They are two subsamples or repeated measures.

Consider an experiment for comparing the abilities of three preservatives to inhibit mold growth on a certain type of cake. The baker makes one cake with each preservative. After nine days of storage, the number of mold spores per cubic centimeter of cake is measured. The baker wanted ten "replications" for the analysis, so he split each cake into ten parts and obtained the spore count on each part. However, those ten measurements did not result from ten independent applications of the preservative. The variation measured by his subsamples is an index of the within-cake variation and not experimental-unit-to-experimental-unit variation. To have ten replications, the baker needs to bake ten cakes with each preservative, each one mixed independently of the other.

Another example of nonreplication involves what some researchers call a *strip trial*. In agronomy, the strip trial consists of planting all the seed of a given variety of plant in one row, each row planted with a different variety. The rows are then partitioned into, say, eight parts, and the parts are called "replications" (see Figure 4.1).

The advantage of using a strip trial instead of eight independent replications is that the researcher need not continually change the seed from planter to planter in the strip trial, whereas in good designs she would have to change the seed several times as dictated by the randomization scheme. If the experimenter analyzes such an experiment as she would a randomized complete block design, her analysis will be incorrect. In fact, the strip trial experiment cannot be used to compare variety differences since there is only one observation of each variety. The row is the experimental unit to which the variety is applied.

In the strip trial, the researcher could have just as easily partitioned the rows into 20 parts or 100 parts; after all, with more "replications," one can detect smaller differences between two means as being signifi-

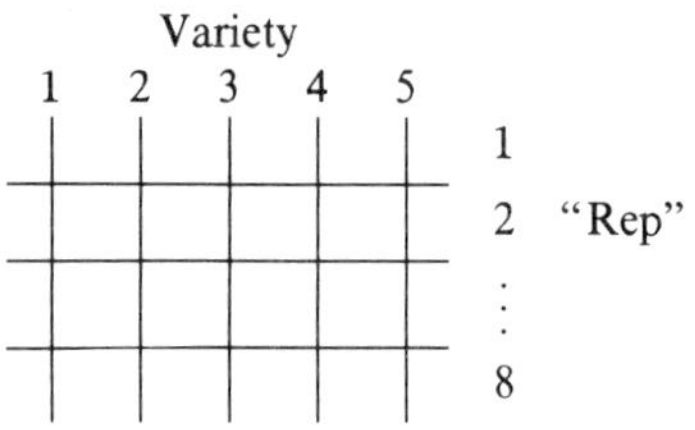

Figure 4.1 Schematic of a strip trial with five varieties arranged in eight pseudo-replications.

cant. (But these are not true replications—they are subsamples. Thus, more and more parts do not aid in detecting differences between the means). One test for determining whether a part or an observation is a true replication is the following: If the researcher could have just as easily obtained more "replications" by splitting, then she is not obtaining true replications but is obtaining subsamples or repeated measures.

It is very important to distinguish between a subsample and a replication, since the error variance estimated from between subsamples is in general considerably smaller than the error variance estimated from between replications or experimental units. Thus results from F-tests constructed by using the error variance computed from subsamples will be much larger than they should be, leading the experimenter to determine more differences as being significant than she should.

4.2 STRUCTURES OF AN EXPERIMENTAL DESIGN

An experimental design consists of two basic structures, and it is very important to be able to identify and distinguish between each structure.

Definition 4.1: The *treatment structure* of an experimental design consists of the set of treatments, treatment combinations, or populations that the experimenter has selected to study and/or compare.

The treatment structure is constructed from those factors or treatments to be compared as measured by their effect on given response variables. The treatment structure could be a set of t treatments, called a one-way treatment structure, or a set of treatment combinations, such as a two-way factorial arrangement or a higher-order factorial arrangement, plus any controls or other standard treatments.

Definition 4.2: The *design structure* of an experimental design consists of the grouping of the experimental units into homogeneous groups or blocks.

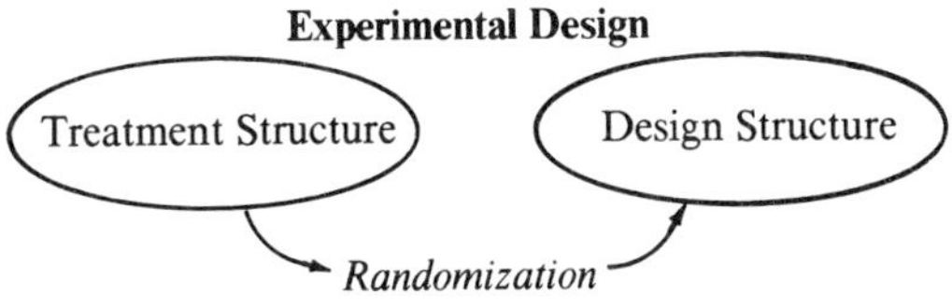

Figure 4.2 Combining the treatment structure with the design structure to form an experimental design.

The design structure of an experiment involves the grouping of experimental units so that the conditions under which the treatments are observed are as uniform as possible. If all the experimental units are very homogeneous, then there need only be one group or block of observations, and the experimental units can be assigned to the treatments completely at random. Such a design structure is called a *completely randomized design*.

If more than one group of experimental units is required so that the units within a group are much more homogeneous than experimental units between groups, then the design structure is some type of a blocked design. Once the treatment structure and design structure have been selected, the experimental design is specified by describing exactly the method of randomly assigning (randomizing) the treatments of the treatment structure to the experimental units in the design structure. Thus, the experimental design involves (1) the choice of the treatment structure, (2) the choice of the design structure, and (3) the method of randomization. Figure 4.2 represents the parts of the experimental design.

The total experimental design dictates the appropriate model to be used to obtain a correct analysis. In constructing the model, two basic assumptions are made. *First*, it is assumed that the components of the design structure are random effects, that is, the blocks used are a random sample from the population of possible blocks of experimental units. *Second*, it is assumed that there is no interaction between the components of the design structure and the components of the treatment structure. In other words, it is assumed that the relationships existing between the treatments will be consistent from block to block (except for random variation), or, stated another way, that the blocks will not influence the relationship between the treatments.

The design structure is selected by using all available knowledge of the experimental units and is chosen independently of the treatment structure (don't let the treatment structure dictate using a poor design structure). Likewise, the experimenter should select the treatment structure without any knowledge of the design structure (don't let the design structure hamper selecting the necessary set of treatments). After the appropriate design structure is specified and the desired treatment structure selected, some compromises may be needed in either one or both of the structures to make them compatible and to enable the experimenter to conduct an effective experiment.

4.2.1 Types of Design Structures

The design structure is determined by the type of blocking or grouping of the experimental units into homogeneous groups. The following are descriptions of some common design structures, each of which will be examined in more detail in Section 4.3.

1. *Completely Randomized Design*. In a completely randomized design structure, all experimental units are assumed to be homogeneous and the treatments are assigned to the experimental units completely at random. Generally, the treatments are assigned to an equal number of experimental units, although this is not required. This design structure may also be used when the experimental units are not homogeneous and the experimenter cannot group them into more homogeneous groups.

2. *Randomized Complete Block Design*. If there are t treatments, then the randomized complete block design structure consists of blocks of t experimental units each. Each treatment is randomly assigned to one experimental unit in each block. If each block consists of $c \times t$ experimental units where c is an integer, then each treatment can be assigned to c experimental units within each block. This is also a randomized complete block design structure. A randomized complete block design structure is any blocking scheme in which the number of experimental units within a block is a multiple of the number of treatments, and thus a complete set of treatments can be assigned completely at random to c experimental units in each block.

3. *Latin Square Design*. The Latin square design structure consists of blocking in two directions. For an experiment involving t treatments, t^2 experimental units are arranged into a $t \times t$ square where the rows are called *row blocks* and the columns are called *column blocks*. Thus the $t \times t$ arrangement of experimental units is blocked in two directions (row blocks and column blocks). To construct a Latin square experimental design, the treatments are randomly assigned to experimental units in the square such that each treatment occurs once and only once in each row block and once and only once in each column block. See Cochran and Cox (1957) for various arrangements of treatments into row and column blocks.

4. *Incomplete Block Designs*. Incomplete block designs occur when the number of treatments exceeds the number of experimental units in a block, so that a complete set of treatments cannot occur within each block.

5. *Various Combinations and Generalizations*. There are various ways to group the experimental units. Sometimes a grouping does not satisfy the above definitions but still provides a valid design structure. An

example is where the block sizes vary from block to block or where some blocks are incomplete while others are complete. In any case, these other blocking schemes can provide an experimenter with very viable design structures.

4.2.2 Types of Treatment Structures

The treatment structure consists of the various treatments or treatment combinations that the experimenter wishes to study. Next, some common treatment structures are described, each of which will be examined in more detail in Section 4.3.

1. *One-Way Treatment Structure*. The one-way treatment structure consists of a set of t treatments or populations where there is no assumed relationship among the treatments.

2. *Two-Way Treatment Structure*. A two-way treatment structure consists of the set of treatments constructed by combining the levels of two different types of treatments. The resulting set of treatments, called *treatment combinations*, is generated by combining each level of one treatment type with each level of the other treatment type. If the first treatment type has s levels and the second treatment type has r levels, they produce sr treatment combinations. Figure 4.3 presents an example of a two-way treatment structure.

3. *Factorial Arrangement Treatment Structure*. A factorial arrangement treatment structure consists of the set of treatment combinations constructed by combining the levels of two or more different types of treatments. The two-way treatment structure is a two-way factorial arrangement. A three-way on up to an n-way treatment structure is also a factorial arrangement. An n-way treatment structure is generated by combining n different types of treatments, each with $s_1, s_2, \ldots, s_n$ levels, respectively, into $s_1 \cdot s_2 \cdots s_n$ treatment combinations.

4. *Fractional Factorial Arrangement Treatment Structures*. A fractional factorial arrangement treatment structure consists of only a part, or fraction, of the possible treatment combinations in a factorial arrangement treatment structure. There are many systematic techniques for

	B_1	B_2	B_3	B_4
A_1	A_1B_1	A_1B_2	A_1B_3	A_1B_4
A_2	A_2B_1	A_2B_2	A_2B_3	A_2B_4
A_3	A_3B_1	A_3B_2	A_3B_3	A_3B_4

Figure 4.3 Two-way treatment structure where factor A has 3 levels and factor B has 4 levels, generating 12 treatment combinations.

selecting an appropriate fraction, most of which depend on the assumptions the experimenter makes about interactions between the various types of treatments in the treatment structure. A Latin square arrangement treatment structure involves a three-way factorial arrangement with n row treatments, n column treatments, and n cell treatments. The Latin square arrangement consists of n^2 of the n^3 possible treatment combinations, or a $1/n$th $= (n^2/n^3)$th fraction of the n^3 possible treatment combinations. One possible use of the Latin square arrangement is when it can be assumed there are no two-way or three-way interactions among the three types of treatments.

5. *Factorial Arrangement with One or More Controls.* Treatment structures can be generated in many different ways and often involve combining some of the above treatment structures. For example, a treatment structure for an experiment could consist of combining a one-way treatment structure of t controls with a two-way factorial arrangement treatment structure.

All of the above treatment structures can always be considered as a one-way treatment structure for analysis purposes. In particular, when the treatment structure is a complex combination of two or more treatment structures, it is usually best to consider the set of treatments as a one-way treatment structure.

Split-plot and repeated measures experimental designs are constructed from incomplete block design structures and factorial arrangement treatment structures involving two or more factors or types of treatments. In effect, the combination of the design structure and the treatment structure for the split-plot and repeated measures designs generates different sizes of experimental units that must be addressed by the analysis. Such designs are discussed in Chapter 5.

4.3 EXAMPLES OF DIFFERENT EXPERIMENTAL DESIGNS

There is a vast amount of published information about various types of experimental designs, for example, see Cochran and Cox (1957), Davies (1954), Federer (1955), Hicks (1973), John (1971), Kirk (1968), and Winer (1971). This section contains several examples that demonstrate the design structures and treatment structures described in Section 4.2. Hopefully, this will help readers to apply these concepts to their own experiments. In most examples, the experimental design is named by specifying the type of design structure and the type of treatment structure. For example, an experimental design could be a two-way treatment structure in a randomized complete block design structure. This method of describing an experimental design differs from that generally used in the literature, but the authors feel that this is the best way to identify the experimental design; in addition, it also helps one to construct an appropriate model and develop the correct analysis. For each experimental situation, the design structure and the treatment structure are specified and the corresponding model and the resulting analysis of variance table with the sources of variation and corresponding degrees of freedom are given. The formulas for computing sums of squares are not given in this section, but some examples with computations are included in other chapters.

EXAMPLE 4.1: Diets

A nutritionist wants to study the effect of five diets on losing weight. The treatment structure of this experiment is a one-way classification involving five treatments (diets). Many different design structures can be selected. If there are 20 homogeneous people, then a completely randomized design structure can be used where each diet is randomly assigned to 4 people. One model for a one-way treatment structure in a completely randomized design structure is

$$y_{ij} = \mu_i + \varepsilon_{ij} \qquad i = 1, 2, \ldots, t, \quad j = 1, 2, \ldots, n, \qquad (4.3.1)$$

where μ_i denotes the mean of the ith treatment (diet) and ε_{ij} denotes the random error. The analysis of variance table for model (4.3.1), listing its sources of variation and degrees of freedom, is given in Table 4.1.

If there are 10 males and 10 females instead of 20 homogeneous people, then sex of person could be used as a blocking factor; a randomized block design structure could be used, each diet being randomly assigned to two males and two females. The model for a one-way treatment structure in a randomized complete block design structure is

$$y_{ij} = \mu_i + b_j + \varepsilon_{ij} \qquad i = 1, 2, \ldots, t, \quad j = 1, 2, \ldots, n \quad (4.3.2)$$

where y_{ij}, μ_i, and ε_{ij} are as defined in (4.3.1) and b_j denotes the effect of block j. The analysis of variance table for model (4.3.2) is given in Table 4.2.

Table 4.1 Analysis of Variance Table for a One-Way Treatment Structure in a Completely Randomized Design Structure

Source of Variation	*df*
Diet	4
Error	15

Table 4.2 Analysis of Variance Table for a One-Way Treatment Structure in a Randomized Complete Block Design Structure

Source of Variation	*df*
Block (Sex of Person)	1
Diet	4
Error	14

In some cases, sex of person is not a good choice for a blocking factor, since it can also be a type of treatment. In that case, the treatment structure is a two-way factorial arrangement or a two-way treatment structure consisting of the ten treatment combinations generated by combining the two levels of sex of person with the five levels of diet. The design structure would be a completely randomized design with each treatment combination observed twice.

One model for a two-way treatment structure in a completely randomized design structure is

$$y_{ijk} = \mu_{ij} + \varepsilon_{ijk} \qquad i = 1, 2, \ldots, a, \quad j = 1, 2, \ldots, b, \quad k = 1, 2, \ldots, n \tag{4.3.3}$$

Table 4.3 Analysis of Variance Table for a Two-Way Treatment Structure in a Completely Randomized Design Structure

Source of Variation	*df*
μ_{ij} Model	
Sex * Diet	9
Error	10
$\mu + \alpha_i + \beta_j + \gamma_{ij}$ Model	
Sex	1
Diet	4
Sex * Diet	4
Error	10

where μ_{ij} is the mean of the combination of sex i and diet j. Sometimes the mean μ_{ij} is expressed as

$$\mu_{ij} = \mu + \alpha_i + \beta_j + \gamma_{ij}$$

where μ is the overall mean, α_i is the effect of the ith sex, β_j is the effect of the jth diet, and γ_{ij} is the interaction effect. The analysis of variance tables for model (4.3.3) for both expressions of μ_{ij} are given in Table 4.3.

Next, suppose that the diets have a structure consisting of a control diet and four diets made up of the four combinations of two protein levels and two carbohydrate levels. The Diet treatment structure is a two-way factorial arrangement with a control that, when crossed with Sex of person, generates a three-way treatment structure with two controls (one for males and one for females). The design structure is completely randomized. The model is

$$y_{ijk} = \mu_{ij} + \varepsilon_{ijk} \qquad i = 1,2, \quad j = 0,1,2,3,4, \quad k = 1,2, \qquad (4.3.4)$$

where μ_{10} and μ_{20} denote the controls and the μ_{ij}'s, $i = 1,2$ and $j = 1,2,3,4$, denote the Sex of person by Diet treatment combinations. The analysis of variance table for model (4.3.4) is in Table 4.4, where

Table 4.4 Analysis of Variance Table for a Three-Way Treatment Structure with Two Controls

Source of Variation	*df*	
Sex	1	
Diet	4	
Control vs. 2^2		1
Protein		1
Carbohydrate		1
Protein * Carbohydrate		1
Sex * Diet	4	
Sex * Control vs. 2^2		1
Sex * Protein		1
Sex * Carbohydrate		1
Sex * Carbohydrate * Protein		1
Error	10	

"Control vs. 2^2" denotes a comparison between the control diet and the average of the four protein × carbohydrate treatment combinations.

EXAMPLE 4.2: House Paint

A paint company wants to compare the abilities of four white house paints to withstand environmental conditions. Four square houses, each with one side facing exactly north, were available for the experiment; thus houses can be used as a blocking factor.

Each side of a house is possibly exposed to different types of weather, thus the sides (indicated here by directions—north, south, east, and west) of the houses can also be used as a blocking factor. Since the number of treatments (the four paints) was the same as both block sizes, a Latin square design structure can be used, where each paint can occur once and only once on each house and once and only once in each direction. One such arrangement of paints to houses and directions is shown in Table 4.5. (The randomization scheme is to randomly assign houses to row blocks and directions to column blocks.)

The experimental design is a one-way treatment structure in a Latin square design structure. A corresponding model is

$$y_{ijk} = \mu_i + h_j + d_k + \varepsilon_{ijk} \quad i = 1,2,3,4, \quad j = 1,2,3,4, \quad k = 1,2,3,4, \tag{4.3.5}$$

Table 4.5 Arrangement of One-Way Treatment Structure in a Latin Square Design Structure

	DIRECTION			
House	*N*	*S*	*E*	*W*
1	*A*	*B*	*C*	*D*
2	*D*	*A*	*B*	*C*
3	*C*	*D*	*A*	*B*
4	*B*	*C*	*D*	*A*

Notes: *A*, *B*, *C*, and *D* denote the four paints. Directions are randomly assigned to columns, and houses are randomly assigned to rows.

Table 4.6 Analysis of Variance Table for One-Way Treatment Structure in a Latin Square Design Structure

Source of Variation	*df*
House	3
Direction	3
Paint	3
Error	6

where μ_i denotes the mean for paint i, h_j denotes the effect of the jth house, d_k denotes the effect of the kth direction, and ε_{ijk} denotes the error. The analysis of variance table for model (4.3.5) is given in Table 4.6.

Next, suppose the paints have a structure as given by (1) base paint, (2) base plus additive I, (3) base plus additive II, (4) base plus additive I plus additive II. This is a two-way treatment structure where one treatment type is additive I with two levels, zero and some, and the second treatment type is additive II with two levels, zero and some. The resulting four treatment combinations are shown in Table 4.7.

One model for a two-way treatment structure in a Latin square design structure is

$$y_{ijkm} = \mu + \gamma_i + \beta_j + (\gamma\beta)_{ij} + h_k + d_m + \varepsilon_{ijkm} \qquad (4.3.6)$$

where γ_i denotes the effect of additive I, β_j denotes the effect of additive II, and $(\gamma\beta)_{ij}$ denotes the interaction between the two additives. The analysis of variance table for model (4.3.6) is given in Table 4.8. The only

Table 4.7 Two-Way Treatment Structure for House Paint Example

	Additive II	
Additive I	*None*	*Some*
None	Base	Base + I
Some	Base + I	Base + I + II

Table 4.8 Analysis of Variance Table for Two-Way Treatment Structure in a Latin Square Design Structure

Source of Variation	*df*	
House	3	
Direction	3	
Paint	3	
I		1
II		1
I∗II		1
Error	6	

difference between analyzing models (4.3.5) and (4.3.6) is that in (4.3.6) the paints have a structure that is used to partition the paint effect into effects due to additive I, additive II, and the interaction of additive I and additive II. The part of the analysis corresponding to the design structure remains unchanged, even though the analysis of the treatment structure was changed.

Finally, suppose eight houses were available so that the experiment could be conducted by using two repeated Latin square design structures. Table 4.9 shows one possible assignment of paints to the house–direction combinations.

Table 4.9 Arrangement Showing a One-Way Treatment Structure in a Repeated Latin Square Design Structure

	House							
	Square 1				Square 2			
Direction	*1*	*2*	*3*	*4*	*5*	*6*	*7*	*8*
N	*C*	*A*	*B*	*D*	*D*	*C*	*A*	*B*
S	*D*	*B*	*C*	*A*	*C*	*B*	*D*	*A*
E	*A*	*C*	*D*	*B*	*A*	*D*	*B*	*C*
W	*B*	*D*	*A*	*C*	*B*	*A*	*C*	*D*

Note: *A*, *B*, *C*, and *D* denote the paints.

Table 4.10 Analysis of Variance Table for a Two-Way Treatment Structure in a Repeated Latin Square Design Structure

Source of Variation	*df*	
House	7	
Square		1
House (Square)		6
Direction	3	
Paint	3	
I		1
II		1
I∗II		1
Error	18	

If the paints have the two-way treatment structure of Table 4.7, then a model is given by

$$y_{ijkmn} = \mu + \gamma_i + \beta_j + (\gamma\beta)_{ij} + s_k + h_{km} + d_n + \varepsilon_{ijkmn} \quad (4.3.7)$$

where s_k denotes square k and h_{km} denotes house m in square k. The analysis of variance table for model (4.3.7) is given in Table 4.10.

EXAMPLE 4.3: Steel Plates

A Latin square design structure is very useful when there is a need to block in two directions, but every Latin square arrangement used by experimenters is not a Latin square design structure. This example demonstrates the consequences of using a Latin square arrangement treatment structure.

Two types of paint additives are to be combined and steel plates are to be painted. The objective of the experiment is to study the ability of the paint combinations to protect steel from heat. There are five levels of each additive and five temperatures at which to check the protecting ability.

This experiment is suited for a Latin square array where the levels of additive I are the rows, the levels of additive II are the columns, and the levels of temperature are assigned to the cells within the square. This arrangement generates 25 treatment combinations. The experimental units are 25 sheets of steel 0.2 cm thick and 1 m^2 in area. The randomization process used should randomly assign one of the 25 treatment combinations to each of the 25 sheets of steel.

In this case, the treatment structure is a $\frac{1}{5}$ fraction of a 5^3 factorial arrangement (as it consists of 25 of the 125 possible treatment combinations), called a Latin square arrangement. The design structure is a completely randomized design, as the treatment combinations are assigned completely at random to the sheets of steel. Since this is a fractional factorial, each main effect is partially aliased (see Cochran and Cox (1957), p. 245) with the two-factor interaction of the other two factors and the three-factor interaction.

In order to properly analyze this experimental design, some assumptions must be made about the parameters in the model. The usual assumptions are that there are no two-way interactions and no three-way interaction. However, one should be very careful not to make such assumptions without having some prior information (which can come from other experiments, existing literature, and so on) showing that the interactions are, in fact, negligible. One such Latin square arrangement is given in Table 4.11, and a model for a Latin square treatment structure in a completely randomized design structure is

$$y_{ijk} = \mu + A1_i + A2_j + T_k + \varepsilon_{ijk} \tag{4.3.8}$$

where $A1_i$ denotes the effect of the ith level of additive I, $A2_j$ denotes the

Table 4.11 Latin Square Arrangement Treatment Structure

Level of Additive I	Level of Additive II				
	1	2	3	4	5
1	T_1	T_2	T_3	T_4	T_5
2	T_5	T_1	T_2	T_3	T_4
3	T_4	T_5	T_1	T_2	T_3
4	T_3	T_4	T_5	T_1	T_2
5	T_2	T_3	T_4	T_5	T_1

Note: T_1, T_2, T_3, T_4, and T_5 denote the temperatures.

Table 4.12 Analysis of Variance Table for Latin Square Treatment Structure in a Completely Randomized Design Structure

Source of Variation	*df*
Additive I	4
Additive II	4
Temperature	4
Residual	12

effect of the jth level of additive II, and T_k denotes the effect of the kth level of temperature.

The analysis of variance table for model (4.3.8) is given in Table 4.12. The term *Residual* is used rather than *Error*, since the corresponding sum of squares involves error plus any interaction effects that may not be zero. If the assumption of zero interactions is not correct, then the Residual mean square will be too large and the resulting F-tests will be too small. Consequently, if there is interaction in the experiment, it cannot be discovered and any other detectable treatment effects may be masked.

EXAMPLE 4.4: Levels of N and K

A model and the resulting analysis consists of three basic components, namely, the model's treatment structure, design structure, and error structure(s). This example demonstrates how the three basic components can be used.

A plant breeder wants to study the effect of three levels of nitrogen and four levels of potassium on his new variety of corn. His treatment structure is a two-way factorial arrangement with 12 (3 levels of $N \times 4$ levels of K) treatment combinations. He has three plots of land called blocks, each of which is partitioned into 12 parts. Each treatment combination is assigned at random to one part in each plot. Thus, the design structure is a randomized complete block, since each treatment combination occurs once in each block. The experimental design is a two-way treatment structure in a randomized complete block design structure. (Blocks in a randomized complete block design are called replications by some authors; however, we prefer to call them blocks or blocked replications in order to distinguish them from replications in the

Table 4.13 Analysis of Variance Table for the General Model

Source of Variation	*df*
Design Structure	(df)
Treatment Structure	(df)
Error Structure(s)	(df['s])

completely randomized design. Example 4.5 discusses differences between blocks and replications.)

The model for this example is

$$y_{ijk} = \mu_{ij} + b_k + \varepsilon_{ijk} \tag{4.3.9}$$

where μ_{ij} is the mean of the ith level of N with the jth level of K, b_k is the effect of the kth block, and ε_{ijk} denotes the random error. In general, when a model is constructed, it involves the sum of three parts:

$$y = \text{Treatment Structure} + \text{Design Structure} + \text{Error Structure(s)} \tag{4.3.10}$$

Likewise, the corresponding analysis of variance table has three parts. The general analysis of variance table for model (4.3.10) is given in Table 4.13. The analysis of variance table for model (4.3.9) is given in Table 4.14.

Table 4.14 Analysis of Variance Table for a Two-Way Treatment Structure in a Randomized Complete Block Design Structure

Source of Variation	*df*	
Design		
Blocks	2	
Treatment	11	
N		2
K		3
$N * K$		6
Error	22	

In general, allowance must be made for the possibility of more than one error term. For example, split-plot and repeated measures models will have more than one error term (see Chapter 5).

EXAMPLE 4.5: Blocks and Replications

In many of the textbooks on experimental design, there is either no distinction made between blocks and replications or, at the very least, there is confusion about the distinction. This example is included to demonstrate the difference between the two concepts.

Suppose the researcher wants to study four treatments in a one-way treatment structure and only has two homogeneous experimental units per block. In this case, the design structure that must be used is an incomplete block design. If there are enough blocks so that every pair of treatments can occur together in a block the same number of times, then it is possible to use a balanced incomplete block design (Cochran and Cox, 1957). For example, the four treatments could be assigned to blocks as shown in Table 4.15. In this case, there are six blocks in the design, and each treatment is replicated three times. This points out that blocks and replications are not the same concept. They are equivalent only for the randomized complete block design structure, where each treatment is observed once and only once in each block. In this example, the design structure is associated with the six blocks (see Table 4.14), not the three replications that just happen to occur. The model for the arrangement in

Table 4.15 Assignment of Treatments to Blocks for Example 4.5

Blocks	*Treatments*
1	1 and 2
2	1 and 3
3	1 and 4
4	2 and 3
5	2 and 4
6	3 and 4

Note: Treatments should be randomly assigned to the experimental units within each block.

Table 4.16 Analysis of Variance Table for a One-Way Treatment Structure in an Incomplete Block Design Structure

Source of Variation	*df*
Block	5
Treatment	3
Error	3

Table 4.15 is

$$y_{ij} = \mu_i + b_j + \varepsilon_{ij} \tag{4.3.11}$$

where the pair (i, j) can take on only those values of Treatment $*$ Block combinations that are observed. The analysis of variance table for model (4.3.11) is given in Table 4.16.

There are many other ways to construct experimental designs by combining various design structures and treatment structures. Hopefully, the above examples will enable the experimenter to construct the desired experimental design, construct an appropriate model, and develop the proper analysis.

CONCLUDING REMARKS

This chapter presented concepts and methods experimenters can use in designing and analyzing experiments. The basic ideas of good experimental design were also introduced. All experimental designs consist of two basic features: the treatment structure and the design structure. These concepts are generally not used in other statistical analysis books. Understanding the difference between these two features of an experimental design will help data analysts select appropriate analyses for their experiments.

5

Experimental Designs Involving Several Sizes of Experimental Units

CHAPTER OUTLINE

Statistical consultants do not always get the chance to design an experiment; instead, they must identify the type of experimental design that an experimenter has employed. One important step in the identification process is to determine if more than one size of experimental unit has been used and, if so, to identify the size of each experimental unit. After the different sizes of the experimental units have been identified, the model for carrying out the analysis can be constructed by using the design structure and treatment structure corresponding to each size of experimental unit.

5.1 IDENTIFYING SIZES OF EXPERIMENTAL UNITS

The experimental designs that have several sizes of experimental units (SSEU) are repeated measures designs, split-plot type designs, some nested type designs, and designs involving various combinations of the above.

Repeated measures and split-plot type designs have identical structures, although the assumptions used to develop the analyses can be different. Split-plot designs evolved from the agricultural sciences (their analyses are discussed in Chapters 24, 25, and 28); repeated measures designs are used extensively in the social and biological sciences (their analyses are presented in Chapters 26, 27, 28, 32, and 33). Nested designs differ from the repeated measures and split-plot designs in treatment structure (the nested model is analyzed in Chapter 30).

SSEU designs have two important characteristics. First, the treatments consist of at least a two-way set of treatment combinations, and the design structure consists of incomplete blocks. The second characteristic, which distinguishes the SSEU designs from those in Chapter 4, is that more than one size of experimental unit is used in an experiment. Each size of experimental unit has its own design structure and treatment structure, and the model can be constructed from the structures for each size of experimental unit. Since there is more than one experimental unit size, there is more than one error term used in the analysis; that is, there is one error term for each size of experimental unit in the experiment, which is also reflected in the model.

This chapter presents several examples that demonstrate the principles needed to properly identify the experimental design used. Once the experimenter is able to use the principles to identify experimental designs, he or she can also use them to design other experiments.

SSEU experimental designs can be structured in many different ways. The next series of examples demonstrates how to look for the different sizes of experimental units and then use this information to construct an appropriate model on which to base the analysis. Each example includes an analysis of variance table that lists the sources of variation and the degrees of freedom. It is important for the experimenter to list the appropriate sources of variation and the corresponding degrees of freedom for an analysis before subjecting the data to computer analysis, because this provides an excellent check on whether the experimenter has given the computer the appropriate instructions.

5.2 SPLIT-PLOT DESIGNS

Split-plot designs are used mainly in agricultural, industrial, and biological research, but they can also be used effectively in other areas. Split-plot designs involve a two- or higher-way treatment structure with an incomplete block design structure and at least two different sizes of experimental units. The feature that distinguishes split-plot designs from repeated measures designs is that the levels of the treatments can be applied to the various sizes of experimental units by using randomization; in contrast, repeated measures designs involve a step where the levels of at least one treatment (usually time) cannot be assigned at random. The examples given below demonstrate the uses of split-plot designs and provide guides for identifying a proper design.

EXAMPLE 5.1: Cooking Beans

An experimenter wants to study how five varieties of beans respond to three cooking methods. The dependent variables of interest are the tenderness of the beans and their flavor. The experimenter has a field consisting of 15 homogeneous rows, which forms a completely randomized design structure. He randomly assigns each one of the five varieties of the one-way treatment structure to three rows, thus generating a one-way treatment structure in a completely randomized design structure. The varieties are assigned to the rows; hence, the rows are the experimental units for varieties. At harvest time, the beans from each row are put into a box. For some measurement made on a box of beans (or a row), the model is

$$y_{ij} = \mu_i + \varepsilon_{ij} \qquad i = 1, 2, 3, 4, 5, \quad j = 1, 2, 3,$$

where μ_i represents the mean of the ith variety and ε_{ij} denotes the error associated with the ith variety being assigned to the jth row. An analysis of variance table to compare varieties is given in Table 5.1.

Table 5.1 Analysis of Variance Table for Comparing Varieties in the Cooking Beans Example

Source of Variation	*df*
Variety	4
Error(Row)	10

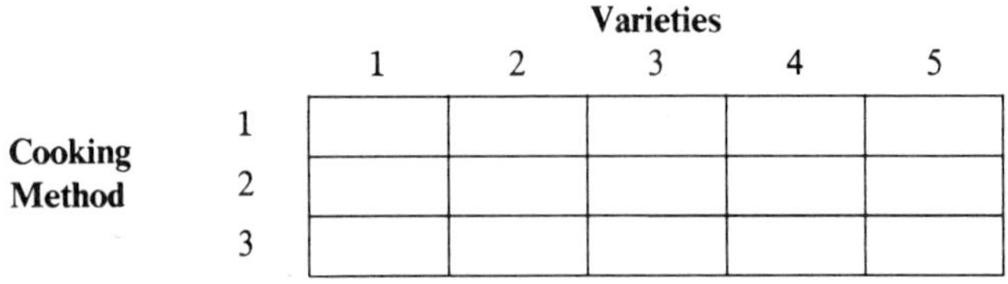

Figure 5.1 Arrangement resulting from assigning cooking methods to boxes of beans from each variety.

Next, the experimenter wants to examine the cooking methods. There are two possible ways to carry out this part of the experiment. First, the experimenter could assign a different cooking method to each of the three rows planted with a given variety. Such an assignment would yield the arrangement shown in Figure 5.1.

The arrangement in Figure 5.1 produces a two-way treatment structure in a completely randomized design structure where a row is the experimental unit. There is only one replication of each treatment combination, and hence there is no measure of the experimental error variance. The resulting analysis of variance table is shown in Table 5.2.

A design with zero degrees of freedom for error is not very desirable (although some analyses can be done by using the two-way nonreplicated experiment techniques discussed in Milliken and Graybill, 1970, and Johnson and Graybill, 1972). However, another way of assigning the cooking methods avoids the zero degrees of freedom problem but does make the experimental design and analysis more complex.

The alternative method is to split each box of beans (a box is obtained from one row) into three batches, and then randomly assign the cooking treatments to the three batches within a row. Since a cooking method is assigned to a batch, the experimental unit for the cooking

Table 5.2 Analysis of Variance Table for Two-Way Treatment Structure Shown in Figure 5.1

Source of Variation	*df*
Variety	4
Cooking Method	2
$C * V$	8
Error(Row)	0

		Variety 1	2	3	4	5
	1	C_1 C_2 C_3	C_2 C_1 C_3	C_1 C_3 C_2	C_3 C_1 C_2	C_1 C_3 C_2
Row	2	C_1 C_2 C_3	C_3 C_2 C_1	C_1 C_2 C_3	C_3 C_2 C_1	C_1 C_2 C_3
	3	C_1 C_2 C_3	C_3 C_1 C_2	C_2 C_3 C_1	C_2 C_3 C_1	C_3 C_1 C_2

The symbols C_1, C_2, and C_3 denote the three cooking methods assigned to the batches within a row.

Figure 5.2 Assignment of cooking treatments to batches within a row.

method is a batch. Thus, there are two sizes of experimental units for this experiment, the row (larger size) for varieties and the batch (smaller size) for cooking treatments. Such an assignment provides the array of data in Figure 5.2.

The treatment structure for the batch experimental units is a one-way set of treatments, and the design structure for a batch is a randomized complete block design where the rows (or boxes) are the blocks. The analysis of variance table for this part of the design is given in Table 5.3.

The sums of squares for rows consists of the sum of squares for Variety plus the sum of squares for Error (Row). Thus the Cooking Method * Row interaction can be partitioned into the Cooking Method * Variety sum of squares plus the Error(Batch) sum of squares. The parts Cooking Method, Variety * Cooking Method, and Error(Batch) are called the within-row, subplot, or batch comparisons, that is, it is the

Table 5.3 Initial Analysis of Variance Table for Batch Experimental Unit

Source of Variation	*df*
Row	14
Cooking Method	2
Cooking Method * Row	28

Table 5.4 Analysis of Variance Table for Cooking Bean Experiment Showing Analysis for Each Size of Experimental Unit

Source of Variation	*df*
Row Analysis	
Variety	4
Error(Row)	10
Batch Analysis	
Row	14
Cooking Method	2
Variety * Cooking Method	8
Error(Batch)	20

analysis for the batch experimental units. A complete analysis of variance table is given in Table 5.4.

The model used to describe data from Figure 5.2 is

$$y_{ijk} = \mu_{ij} + r_{ik} + \varepsilon_{ijk} \tag{5.2.1}$$

where μ_{ij} denotes the mean of variety i with cooking method j, r_{ik} denotes the random effect of the kth row of variety i that is assumed to be normally distributed as $N(0, \sigma_R^2)$, ε_{ijk} denotes the random effect of the jth batch of the ikth row, which is assumed to be normally distributed as $N(0, \sigma_B^2)$. It is also assumed that r_{ik} and ε_{ijk} are independent random variables.

The mean μ_{ij} can be expressed as

$$\mu_{ij} = \mu + \tau_i + \alpha_j + \gamma_{ij}$$

where μ denotes the overall mean, τ_i denotes the effect of variety i, α_j denotes the effect of cooking method j, and γ_{ij} denotes the interaction between Variety and Cooking Method. The model of equation (5.2.1) can then be reexpressed in terms of each size of experimental unit:

$$y_{ijk} = \mu + \tau_i + r_{ik} \quad] \text{ Row Part of Model}$$

$$+ \alpha_j + \gamma_{ij} + \varepsilon_{ijk}] \text{ Batch Part of Model}$$

For balanced data, it may not be economical to analyze these data in two parts via the computer, but it is instructive to do so.

EXAMPLE 5.2: The Simplest Split-Plot Experimental Design

The simplest split-plot experimental design involves a treatment structure with two factors and an incomplete block design structure. There are two sizes of experimental units: The larger experimental units are called the *whole plots*, and the smaller experimental units are called the *subplots*. To construct such a design, the smaller experimental units are grouped into several blocks of size b where b is the number of levels of one of the factors in the treatment structure, say B, which should be the factor of primary importance to the experimenter. The number of blocks is a multiple of a where a is the number of levels of the other factor in the treatment structure, say A. If the blocks of size b are homogeneous, then the levels of A are assigned completely at random to these blocks where all experimental units in a block receive the same level of A.

With this method of assigning the levels of A to blocks, the blocks become the experimental units associated with the levels of A and are the whole plots. The whole plot experimental design consists of a completely randomized design structure and a one-way treatment structure made up of the levels of factor A. Factor A is called the *whole plot treatment*.

The experimental units within a block are the subplots, and the levels of B are randomly assigned to the experimental units (subplots) within each block. Thus, the subplot treatment structure is a one-way set of treatments and the subplot design structure is a randomized complete block design. Note that the whole plot design structure may or may not involve blocking.

For a specific example, suppose factor A has four levels, factor B has three levels, and there are 12 homogeneous whole plots (blocks) of three subplot experimental units each. The levels of factor A are randomly assigned to blocks, as shown in Figure 5.3. The whole plot experimental design is a one-way treatment structure in a completely randomized design structure. Next, randomly assign the levels of B to the experimental units within the whole plots, as shown in Figure 5.4. The subplot design structure is a randomized complete block design, and the subplot treatment structure is a one-way set of treatments involving three treatments (the levels of B).

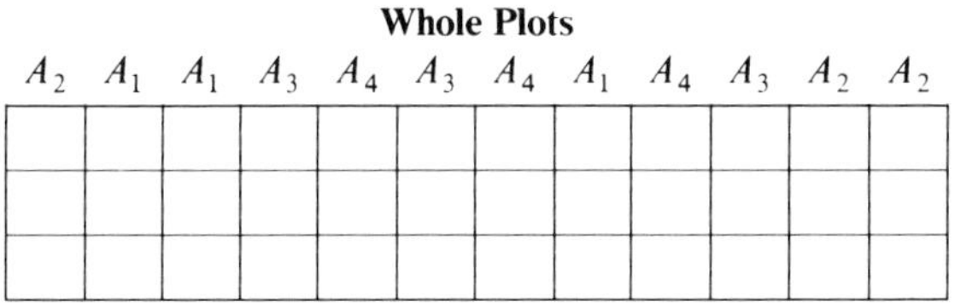

Figure 5.3 Assignment of the levels of factor A to the whole plots.

Whole Plots or Blocks

A_2	A_1	A_1	A_3	A_4	A_3	A_4	A_1	A_4	A_3	A_2	A_2
B_1	B_2	B_3	B_1	B_2	B_3	B_1	B_2	B_1	B_1	B_3	B_2
B_3	B_3	B_1	B_2	B_1	B_1	B_3	B_1	B_2	B_3	B_2	B_3
B_2	B_1	B_2	B_3	B_3	B_2	B_2	B_3	B_3	B_2	B_1	B_1

Figure 5.4 Assignment of the levels of *B* to the subplots.

After running the experiment, the analysis can be done in two parts. First, compute the whole plot means (or totals) and do a whole plot analysis of variance on these means. For this simple split-plot design, the whole plot model is

$$\bar{y}_{ij\cdot} = \mu_i + \varepsilon^*_{ij\cdot}$$

The sources of variation and the degrees of freedom in the analysis of variance table are shown in Table 5.5. Since the analysis is conducted on the means, the sums of squares need to be multiplied by n, the number of observations in each mean. The initial analysis of variance table for the subplot analysis is shown in Table 5.6. The subplot residual sum of squares includes the interaction sum of squares between factors A and B as well as the subplot error. The interaction sum of squares must be

Table 5.5 Analysis of Variance Table for Whole Plot Experimental Units

Source of Variation	*df*
A	3
Error(Whole Plot)	8

Table 5.6 Initial Analysis of Variance Table for Subplot Experimental Units

Source of Variation	*df*
Block	11
B	2
Subplot Residual	22

Table 5.7 Final Analysis of Variance Table for Subplot Experimental Units

Source of Variation	*df*
Blocks(Whole Plot)	11
B	2
$A * B$	6
Error(Subplot)	16

removed before the subplot error can be estimated. The final subplot analysis of variance table is shown in Table 5.7.

A model for this split-plot experimental design is

$$y_{ijk} = \mu_{ik} + \varepsilon_{ij} + e_{ijk}$$

or

$$y_{ijk} = \mu + A_i + \varepsilon_{ij} + B_k + (AB)_{ik} + e_{ijk}$$

where $\mu + A_i + \varepsilon_{ij}$ is the whole plot part of the model and at the same time represents the blocking structure for the subplot part of the model. The subplot part of the model is $B_k + (AB)_{ik} + e_{ijk}$. The complete analysis of variance table is shown in Table 5.8.

Table 5.8 Complete Analysis of Variance Table for Simplest Split-Plot Experimental Design

Source of Variation	*df*
Whole Plot Analysis	
A	3
Error(Whole Plot)	8
Subplot Analysis	
Block or Whole Plot	11 ← (3 + 8)
B	2
$A * B$	6
Error(Subplot)	16

Groups of Whole Plots

A_2	A_1	A_3	A_4	A_1	A_2	A_4	A_3	A_4	A_1	A_2	A_3
B_1	B_2	B_1	B_2	B_3	B_3	B_1	B_3	B_1	B_2	B_2	B_1
B_3	B_3	B_2	B_1	B_1	B_2	B_3	B_1	B_2	B_1	B_3	B_3
B_2	B_1	B_3	B_3	B_2	B_1	B_2	B_2	B_3	B_3	B_1	B_2

Figure 5.5 An example of the treatment assignments in the usual split-plot experimental design.

EXAMPLE 5.3: Usual Split-Plot Experimental Design

Most authors introduce the split-plot experimental design so that the design structure for the whole plot experimental units is a randomized complete block design. To illustrate this design, suppose the 12 blocks of Example 5.2 can be further grouped into three groups of four homogeneous blocks. Then randomly assign the levels of A to the blocks within each group of four blocks; finally, randomly assign levels of B to the subplots as before. The resulting whole plot experimental design is changed, but the subplot experimental design is unchanged. One such grouping of the blocks and the assignment of the levels of A and the levels of B is given in Figure 5.5.

A model for this example is

$$y_{ijk} = \mu + A_i + r_j + \varepsilon_{ij} + B_k + (AB)_{ik} + e_{ijk}$$

The corresponding analysis of variance table has an effect for the groups

Table 5.9 Analysis of Variance Table for Usual Split-Plot Experimental Design

Source of Variation	*df*
Whole Plot Analysis	
Rep or Group	2
A	3
Error(Whole Plot)	6
Subplot Analysis	
Block or Subplot	11
B	2
$A * B$	6
Error(Subplot)	16

of four blocks, that has been removed from the whole plot error. The groups in this case are equivalent to replications. The resulting analysis of variance table is shown in Table 5.9.

As indicated by the above table, the interaction between the levels of a whole plot treatment and the levels of a subplot treatment are within-block comparisons and thus belong in the subplot analysis. When comparisons are made between pairs of treatments, the different error terms must be taken into account. Methods to construct multiple comparisons and test contrasts are discussed in Chapter 24.

EXAMPLE 5.4: Meat in Display Case (A Complex Split-Plot Design)

A meat scientist wants to study the effect of temperature (T) with three levels, types of packaging (P) with two levels, types of lighting (L) with four levels, and intensity of light (I) with four levels on the color of meat stored in a meat cooler for seven days. Six coolers are available for the experiment, and the three temperatures (34°F, 40°F, and 46°F) are each assigned at random to two coolers, as shown in Figure 5.6.

Each cooler is partitioned into 16 compartments on a 4×4 grid (Figure 5.7). Because the light intensities are regulated by distance, all partitions in a column are assigned the same light intensity. The types of light are randomly assigned to each partition within a column. Then the two types of packaging are assigned to the steaks, and both types of packaging are put into each partition. Figure 5.7 shows how one such cooler is arranged.

In order to analyze this experiment, one must first identify the different sizes of experimental units. The experimental unit for the temperature factor is a cooler. The cooler experimental design is a one-way treatment structure (levels of T) in a completely randomized design structure. The experimental unit for the intensity factor is a column of four partitions in a cooler. The column experimental design consists of a one-way treatment structure (levels of I) in a randomized complete block design structure with 6 blocks (coolers). The experimental unit for the factor lighting type is a partition of a column. The partition experimental design is a one-way treatment structure (levels of L) in a randomized complete block design structure with 24 blocks. Finally, the experimental unit for packaging is half a partition (or steak). The

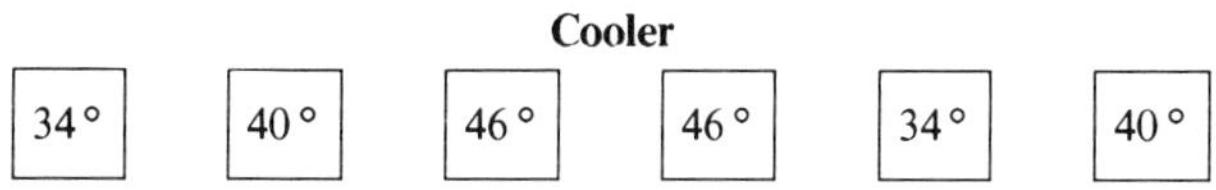

Figure 5.6. Assignment of temperatures to coolers.

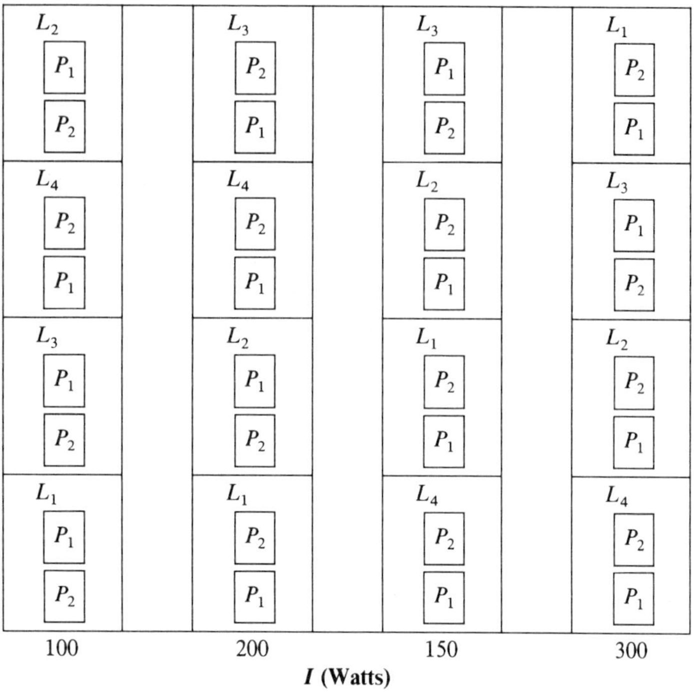

Figure 5.7 Assignment of the various treatments to the 16 compartments in each cooler.

half-partition experimental design is a one-way treatment structure (levels of P) in a randomized complete block design structure with 96 blocks.

A model for this split-split-split-plot experiment consisting of four sizes of experimental units involves four error terms and can be expressed as

$$y_{ijkmn} = \mu + T_i + a_{ij} \} \quad \text{Cooler Part of Model}$$

$$+ I_k + (TI)_{ik} + b_{ijk} \} \quad \text{Column Part of Model}$$

$$+ L_m + (TL)_{im} + (IL)_{km} (TIL)_{ikm} + c_{ijkm} \} \quad \text{Partition Part of Model}$$

$$\left.\begin{array}{l} + P_n + (TP)_{in} + (IP)_{kn} + (TIP)_{ikn} \\ + (LP)_{mn} + (TLP)_{imn} + (ILP)_{kmn} \\ + (TILP)_{ikmn} + d_{ijkmn} \end{array}\right\} \quad \text{Half-Partition Part of Model}$$

Table 5.10 Analysis of Variance Table for Meat in Cooler Example

Source of Variation	*df*
Cooler Analysis	
T	2
Error(Cooler)	3
Column Analysis	
Block	5
I	3
$I * T$	6
Error(Column)	9
Partition Analysis	
Block	23
L	3
$L * T$	6
$L * I$	9
$L * I * T$	18
Error(Partition)	36
Half-Partition Analysis	
Block	95
P	1
$P * T$	2
$P * I$	3
$P * I * T$	6
$P * L$	3
$P * L * T$	6
$P * L * I$	9
$P * L * I * T$	18
Error(Half-Partition)	48

Note: The brackets and arrows indicate which effects form the blocks for the next smaller size of experimental unit.

where

$a_{ij} \sim$ i.i.d. $N(0, \sigma_a^2)$ denotes the cooler error term,

$b_{ijk} \sim$ i.i.d. $N(0, \sigma_b^2)$ denotes the column error term,

$c_{ijkm} \sim$ i.i.d. $N(0, \sigma_c^2)$ denotes the partition error term, and

$d_{ijkmn} \sim$ i.i.d. $N(0, \sigma_d^2)$ denotes the half-partition error term.

The corresponding analysis of variance table is shown in Table 5.10.

5.3 REPEATED MEASURES DESIGNS

Repeated measures designs are effectively being used in many areas of study. These designs involve a treatment structure with at least two factors, an incomplete block design structure and at least two sizes of experimental units. A repeated measures type of design differs from a split-plot type of design in that time or sequence is usually one of the treatments in the former design and cannot be randomly assigned to its experimental unit. Thus, repeated measures involve a step or steps where there is no randomization of treatment levels, whereas in split-plot type designs there is randomization at each step. The examples below demonstrate some uses of repeated measures designs and provide a guide for identifying proper experimental designs.

EXAMPLE 5.5: Horse Feet

This particular experiment will lead to the idea of a repeated measures design where two different sizes of experimental units are used. However, if the experimenter is not careful, he or she may inadvertently miss the fact that there are two different sizes of experimental units.

A veterinarian has two techniques of fusing the joint of a horse's foot after it is broken, and she wishes to determine if one technique is better than the other. The experiment consists of taking some horses, breaking a joint on each horse, repairing it with one of the two techniques, and determining the strength of the fused joint four months later. She also wants to determine if the same techniques work equally well for front feet and back feet. Because the horses are scarce and expensive, she plans to break the joint on a front foot, wait until it heals, and then break the joint on a rear foot, or vice versa. This healing process also introduces a time factor into the design. Thus, the treatment structure is a 2^3 factorial arrangement generated from two techniques (F), two positions (P), and two times (T). The design structure is an incomplete block design where each horse is a block and there are two observations per block (horse). Since the blocks are incomplete, some of the treatment structure information will be confounded with block or horse effects (Cochran and Cox, 1957).

There are various ways of assigning the treatment combinations to the two feet of a horse. Suppose there are four horses and one technique is assigned to each horse, as in Table 5.11.

Let y_{ijkm} denote the observation obtained from fusion technique i, position j, time k, and horse m. The model used to describe the data is

$$y_{ijkm} = \mu_{ijk} + h_m + \varepsilon_{ijkm} \tag{5.3.1}$$

where h_m denotes the effect of the mth horse and ε_{ijkm} denotes the response error. There are two types of comparisons that can be made among the μ_{ijk}'s, intrahorse (or within-horse) comparisons and interhorse (or between-horse) comparisons. The factorial effects are defined as

$$\begin{aligned}
\text{Mean} &= \bar{\mu}_{\cdots}, \\
F &= \bar{\mu}_{1\cdot\cdot} - \bar{\mu}_{2\cdot\cdot}, \quad P = \bar{\mu}_{\cdot 1 \cdot} - \bar{\mu}_{\cdot 2 \cdot}, \quad T = \bar{\mu}_{\cdot\cdot 1} - \bar{\mu}_{\cdot\cdot 2}, \\
F * P &= \bar{\mu}_{11\cdot} - \bar{\mu}_{12\cdot} - \bar{\mu}_{21\cdot} + \bar{\mu}_{22\cdot}, \\
F * T &= \bar{\mu}_{1\cdot 1} - \bar{\mu}_{1\cdot 2} - \bar{\mu}_{2\cdot 1} + \bar{\mu}_{2\cdot 2}, \\
P * T &= \bar{\mu}_{\cdot 11} - \bar{\mu}_{\cdot 12} - \bar{\mu}_{\cdot 21} + \bar{\mu}_{\cdot 22}, \quad \text{and} \\
F * P * T &= \mu_{111} - \mu_{112} - \mu_{121} + \mu_{122} - \mu_{211} + \mu_{212} + \mu_{221} - \mu_{222}.
\end{aligned}$$

The best estimator of μ_{ijk} is $\bar{y}_{ijk\cdot}$. The best estimator of P can be expressed as

$$\begin{aligned}
\hat{P} &= \bar{y}_{\cdot 1\cdot\cdot} - \bar{y}_{\cdot 2\cdot\cdot} \\
&= \tfrac{1}{4}\left[(y_{1111} - y_{1221}) + (y_{1122} - y_{1212}) + (y_{2113} - y_{2223}) \right. \\
&\quad \left. + (y_{2124} - y_{2214})\right],
\end{aligned}$$

Table 5.11 First Assignment of Treatment Combinations for Horse Feet Experiment

HORSE			
1	*2*	*3*	*4*
$F_1P_1T_1$	$F_1P_1T_2$	$F_2P_1T_1$	$F_2P_1T_2$
$F_1P_2T_2$	$F_1P_2T_1$	$F_2P_2T_2$	$F_2P_2T_1$

Note: The two fusion techniques are F_1 and F_2, the two times are T_1 and T_2, and the two positions are P_1 and P_2.

which is an intrahorse or within-horse comparison. This is easily seen by substituting the right-hand side of the above model for y_{ijkm} in $\hat{P}$, which gives

$$\hat{P} = \bar{\mu}_{\cdot 1 \cdot} - \bar{\mu}_{\cdot 2 \cdot} + \tfrac{1}{4}(\varepsilon_{1111} - \varepsilon_{1221} + \varepsilon_{1122} - \varepsilon_{1212} + \varepsilon_{2113} - \varepsilon_{2223} + \varepsilon_{2124} - \varepsilon_{2214}),$$

that is, the horse effects, h_m, subtract out of the expression.

The best estimator of F can be expressed as

$$\begin{aligned}\hat{F} &= \bar{y}_{1\cdots} - \bar{y}_{2\cdots}\\ &= \tfrac{1}{4}[(y_{1111} + y_{1221}) + (y_{1122} + y_{1212}) - (y_{2113} + y_{2223}) - (y_{2124} + y_{2214})]\end{aligned}$$

This estimator is a between-horse comparison; that is, it is a comparison of horses 1 and 2 with horses 3 and 4, and thus depends on the horse effects. This is seen by expressing $\hat{F}$ in terms of the right-hand side of the model as

$$\hat{F} = \bar{\mu}_{1\cdot\cdot} - \bar{\mu}_{2\cdot\cdot} + \tfrac{1}{4}(\varepsilon_{1111} + \varepsilon_{1221} + \varepsilon_{1122} + \varepsilon_{1212} - \varepsilon_{2113} - \varepsilon_{2223} - \varepsilon_{2124} - \varepsilon_{2214}) + \tfrac{1}{2}(h_1 + h_2 - h_3 - h_4),$$

which involves the h_m's, whereas $\hat{P}$ did not. Similarly, we can show that P, T, $F*T$, and $F*P$ are within-horse effects while F, $P*T$, and $F*T*P$ are between-horse effects. Since the between-horse effects involve the h_m's, they are said to be confounded with the horse effects; that is, F, $P*T$, and $F*P*T$ are confounded with horse effects.

This experimental design consists of a three-way treatment structure and an incomplete block design structure where each horse is a block. Since some of the comparisons are comparisons between horses and others are comparisons between feet "within" a horse, this experiment involves two sizes of experimental units. The feet are the small experimental units, while the horses are the large experimental units. The h_m term in the model represents the horse error (that is, variation due to differences between horses), while the ε_{ijkm} term represents the foot error (that is, variation due to differences between feet on the same horse). When designing experiments involving different sizes of experimental units, it is desirable to choose those effects that are most important so that they involve comparisons between the smaller experimental units, and to let those effects that are least important involve comparisons between larger experimental units. However, this arrangement is not always possible. For example, if the experiment here also involved two types of horses (say, racing and working), it would be impossible for types of horse to be other than a between-horse comparison. In our experiment, which has only three factors, the experimenter is most interested in comparing the two fusion techniques. The design given in Table 5.11 is such that the F effect (fusion) is confounded with horses,

Table 5.12 Second Assignment of Treatment Combinations for Horse Feet Experiment

HORSE			
1	*2*	*3*	*4*
$F_1P_1T_1$	$F_2P_1T_1$	$F_1P_1T_2$	$F_2P_1T_2$
$F_2P_2T_2$	$F_1P_2T_2$	$F_2P_2T_1$	$F_1P_2T_1$

resulting in less precision for fusion than is desired. The design given in Table 5.12 yields a fusion comparison that is not confounded with horses and thus achieves the goal of having the most important effect being compared on the smaller experimental unit, that is, the feet.

Model (5.3.1) can be used to represent this second assignment of treatment combinations. Using the same techniques as for the first assignment, it can be shown that the F, P, T, and $F*P*T$ effects are within-horse or intrahorse comparisons and that the $F*T$, $F*P$, and $P*T$ effects are interhorse comparisons.

Neither of the above two designs yield enough observations to provide any degrees of freedom for estimating the two error terms. In order to obtain some degrees of freedom for the two types of error variances, one could repeat the design in Table 5.12 by using eight

Table 5.13 Between-Horse and Within-Horse Analysis of Variance Table

Source of Variation	*df*	
Between-Horse	7	
$F*P$	1	Horse is experimental unit.
$F*T$	1	
$P*T$	1	
Error(Horse)	4	
Within-Horse	8	
F	1	Foot of horse is experimental unit.
T	1	
P	1	
$F*P*T$	1	
Error(Foot)	4	

horses, where two horses are randomly assigned each set of treatment combinations. The analysis would consist of two parts, a between-horse analysis and a within-horse or feet-within-horse analysis. There would be eight within-horse comparisons (one from each horse), which can be partitioned into estimates of the F, T, P, and $F*P*T$ effects and an estimate of the within-horse error variance denoted by Error(Feet). There are seven between-horse comparisons, which can be partitioned into estimates of the $F*T$, $F*P$, and $T*P$ effects and an estimate of the between-horse error variance, denoted by Error(Horse). The resulting analysis of variance is displayed in Table 5.13.

This experimental design falls into the class of repeated measures designs, since there are two measurements (repeated) on each horse; that is, a front foot and a rear foot are measured on each horse, and this factor cannot be randomly assigned.

EXAMPLE 5.6: Comfort Study (Repeated Measures Design)

An experimenter wants to study the effect of six environmental conditions on the comfort of people. He has six environmental chambers, and each can be set with a different environment. The experiment consists of putting one person in a chamber; after one hour, two hours, and three hours, an instrument is used to measure the subject's comfort. There are 36 subjects in the study, and 6 subjects are assigned at random to each environmental chamber.

The data arrangement is shown in Figure 5.8. Each rectangle in Figure 5.8 represents a person, and T_1, T_2, and T_3 represent the comfort measurements at the respective times. The experimental unit for environments is a person, and the experimental unit for time is a one-hour interval "within" that person. In effect, the person is "split" into three

Environment

1	2	3	4	5	6
(1) $T_1T_2T_3$	(7) $T_1T_2T_3$	(13) $T_1T_2T_3$	(19) $T_1T_2T_3$	(25) $T_1T_2T_3$	(31) $T_1T_2T_3$
(2) $T_1T_2T_3$	(8) $T_1T_2T_3$	(14) $T_1T_2T_3$	(20) $T_1T_2T_3$	(26) $T_1T_2T_3$	(32) $T_1T_2T_3$
(3) $T_1T_2T_3$	(9) $T_1T_2T_3$	(15) $T_1T_2T_3$	(21) $T_1T_2T_3$	(27) $T_1T_2T_3$	(33) $T_1T_2T_3$
(4) $T_1T_2T_3$	(10) $T_1T_2T_3$	(16) $T_1T_2T_3$	(22) $T_1T_2T_3$	(28) $T_1T_2T_3$	(34) $T_1T_2T_3$
(5) $T_1T_2T_3$	(11) $T_1T_2T_3$	(17) $T_1T_2T_3$	(23) $T_1T_2T_3$	(29) $T_1T_2T_3$	(35) $T_1T_2T_3$
(6) $T_1T_2T_3$	(12) $T_1T_2T_3$	(18) $T_1T_2T_3$	(24) $T_1T_2T_3$	(30) $T_1T_2T_3$	(36) $T_1T_2T_3$

Figure 5.8 Data arrangement for the comfort study.

Table 5.14 Analysis of Variance Table for Comfort Study

Source of Variation	*df*
Person Analysis	
Environment	5
Error(Person)	30
Hour-Interval Analysis	
People	35
Time	2
Time * Environment	10
Error(Interval)	60

parts, but this design is considered a repeated measures design rather than a split-plot design since time cannot be randomized. This repeated measures experimental design consists of a one-way whole plot (or person) treatment structure in a completely randomized whole plot design structure and a one-way subplot (or hour) treatment structure in a randomized complete block design structure. The analysis of variance is shown in Table 5.14.

An alternative to assigning six people to an environmental chamber six different times is to assign the six people to the environmental chamber all at one time. This seems like a good idea, but the resulting data do not provide any measure of error to assess the effects of environment on comfort. This is because the six people form the experimental unit for chambers, and thus there is only one observation of each environment.

EXAMPLE 5.7: Crossover or Change-Over Designs

A useful method for comparing treatments to be administered to animals (experimental units), including humans, is to assign treatment A to an animal, observe the response of the animal after a given period of time, allow a certain amount of time for the animal to recover from treatment A, assign treatment B to the same animal, and then observe the response of the animal after a given period of time. (This approach can also be used on plants, plots of land, or other types of processes that can at least partially recover from the effects of the first treatment.)

In this method of comparing treatments, there are two sequences of treatment assignments for an animal—A followed by B and B followed by A. The treatment structure is a one-way set of treatments with two treatments (A and B), but since the treatments are applied in sequence, we have generated another type of treatment, called a *sequence*. Thus the experimental design involves a two-way treatment structure with treatments crossed with sequences.

A repeated measures design with two sizes of experimental units is generated, since each treatment is observed on the same animal. The treatment sequence is assigned to an animal, which is the large experimental unit. The time periods or times during which the treatments are being observed are the small experimental units. The treatment structure for the large experimental units is a one-way set of treatments with two levels (the two possible sequences).

Although any design structure can be used, a completely randomized design is generally used, where the animals are assigned to the treatment sequences completely at random.

The experimental design for the small experimental units (time) is a one-way treatment structure (levels A and B) in a randomized complete block design structure where the animals are the blocks. The data can be arranged as shown in Table 5.15. The model used to describe the array of data is

$$y_{ijk} = \mu_{ik} + \varepsilon_{ij} + e_{ijk} \tag{5.3.2}$$

where μ_{ik} denotes the effect of the kth treatment in the ith sequence, ε_{ij} denotes the random effect of animal j in sequence i and is assumed to be

Table 5.15 Data Arrangement for Crossover Design

	ANIMAL			
Sequence 1	*1*	*2*	. . .	n_1
A	y_{11A}	y_{12A}	. . .	y_{1n_1A}
B	y_{11B}	y_{12B}	. . .	y_{1n_1B}
	ANIMAL			
Sequence 2	*1*	*2*	. . .	n_2
B	y_{21B}	y_{22B}	. . .	y_{2n_2B}
A	y_{21A}	y_{22A}	. . .	y_{2n_2A}

distributed i.i.d. $N(0, \sigma_\varepsilon^2)$, and the e_{ijk} denotes the random error of a measurement and is assumed to be distributed i.i.d. $N(0, \sigma_e^2)$. It is also assumed that ε_{ij} and e_{ijk} are independent. An analysis of crossover designs is presented in Chapter 32.

5.4 DESIGNS INVOLVING NESTED FACTORS

In a given experimental design, it is possible to have nested effects in the design structure, the treatment structure, or both. These situations are examined here. Nesting occurs most often in the design structure of an experiment, where a smaller experimental unit is nested within a larger one. One size of experimental unit is nested within a larger size if the smaller experimental units are different for each large experimental unit. In a treatment structure, nesting occurs when the levels of one factor occur with only one level of a second factor. In that case, the levels of the first factor are said to be nested within the level of the second factor.

Table 5.16 illustrates nesting in the design structure for example 4.2; in this case, houses are nested within squares (an "X" indicates that a house is in the indicated square). Each square is the large experimental unit, and the house is the small experimental unit. Since the houses for the first square are different from the houses for the second square, houses are nested within squares. Such a nested effect is often expressed in a model by $s_k + h_{m(k)}$ where s_k denotes the effect of the kth square and $h_{m(k)}$ denotes the effect of the mth house in the kth square. The sums of squares for squares and houses within squares are denoted by SSSQUARE and SSHOUSES(SQUARES), respectively. If houses but not squares are included in the model, then there is only a sum of squares due to houses (SSHOUSES). The sum of squares due to houses is partitioned as

$$\text{SSHOUSES} = \text{SSSQUARES} + \text{SSHOUSES(SQUARES)}.$$

Table 5.16 Diagram Showing Houses Nested within Squares

	HOUSES							
Square	*1*	*2*	*3*	*4*	*5*	*6*	*7*	*8*
1	X	X	X	X				
2					X	X	X	X

EXAMPLE 5.8: Animal Genetics

An animal scientist wants to study the growth rate of lambs. She has 4 males (sires) and 12 females (dams). The breeding structure is shown in Table 5.17 (an "X" denotes a mating). For this example, each sire is mated to three dams, the three dams being different for each sire. Thus, Dam is called a nested effect, where Dam is nested within Sire.

When nesting occurs in the treatment structure, the treatment structure must consist of at least two factors. In this case, each level of the nested factor occurs just once with a level or levels of the other factor. The next example demonstrates nesting in a treatment structure.

Table 5.17 Breeding Structure Showing Dams Nested Within Sires

	DAMS											
Sires	*1*	*2*	*3*	*4*	*5*	*6*	*7*	*8*	*9*	*10*	*11*	*12*
1	X	X	X									
2				X	X	X						
3							X	X	X			
4										X	X	X

EXAMPLE 5.9: Engines on Aircraft

An aircraft company wants to evaluate the performance of seven engine types with three aircraft types. Because of certain mechanical characteristics, only certain engines can be used with each type of aircraft. Table 5.18 shows the possible engine–aircraft configurations (marked by an "X").

As seen from Table 5.18, the levels of engine types are nested within aircraft types. The aircraft company made three aircraft for each of the seven treatment combinations. Let y_{ijk} denote the performance measure of the kth airplane made from aircraft type i with engine type j. The model is

$$y_{ijk} = \mu_{ij} + \varepsilon_{ijk} \qquad (5.4.1)$$

Table 5.18 Observable Engine–Aircraft Configurations, Which Show Nested Treatment Structure

	ENGINE TYPE						
Aircraft Type	*A*	*B*	*C*	*D*	*E*	*F*	*G*
I	X	X	X				
II				X	X		
III						X	X

or

$$y_{ijk} = \mu + A_i + E_{j(i)} + \varepsilon_{ijk} \tag{5.4.2}$$

An analysis of variance table for model (5.4.1) is shown in Table 5.19, where F_A tests $\bar{\mu}_{1\cdot} = \bar{\mu}_{2\cdot} = \bar{\mu}_{3\cdot}$ and $F_{E(A)}$ tests $\mu_{11} = \mu_{12} = \mu_{13}$, $\mu_{21} = \mu_{22}$, and $\mu_{31} = \mu_{32}$.

One has to be quite careful when nesting occurs in the treatment structure. In the airplane example, there is only one size of experimental unit, an airplane. Thus, there is only one error term in the model. When the nesting occurs in the design structure, there is more than one size of experimental unit and thus more than one error term in the model. The next example illustrates nesting in the design structure.

Table 5.19 Analysis of Variance Table for Two-Way Nested Treatment Structure in a Completely Randomized Design Structure

Source of Variation	*df*	*MS*	*F*
Aircraft	2	MSAIRCRAFT	$F_A = \dfrac{\text{MSAIRCRAFT}}{\text{MSERROR}}$
Engine(Aircraft)	4	MSENGINES(AIRCRAFT)	$F_{E(A)} = \dfrac{\text{MSENGINES(AIRCRAFT)}}{\text{MSERROR}}$
Error	14	MSERROR	

EXAMPLE 5.10: Simple Comfort Experiment

A comfort experiment was conducted to study the effects of temperature (three levels—65 °F, 70 °F, and 75 °F) and sex of person (two

Temperature	**Sex**	(1)	(2)	(3)
65 °F	M	1 2	3 4	5 6
	F	1 2	3 4	5 6
		(4)	(5)	(6)
70 °F	M	7 8	9 10	11 12
	F	7 8	9 10	11 12
		(7)	(8)	(9)
75 °F	M	13 14	15 16	17 18
	F	13 14	15 16	17 18

Numbers in parentheses denote chamber numbers. Number within boxes denote person numbers of the given sex.

Figure 5.9 Assignment of people and temperatures to chambers for Example 5.10.

levels—male [M] and female [F]) in a two-way treatment structure on a person's comfort. The three temperatures were each randomly assigned to three of the nine available environmental chambers. A chamber is the experimental unit for temperature; the chamber experimental design is a one-way treatment structure in a completely randomized design structure.

Eighteen males and eighteen females were randomly assigned to chambers so that two males and two females were assigned to each of the nine chambers. The experimental unit for sex of person is a person, and the person experimental design is a one-way treatment structure in a randomized complete block design structure where the chambers are the blocks.

There are two observations on each treatment (sex) in each block. Figure 5.9 shows how the people and the temperatures were assigned to the chambers.

After the people were subjected to the environmental condition for three hours, their comfort was measured. A model to describe these data is

$$y_{ijkm} = \mu_{ik} + c_{j(i)} + p_{m(ijk)}$$

where μ_{ik} is the mean response at temperature i and sex k, $c_{j(i)}$ is the effect of the jth chamber assigned temperature i, and $p_{m(ijk)}$ denotes the effect of the mth person of the kth sex assigned to the jth chamber with temperature i. The $c_{j(i)}$'s are assumed to be distributed i.i.d. $N(0, \sigma_c^2)$, the

Table 5.20 Analysis of Variance Table for Simple Comfort Experiment

Source of Variation	*df*	*EMS*[1]
Temperature	2	
Chamber(Temperature)[2]	6	$\sigma_p^2 + 4\sigma_c^2$
Sex	1	
Sex * Temperature	2	
Error(Person)	24	σ_p^2

[1]"EMS" denotes expected mean squares; see Chapters 18–21.
[2]Chamber(Temperature) = Error(Chamber).

$p_{m(ijk)}$'s are assumed to be distributed i.i.d. $N(0, \sigma_p^2)$, and $c_{j(i)}$ is assumed to be independent of $p_{m(ijk)}$.

There are two types of nesting involved in this experiment. First, environmental chambers are nested within temperatures. Second, people of the same sex are nested within chambers. The analysis of variance table for this experiment is shown in Table 5.20. The effects of Sex and the Sex * Temperature interaction are between-person comparisons, and thus the person error term is used for those comparisons. The chamber error term is used to make comparisons between temperatures.

The designs discussed in this and the previous two sections can be combined into some very complex experimental designs. In such cases, the key to determining the type of model needed is to identify the sizes of the experimental units and to identify the treatment structure and design structure for each size of experimental unit. Analyses of designs involving more than one size of experimental unit are presented in Chapters 24 through 32, where the assumptions concerning the models are discussed.

CONCLUDING REMARKS

In this chapter, design structures involving several different sizes of experimental units were considered. Design types discussed included split-plot designs, nested designs, repeated measures designs, and several variations and combinations of these. The emphasis was on recognizing such designs and on when and how to use them. These designs shall be analyzed in later chapters.

6

Matrix Form of the Model

CHAPTER OUTLINE

When one works with unbalanced models, random models, or mixed models, summation notation becomes very laborious and sometimes nearly impossible to use. This problem can be solved by using a matrix form of the model. This chapter discusses the construction of the matrix form and how to compute with matrices to obtain least squares estimators, to test hypotheses, and to construct confidence intervals. The concept of estimability is discussed in Section 6.3.

6.1 BASIC NOTATION

The matrix form of a model can be expressed as

$$\underset{n\times 1}{\mathbf{y}} = \underset{n\times p}{\mathbf{X}}\ \underset{p\times 1}{\boldsymbol{\beta}} + \underset{n\times 1}{\boldsymbol{\varepsilon}} \tag{6.1.1}$$

where $\mathbf{y}$ denotes the $n \times 1$ vector of observations; $\mathbf{X}$ denotes the $n \times p$ matrix of known constants, called the *design matrix*; $\boldsymbol{\beta}$ denotes the $p \times 1$ vector of unknown parameters; and $\boldsymbol{\varepsilon}$ denotes the $n \times 1$ vector of unobserved random errors. The model for the ith observation (ith element of $\mathbf{y}$) is of the form

$$y_i = \beta_0 + \beta_1 X_{i1} + \cdots + \beta_{p-1} X_{ip-1} + \varepsilon_i. \tag{6.1.2}$$

The vectors and matrices used to represent model (6.1.2) as the matrix model (6.1.1) are

$$\mathbf{y} = \begin{bmatrix} y_1 \\ y_2 \\ \vdots \\ y_n \end{bmatrix}, \quad \mathbf{X} = \begin{bmatrix} 1 & x_{11} & \cdots & x_{1p-1} \\ 1 & x_{21} & \cdots & x_{2p-1} \\ \vdots & \vdots & & \vdots \\ 1 & x_{n1} & \cdots & x_{np-1} \end{bmatrix}, \quad \boldsymbol{\beta} = \begin{bmatrix} \beta_0 \\ \beta_1 \\ \vdots \\ \beta_{p-1} \end{bmatrix}, \quad \text{and}$$

$$\boldsymbol{\varepsilon} = \begin{bmatrix} \varepsilon_1 \\ \varepsilon_2 \\ \vdots \\ \varepsilon_n \end{bmatrix}. \tag{6.1.3}$$

Matrices of the type in (6.1.3) can be used to represent any type of model, such as design models (one-way, two-way, and block treatment), regression models, covariance models, random models, split-plot models, mixed models, and random coefficient regression models, by specifying the appropriate elements for $\mathbf{X}$ and the appropriate assumptions on $\boldsymbol{\beta}$ and $\boldsymbol{\varepsilon}$. We shall now present some matrix models for various experimental situations.

6.1.1 Simple Linear Regression Model

The simple linear regression model, $y_i = \beta_0 + \beta_1 x_i + \varepsilon_i$, $i = 1, 2, \ldots, n$, can be represented in matrix form as

$$\begin{bmatrix} y_1 \\ y_2 \\ \vdots \\ y_n \end{bmatrix} = \begin{bmatrix} 1 & x_1 \\ 1 & x_2 \\ \vdots & \vdots \\ 1 & x_n \end{bmatrix} \begin{bmatrix} \beta_0 \\ \beta_1 \end{bmatrix} + \boldsymbol{\varepsilon}.$$

6.1.2 One-Way Treatment Structure Model

To represent the model for a one-way treatment structure in a completely randomized design structure, let the independent variables x_{ij} be defined by

$$x_{ij} = \begin{cases} 0 & \text{if the } j\text{th observation is not from the } i\text{th treatment} \\ 1 & \text{if the } j\text{th observation is from the } i\text{th treatment} \end{cases}$$

for $i = 1, 2, \ldots, t$ and $j = 1, 2, \ldots, n_i$. The variable x_{ij} is called an *indicator variable*, as $x_{ij} = 1$ indicates which observations belong to treatment i and $x_{ij} = 0$ indicates which observations do not. Then the model $y_{ij} = \mu + \tau_i + \varepsilon_{ij}$, $i = 1, 2, \ldots, t$; $j = 1, 2, \ldots, n_i$, can be expressed in matrix form as

$$\begin{bmatrix} y_{11} \\ y_{12} \\ \vdots \\ y_{1n_1} \\ y_{21} \\ \vdots \\ y_{2n_2} \\ \vdots \\ y_{t1} \\ \vdots \\ y_{tn_t} \end{bmatrix} = \begin{bmatrix} 1 & 1 & 0 & \cdots & 0 \\ 1 & 1 & 0 & \cdots & 0 \\ \vdots & \vdots & \vdots & & \vdots \\ 1 & 1 & 0 & \cdots & 0 \\ 1 & 0 & 1 & \cdots & 0 \\ \vdots & \vdots & \vdots & & \vdots \\ 1 & 0 & 1 & \cdots & 0 \\ \vdots & \vdots & \vdots & & \vdots \\ 1 & 0 & 0 & \cdots & 1 \\ \vdots & \vdots & \vdots & & \vdots \\ 1 & 0 & 0 & \cdots & 1 \end{bmatrix} \begin{bmatrix} \mu \\ \tau_1 \\ \vdots \\ \tau_t \end{bmatrix} + \boldsymbol{\varepsilon}. \tag{6.1.4}$$

6.1.3 Two-Way Treatment Structure Model

A form of the model used for a two-way treatment structure in a completely randomized design structure is

$$y_{ijk} = \mu_{ij} + \varepsilon_{ijk} \qquad i = 1, 2, \ldots, t, \quad j = 1, 2, \ldots, b, \quad k = 1, 2, \ldots, n_{ij}. \tag{6.1.5}$$

The model used in (6.1.5), called the *means model*, can be represented in matrix form as

$$
\begin{bmatrix} y_{111} \\ y_{112} \\ \vdots \\ y_{11n_{11}} \\ y_{121} \\ \vdots \\ y_{12n_{12}} \\ \vdots \\ y_{1b1} \\ \vdots \\ y_{1bn_{1b}} \\ y_{211} \\ \vdots \\ y_{21n_{21}} \\ \vdots \\ y_{tb1} \\ \vdots \\ y_{tbn_{tb}} \end{bmatrix}
=
\begin{bmatrix}
1 & 0 & \cdots & 0 & 0 & \cdots & 0 \\
1 & 0 & \cdots & 0 & 0 & \cdots & 0 \\
\vdots & \vdots & & \vdots & \vdots & & \vdots \\
1 & 0 & \cdots & 0 & 0 & \cdots & 0 \\
0 & 1 & \cdots & 0 & 0 & \cdots & 0 \\
\vdots & \vdots & & \vdots & \vdots & & \vdots \\
0 & 1 & \cdots & 0 & 0 & \cdots & 0 \\
\vdots & \vdots & & \vdots & \vdots & & \vdots \\
0 & 0 & \cdots & 1 & 0 & \cdots & 0 \\
\vdots & \vdots & & \vdots & \vdots & & \vdots \\
0 & 0 & \cdots & 1 & 0 & \cdots & 0 \\
0 & 0 & \cdots & 0 & 1 & \cdots & 0 \\
\vdots & \vdots & & \vdots & \vdots & & \vdots \\
0 & 0 & \cdots & 0 & 1 & \cdots & 0 \\
\vdots & \vdots & & \vdots & \vdots & & \vdots \\
0 & 0 & \cdots & 0 & 0 & \cdots & 1 \\
\vdots & \vdots & & \vdots & \vdots & & \vdots \\
0 & 0 & \cdots & 0 & 0 & \cdots & 1
\end{bmatrix}
\begin{bmatrix} \mu_{11} \\ \mu_{12} \\ \vdots \\ \mu_{1b} \\ \mu_{21} \\ \vdots \\ \mu_{tb} \end{bmatrix}
+ \boldsymbol{\varepsilon}.
$$

This book emphasizes models that correspond to experimental design situations. The next two examples demonstrate how to construct such models.

EXAMPLE 6.1: Means Model for Two-Way Treatment Structure

The information in Figure 6.1 represents data from a two-way treatment structure in a completely randomized design structure. The matrix form

		Column Treatment 1	2	3
Row Treatment	1	3, 6	9	10
	2	2	5, 3	8
	3	4	2	6

Figure 6.1 Data for Example 6.1.

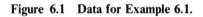

of the model corresponding to model (6.1.5) for the data in Figure 6.1 is

$$\begin{bmatrix} 3 \\ 6 \\ 9 \\ 10 \\ 2 \\ 5 \\ 3 \\ 8 \\ 4 \\ 2 \\ 6 \end{bmatrix} = \begin{bmatrix} 1 & 0 & 0 & 0 & 0 & 0 & 0 & 0 & 0 \\ 1 & 0 & 0 & 0 & 0 & 0 & 0 & 0 & 0 \\ 0 & 1 & 0 & 0 & 0 & 0 & 0 & 0 & 0 \\ 0 & 0 & 1 & 0 & 0 & 0 & 0 & 0 & 0 \\ 0 & 0 & 0 & 1 & 0 & 0 & 0 & 0 & 0 \\ 0 & 0 & 0 & 0 & 1 & 0 & 0 & 0 & 0 \\ 0 & 0 & 0 & 0 & 1 & 0 & 0 & 0 & 0 \\ 0 & 0 & 0 & 0 & 0 & 1 & 0 & 0 & 0 \\ 0 & 0 & 0 & 0 & 0 & 0 & 1 & 0 & 0 \\ 0 & 0 & 0 & 0 & 0 & 0 & 0 & 1 & 0 \\ 0 & 0 & 0 & 0 & 0 & 0 & 0 & 0 & 1 \end{bmatrix} \begin{bmatrix} \mu_{11} \\ \mu_{12} \\ \mu_{13} \\ \mu_{21} \\ \mu_{22} \\ \mu_{23} \\ \mu_{31} \\ \mu_{32} \\ \mu_{33} \end{bmatrix} + \boldsymbol{\varepsilon}.$$

In Example 6.1, the means or μ_{ij} model was used for the two-way treatment structure. The next example expresses the data in Figure 6.1 in matrix form using the *classical* or *effects model*,

$$y_{ijk} = \mu + \tau_i + \beta_j + \gamma_{ij} + \varepsilon_{ijk}. \tag{6.1.6}$$

EXAMPLE 6.2: Effects or Classical Model for Two-Way Treatment Structure

The matrix form of the model that represents the data in Table 6.1 as the effects model (6.1.6) is

$$\begin{bmatrix} 3 \\ 6 \\ 9 \\ 10 \\ 2 \\ 5 \\ 3 \\ 8 \\ 4 \\ 2 \\ 6 \end{bmatrix} = \begin{bmatrix} 1 & 1 & 0 & 0 & 1 & 0 & 0 & 1 & 0 & 0 & 0 & 0 & 0 & 0 & 0 & 0 \\ 1 & 1 & 0 & 0 & 1 & 0 & 0 & 1 & 0 & 0 & 0 & 0 & 0 & 0 & 0 & 0 \\ 1 & 1 & 0 & 0 & 0 & 1 & 0 & 0 & 1 & 0 & 0 & 0 & 0 & 0 & 0 & 0 \\ 1 & 1 & 0 & 0 & 0 & 0 & 1 & 0 & 0 & 1 & 0 & 0 & 0 & 0 & 0 & 0 \\ 1 & 0 & 1 & 0 & 1 & 0 & 0 & 0 & 0 & 0 & 1 & 0 & 0 & 0 & 0 & 0 \\ 1 & 0 & 1 & 0 & 0 & 1 & 0 & 0 & 0 & 0 & 0 & 1 & 0 & 0 & 0 & 0 \\ 1 & 0 & 1 & 0 & 0 & 1 & 0 & 0 & 0 & 0 & 0 & 1 & 0 & 0 & 0 & 0 \\ 1 & 0 & 1 & 0 & 0 & 0 & 1 & 0 & 0 & 0 & 0 & 0 & 1 & 0 & 0 & 0 \\ 1 & 0 & 0 & 1 & 1 & 0 & 0 & 0 & 0 & 0 & 0 & 0 & 0 & 1 & 0 & 0 \\ 1 & 0 & 0 & 1 & 0 & 1 & 0 & 0 & 0 & 0 & 0 & 0 & 0 & 0 & 1 & 0 \\ 1 & 0 & 0 & 1 & 0 & 0 & 1 & 0 & 0 & 0 & 0 & 0 & 0 & 0 & 0 & 1 \end{bmatrix} \begin{bmatrix} \mu \\ \tau_1 \\ \tau_2 \\ \tau_3 \\ \beta_1 \\ \beta_2 \\ \beta_3 \\ \gamma_{11} \\ \gamma_{12} \\ \gamma_{13} \\ \gamma_{21} \\ \gamma_{22} \\ \gamma_{23} \\ \gamma_{31} \\ \gamma_{32} \\ \gamma_{33} \end{bmatrix} + \boldsymbol{\varepsilon}$$

(columns: $\mathbf{j}$ | $\mathbf{X}_1$ | $\mathbf{X}_2$ | $\mathbf{X}_3$)

$$= \mathbf{j}\mu + \mathbf{X}_1\boldsymbol{\tau} + \mathbf{X}_2\boldsymbol{\beta} + \mathbf{X}_3\boldsymbol{\gamma} + \boldsymbol{\varepsilon}$$

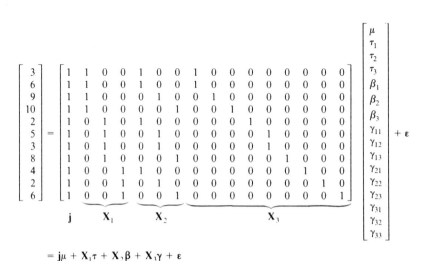

where $\mathbf{j}$ is an $n \times 1$ vector of ones, and $\mathbf{X}_1$, $\mathbf{X}_2$ and $\mathbf{X}_3$ are partitions of the above matrix.

Once the model is specified in matrix form, the next step in the analysis is to obtain the least squares estimator of the parameter vector $\boldsymbol{\beta}$.

6.2 LEAST SQUARES ESTIMATION

The method of least squares can be used to estimate the parameters of the model. To use this method, assume that the model can be expressed as

$$y_i = f(\mathbf{X}_i; \boldsymbol{\beta}) + \varepsilon_i \tag{6.2.1}$$

where $f(\mathbf{X}_i; \boldsymbol{\beta})$ is a function of the design variables that also depends on the parameter vector $\boldsymbol{\beta}$. The least squares estimator of $\boldsymbol{\beta}$ is the value of $\boldsymbol{\beta}$, usually denoted by $\hat{\boldsymbol{\beta}}$, that minimizes the sum of squares

$$\mathrm{SS}(\boldsymbol{\beta}) = \sum_{i=1}^{n} [y_i - f(\mathbf{X}_i; \boldsymbol{\beta})]^2. \tag{6.2.2}$$

If, in addition to assuming the model is of the form (6.2.1), one assumes that $\varepsilon_i \sim$ i.i.d. $N(0, \sigma^2)$, $i = 1, 2, \ldots, n$, then the least squares estimator is also a maximum likelihood estimator.

For example, the model function for the two-way treatment structure in a completely randomized design for Example 6.1 is

$$f(\mathbf{X}_i; \boldsymbol{\beta}) = \mu_{ij}.$$

The least squares estimator of μ_{ij} is that value, say $\hat{\mu}_{ij}$, which minimizes

$$\mathrm{SS}(\boldsymbol{\mu}) = \sum_{i=1}^{3} \sum_{j=1}^{3} \sum_{k=1}^{n_{ij}} (y_{ijk} - \mu_{ij})^2.$$

For the effects model of Example 6.2, the model function is

$$f(x_{ij}; \boldsymbol{\beta}) = \mu + \tau_i + \beta_j + \gamma_{ij}$$

and the least squares estimators of μ, $\boldsymbol{\tau}$, $\boldsymbol{\beta}$, and $\boldsymbol{\gamma}$ are obtained by minimizing

$$\mathrm{SS}(\mu, \boldsymbol{\tau}, \boldsymbol{\beta}, \boldsymbol{\gamma}) = \sum_{i=1}^{3} \sum_{j=1}^{3} \sum_{k=1}^{n_{ij}} (y_{ijk} - \mu - \tau_i - \beta_j - \gamma_{ij})^2.$$

In general, the model can be written in the matrix form of (6.1.1), and the least squares estimator of $\boldsymbol{\beta}$ is the value $\hat{\boldsymbol{\beta}}$ that minimizes the sum of squares

$$\mathrm{SS}(\boldsymbol{\beta}) = (\mathbf{y} - \mathbf{X}\boldsymbol{\beta})'(\mathbf{y} - \mathbf{X}\boldsymbol{\beta}). \tag{6.2.3}$$

6.2.1 Least Squares Equations

When the sum of squares of equation (6.2.3) is minimized with respect to $\boldsymbol{\beta}$, one obtains a set of equations which $\hat{\boldsymbol{\beta}}$ must satisfy, called the *least squares equations* or the *normal equations* for the model. The normal equations for model (6.1.1) are given by

$$\mathbf{X}'\mathbf{X}\hat{\boldsymbol{\beta}} = \mathbf{X}'\mathbf{y}. \tag{6.2.4}$$

Any vector $\hat{\boldsymbol{\beta}}$ that satisfies the normal equations is a least squares estimator of $\boldsymbol{\beta}$. The least squares estimator need not be unique. To help the reader become more familiar with the normal equations, the normal equations for the models discussed in Examples 6.1 and 6.2 are given below.

The normal equations for the means model in Example 6.1 are

$$\begin{bmatrix} 2 & 0 & 0 & 0 & 0 & 0 & 0 & 0 & 0 \\ 0 & 1 & 0 & 0 & 0 & 0 & 0 & 0 & 0 \\ 0 & 0 & 1 & 0 & 0 & 0 & 0 & 0 & 0 \\ 0 & 0 & 0 & 1 & 0 & 0 & 0 & 0 & 0 \\ 0 & 0 & 0 & 0 & 2 & 0 & 0 & 0 & 0 \\ 0 & 0 & 0 & 0 & 0 & 1 & 0 & 0 & 0 \\ 0 & 0 & 0 & 0 & 0 & 0 & 1 & 0 & 0 \\ 0 & 0 & 0 & 0 & 0 & 0 & 0 & 1 & 0 \\ 0 & 0 & 0 & 0 & 0 & 0 & 0 & 0 & 1 \end{bmatrix} \begin{bmatrix} \hat{\mu}_{11} \\ \hat{\mu}_{12} \\ \hat{\mu}_{13} \\ \hat{\mu}_{21} \\ \hat{\mu}_{22} \\ \hat{\mu}_{23} \\ \hat{\mu}_{31} \\ \hat{\mu}_{32} \\ \hat{\mu}_{33} \end{bmatrix} = \begin{bmatrix} y_{11\cdot} \\ y_{12\cdot} \\ y_{13\cdot} \\ y_{21\cdot} \\ y_{22\cdot} \\ y_{23\cdot} \\ y_{31\cdot} \\ y_{32\cdot} \\ y_{33\cdot} \end{bmatrix}$$

where

$$y_{ij\cdot} = \sum_{k=1}^{n_{ij}} y_{ijk}.$$

When the $\mathbf{X}'\mathbf{X}$ matrix is of full rank (Graybill, 1976), that is, nonsingular, its inverse exists and the least squares estimator for $\boldsymbol{\beta}$ (the solution for $\hat{\boldsymbol{\beta}}$ in 6.2.4) is

$$\hat{\boldsymbol{\beta}} = (\mathbf{X}'\mathbf{X})^{-1}(\mathbf{X}'\mathbf{y}). \tag{6.2.5}$$

When $\mathbf{X}'\mathbf{X}$ is of full rank, the least squares estimator is unique.

Computing the inverse of $\mathbf{X}'\mathbf{X}$ is generally not an easy task. However, when $\mathbf{X}'\mathbf{X}$ has certain patterns, the inverse can be computed easily. For the normal equations of Example 6.1, the $\mathbf{X}'\mathbf{X}$ matrix is diagonal and the inverse of $\mathbf{X}'\mathbf{X}$ is obtained by simply replacing each diagonal element by its inverse. Thus, the least squares estimator of μ_{ij} is

$$\hat{\mu}_{ij} = \frac{y_{ij\cdot}}{n_{ij}} = \bar{y}_{ij\cdot} \qquad \text{for all } i \text{ and } j.$$

The normal equations for the effects model used in Example 6.2 are

$$\begin{bmatrix} 11 & 4 & 4 & 3 & 4 & 4 & 3 & 2 & 1 & 1 & 1 & 2 & 1 & 1 & 1 & 1 \\ 4 & 4 & 0 & 0 & 2 & 1 & 1 & 2 & 1 & 1 & 0 & 0 & 0 & 0 & 0 & 0 \\ 4 & 0 & 4 & 0 & 1 & 2 & 1 & 0 & 0 & 0 & 1 & 2 & 1 & 0 & 0 & 0 \\ 3 & 0 & 0 & 3 & 1 & 1 & 1 & 0 & 0 & 0 & 0 & 0 & 0 & 1 & 1 & 1 \\ 4 & 2 & 1 & 1 & 4 & 0 & 0 & 2 & 0 & 0 & 1 & 0 & 0 & 1 & 0 & 0 \\ 4 & 1 & 2 & 1 & 0 & 4 & 0 & 0 & 1 & 0 & 0 & 2 & 0 & 0 & 1 & 0 \\ 3 & 1 & 1 & 1 & 0 & 0 & 3 & 0 & 0 & 1 & 0 & 0 & 1 & 0 & 0 & 1 \\ 2 & 2 & 0 & 0 & 2 & 0 & 0 & 2 & 0 & 0 & 0 & 0 & 0 & 0 & 0 & 0 \\ 1 & 1 & 0 & 0 & 0 & 1 & 0 & 0 & 1 & 0 & 0 & 0 & 0 & 0 & 0 & 0 \\ 1 & 1 & 0 & 0 & 0 & 0 & 1 & 0 & 0 & 1 & 0 & 0 & 0 & 0 & 0 & 0 \\ 1 & 0 & 1 & 0 & 1 & 0 & 0 & 0 & 0 & 0 & 1 & 0 & 0 & 0 & 0 & 0 \\ 2 & 0 & 2 & 0 & 0 & 2 & 0 & 0 & 0 & 0 & 0 & 2 & 0 & 0 & 0 & 0 \\ 1 & 0 & 1 & 0 & 0 & 0 & 1 & 0 & 0 & 0 & 0 & 0 & 1 & 0 & 0 & 0 \\ 1 & 0 & 0 & 1 & 1 & 0 & 0 & 0 & 0 & 0 & 0 & 0 & 0 & 1 & 0 & 0 \\ 1 & 0 & 0 & 1 & 0 & 1 & 0 & 0 & 0 & 0 & 0 & 0 & 0 & 0 & 1 & 0 \\ 1 & 0 & 0 & 1 & 0 & 0 & 1 & 0 & 0 & 0 & 0 & 0 & 0 & 0 & 0 & 1 \end{bmatrix} \begin{bmatrix} \mu \\ \tau_1 \\ \tau_2 \\ \tau_3 \\ \beta_1 \\ \beta_2 \\ \beta_3 \\ \gamma_{11} \\ \gamma_{12} \\ \gamma_{13} \\ \gamma_{21} \\ \gamma_{22} \\ \gamma_{23} \\ \gamma_{31} \\ \gamma_{32} \\ \gamma_{33} \end{bmatrix} = \begin{bmatrix} y_{\cdot\cdot\cdot} \\ y_{1\cdot\cdot} \\ y_{2\cdot\cdot} \\ y_{3\cdot\cdot} \\ y_{\cdot 1\cdot} \\ y_{\cdot 2\cdot} \\ y_{\cdot 3\cdot} \\ y_{11\cdot} \\ y_{12\cdot} \\ y_{13\cdot} \\ y_{21\cdot} \\ y_{22\cdot} \\ y_{23\cdot} \\ y_{31\cdot} \\ y_{32\cdot} \\ y_{33\cdot} \end{bmatrix}$$

where

$$y_{\cdot\cdot\cdot} = \sum_i \sum_j \sum_k y_{ijk}, \quad y_{i\cdot\cdot} = \sum_j \sum_k y_{ijk}, \quad \text{and} \quad y_{\cdot j\cdot} = \sum_i \sum_k y_{ijk}.$$

Unlike the normal equations in Example 6.1, $\mathbf{X}'\mathbf{X}$ is singular, and the inverse of $\mathbf{X}'\mathbf{X}$ for the effects model does not exist. In this case there are many solutions to the normal equations. The effects model is called an *overspecified model* in that the model has more parameters than can be uniquely estimated from the data collected.

Overspecified models are commonly used, and there are several ways to solve their normal equations. The following discussion addresses two-way treatment structures in a completely randomized design structure, but similar techniques can be used in other cases.

Theoretically a generalized inverse can be used to solve the normal equations of $\hat{\boldsymbol{\beta}}$ (Graybill, 1976), but the most commonly used method for solving the normal equations of an overspecified model is to restrict the parameters of the model (which in effect generates a g-inverse solution). There are several ways to do this, two of which are considered here.

6.2.2 Sum-to-Zero Restrictions

The most common technique is to require the sums of certain parameters to be equal to zero. This procedure has been used to solve normal equations from the very beginning of the analysis of variance. For the model in Example 6.2, the sum-to-zero restrictions are

$$\sum_{i=1}^{3} \tau_i = 0, \quad \sum_{j=1}^{3} \beta_j = 0, \quad \sum_{i=1}^{3} \gamma_{i1} = 0, \quad \sum_{i=1}^{3} \gamma_{i2} = 0, \quad \sum_{i=1}^{3} \gamma_{i3} = 0,$$

$$\sum_{j=1}^{3} \gamma_{1j} = 0, \quad \sum_{j=1}^{3} \gamma_{2j} = 0, \quad \text{and} \quad \sum_{j=1}^{3} \gamma_{3j} = 0.$$

Next, these restrictions are incorporated into the model by solving for some of the parameters in terms of others with the restrictions being observed and then substituting the expressions back into the model. For example, the parameters that can be replaced are

$$\begin{aligned}
\tau_3 &= -\tau_1 - \tau_2 & \gamma_{33} &= -\gamma_{31} - \gamma_{32} \\
\beta_3 &= -\beta_1 - \beta_2 & \gamma_{31} &= -\gamma_{11} - \gamma_{21} \\
\gamma_{13} &= -\gamma_{11} - \gamma_{12} & \gamma_{32} &= -\gamma_{12} - \gamma_{22} \\
\gamma_{23} &= -\gamma_{21} - \gamma_{22} & \gamma_{33} &= -\gamma_{13} - \gamma_{23} = \gamma_{11} + \gamma_{12} + \gamma_{21} + \gamma_{22}.
\end{aligned}$$

Thus, replace τ_3, β_3, γ_{13}, γ_{23}, γ_{33}, γ_{31}, and γ_{32} in the model to obtain a reparameterized model

$$\begin{bmatrix} y_{111} \\ y_{112} \\ y_{121} \\ y_{131} \\ y_{211} \\ y_{221} \\ y_{222} \\ y_{231} \\ y_{311} \\ y_{321} \\ y_{331} \end{bmatrix} = \begin{bmatrix}
1 & 1 & 0 & 1 & 0 & 1 & 0 & 0 & 0 \\
1 & 1 & 0 & 1 & 0 & 1 & 0 & 0 & 0 \\
1 & 1 & 0 & 0 & 1 & 0 & 1 & 0 & 0 \\
1 & 1 & 0 & -1 & -1 & -1 & -1 & 0 & 0 \\
1 & 0 & 1 & 1 & 0 & 0 & 0 & 1 & 0 \\
1 & 0 & 1 & 0 & 1 & 0 & 0 & 0 & 1 \\
1 & 0 & 1 & 0 & 1 & 0 & 0 & 0 & 1 \\
1 & 0 & 1 & -1 & -1 & 0 & 0 & -1 & -1 \\
1 & -1 & -1 & 1 & 0 & -1 & 0 & -1 & 0 \\
1 & -1 & -1 & 0 & 1 & 0 & -1 & 0 & -1 \\
1 & -1 & -1 & -1 & -1 & 1 & 1 & 1 & 1
\end{bmatrix} \begin{bmatrix} \mu \\ \tau_1 \\ \tau_2 \\ \beta_1 \\ \beta_2 \\ \gamma_{11} \\ \gamma_{12} \\ \gamma_{21} \\ \gamma_{22} \end{bmatrix} + \boldsymbol{\varepsilon},$$

which we shall write as

$$\mathbf{y} = \mathbf{X}^*\boldsymbol{\beta}^* + \boldsymbol{\varepsilon}.$$

The solution to the normal equations corresponding to the sum-to-zero restrictions is

$$\hat{\boldsymbol{\beta}}^* = (\mathbf{X}^{*\prime}\mathbf{X}^*)^{-1}\mathbf{X}^{*\prime}\mathbf{y},$$

where

$$\hat{\boldsymbol{\beta}}^{*\prime} = \left[\hat{\mu}^*, \hat{\tau}_1^*, \hat{\tau}_2^*, \hat{\beta}_1^*, \hat{\beta}_2^*, \hat{\gamma}_{11}^*, \hat{\gamma}_{12}^*, \hat{\gamma}_{21}^*, \hat{\gamma}_{22}^*\right]$$

$$= [5.500, 2.333, -.833, -2.000, -.500, -1.333, 1.667, -.667, -.167].$$

The estimators for the remaining elements of $\boldsymbol{\beta}$ are obtained from the restrictions as follows:

$$\hat{\tau}_3^* = -\hat{\tau}_1^* - \hat{\tau}_2^* = 2.50 \qquad \hat{\gamma}_{23}^* = -\hat{\gamma}_{21}^* - \hat{\gamma}_{22}^* = .84$$

$$\hat{\beta}_3^* = -\hat{\beta}_1^* - \hat{\beta}_2^* = -1.50 \qquad \hat{\gamma}_{31}^* = -\hat{\gamma}_{11}^* - \hat{\gamma}_{21}^* = 2.06$$

$$\hat{\gamma}_{13}^* = -\hat{\gamma}_{11}^* - \hat{\gamma}_{12}^* = -.34 \qquad \hat{\gamma}_{32}^* = -\hat{\gamma}_{12}^* - \hat{\gamma}_{22}^* = -1.50$$

$$\hat{\gamma}_{33}^* = \hat{\gamma}_{11}^* + \hat{\gamma}_{12}^* + \hat{\gamma}_{21}^* + \hat{\gamma}_{22}^* = -.50$$

In relation to the means model, the parameters μ_i, τ_i, β_j, and γ_{ij} can be selected to satisfy the sum-to-zero restrictions by defining

$$\mu = \bar{\mu}_{..} \qquad \beta_j = \bar{\mu}_{.j} - \bar{\mu}_{..}$$

$$\tau_i = \bar{\mu}_{i.} - \bar{\mu}_{..} \qquad \gamma_{ij} = \mu_{ij} - \bar{\mu}_{i.} - \bar{\mu}_{j} + \bar{\mu}_{...}$$

6.2.3 Set-to-Zero Restrictions

Another reparameterization technique often used to solve normal equations uses restrictions that set the last parameter in each set equal to zero (the last parameter is selected for convenience; one could also use the first, or second, or any other). For the effects model in Example 6.2, the restrictions are

$$\tau_3 = 0, \;\; \beta_3 = 0, \;\; \gamma_{31} = 0, \;\; \gamma_{32} = 0, \;\; \gamma_{33} = 0, \;\; \gamma_{13} = 0, \;\; \text{and} \;\; \gamma_{23} = 0.$$

The resulting reparameterized model obtained by incorporating the above restrictions into the model is

$$\begin{bmatrix} y_{111} \\ y_{112} \\ y_{121} \\ y_{131} \\ y_{211} \\ y_{221} \\ y_{222} \\ y_{231} \\ y_{311} \\ y_{321} \\ y_{331} \end{bmatrix} = \begin{bmatrix} 1 & 1 & 0 & 1 & 0 & 1 & 0 & 0 & 0 \\ 1 & 1 & 0 & 1 & 0 & 1 & 0 & 0 & 0 \\ 1 & 1 & 0 & 0 & 1 & 0 & 1 & 0 & 0 \\ 1 & 1 & 0 & 0 & 0 & 0 & 0 & 0 & 0 \\ 1 & 0 & 1 & 1 & 0 & 0 & 0 & 1 & 0 \\ 1 & 0 & 1 & 0 & 1 & 0 & 0 & 0 & 1 \\ 1 & 0 & 1 & 0 & 1 & 0 & 0 & 0 & 1 \\ 1 & 0 & 1 & 0 & 0 & 0 & 0 & 0 & 0 \\ 1 & 0 & 0 & 1 & 0 & 0 & 0 & 0 & 0 \\ 1 & 0 & 0 & 0 & 1 & 0 & 0 & 0 & 0 \\ 1 & 0 & 0 & 0 & 0 & 0 & 0 & 0 & 0 \end{bmatrix} \begin{bmatrix} \mu \\ \tau_1 \\ \tau_2 \\ \beta_1 \\ \beta_2 \\ \gamma_{11} \\ \gamma_{12} \\ \gamma_{21} \\ \gamma_{22} \end{bmatrix} + \boldsymbol{\varepsilon},$$

which is written as $\mathbf{y} = \mathbf{X}^+\boldsymbol{\beta}^+ + \boldsymbol{\varepsilon}$. The matrix $\mathbf{X}^+$ is obtained from $\mathbf{X}$ by deleting the columns corresponding to β_3, τ_3, γ_{13}, γ_{23}, γ_{33}, γ_{31}, and γ_{32}. The solution to the normal equations corresponding to the set-to-zero restrictions is

$$\hat{\boldsymbol{\beta}}^+ = (\mathbf{X}^{+\prime}\mathbf{X}^+)^{-1}\mathbf{X}^{+\prime}\mathbf{y}$$

or

$$\hat{\boldsymbol{\beta}}^+ = [\hat{\mu}^+, \hat{\tau}_1^+, \hat{\tau}_2^+, \hat{\beta}_1^+, \hat{\beta}_2^+, \hat{\gamma}_{11}^+, \hat{\gamma}_{12}^+, \hat{\gamma}_{21}^+, \hat{\gamma}_{22}^+]'$$
$$= [6.0, 4.0, 2.0, -2.0, -4.0, -3.5, 3.0, -4.0, 0.0]'.$$

The estimates of the remaining parameters are zero, that is,

$$\hat{\tau}_3^+ = \hat{\beta}_3^+ = \hat{\gamma}_{13}^+ = \hat{\gamma}_{23}^+ = \hat{\gamma}_{33}^+ = \hat{\gamma}_{31}^+ = \hat{\gamma}_{32}^+ = 0$$

since they are specified by the zero restrictions. To relate the set-to-zero restrictions to the mean model, define μ, β_i, τ_j, and γ_{ij} as

$$\mu = \mu_{tb} \qquad \beta_j = \mu_{tj} - \mu_{tb}$$
$$\tau_i = \mu_{ib} - \mu_{tb} \qquad \gamma_{ij} = \mu_{ij} - \mu_{tj} - \mu_{ib} + \mu_{tb}.$$

Thus there are several possible solutions to the normal equations when $\mathbf{X}'\mathbf{X}$ is not of full rank (that is, singular). This occurs because the model is overparameterized; that is, there are more parameters in the model (16 in this case) than can be uniquely estimated from the available data (9 in this case). The number of parameters that can be estimated uniquely might be called the number of *essential* parameters. To cope with the overparameterized model and nonunique least squares solutions, the concept of estimability must be considered, which is the topic of the next section. The next example obtains two possible solutions for a one-way treatment structure.

EXAMPLE 6.3: A One-Way Treatment Structure

This is an example of a one-way treatment structure with four treatments in a completely randomized design structure. The data are shown in Figure 6.2, and the $\mathbf{X}^*$ matrix comes from the reparameterized model by using the sum-to-zero restrictions; that is, it was assumed that

$$\tau_1^* + \tau_2^* + \tau_3^* + \tau_4^* = 0.$$

The resulting normal equations are also shown in Figure 6.2, and the corresponding least squares estimators are

$$\hat{\boldsymbol{\beta}}^* = \begin{bmatrix} \hat{\mu}^* \\ \hat{\tau}_1^* \\ \hat{\tau}_2^* \\ \hat{\tau}_3^* \end{bmatrix} = (\mathbf{X}^{*\prime}\mathbf{X}^*)^{-1}\mathbf{X}^{*\prime}\mathbf{y} = \begin{bmatrix} 2.150 \\ .020 \\ .569 \\ -.451 \end{bmatrix}$$

and $\hat{\tau}_4^* = -\hat{\tau}_1^* - \hat{\tau}_2^* - \hat{\tau}_3^* = -.138$. By using the set-to-zero restric-

$$
\mathbf{X}^* = \begin{bmatrix}
\mu^* & \tau_1^* & \tau_2^* & \tau_3^* \\
1 & 1 & 0 & 0 \\
1 & 1 & 0 & 0 \\
1 & 1 & 0 & 0 \\
1 & 1 & 0 & 0 \\
1 & 1 & 0 & 0 \\
1 & 1 & 0 & 0 \\
1 & 1 & 0 & 0 \\
1 & 0 & 1 & 0 \\
1 & 0 & 1 & 0 \\
1 & 0 & 1 & 0 \\
1 & 0 & 1 & 0 \\
1 & 0 & 1 & 0 \\
1 & 0 & 0 & 1 \\
1 & 0 & 0 & 1 \\
1 & 0 & 0 & 1 \\
1 & 0 & 0 & 1 \\
1 & 0 & 0 & 1 \\
1 & 0 & 0 & 1 \\
1 & -1 & -1 & -1 \\
1 & -1 & -1 & -1 \\
1 & -1 & -1 & -1 \\
1 & -1 & -1 & -1 \\
1 & -1 & -1 & -1 \\
1 & -1 & -1 & -1 \\
1 & -1 & -1 & -1 \\
1 & -1 & -1 & -1
\end{bmatrix}
\qquad
\begin{bmatrix}
Y \\
2.2 \\ 2.4 \\ 2.5 \\ 2.3 \\ 2.0 \\ 1.9 \\ 1.9 \\ 2.4 \\ 2.6 \\ 3.0 \\ 3.1 \\ 2.5 \\ 1.8 \\ 1.7 \\ 1.8 \\ 1.6 \\ 1.4 \\ 1.9 \\ 1.9 \\ 2.0 \\ 2.3 \\ 2.1 \\ 1.9 \\ 2.0 \\ 2.4 \\ 1.5
\end{bmatrix}
$$

$$
\mathbf{X}^{*\prime}\mathbf{X}^*\boldsymbol{\beta}^* = \begin{bmatrix} 26 & -1 & -3 & -2 \\ -1 & 15 & 8 & 8 \\ -3 & 8 & 13 & 8 \\ -2 & 8 & 8 & 14 \end{bmatrix} \begin{bmatrix} \hat{\mu}^* \\ \hat{\tau}_1^* \\ \hat{\tau}_2^* \\ \hat{\tau}_3^* \end{bmatrix} = \begin{bmatrix} 55.1 \\ -\ .9 \\ -\ 2.5 \\ -\ 3.9 \end{bmatrix} = \mathbf{X}^{*\prime}\mathbf{y}
$$

Figure 6.2 Data, design matrix, and normal equations for Example 6.3.

tions, that is, $\tau_4 = 0$, the least squares solution to the normal equations is

$$
\hat{\boldsymbol{\beta}}^+ = \begin{bmatrix} 2.012 \\ .157 \\ .707 \\ -.313 \end{bmatrix} = \begin{bmatrix} \hat{\mu}^+ \\ \hat{\tau}_1^+ \\ \hat{\tau}_2^+ \\ \hat{\tau}_3^+ \end{bmatrix}
$$

and $\hat{\tau}_4^+ = 0$.

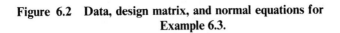

The last parameter that needs to be estimated is the population variance σ^2. An estimate of σ^2 based on the least squares solution for $\boldsymbol{\beta}$ is

$$\hat{\sigma}^2 = \frac{1}{n-p}(\mathbf{y} - \mathbf{X}\hat{\boldsymbol{\beta}})'(\mathbf{y} - \mathbf{X}\hat{\boldsymbol{\beta}})$$

$$= \frac{1}{n-p}\sum_{i=1}^{n}\left(y_i - \hat{\beta}_0 - \hat{\beta}_1 X_{i1} - \cdots - \hat{\beta}_{p-1}X_{ip-1}\right)^2. \qquad (6.2.6)$$

If the errors are assumed to be independently and identically distributed with the first four moments equal to the first four moments of a normal distribution, then $\hat{\sigma}^2$ is the best quadratic unbiased estimate of σ^2. If the errors are also normally distributed, then $\hat{\sigma}^2$ is the best unbiased estimator of σ^2, and the sampling distribution of $(n-p)\hat{\sigma}^2/\sigma^2$ is a central chi-square distribution with $n-p$ degrees of freedom.

6.3 ESTIMABILITY AND CONNECTED DESIGNS

When an overspecified model or a less than full rank model is used for an experimental situation, there are many different least squares solutions (in fact, there is an infinite number). If two researchers analyzed the above two data sets, one using the sum-to-zero restriction and the other using the set-to-zero restriction, they might appear to obtain two different conclusions. For the two-way example in the last section, $\hat{\tau}_2^* = -.83$, while $\hat{\tau}_2^+ = 2.0$; thus one researcher might say τ_2 is most likely to be negative while the other might say τ_2 is most likely to be positive—and both statements may actually be incorrect.

6.3.1 Estimable Functions

Since both researchers are analyzing the same data set, it seems that they should consider only parameters or functions of the parameters that have identical estimates for both reparameterized models. Such functions of the parameters are called *estimable functions of the parameters*.

Definition 6.3.1 A parameter or function of the parameters $f(\boldsymbol{\beta})$ is *estimable* if and only if the estimate of the parameter or function of parameters is invariant with respect to the choice of a least squares solution; that is, the value of the estimate is the same regardless of which solution to the normal equations is used.

If two researchers obtain an estimate of an estimable function of the parameters, they both will obtain the same estimate even if they have two different least squares solutions. Therefore, they will make the same decisions about estimable functions of the parameters. For matrix models, the linear estimable functions of $\boldsymbol{\beta}$ take on the form of linear combinations of the parameter vector such as $\mathbf{a}'\boldsymbol{\beta}$ where $\mathbf{a}$ is a $p \times 1$ vector of constants. A linear function of $\mathbf{a}'\boldsymbol{\beta}$ is estimable if and only if there exists a vector $\mathbf{r}$ such that $\mathbf{a} = \mathbf{X}'\mathbf{X}\mathbf{r}$.

Consider the two solutions obtained for the one-way example in Section 6.2. Since there are two different solutions for each of the parameters μ, τ_1, τ_2, τ_3, and τ_4, these parameters are considered to be nonestimable. But by computing the estimate of $\mu + \tau_i$ from each method, it is seen that

$$\hat{\mu}^* + \hat{\tau}_i^* = \hat{\mu}^+ + \hat{\tau}_i^+ \qquad i = 1, 2, 3, 4.$$

Hence, $\mu + \tau_i$ is an estimable function of the parameters.

All contrasts of τ_i, such as the differences $\tau_1 - \tau_2$, $\tau_2 - \tau_3$, and so on, can also be shown to be estimable functions for the one-way model.

For the two-way effects model, some estimable functions are

$$\begin{aligned}&\mu + \tau_i + \beta_j + \gamma_{ij},\\ &\gamma_{ij} - \gamma_{i'j} - \gamma_{ij'} + \gamma_{i'j'},\\ &\tau_i - \tau_{i'} + \bar{\gamma}_{i\cdot} - \bar{\gamma}_{i'\cdot}, \quad \text{and}\\ &\beta_j - \beta_{j'} + \bar{\gamma}_{\cdot j} - \bar{\gamma}_{\cdot j'}.\end{aligned}$$

Estimable functions are discussed in more detail in Chapter 10. The important thing to remember here is the definition of an estimable function. In making inferences from a data set, one must consider only functions of the parameters that are estimable, since they are the functions of the parameter with estimates that do not depend on which least squares solution is chosen.

6.3.2 Connectedness

Another concept related to estimable functions is the connectedness of a two-way treatment structure. If it can be assumed that the levels of the two treatments do not interact, then the treatment combination mean can be modeled as

$$\mu_{ij} = \mu + \tau_i + \beta_j. \qquad (6.3.1)$$

I. Connected

Column Treatments (rows: Row Treatments)

X				X
	X			X
		X		X
			X	X

II. Not Connected

Column Treatments (rows: Row Treatments)

X		X		X
	X		X	
X		X		X
	X		X	
X		X		X

"X" denotes an observed treatment combination.

Figure 6.3 Connected and unconnected two-way treatment structures.

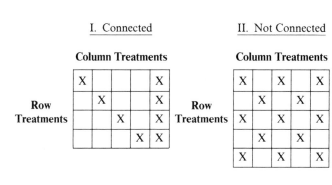

A two-way treatment structure is said to be connected if and only if data occur in the two-way cells in such a way that $\tau_i - \tau_{i'}$ and $\beta_j - \beta_{j'}$ are estimable for all (i, i') and (j, j') for model (6.3.1). Arrangement I in Figure 6.3 is connected, while arrangement II is not.

The next section discusses the testing of hypotheses about estimable functions of the parameters.

6.4 TESTING HYPOTHESES ABOUT LINEAR MODEL PARAMETERS

There are several ways to develop appropriate statistics for testing hypotheses about linear functions of the parameters of a linear model. The method used here, which is expressed in matrix notation, is equivalent to the principle of conditional error (Chapter 1) and the likelihood ratio statistic. Only hypotheses about estimable functions of the parameters are tested. In particular, consider testing the hypothesis

$$\text{H}_0\colon \mathbf{H}\boldsymbol{\beta} = \mathbf{h} \text{ versus } \text{H}_a\colon \mathbf{H}\boldsymbol{\beta} \neq \mathbf{h} \tag{6.4.1}$$

where the linear combinations $\mathbf{H}\boldsymbol{\beta}$ are estimable functions of $\boldsymbol{\beta}$ and $\mathbf{H}$ is a $q \times p$ matrix of rank q. The corresponding test statistic is

$$F_c = \frac{\text{SSH}_0/q}{\hat{\sigma}^2} \tag{6.4.2}$$

where $\hat{\sigma}^2$ is given by (6.2.6) and

$$\text{SSH}_0 = (\mathbf{H}\hat{\boldsymbol{\beta}} - \mathbf{h})'[\mathbf{H}(\mathbf{X}'\mathbf{X})^{-}\mathbf{H}']^{-1}(\mathbf{H}\hat{\boldsymbol{\beta}} - \mathbf{h}) \tag{6.4.3}$$

which is called the *sum of squares due to deviations from the null hypothesis* (the notation "$(\mathbf{X}'\mathbf{X})^{-}$" denotes a generalized inverse of the matrix $\mathbf{X}'\mathbf{X}$, Graybill, 1976). Under the assumption that the elements of the error vector are i.i.d. $N(0, \sigma^2)$, F_c is distributed as an F-distribution with q and $n - p$ degrees of freedom.

The hypothesis in (6.4.1) can always be equivalently stated in terms of a reparameterized model $\mathbf{y} = \mathbf{X}^*\boldsymbol{\beta}^* + \boldsymbol{\varepsilon}$ where $\mathbf{X}^{*\prime}\mathbf{X}^*$ is nonsingular, as

$$\text{H}_0\colon \mathbf{H}^*\boldsymbol{\beta}^* = \mathbf{h}^* \quad \text{versus} \quad \text{H}_a\colon \mathbf{H}^*\boldsymbol{\beta}^* \neq \mathbf{h}^*.$$

Then the SSH_0 of (6.4.3) can be computed as

$$\text{SSH}_0 = (\mathbf{H}^*\hat{\boldsymbol{\beta}}^* - \mathbf{h}^*)'[\mathbf{H}^*(\mathbf{X}^{*\prime}\mathbf{X}^*)^{-1}\mathbf{H}^{*\prime}]^{-1}(\mathbf{H}^*\hat{\boldsymbol{\beta}}^* - \mathbf{h}^*).$$

For a one-way model, testing the hypothesis

$$\text{H}_0\colon \tau_1 = \tau_2 = \cdots = \tau_t \quad \text{versus} \quad \text{H}_a\colon \tau_i \neq \tau_{i'} \text{ for some } i \neq i'$$

for the effects model is equivalent to testing

$$\text{H}_0\colon \tau_1^* = \tau_2^* = \cdots = \tau_{t-1}^* = 0 \quad \text{versus} \quad \text{H}_a\colon \tau_i^* \neq 0 \text{ for some } i$$

in the sum-to-zero reparameterized model. The null hypothesis in terms of $\boldsymbol{\beta}^*$ for Example 6.2 is

$$\mathrm{H}_0\colon \begin{bmatrix} 0 & 1 & 0 & 0 \\ 0 & 0 & 1 & 0 \\ 0 & 0 & 0 & 1 \end{bmatrix} \begin{bmatrix} \mu^* \\ \tau_1^* \\ \tau_2^* \\ \tau_3^* \end{bmatrix} = \mathbf{0} \quad \text{or} \quad \mathbf{H}^*\boldsymbol{\beta}^* = \mathbf{0},$$

and then the sum of squares due to deviations from H_0 is

$$\mathrm{SSH}_0 = \left(\hat{\tau}_1^*, \hat{\tau}_2^*, \hat{\tau}_3^*\right)\mathbf{Z}^{-1} \cdot \begin{bmatrix} \hat{\tau}_1^* \\ \hat{\tau}_2^* \\ \hat{\tau}_3^* \end{bmatrix}$$

where $\mathbf{Z} = \mathbf{H}^*(\mathbf{X}^{*\prime}\mathbf{X}^*)^{-1}\mathbf{H}^{*\prime}$, which is the portion of $(\mathbf{X}^{*\prime}\mathbf{X}^*)^{-1}$ corresponding to the rows and columns associated with $(\tau_1^*, \tau_2^*, \tau_3^*)$.

A $(1 - \alpha)100\%$ confidence interval about an estimable function $\mathbf{a}'\boldsymbol{\beta}$ is

$$\mathbf{a}'\hat{\boldsymbol{\beta}} - \left[t_{\alpha/2}, n - p\right] S_{\mathbf{a}'\hat{\boldsymbol{\beta}}} \le \mathbf{a}'\boldsymbol{\beta} \le \mathbf{a}'\hat{\boldsymbol{\beta}} + \left[t_{\alpha/2}, n - p\right] S_{\mathbf{a}'\hat{\boldsymbol{\beta}}}$$

where $S^2_{\mathbf{a}'\hat{\boldsymbol{\beta}}} = \hat{\sigma}^2\mathbf{a}'(\mathbf{X}'\mathbf{X})^{-}\mathbf{a}$. Simultaneous confidence intervals can be constructed about several estimable functions by using a Bonferroni or Scheffé procedure for making multiple comparisons (see Sections 3.5 and 3.7).

6.5 POPULATION MARGINAL MEANS

After analyzing a cross-classified data set via an analysis of variance, the experimenter is interested in estimating the means of particular effects or cells. The population marginal mean is a linear combination of the parameters averaged over specified classes as if there were one observation in each cell (Searle, Speed, and Milliken, 1980). If every cell has at least one observation, then all population marginal means are estimable, whereas they are not necessarily estimable if some cells are empty. This definition does not depend on the sample sizes in the cells. If the data represent a proportional sampling of cells, then the experimenter might want to consider a weighted average of the cell means where the weights are given by the sample sizes (see Chapter 10).

For a one-way treatment structure in a completely randomized design, the population marginal mean for the ith treatment is $\mu + \tau_i = \mu_i$ and has estimator

$$\widehat{\mu + \tau_i} = \hat{\mu} + \hat{\tau}_i$$

where $\hat{\mu}$ and $\hat{\tau}_i$ are obtained from any solution to the normal equations. These estimated values are called *estimated population marginal means*.

For a two-way treatment structure in a completely randomized design, the population marginal mean for cell (i, j) is $\mu_{ij} = \mu + \tau_i + \beta_j + \gamma_{ij}$. The population marginal mean for row i is the average of the μ_{ij}'s in that row, or

$$\bar{\mu}_{i\cdot} = \sum_{j=1}^{b} \frac{\mu_{ij}}{b} = \mu + \tau_{\cdot} + \beta_{\cdot} + \bar{\gamma}_{i\cdot}.$$

The population marginal mean for column j is the average of the μ_{ij}'s in that column, or

$$\bar{\mu}_{\cdot j} = \sum_{i=1}^{t} \frac{\mu_{ij}}{t_b} = \mu + \tau_{\cdot}\beta_j + \bar{\gamma}_{\cdot j}.$$

The estimates of the population marginal means are

$$\hat{\mu}_{ij} = \hat{\mu} + \hat{\tau}_i + \hat{\beta}_j + \hat{\gamma}_{ij},$$

$$\hat{\bar{\mu}}_{i\cdot} = \hat{\mu} + \hat{\tau}_i + \frac{1}{b}\sum_{j=1}^{b} \hat{\beta}_j + \frac{1}{b}\sum_{j=1}^{b} \hat{\gamma}_{ij}, \quad \text{and}$$

$$\hat{\bar{\mu}}_{\cdot j} = \hat{\mu} + \frac{1}{t}\sum_{i=1}^{t} \hat{\tau}_i + \hat{\beta}_j + \frac{1}{t}\sum_{i=1}^{t} \hat{\gamma}_{ij}.$$

The estimate of the population marginal mean $\bar{\mu}_{1\cdot}$ for the two-way example in Section 6.2, computed from the sum-to-zero restricted model, is

$$\begin{aligned}\hat{\bar{\mu}}_{1\cdot} &= \hat{\mu}^* + \hat{\tau}_1^* + \frac{\hat{\beta}_1^* + \hat{\beta}_2^* + \hat{\beta}_3^*}{3} + \frac{\hat{\gamma}_{11}^* + \hat{\gamma}_{12}^* + \hat{\gamma}_{13}^*}{3} \\ &= 5.5 + 2.33 + \frac{-2.0 - .5 + 2.5}{3} + \frac{-1.333 + 1.667 - .34}{3} \\ &= 7.83.\end{aligned}$$

We obtain the same value for $\hat{\bar{\mu}}_{1\cdot}$ when the set-to-zero least squares solution is used as expected, since $\bar{\mu}_{1\cdot}$ is estimable for this example.

When there are no empty cells, then all population marginal means are estimable. If there are empty cells, then any population marginal mean involving one or more of the missing cells is not estimable. For example, if the (2, 2) cell is missing in a 2×2 treatment structure, there is no information about μ_{22} and hence μ_{22} is not estimable. The population marginal mean for column 2 is

$$\bar{\mu}_{\cdot 2} = \frac{\mu_{12} + \mu_{22}}{2}.$$

Since $\bar{\mu}_{\cdot 2}$ depends on μ_{22}, it follows that $\bar{\mu}_{\cdot 2}$ is not estimable ($\bar{\mu}_{2\cdot}$ is not estimable either).

Any population marginal mean that is estimable can be expressed as a linear combination of the elements of the parameter vector of a reparameterized model; that is, the population marginal mean can be expressed as $\mathbf{a}'\boldsymbol{\beta}^*$ for a proper choice of $\mathbf{a}$. The variance of the estimated population marginal mean is

$$\text{Var}(\mathbf{a}'\hat{\boldsymbol{\beta}}^*) = \sigma^2\mathbf{a}'(\mathbf{X}^{*\prime}\mathbf{X}^*)^{-1}\mathbf{a},$$

and the estimated standard error of $\mathbf{a}'\hat{\boldsymbol{\beta}}^*$ is

$$\text{SE}(\mathbf{a}'\hat{\boldsymbol{\beta}}^*) = \sigma^2\sqrt{\mathbf{a}'(\mathbf{X}^{*\prime}\mathbf{X}^*)^{-1}\mathbf{a}}\,.$$

For the one-way model with the set-to-zero restriction, the estimated population marginal means are

$$\widehat{\mu + \tau_i} = \hat{\mu}^* + \hat{\tau}_i^* \qquad i = 1, 2, \ldots, t-1, \quad \text{and}$$

$$\widehat{\mu + \tau_t} = \hat{\mu}^*.$$

The variance of these estimated population marginal means are

$$\text{Var}(\widehat{\mu + \tau_i}) = \text{Var}(\hat{\mu}^*) + 2\,\text{Cov}(\hat{\mu}^*, \hat{\tau}_i^*) + \text{Var}(\hat{\tau}_i^*) \qquad i = 1, 2, \ldots, t-1,$$

and

$$\text{Var}(\widehat{\mu + \tau_t}) = \text{Var}(\hat{\mu}^*).$$

CONCLUDING REMARKS

This chapter, required only for those interested in a theoretical background, introduced least squares estimation procedures and discussed the important concept of estimability. Also discussed were the definition and estimation of population marginal means. This chapter provided general formulas for those who want to develop statistical software for their own personal computer.

7

Balanced Two-Way Treatment Structures

CHAPTER OUTLINE

In Chapters 4 and 5 we discussed how to analyze the design structure of an experiment. In this chapter we discuss methods of analyzing treatment structures, specifically two-way treatment structures. It is assumed that there are two sets of treatments $T_1, T_2, \ldots, T_t$ and $B_1, B_2, \ldots, B_b$. Each one of the T-treatments is to be combined with each one of the B-treatments and assigned to an experimental unit. Thus, a total of bt populations are sampled in a cross-classified treatment structure.

For convenience, we assume that the experimental units are assigned to the treatments completely at random, thus giving a completely randomized design. Analyzing the treatment structure and analyzing the design structure of an experiment are usually performed independently except for split-plot and repeated measures designs and their variations, which are considered in Chapters 24 through 32. Thus, it makes little difference whether the experimental units are grouped into complete blocks, balanced incomplete blocks, Latin squares, or some other grouping—the analysis of the treatment structure is similar for most standard designs. Since bt populations are sampled, there are $bt - 1$ degrees of freedom in the sum of squares for testing the hypothesis of equal treatment means. We shall consider different partitions of this treatment sum of squares to test the different kinds of hypotheses that are usually of interest to experimenters.

7.1 MODEL DEFINITION AND ASSUMPTIONS

7.1.1 Means Model

The *means model* is defined by

$$Y_{ijk} = \mu_{ij} + \varepsilon_{ijk} \qquad i = 1,2,\ldots,t, \quad j = 1,2,\ldots,b, \quad k = 1,2,\ldots,n, \tag{7.1.1}$$

where it is assumed that $\varepsilon_{ijk} \sim$ i.i.d. $N(0, \sigma^2)$, $i = 1, 2, \ldots, t$; $j = 1, 2, \ldots, b$; $k = 1, 2, \ldots, n$, and μ_{ij} is the response expected when treatment combination (T_i, B_j) is assigned to a randomly selected experimental unit.

7.1.2 Effects Model

Another common model in experiments having two-way treatment structures is known as the *effects model*, which is defined by (7.1.1) with μ_{ij} replaced by

$$\mu_{ij} = \mu + \tau_i + \beta_j + \gamma_{ij} \qquad i = 1,2,\ldots,t, \quad j = 1,2,\ldots,b. \tag{7.1.2}$$

Philosophically, such a model is intuitively appealing because it might be motivated by assuming that μ represents some overall mean effect, τ_i represents the effect on the mean as a result of assigning T_i to the experimental unit, β_j represents the effect of assigning B_j to the experimental unit, and γ_{ij} represents any additional effects that might result from using T_i and B_j at the same time on the experimental unit.

However, such a model presents some estimability problems, as was pointed out in Chapter 6. We discuss in detail the analysis of both model forms below.

7.2 PARAMETER ESTIMATION

In most experiments, the estimate of σ^2 is derived solely from the design structure of the experiment. However, in some instances the experimenter may be able to make certain assumptions about the μ_{ij}'s or the treatment effects that might provide some additional information about σ^2. For example, if the experimenter can assume that $\mu_{11} = \mu_{12}$, then the samples from the populations with means μ_{11} and μ_{12} could be combined to provide an additional degree of freedom for estimating σ^2. In this case, the corresponding single-degree-of-freedom sum of squares is given by $n(\bar{y}_{11\cdot} - \bar{y}_{12\cdot})^2/2$.

Most experimenters would not feel that this assumption is appropriate, but there are other assumptions that might be appropriate at certain times. One such assumption is that the effects of the two sets of treatments are additive, which if true would provide additional information for estimating σ^2.

All of the discussion for the one-way treatment structure applies to the two-way treatment structure if we consider the bt combinations as bt different treatments; that is,

$$[\mu_{11}, \mu_{12}, \ldots, \mu_{1b}, \ldots, \mu_{t1}, \mu_{t2}, \ldots, \mu_{tb}] = [\mu_1, \mu_2, \ldots, \mu_{bt}].$$

Thus, the best estimates of the parameters in the model are

$$\hat{\mu}_{ij} = \frac{1}{n} \sum_{k=1}^{n} y_{ijk} = \bar{y}_{ij\cdot} \qquad i = 1, 2, \ldots, t, \quad j = 1, 2, \ldots, b,$$

and

$$\hat{\sigma}^2 = \frac{1}{N - bt} \sum_{ijk} (y_{ijk} - \bar{y}_{ij\cdot})^2$$

where $N = nbt$. The sampling distribution of $\hat{\mu}_{ij}$ is $N(\mu_{ij}, \sigma^2/n)$ where $i = 1, 2, \ldots, t$, and $j = 1, 2, \ldots, b$, and the sampling distribution of $(N - bt)\hat{\sigma}^2/\sigma^2$ is $\chi^2(N - bt)$. In addition, $\hat{\mu}_{11}, \hat{\mu}_{12}, \ldots, \hat{\mu}_{tb}$ and $\hat{\sigma}^2$ are all independently distributed statistics.

Most often the experimenter will want to answer the following questions:

1. How do the T-treatments affect the response?
2. How do the B-treatments affect the response?

In order to give good answers to these questions, the experimenter must first determine whether these two sets of treatments interact.

7.3 INTERACTIONS AND THEIR IMPORTANCE

The interaction hypothesis can be stated in many different ways, all of which are equivalent. Two such ways are given below:

1. $H_0: \mu_{ij} - \mu_{ij'} - \mu_{i'j} + \mu_{i'j'} = 0$ for all i, i', j, j'; (7.3.1)
2. $H_0: \mu_{ij} = \mu + \tau_i + \beta_j$ for all i and j for some set of parameters

$$\mu\ \tau_1, \tau_2, \ldots, \tau_t, \quad \text{and} \quad \beta_1, \beta_2, \ldots, \beta_b. \tag{7.3.2}$$

Both (1) and (2) imply that there is no interaction between the treatment effects. The interpretation of (1) is that the difference between the effects of B_j and $B_{j'}$ is the same regardless of which T-treatment they are combined with. Equivalently, the difference between the effects of T_i and $T_{i'}$ is the same regardless of which B-treatment they are combined with. The interpretation corresponding to (2) is that the effects of the two sets of treatments are additive.

If the two sets of treatments do not interact, then the effects of each set of treatments can best be compared after averaging over the effects of the second set of treatments. Such a comparison is best in the sense that averaging provides more power for comparing the effects of two or more treatments, or equivalently, averaging gives the shortest possible confidence intervals on effect differences. If the two sets of treatments interact, then differences between the effects of one set of treatments depend on the level of the second treatment set with which they are combined, and the analysis of the experiment is slightly more complex.

7.4 MAIN EFFECTS

If the experimenter concludes that the two sets of treatments do not interact, then hypotheses about the main effects can be tested. These hypotheses can be written as:

$$H_{01}: \bar{\mu}_{1\cdot} = \bar{\mu}_{2\cdot} = \cdots = \bar{\mu}_{t\cdot} \quad \text{and}$$
$$H_{02}: \bar{\mu}_{\cdot 1} = \bar{\mu}_{\cdot 2} = \cdots = \bar{\mu}_{\cdot b}.$$

Even if there is interaction in the experiment, the above two hypotheses can still be tested. However, the interpretations of the results of the tests in these two situations will be quite different.

7.5 COMPUTER ANALYSES

Most statistical analysis packages automatically give the tests for the three hypotheses given in the preceding two sections provided that one specifies a model of the form

$$Y = \quad T \quad B \quad T * B.$$

Most also have an option that allows the user to compare the main effect means by using one or more of the multiple comparison procedures discussed in Chapter 3. Most packages also allow the user to specify and test contrasts of the user's own choosing.

If it is determined that the two sets of treatments interact, then the experimenter may want to compare the effects of all *bt* treatment combinations. This can be done by hand if such comparisons cannot be made by the statistical package being used. Alternatively, if the statistical analysis package does not allow multiple comparisons of the $T * B$ cell means, it can often be tricked into doing so. To do this, one includes a new identification variable on the data cards so that the new variable takes on *bt* different values, one for each of the *bt* treatment combinations. This new variable can be used to reanalyze the data as a one-way treatment structure, thus yielding multiple comparisons on the two-way cell means.

In the next chapter, a case study is considered that illustrates the concepts discussed in this chapter.

CONCLUDING REMARKS

There are three basic preliminary hypotheses that are often tested when the treatments are arranged in a two-way treatment structure. The most important of these is the interaction hypothesis. If there is no interaction, then the main effects of each of the treatments can best be compared by averaging over the levels of the other treatment. If there is interaction, then the experimenter must be careful to determine whether it makes sense to average over the levels of the second treatment when comparing the effects of the first treatment. Often it does not make sense.

8

Case Study: Complete Analyses of Balanced Two-Way Experiments

CHAPTER OUTLINE

In the preceding chapter, it was assumed that when the T-treatments and the B-treatments interact in an experiment, the experimenter will want to compare the effects of the T-treatments at each level of the B-treatments, or vice versa. In many instances, interaction does not occur everywhere in the experiment—often just one or two treatment combinations cause most of the interaction. In other cases, one level of one of the treatments may interact with the levels of the second treatment, while all other levels of the first treatment do not interact with any levels of the second.

In order to conduct a more complete analysis of data with interaction, it is helpful to determine where the interaction occurs in the data. For example, if it is known that all of the interaction in an experiment is caused by only one level of the T-treatments, then all of the other levels of the T-treatments could still be compared after averaging over all of the levels of B-treatments. This provides more power for comparing the remaining levels of the T-treatments.

8.1 CONTRASTS OF MAIN EFFECTS

Very often the structure of the T-treatments and the structure of the B-treatments suggest main-effect contrasts of particular interest to the experimenter. These main-effect contrasts can give rise to special types of interaction contrasts that should also be of interest to the experimenter and that should be easy to interpret.

Next we define contrasts in the main effects.

Definition 8.1 A linear combination of the $\bar{\mu}_{i\cdot}$'s, $\Sigma c_i \bar{\mu}_{i\cdot}$, is called a *contrast in the T-effects* if $\Sigma c_i = 0$. A linear combination of the $\bar{\mu}_{\cdot j}$'s, $\Sigma d_j \bar{\mu}_{\cdot j}$, is called a *contrast in the B-effects* if $\Sigma d_j = 0$.

Now we define orthogonal contrasts in the main effects.

Definition 8.2 Two contrasts, $\Sigma c_i \bar{\mu}_{i\cdot}$ and $\Sigma c_i' \bar{\mu}_{i\cdot}$, are called *orthogonal contrasts in the T-effects* if $\Sigma c_i c_i' = 0$. Two contrasts, $\Sigma d_j \bar{\mu}_{\cdot j}$ and $\Sigma d_j' \bar{\mu}_{\cdot j}$, are called *orthogonal contrasts in the B-effects* if $\Sigma d_j d_j' = 0$.

Now suppose that

$$S_1 = \left\{ \sum_i c_{i1} \bar{\mu}_{i\cdot}, \sum_i c_{i2} \bar{\mu}_{i\cdot}, \ldots, \sum_i c_{it-1} \bar{\mu}_{i\cdot} \right\}$$

is an orthogonal set of $t - 1$ contrasts in the T-effects and that

$$S_2 = \left\{ \sum_j d_{j1} \bar{\mu}_{\cdot j}, \sum_j d_{j2} \bar{\mu}_{\cdot j}, \ldots, \sum_j d_{jb-1} \bar{\mu}_{\cdot j} \right\}$$

is an orthogonal set of $b - 1$ contrasts in the B-effects.

Each of the sets S_1 and S_2 suggests a partitioning of the two main-effect sums of squares. That is,

$$S_1^* = \left\{ Q_p^2 = \frac{nb(\Sigma_i c_{ip}\bar{y}_{i\cdot\cdot})^2}{\Sigma_i c_{ip}^2} \qquad p = 1, 2, \ldots, t-1 \right\} \tag{8.1.1}$$

defines a partitioning of the sum of squares for T, and

$$S_2^* = \left\{ Q_q^2 = \frac{nt(\Sigma_j d_{jq}\bar{y}_{\cdot j\cdot})^2}{\Sigma_j d_{jq}^2} \qquad q = 1, 2, \ldots, b-1 \right\} \tag{8.1.2}$$

defines a partitioning of the sum of squares for B. That is, each Q_p^2 in S_1^* is a single-degree-of-freedom sum of squares used for testing whether the corresponding contrast of the main-effect means is zero, and the sum of all Q_p^2 in S_1^* is equal to the sum of squares for testing H_0: $\bar{\mu}_{1\cdot} = \bar{\mu}_{2\cdot} = \cdots = \bar{\mu}_{t\cdot}$. A similar situation exists for the elements of S_2^*.

We do not wish to overemphasize the desirability of obtaining orthogonal partitions of the basic sums of squares. Orthogonal partitions are nice from a mathematical point of view, and quite often a well-chosen set of orthogonal contrasts will enable the experimenter to interpret his or her data wisely, clearly, and completely. However, we believe that the experimenter should consider any and all contrasts that may be meaningful and should not be overly concerned about whether his or her selected contrasts are orthogonal.

8.2 CONTRASTS OF INTERACTION EFFECTS

We begin with a definition of an interaction contrast.

Definition 8.3 A linear combination of the μ_{ij}'s,

$$\sum_i \sum_j \omega_{ij}\mu_{ij},$$

is called an *interaction contrast* if $\Sigma_i \omega_{ij} = 0$ for $j = 1, 2, \ldots, b$ and $\Sigma_j \omega_{ij} = 0$ for $i = 1, 2, \ldots, t$.

Contrasts in the main effects of a two-way experiment give rise to special types of interaction contrasts. Suppose that $\Sigma_i c_i \bar{\mu}_{i\cdot}$ is a contrast in the T-effects and $\Sigma_j d_j \bar{\mu}_{\cdot j}$ is a contrast in the B-effects. Then $\Sigma_i \Sigma_j c_i d_j \mu_{ij}$ is an interaction contrast.

Two interaction contrasts $\Sigma_{ij}\omega_{ij}\mu_{ij}$ and $\Sigma_{ij}\omega'_{ij}\mu_{ij}$ are called *orthogonal contrasts in the interaction effects* if $\Sigma_{ij}\omega_{ij}\omega'_{ij} = 0$. Orthogonal contrasts in the two sets of main effects give rise to orthogonal contrasts in the interaction effects. Suppose that $\Sigma_i c_i \bar{\mu}_{i\cdot}$ and $\Sigma_i c'_i \bar{\mu}_{i\cdot}$ are two contrasts in the $\bar{\mu}_{i\cdot}$'s and that $\Sigma_j d_j \bar{\mu}_{\cdot j}$ and $\Sigma_j d'_j \bar{\mu}_{\cdot j}$ are two contrasts in the $\bar{\mu}_{\cdot j}$'s. Then $\Sigma_{ij} c_i d_j \mu_{ij}$ and $\Sigma_{ij} c'_i d'_j \mu_{ij}$ are orthogonal contrasts in the μ_{ij}'s if either $\Sigma_i c_i c_i' = 0$ or $\Sigma_j d_j d_j' = 0$, that is, if at least one of the pairs of main-effect contrasts is orthogonal.

Suppose S_1 and S_2 are as defined in Section 8.1. Then

$$S_3 = \left\{ \sum_{ij} c_{i1} d_{j1} \mu_{ij}, \sum_{ij} c_{i1} d_{j2} \mu_{ij}, \ldots, \sum_{ij} c_{it-1} d_{jb-1} \mu_{ij} \right\}$$

is an orthogonal set of $(t-1)(b-1)$ contrasts in the $T * B$ interaction effect.

Next let

$$Q_{pq}^2 = \frac{n\left(\Sigma_{ij} c_{ip} d_{jq} \bar{y}_{ij\cdot}\right)^2}{\Sigma_i c_{ip}^2 \Sigma_j d_{jq}^2} \qquad p = 1, 2, \ldots, t-1, \quad q = 1, 2, \ldots, b-1. \tag{8.2.1}$$

The set $S_3^* = \{Q_{11}^2, Q_{12}^2, \ldots, Q_{t-1b-1}^2\}$ defines a partitioning of the sum of squares for interaction. That is,

$$\sum_p \sum_q Q_{pq}^2 = T * B \text{ SS},$$

and Q_{pq}^2 and $Q_{p'q'}^2$ are independent single-degree-of-freedom sum of squares if $p \neq p'$ or $q \neq q'$.

In the next section, an example is discussed that illustrates the ideas described in this and the preceding section.

8.3 PAINT–PAVING EXAMPLE

Consider the experiment in Table 8.1, which gives the means of three observations. This experiment was conducted to compare the lifetimes measured in weeks of two colors of paint manufactured by two different companies on three types of paving surfaces. The error sum of squares for this experiment was 455.04 with 24 degrees of freedom so that $\hat{\sigma}^2 = 18.96$. The usual analysis of variance table for this experiment is given in Table 8.2.

Table 8.1 Paint–Paving Cell Means

Paint	*Asphalt I*	*Asphalt II*	*Concrete*	*Mean*
Yellow I	15	17	32	21.333
Yellow II	27	30	20	25.667
White I	30	28	29	29.0
White II	34	35	36	35.0
Mean	26.5	27.5	29.25	27.75

Table 8.2 Analysis of Variance Table for Paint–Paving Data

Source of Variation	*df*	*SS*	*MS*	*F*	*p*
Total	35	2,039.79			
Paint	3	896.75	298.92	15.76	< .001
Paving	2	46.50	23.25	1.25	n.s.
Paint * Paving	6	641.50	106.42	5.64	< .001
Error	24	455.04	18.96		

The structure of the treatment combinations in this experiment gives rise to a set of orthogonal contrasts on the two sets of main effects that might be of interest. These are given in Table 8.3.

These two sets of orthogonal contrasts in main effects suggest six orthogonal contrasts in the interaction effects. These are given in Table 8.4.

A more complete analysis of these data using the partitioning suggested in Tables 8.3 and 8.4 is given in Table 8.5.

We next show the details of the computations necessary to obtain the sum of squares for some selected contrasts from Table 8.5.

The single-degree-of-freedom sum of squares for comparing White is, from (8.1.1),

$$\frac{3 \cdot 3[(1)29.0 + (-1)(35.0)]^2}{1^2 + (-1)^2} = 162.0.$$

The single-degree-of-freedom sum of squares for comparing Type is,

Table 8.3 Main-Effect Hypotheses for Paint–Paving Data

Comparison	*Hypothesis*
Paints	
Yellow I vs. Yellow II	$\bar{\mu}_{1\cdot} - \bar{\mu}_{2\cdot} = 0$
White I vs. White II	$\bar{\mu}_{3\cdot} - \bar{\mu}_{4\cdot} = 0$
Yellow vs. White	$\bar{\mu}_{1\cdot} + \bar{\mu}_{2\cdot} - \bar{\mu}_{3\cdot} - \bar{\mu}_{4\cdot} = 0$
Paving	
Asphalt I vs. Asphalt II	$\bar{\mu}_{\cdot 1} - \bar{\mu}_{\cdot 2} = 0$
Asphalt vs. Concrete	$\bar{\mu}_{\cdot 1} + \bar{\mu}_{\cdot 2} - 2\bar{\mu}_{\cdot 3} = 0$

Table 8.4 Interaction Hypotheses for Paint–Paving Data

Comparison	*Hypothesis*
Yellow * Asphalt	$\mu_{11} - \mu_{12} - \mu_{21} + \mu_{22} = 0$
White * Asphalt	$\mu_{31} - \mu_{32} - \mu_{41} + \mu_{42} = 0$
Color * Asphalt	$\mu_{11} + \mu_{21} - \mu_{12} - \mu_{22} - \mu_{31} - \mu_{41} + \mu_{32} + \mu_{42} = 0$
Yellow * Type	$\mu_{11} + \mu_{12} - 2\mu_{13} - \mu_{21} - \mu_{22} + 2\mu_{23} = 0$
White * Type	$\mu_{31} + \mu_{32} - 2\mu_{33} - \mu_{41} - \mu_{42} + 2\mu_{43} = 0$
Color * Type	$\mu_{11} + \mu_{12} + \mu_{21} + \mu_{22} - 2\mu_{13} - 2\mu_{23} - \mu_{31} - \mu_{32}$ $- \mu_{41} - \mu_{42} + 2\mu_{33} + 2\mu_{43} = 0$

Note: "Type" refers to Asphalt versus Concrete.

Table 8.5 Analysis of Variance Table for Paint–Paving Data Including Single-Degree-of-Freedom Tests

Source of Variation	*df*		*SS*	*MS*	*F*	*p*
Total	35		2,039.79			
Paint	3		896.92	298.97	15.77	< .0001
Yellow		1	84.5	84.5	4.46	< .05
White		1	162.0	162.0	8.54	< .01
Color		1	650.25	650.25	34.30	< .0001
Paving	2		46.5	23.25	1.23	n.s.
Asphalt		1	6.0	6.0	.32	n.s.
Type		1	40.5	40.5	2.14	n.s.
*Paint * Paving*	6		641.5	106.92	5.64	< .001
Yellow * Asphalt		1	.75	.75	.04	n.s.
White * Asphalt		1	6.75	6.75	.36	n.s.
Color * Asphalt		1	13.5	13.5	.71	n.s.
Yellow * Type		1	600.25	600.25	31.66	< .0001
White * Type		1	2.25	2.25	.12	n.s.
Color * Type		1	18.0	18.0	.95	n.s.
Error	24		455.04	18.96		

from (8.1.2),

$$\frac{3 \cdot 4[(1)(26.5) + (1)(27.5) + (-2)(29.5)]^2}{1^2 + 1^2 + (-2)^2} = 40.5.$$

The single-degree-of-freedom sum of squares for comparing the White * Type interaction is

$$\frac{3[(1)(30) + (1)(28) + (-2)(29) + (-1)(34) + (-1)(35) + 2(36)]^2}{\left(1^2 + (-1)^2\right)\left(1^2 + 1^2 + (-2)^2\right)}$$

$$= 2.25$$

From examining the analysis in Table 8.5, one can make the following conclusions:

1. All of the interaction in the experiment is caused by the two yellow paints acting differently on the two types of surfaces, since this interaction contrast is the only single-degree-of-freedom sum of squares for interaction that is significant.
2. Because we now know where the interaction exists in the data, we can make the following observations:
   a. Since there is no interaction between Asphalt and Paint, the two asphalts can be compared after averaging across all paints. The value of the F-statistic for this comparison is $F = .32$; thus, there is no significant difference between Asphalts I and II.
   b. Since there is no interaction between the white paints and the three pavings, the two white paints can be compared after averaging across all pavings. The value of the F-statistic for this comparison is $F = 8.54$, which indicates that White Paint II is significantly different from White Paint I. From Table 8.1, we see that White Paint II lasts longer.
   c. Although the statistic for comparing Yellow I versus Yellow II is significant ($F = 4.45$), one must be careful when making an interpretation because of the significant interaction between the brands of yellow paint and the type of paving.
   d. Even though the comparison for Asphalt versus Concrete is not significant ($F = 2.14$), one must again be careful when making an interpretation because of the significant interaction between the brands of paint and the types of paving.
3. To complete the analysis of these data, we should yet examine:
   a. Yellow I versus Yellow II on Asphalt (that is, $\mu_{11} + \mu_{12} - \mu_{21} - \mu_{22} = 0$),
   b. Yellow I versus Yellow II on Concrete,
   c. Concrete versus Asphalt for Yellow I,
   d. Concrete versus Asphalt for Yellow II, and
   e. the three pavings for white paints.

   The results are given in Table 8.6.

Table 8.6 Tests of Hypotheses in Conclusion 3

Comparison	*df*	*SS*	*MS*	*F*	*p*
a.	1	468.75	468.75	24.72	< .0001
b.	1	216.00	216.00	11.39	< .005
c.	1	512.00	512.00	27.00	< .0001
d.	1	144.50	144.50	7.62	< .02
e.	2	3.00	1.50	.08	n.s.

Examination of the results in Table 8.6 and the means in Table 8.1 reveals that (a) Yellow II is significantly better than Yellow I on asphalt, but (b) Yellow I is significantly better than Yellow II on concrete. (c) Yellow I lasts significantly longer on concrete than on asphalt; (d) Yellow II lasts significantly longer on asphalt than on concrete; and finally, (e) the white paint lasts about the same length of time on all three pavings.

All of the results obtained for our analysis of this example can be obtained using SAS®-GLM, SPSS-ANOVA, and BMD-P1V or BMD-P4V by using their contrast options. The latter three require that the 12 treatment combinations be considered as a one-way treatment structure, while SAS® allows either possibility.

8.4 ANALYZING QUANTITATIVE TREATMENT FACTORS

In this section it is assumed that the levels of both factors of an experiment are quantitative. In this case, we can define contrasts that measure curvilinear trends in each set of main effect treatment means. Trends of interest are often linear, quadratic, cubic, and so on.

The corresponding orthogonal contrasts that partition the main-effect sums of squares into effects that measure linear, quadratic, cubic, and so on, trends can then be used to construct orthogonal contrasts in the interaction effects. The resulting contrasts are called "Lin $T*$Lin B" (linear effect of T by linear effect of B), "Lin $T*$Quad B," and so on.

For a 3×4 experiment where each of the two treatments has equally spaced levels, the Lin $T*$Quad B contrast is $\Sigma_{ij} c_i d_j \mu_{ij}$ where

$$c_1 = -1, \quad c_2 = 0, \quad c_3 = 1, \quad d_1 = 1, \quad d_2 = -1, \quad d_3 = -1, \quad d_4 = 1.$$

The values of the coefficients for these particular contrasts can be obtained from a table of orthogonal polynomials. See Beyer (1968).

Let $x_1, x_2, \ldots, x_t$ represent the levels of factor T and let $z_1, z_2, \ldots, z_b$ represent the levels of factor B. There always exist parameters α_{kh}, $k = 0, 1, \ldots, t-1$, $h = 0, 1, \ldots, b-1$, such that the means μ_{ij} can be represented as a polynomial function of x_i and z_j. That is,

$$\begin{aligned}\mu_{ij} &= \sum_{k=0}^{t-1}\sum_{h=0}^{b-1} \alpha_{kh} x_i^k z_j^h \\ &= \alpha_{00} + \alpha_{10}x_i + \alpha_{20}x_i^2 + \cdots + \alpha_{t-10}x_i^{t-1} \\ &\quad + \alpha_{01}z_j + \alpha_{02}z_j^2 + \cdots + \alpha_{0b-1}z_j^{b-1} \\ &\quad + \alpha_{11}x_iz_j + \alpha_{12}x_iz_j^2 + \cdots + \alpha_{t-1b-1}x_i^{t-1}z_j^{b-1}. \qquad (8.4.1)\end{aligned}$$

If $\Sigma c_i = \Sigma d_j = 0$, then

$$\sum_{ij} c_i d_j \mu_{ij} = \sum_{ij} c_i d_j \sum_{k=1}^{t-1}\sum_{h=1}^{b-1} \alpha_{kh} x_i^k z_j^h.$$

Table 8.7 gives the expected values of main-effect and interaction contrasts for a 3×4 experiment in terms of the coefficients in model (8.4.1). In constructing the table, it was assumed that the three levels of the x's were coded to -1, 0, and 1 and that the four levels of the z's were coded to -3, -1, 1, and 3.

Our purpose in providing Table 8.7 is to point out the hypotheses that are being tested when contrasts in the interaction effects are investigated. The Lin∗Lin effect tests the hypothesis that $40\alpha_{11} + 328\alpha_{13} = 0$.

Table 8.7 Expected Values of Orthogonal Polynomials in a 3 × 4 Experiment

Effect	*Expected Value*
Lin T	$2\alpha_{10} + 10\alpha_{12}$
Quad T	$2\alpha_{20} + 10\alpha_{22}$
Lin B	$20\alpha_{01} + \frac{40}{3}\alpha_{21} + 164\alpha_{03} + \frac{328}{3}\alpha_{23}$
Quad B	$16\alpha_{02} + \frac{32}{3}\alpha_{22}$
Cubic B	$48\alpha_{03} + 32\alpha_{23}$
Lin∗Lin	$40\alpha_{11} + 328\alpha_{13}$
Lin∗Quad	$32\alpha_{12}$
Lin∗Cubic	$96\alpha_{13}$
Quad∗Lin	$40\alpha_{21} + 328\alpha_{23}$
Quad∗Quad	$32\alpha_{22}$
Quad∗Cubic	$96\alpha_{23}$

Note: Expected values are not normalized.

Thus, if this effect is significant, it could be because either α_{11} or α_{13} is nonzero, and not only because α_{11} is nonzero (as many data analysts might believe).

If one is going to examine orthogonal polynomials, we recommend that one look at the coefficients of the highest-degree term first and consider the remaining terms in descending order of degree. Once a term is determined to be in the model, then all terms whose two components both have degrees lower than that of the significant term should also be included in the model. For example, if one decides that x^2z^2 should be in the model, then the model should also include xz^2, z^2, x^2z, x^2, xz, x, and z. Our reasoning for this is as follows: The orthogonal polynomials always refer to coded values of the quantitative variables. Thus, if α_{22} is nonzero, it really implies that

$$\left(\frac{x - h_1}{c_1}\right)^2 \left(\frac{z - h_2}{c_2}\right)^2$$

belongs in the model where $(x - h_1)/c_1$ and $(z - h_2)/c_2$ are the coded values of x and z. Expansion of

$$\left(\frac{x - h_1}{c_1}\right)^2 \left(\frac{z - h_2}{c_2}\right)^2$$

demonstrates that the terms xz^2, z^2, x^2z, x^2, xz, x, and z are also in the model, even though other lower-degree orthogonal polynomials may not be significant.

8.5 MULTIPLE COMPARISONS

Any of the multiple comparison procedures discussed in Chapter 3 can be used with only some very minor adjustments for making multiple comparisons on the main effects of a two-factor experiment. The adjustments require that the n's and the n_i's be replaced by the total number of observations that were averaged to estimate the main-effect means being compared. In this chapter, the sample sizes are nt for the B-main-effect means and nb for the T-main-effect means. Our recommendations for multiple comparisons on main-effect means are the same as those given in Section 3.2.

The only procedures given in Chapter 3 that are easily generalized to contrasts in the interaction effects are the LSD procedure, Bonferroni's method, the multivariate t-method, and Scheffé's procedure. We have found Scheffé's procedure not very satisfactory because the required critical point is much too large and the procedure is much too conservative. Our recommendations for multiple comparisons of interaction contrasts are as follows:

1. Conduct an F-test for interaction.
2. If the F-statistic is significant, make any planned comparisons by using the LSD procedure (or equivalently, the contrast procedure given in the preceding section). For data snooping and unplanned

comparisons, use the procedure given by Johnson (1976), which is not discussed here.

3. If the F-test for interaction is not significant, the experimenter should still examine any individual interaction contrasts that he or she had planned to consider but by using the multivariate t-method or Bonferroni's method. The multivariate T-method is used whenever the selected contrasts are linearly independent; otherwise, Bonferroni's method should be used.

CONCLUDING REMARKS

In this chapter, we introduced, by giving examples, methods for obtaining a maximum amount of information from an experiment. Included were methods for discovering where interaction occurs in an experiment. Knowing where interaction occurs in an experiment is valuable in determining the best answers to questions that may be raised. The techniques introduced in this chapter should help experimenters do a better job of analyzing their experiments.

The analysis of quantitative treatment factors was also considered, including how to determine what kinds of trends might be related to the different levels of the treatment factors.

9

Using the Means Model to Analyze Balanced Two-Way Treatment Structures with Unequal Subclass Numbers

CHAPTER OUTLINE

In Chapters 7 and 8 we considered the equal-sample-size case, where each treatment combination is observed an equal number of times. In Chapters 13 through 15 we consider cases where some treatment combinations are missing, but in this chapter as well as Chapters 10 through 12, we assume that every treatment combination is observed and at least one combination is observed more than once.

9.1 MODEL DEFINITIONS AND ASSUMPTIONS

As in Section 7.1.1, let μ_{ij} be the expected response when level i of a treatment T and level j of a treatment factor B are both applied to the same experimental unit. In this chapter, we assume that the observed response can be modeled by

$$y_{ijk} = \mu_{ij} + e_{ijk}$$
$$i = 1,2,\ldots,t, \quad j = 1,2,\ldots,b, \quad k = 1,2,\ldots,n_{ij} \quad (9.1.1)$$

where $e_{ijk} \sim$ i.i.d. $N(0,\sigma^2)$ and $n_{ij} > 0$ for every i and j.

9.2 PARAMETER ESTIMATION

All of the discussion for the one-way treatment structure in Chapters 1 through 3 applies to the two-way treatment structure as well if we consider the bt treatment combinations as bt different treatments. For unbalanced data problems this is a good way, and often the best way, to analyze the data. The best estimates of the parameters in the means model are

$$\hat{\mu}_{ij} = \frac{1}{n_{ij}} \sum_{k=1}^{n_{ij}} y_{ijk} = \bar{y}_{ij\cdot}$$
$$i = 1,2,\ldots,t, \quad j = 1,2,\ldots,b, \quad (9.2.1)$$

and

$$\hat{\sigma}^2 = \frac{1}{N - tb} \sum_{ijk} (y_{ijk} - \bar{y}_{ij\cdot})^2 \quad \text{where} \quad N = n_{\cdot\cdot\cdot} \quad (9.2.2)$$

We note that the sampling distributions of the $\hat{\mu}_{ij}$'s and $\hat{\sigma}^2$ are

$$\hat{\mu}_{ij} \sim N\left(\mu_{ij}, \frac{\sigma^2}{n_{ij}}\right) \qquad i = 1,2,\ldots,t, \quad j = 1,2,\ldots,b,$$

and

$$\frac{(N - tb)\hat{\sigma}^2}{\sigma^2} \sim \chi^2(N - tb),$$

and that $\hat{\mu}_{11}, \hat{\mu}_{12}, \ldots, \hat{\mu}_{tb}$, and $\hat{\sigma}^2$ are all independently distributed as before. The experimenter will usually want to answer the same questions when the data are unbalanced as when they are balanced. We recall that those questions are:

1. Do the two sets of treatments interact?
2. How do the T-treatments affect the response?
3. How do the B-treatments affect the response?

These questions can be stated as hypotheses in terms of the parameters of the model. These hypotheses are:

$$H_{01}: \mu_{ij} - \mu_{i'j} - \mu_{ij'} + \mu_{i'j'} = 0 \quad \text{for all} \quad i \neq i' \quad \text{and} \quad j \neq j,$$
$$H_{02}: \bar{\mu}_{1\cdot} = \bar{\mu}_{2\cdot} = \cdots = \bar{\mu}_{t\cdot}, \quad \text{and}$$
$$H_{03}: \bar{\mu}_{\cdot 1} = \bar{\mu}_{\cdot 2} = \cdots = \bar{\mu}_{\cdot b}.$$

Formulating the above hypotheses should be considered as a first step in analyzing any experiment. There will usually be well-defined contrasts that directly address the questions of interest to the researcher. The hypotheses H_{01}, H_{02}, and H_{03} are tested to help choose an appropriate multiple comparison procedure for addressing these questions.

As in (7.1.2), there always exist parameters

$$\mu, \quad \tau_1, \tau_2, \ldots, \tau_t, \quad \beta_1, \beta_2, \ldots, \beta_b, \quad \gamma_{11}, \gamma_{12}, \ldots, \gamma_{tb}$$

such that μ_{ij} can be expressed in an effects model as

$$\mu_{ij} = \mu + \tau_i + \beta_j + \gamma_{ij} \qquad i = 1,2,\ldots,t, \quad j = 1,2,\ldots,b.$$

Many experimenters prefer to look at a representation of the treatment combination means like the one above. This may be because statisticians have encouraged experimenters to consider such models. As a result, much of the existing computer software leads us to use this representation. We shall consider both types of models, the means model in this chapter and the effects model in Chapter 10.

In Sections 1.5 and 1.6, we introduced two different procedures for developing test statistics, both of which gave rise to the same test statistics in the one-way case. This is, in fact, always the case for well-balanced data sets. However, it is not the case with unbalanced data sets. By "well-balanced" we mean that there are equal numbers of observations on each treatment combination. The matrix procedure is used in this chapter, and the model-fitting procedure is used in Chapter 10 to obtain test statistics.

9.3 TESTING WHETHER ALL MEANS ARE EQUAL

Consider the data in Table 9.1. The data is from a small two-way treatment structure experiment conducted in a completely randomized design structure.

To begin, we compute the two-way cell means and the marginal means for the data in Table 9.1. Table 9.2 gives these means where a row marginal mean is defined as the mean of the cell means in the given row,

Table 9.1 An Unbalanced Two-Way Experiment

	B_1	B_2	B_3	*Total*
T_1	19 20 21	24 26	22 25 25	182
	60	50	72	
T_2	25 27	21 24 24	31 32 33	217
	52	69	96	
Total	112	119	168	399

and a column marginal mean is the mean of the cell means in the given column.

The error sum of squares for the data in Table 9.1 is

$$\text{ESS} = \sum_{ijk} y_{ijk}^2 - \sum_{ij} \frac{y_{ij\cdot}^2}{n_{ij}} = 20$$

with $N - bt = 10$ degrees of freedom. The best estimate of σ^2 is $\hat{\sigma}^2 = \frac{20}{10} = 2$.

We now consider the experiment as a one-way treatment structure with six treatments and test H_0: $\mu_{11} = \mu_{12} = \mu_{13} = \mu_{21} = \mu_{22} = \mu_{23}$. Using the results in (1.5.1), we get

$$\begin{aligned} SS_{H_0} &= \frac{60^2}{3} + \frac{50^2}{2} + \frac{72^2}{3} + \frac{52^2}{2} + \frac{69^2}{3} + \frac{96^2}{3} - \frac{399^2}{16} \\ &= 238.9375, \end{aligned}$$

Table 9.2 Cell Means for Data in Table 9.1

	B_1	B_2	B_3	*Marginal Mean*
T_1	20	25	24	23
T_2	26	23	32	27
Marginal Mean	23	24	28	25

which is based on 5 degrees of freedom. The F-statistic for testing H_0 is

$$F_c = \frac{238.9375/5}{2} = 23.89$$

which is significant at the $\hat{\alpha} = .00003$ level. Thus, we conclude that there are differences between the means of the six different treatment combinations.

9.4 INTERACTION AND MAIN-EFFECT HYPOTHESES

In the previous section, it was determined that there are significant differences among the six treatment combination means. Now it is necessary to see where differences occur. As a first step, we determine whether there is significant interaction among the data given in Table 9.1. We test

$$H_{01}: \mu_{ij} - \mu_{i'j} - \mu_{ij'} + \mu_{i'j'} = 0$$

for all $i \neq i'$ and $j \neq j'$. We do this by utilizing the matrix procedure for developing test statistics that was discussed in Section 1.4. The hypothesis H_{01} will be true if and only if $\mu_{11} - \mu_{12} - \mu_{21} + \mu_{22} = 0$ and $\mu_{11} - \mu_{13} - \mu_{21} + \mu_{23} = 0$. In turn, these statements are true if and only if $\mathbf{C}\boldsymbol{\mu} = \mathbf{0}$ where

$$\mathbf{C} = \begin{bmatrix} 1 & -1 & 0 & -1 & 1 & 0 \\ 1 & 0 & -1 & -1 & 0 & 1 \end{bmatrix}$$

and

$$\boldsymbol{\mu}' = [\mu_{11} \quad \mu_{12} \quad \mu_{13} \quad \mu_{21} \quad \mu_{22} \quad \mu_{23}].$$

Then, from (1.4.4),

$$\begin{aligned} SS_{H_{01}} &= [\mathbf{C}\hat{\boldsymbol{\mu}}]'[\mathbf{CDC}']^{-1}[\mathbf{C}\hat{\boldsymbol{\mu}}] \\ &= [-8 \quad 2]\begin{bmatrix} \frac{10}{6} & \frac{5}{6} \\ \frac{5}{6} & \frac{9}{6} \end{bmatrix}^{-1}\begin{bmatrix} -8 \\ 2 \end{bmatrix} \end{aligned}$$

since $\mathbf{D} = \text{Diag}(\frac{1}{3}, \frac{1}{2}, \frac{1}{3}, \frac{1}{2}, \frac{1}{3}, \frac{1}{3})$. Thus,

$$\begin{aligned} SS_{H_{01}} &= [-8 \quad 2]\left(\frac{6}{65}\begin{bmatrix} 9 & -5 \\ -5 & 10 \end{bmatrix}\right)\begin{bmatrix} -8 \\ 2 \end{bmatrix} \\ &= 776\left(\frac{6}{65}\right) = 71.631 \end{aligned}$$

and is based on 2 degrees of freedom. The F-statistic is

$$F_c = \frac{71.631/2}{2} = 17.91,$$

which is based on 2 and 10 degrees of freedom and is significant at the $\hat{\alpha} = .0005$ level. Other matrices **C** could also be used, but all produce the same test statistic.

Next we test the equality of the expected row marginal means for illustration purposes. The appropriate hypothesis is H_{02}: $\bar{\mu}_{1\cdot} = \bar{\mu}_{2\cdot}$. We note that H_{02} is true if and only if $\mathbf{C\mu} = \mathbf{0}$ where

$$\mathbf{C} = \begin{bmatrix} 1 & 1 & 1 & -1 & -1 & -1 \end{bmatrix}.$$

Using (1.4.4) we get

$$\begin{aligned} SS_{H_{02}} &= [\mathbf{C}\hat{\boldsymbol{\mu}}]'[\mathbf{CDC}]^{-1}[\mathbf{C}\hat{\boldsymbol{\mu}}] \\ &= [-12]\left[\frac{14}{6}\right]^{-1}[-12] = 61.714, \end{aligned}$$

which is based on 1 degree of freedom, and the corresponding F-statistic is

$$F_c = \frac{61.714/1}{2} = 30.857$$

which is significant at the $\hat{\alpha} = .00024$ level. One could also take

$$\mathbf{C} = \begin{bmatrix} \frac{1}{3} & \frac{1}{3} & \frac{1}{3} & -\frac{1}{3} & -\frac{1}{3} & -\frac{1}{3} \end{bmatrix}.$$

The reader should verify that this second choice of **C** leads to the same test statistic.

Finally, we test the equality of the expected column marginal means by testing H_{03}: $\bar{\mu}_{\cdot 1} = \bar{\mu}_{\cdot 2} = \bar{\mu}_{\cdot 3}$. We first note that H_{03} is true if and only if $\mathbf{C\mu} = \mathbf{0}$ where

$$\mathbf{C} = \begin{bmatrix} 1 & -1 & 0 & 1 & -1 & 0 \\ 1 & 0 & -1 & 1 & 0 & -1 \end{bmatrix}.$$

Using (1.4.4), we get

$$\begin{aligned} SS_{H_{03}} &= [-2 \quad -10]\begin{bmatrix} \frac{10}{6} & \frac{5}{6} \\ \frac{5}{6} & \frac{9}{6} \end{bmatrix}^{-1}\begin{bmatrix} -2 \\ -10 \end{bmatrix} \\ &= [-2 \quad -10]\left(\frac{6}{65}\begin{bmatrix} 9 & -5 \\ -5 & 10 \end{bmatrix}\right)\begin{bmatrix} -2 \\ -10 \end{bmatrix} \\ &= 77.169 \end{aligned}$$

which is based on 2 degrees of freedom. Hence, the corresponding F-statistic is $F_c = 19.29$, which is significant at the $\hat{\alpha} = .0037$ level.

The above tests are summarized in the analysis of variance table given in Table 9.3.

At this time we want to point out some differences between the results in this analysis and those obtained for the balanced case, discussed in Chapter 7.

Table 9.3 Analysis of Variance Table for Data in Table 9.1

Source of Variation	*df*	*SS*	*MS*	*F*	$\hat{p}$
Total	15	258.938			
$\mu_{11} = \cdots = \mu_{23}$	5	238.938	47.79	23.89	.00003
T	1	61.714	61.71	30.86	.00024
B	2	77.169	38.58	19.29	.00037
$T*B$	2	71.631	35.81	17.91	.0005
Error	10	20	2		

1. For balanced data, it is always true that

$$SS_T + SS_B + SS_{T*B} = SS_{\mu_{11}=\mu_{12}=\cdots=\mu_{23}};$$

this is not generally true for unbalanced data.

2. For the balanced case, SS_T, SS_B, and SS_{T*B} are statistically independent; this is not generally true for unbalanced data.

There are other sums of squares that are often associated with analyses of two-way treatment structures. Two of these are examined in Chapter 10.

9.5 POPULATION MARGINAL MEANS

Often the experimenter is interested in making comparisons about and between the levels of each main effect. In the balanced case, one may compare $\bar{\mu}_{1\cdot}, \bar{\mu}_{2\cdot}, \ldots, \bar{\mu}_{t\cdot}$ with each other and compare $\bar{\mu}_{\cdot 1}, \bar{\mu}_{\cdot 2}, \ldots, \bar{\mu}_{\cdot b}$ with each other. These means are called the *population marginal means* for both the balanced and unbalanced case. The best estimate of $\bar{\mu}_{i\cdot}$ is

$$\hat{\bar{\mu}}_{i\cdot} = \frac{1}{b} \sum_{j=1}^{b} \hat{\mu}_{ij}, \qquad i = 1, 2, \ldots, t. \tag{9.5.1}$$

The estimated standard error of $\hat{\bar{\mu}}_{i\cdot}$ is

$$\widehat{\text{s.e.}}(\hat{\bar{\mu}}_{i\cdot}) = \frac{\hat{\sigma}}{b} \sqrt{\sum_{j=1}^{b} \frac{1}{n_{ij}}}. \tag{9.5.2}$$

The best estimate of $\hat{\bar{\mu}}_{\cdot j}$ is

$$\hat{\bar{\mu}}_{\cdot j} = \frac{1}{t} \sum_{i=1}^{t} \hat{\mu}_{ij} \qquad j = 1, 2, \ldots, b, \tag{9.5.3}$$

and the estimated standard error of $\hat{\bar{\mu}}_{.j}$ is

$$\widehat{\text{s.e.}}(\hat{\bar{\mu}}_{.j}) = \frac{\hat{\sigma}}{t}\sqrt{\sum_{i=1}^{t}\frac{1}{n_{ij}}}. \tag{9.5.4}$$

It should be noted that in unbalanced data problems, it is generally the case that $\hat{\bar{\mu}}_{i.}$ will be different from $\bar{y}_{i..}$ and that $\hat{\bar{\mu}}_{.j}$ will be different from $\bar{y}_{.j.}$.

The estimators $\hat{\bar{\mu}}_{i.}$ and $\hat{\bar{\mu}}_{.j}$ are unbiased estimates of $\bar{\mu}_{i.}$ and $\bar{\mu}_{.j}$, respectively, while $\bar{y}_{i..}$ is an unbiased estimate of $(\Sigma_j n_{ij}\mu_{ij})/n_{i.}$ and $\bar{y}_{.j.}$ is an unbiased estimate of $(\Sigma_i n_{ij}\mu_{ij})/n_{.j}$. When using computing packages to analyze data, it is extremely important to determine whether estimates of the main effects are calculated as $\hat{\bar{\mu}}_{i.}$ and $\hat{\bar{\mu}}_{.j}$ or as $\bar{y}_{i..}$ and $\bar{y}_{.j.}$.

For the data in Table 9.1, the estimated population marginal means are shown in Table 9.2. We have $\hat{\bar{\mu}}_{1.} = 23$ and $\hat{\bar{\mu}}_{2.} = 27$, and $\hat{\bar{\mu}}_{.1} = 23$, $\hat{\bar{\mu}}_{.2} = 24$, and $\hat{\bar{\mu}}_{.3} = 28$. The standard errors of these estimates are as follows:

$$\widehat{\text{s.e.}}(\hat{\bar{\mu}}_{1.}) = \frac{1.414}{3}\sqrt{\frac{1}{3}+\frac{1}{2}+\frac{1}{3}} = .51,$$

$$\widehat{\text{s.e.}}(\hat{\bar{\mu}}_{2.}) = \frac{1.414}{3}\sqrt{\frac{1}{2}+\frac{1}{3}+\frac{1}{3}} = .51,$$

$$\widehat{\text{s.e.}}(\hat{\bar{\mu}}_{.1}) = \frac{1.414}{2}\sqrt{\frac{1}{3}+\frac{1}{2}} = .65,$$

$$\widehat{\text{s.e.}}(\hat{\bar{\mu}}_{.2}) = \frac{1.414}{2}\sqrt{\frac{1}{2}+\frac{1}{3}} = .65, \quad \text{and}$$

$$\widehat{\text{s.e.}}(\hat{\bar{\mu}}_{.3}) = \frac{1.414}{2}\sqrt{\frac{1}{3}+\frac{1}{3}} = .58.$$

To make inferences about linear combinations of the population marginal means, say $\Sigma c_i\bar{\mu}_{i.}$ or $\Sigma d_j\bar{\mu}_{.j}$, we note that

$$\frac{\sum c_i\hat{\bar{\mu}}_{i.} - \sum c_i\bar{\mu}_{i.}}{\frac{\hat{\sigma}}{b}\sqrt{\sum_i\left(c_i^2\sum_j\frac{1}{n_{ij}}\right)}} \sim t(\nu), \tag{9.5.5}$$

and that

$$\frac{\sum d_j\hat{\bar{\mu}}_{.j} - \sum_j d_j\bar{\mu}_{.j}}{\frac{\hat{\sigma}}{t}\sqrt{\sum_j\left(d_j^2\sum_i\frac{1}{n_{ij}}\right)}} \sim t(\nu). \tag{9.5.6}$$

The formulas in (9.5.5) and (9.5.6) can be obtained as special cases of (1.3.1). For example, the t-statistic that tests $\bar{\mu}_{i\cdot} = \bar{\mu}_{i'\cdot}$ is

$$t_c = \frac{\hat{\bar{\mu}}_{i\cdot} - \hat{\bar{\mu}}_{i'}}{\frac{\hat{\sigma}}{b}\sqrt{\sum_j \frac{1}{n_{ij}} + \sum_j \frac{1}{n_{i'j}}}}.$$

For the data in Table 9.1, to test $\bar{\mu}_{1\cdot} = \bar{\mu}_{2\cdot}$, we get

$$t_c = \frac{23 - 27}{\frac{1.414}{3}\sqrt{\left(\frac{1}{3} + \frac{1}{2} + \frac{1}{3}\right) + \left(\frac{1}{2} + \frac{1}{3} + \frac{1}{3}\right)}}$$

$$= \frac{-4}{\frac{1.414}{3}\sqrt{\frac{7}{3}}} = \frac{-4}{.72} = -5.55$$

which is significant at the $\hat{\alpha} = .0002$ level. A 95% confidence interval for $\bar{\mu}_{\cdot 1} - \bar{\mu}_{\cdot 2}$ constructed from the data in Table 9.1 is

$$\hat{\bar{\mu}}_{\cdot 1} - \hat{\bar{\mu}}_{\cdot 2} \pm t_{\alpha/2,\nu} \cdot \frac{\hat{\sigma}}{t}\sqrt{\sum_i \frac{1}{n_{i1}} + \sum_i \frac{1}{n_{i2}}}$$

$$= 23 - 24 \pm t_{.025,10} \cdot \frac{1.414}{2}\sqrt{\left(\frac{1}{3} + \frac{1}{2}\right) + \left(\frac{1}{2} + \frac{1}{3}\right)}$$

$$= -1 \pm (2.228)(.91)$$

$$= -1 \pm 2.03.$$

9.6 SIMULTANEOUS INFERENCES AND MULTIPLE COMPARISONS

There are few good procedures available for making multiple comparisons in two-way experiments in which there are unequal numbers of observations per treatment combination. If one wants to compare all pairs of two-way cell means, then any of the techniques discussed in Chapter 3 can be used simply by considering the two-way experiment as a one-way treatment structure experiment. In this case the reader should see the recommendations given in Section 3.2.

If one wishes to make multiple comparisons on the population marginal means, we recommend using t-tests based on (9.5.5) and (9.5.6). Use the given significance levels if the corresponding F-test for comparing the corresponding marginal means is significant. If the F-test is not significant, we still recommend using these t-tests. However, in this case we would use Bonferroni's method and claim two population marginal means to be significantly different only when the calculated $\hat{\alpha}$ is less than α/p where α is our selected experimentwise error rate and p is the number of comparisons we had planned to make prior to conducting the analysis.

If we determine that there is interaction in the data, then we may want to compare the effects of one of the treatments at each level of the other treatment. That is, we may want to compare within the sets

$$\{\mu_{11}, \mu_{12}, \ldots, \mu_{1b}\}, \{\mu_{21}, \mu_{22}, \ldots, \mu_{2b}\}, \ldots, \{\mu_{t1}, \mu_{t2}, \ldots, \mu_{tb}\}$$

or within the sets

$$\{\mu_{11}, \mu_{21}, \ldots, \mu_{t1}\}, \{\mu_{12}, \mu_{22}, \ldots, \mu_{t2}\}, \ldots, \{\mu_{1b}, \mu_{2b}, \ldots, \mu_{tb}\}.$$

There are

$$\frac{b(b+1)}{2} \cdot t$$

pairwise comparisons required in the first group and

$$\frac{t(t+1)}{2} \cdot b$$

comparisons required in the second group. To make comparisons in the first group, we can use the two-way cell means and declare two cell means to be significantly different if the significance level given by the t-test is less than α/p where

$$p = \frac{b(b+1)}{2} \cdot t,$$

and similarly for the second group. To make comparisons in both groups simultaneously, we would take

$$p = \frac{b(b+1)}{2} \cdot t + \frac{t(t+1)}{2} \cdot b.$$

These obviously give Bonferroni tests for the two-way cell means. If the F-test comparing all means is significant, we have no objections to using the actual significance levels. This use is the equivalent of a Fisher's LSD procedure.

For data snooping and unplanned comparisons, we should use Scheffé's procedure. See Johnson (1973) to see how Scheffé's procedure can be applied in messy data situations.

9.7 COMPUTER ANALYSES

The test statistics described in Section 9.4 can be obtained automatically with many statistical computing packages. Since these packages employ the effects model, the interested reader should see Section 10.7.

CONCLUDING REMARKS

This chapter is the first of four considering the analysis of two-way treatment structures with unequal subclass numbers. The analyses presented in this chapter were obtained by using the means model. An important assumption made was that all treatment combinations were observed. Procedures for testing main-effect and interaction hypotheses were obtained as special cases of the general techniques introduced in Chapter 1. Procedures for making inferences on the population marginal means were also given.

In Chapter 10, similar kinds of questions are answered by utilizing the effects model; however, we prefer the means model because of its simplicity.

10

Using the Effects Model to Analyze Balanced Two-Way Treatment Structures with Unequal Subclass Numbers

CHAPTER OUTLINE

In Chapter 9 we discussed using the means model to analyze two-way treatment structures having unequal subclass numbers. This chapter considers using the effects model in the same situation. All questions that can be answered by using the effects model can also be answered by using the means model, and vice versa. We shall discuss the effects model because it is often an important tool in using statistical computing packages. These packages can deliver automatically printed test statistics, that answer questions frequently encountered when analyzing two-way treatment structures.

10.1 MODEL DEFINITION

The effects model corresponding to the means model (9.1.1) is defined by

$$y_{ijk} = \mu + \tau_i + \beta_j + \gamma_{ij} + \varepsilon_{ijk} \qquad i = 1, 2, \ldots, t, \quad j = 1, 2, \ldots, b, \quad k = 1, 2, \ldots, n_{ij}, \tag{10.1.1}$$

where $\varepsilon_{ijk} \sim$ i.i.d. $N(0, \sigma^2)$.

10.2 PARAMETER ESTIMATES AND TYPE I ANALYSIS

Other sums of squares are often associated with analyses of two-way treatment structures having unequal subclass numbers besides those introduced in Chapter 9. We shall examine two sets of these. The first set involves fitting the two-way data in a sequential manner by using generalizations of the model comparison method described in Section 1.6. One sequence of steps often used is the following:

Step 1. Fit $y_{ijk} = \mu + \varepsilon_{ijk}$ and obtain RSS_1.

Step 2. Fit $y_{ijk} = \mu + \tau_i + \varepsilon_{ijk}$ and obtain RSS_2.

Step 3. Fit $y_{ijk} = \mu + \tau_i + \beta_j + \varepsilon_{ijk}$ and obtain RSS_3.

Step 4. Fit $y_{ijk} = \mu + \tau_i + \beta_j + \gamma_{ij} + \varepsilon_{ijk}$ and obtain RSS_4.

The quantity RSS_i is the residual sum of squares after fitting the model in the ith step.

The difference between RSS_1 and RSS_2, denoted by $R(\tau|\mu)$, is called the *reduction due to* τ *adjusted for* μ; that is, $R(\tau|\mu) = \text{RSS}_1 - \text{RSS}_2$. This reduction gives the amount by which one can reduce the residual sum of squares of the model in step 1 by considering a model with τ_i included as well. The larger the value of $R(\tau|\mu)$, the more important it is to have τ_i in the model. Thus, $R(\tau|\mu)$ is a measure of the effect of different levels of treatment T.

The quantity $R(\beta|\mu, \tau) = \text{RSS}_2 - \text{RSS}_3$ is called the *reduction due to* β *adjusted for* μ *and* τ. It gives the additional amount by which one can reduce the residual sum of squares of the model in step 2 by also including β_j in the model. $R(\beta|\mu, \tau)$ is a measure of the effect of different levels of treatment B above and beyond the effect of treatment T.

Finally, the quantity $R(\gamma|\mu, \tau, \beta) = \text{RSS}_3 - \text{RSS}_4$ is called the *reduction due to* γ *adjusted for* μ, τ, *and* β. It gives the additional amount

Table 10.1 Analysis of Variance Table for a Sequential Analysis (Type I Analysis)

Source of Variation	*df*	*SS*	*MS*	*F*
Total	$N-1$	RSS_1		
T	$t-1$	$R(\tau\vert\mu)$	$\dfrac{R(\tau\vert\mu)}{t-1}$	$\dfrac{T\text{MS}}{\hat{\sigma}^2}$
B	$b-1$	$R(\beta\vert\mu,\tau)$	$\dfrac{R(\beta\vert\mu,\tau)}{b-1}$	$\dfrac{B\text{MS}}{\hat{\sigma}^2}$
$T*B$	$(b-1)(t-1)$	$R(\gamma\vert\mu,\tau,\beta)$	$\dfrac{R(\gamma\vert\mu,\tau,\beta)}{(t-1)(b-1)}$	$\dfrac{(T*B)\text{MS}}{\hat{\sigma}^2}$
Error	$N-bt$	RSS_4	$\hat{\sigma}^2$	

by which one can reduce the residual sum of squares of the model in step 3 by adding an interaction term, γ_{ij}, to the model. Clearly, $R(\gamma|\mu,\tau,\beta)$ is a measure of interaction, since the model in step 3 is an additive model that holds if and only if there is no interaction.

An analysis of variance table corresponding to this sequential analysis is given in Table 10.1. This analysis is called a *Type I analysis*.

The sums of squares in the last four lines of Table 10.1 are statistically independent, and the ratios of the T, B, and $T*B$ mean squares to $\hat{\sigma}^2$ all have F-distributions. It is quite interesting and informative to determine exactly what hypotheses each of the F-statistics in Table 10.1 is testing. This determination is made in Section 10.4.

To illustrate, we fit each of the four models required for the Type I analysis to the data in Table 9.1. An understanding of Chapter 6 is required to follow the computations made here. However, such an understanding is not necessary for readers interested in the sequential approach who are willing to let statistical computing packages do the required computations.

The model given in step 1 is $y_{ijk} = \mu + \varepsilon_{ijk}$. The best estimate of μ in this model is $\hat{\mu} = \bar{y}_{...} = 24.9375$, and the residual sum of squares is

$$\text{RSS}_1 = \sum_{ijk}(y_{ijk} - \bar{y}_{...})^2 = \sum_{ijk} y_{ijk}^2 - (n_{..})\bar{y}_{...}^2 = 258.9375,$$

which is based on $n_{..} - 1 = 15$ degrees of freedom.

The normal equations for the model defined in step 2 are

$$\begin{bmatrix} 16 & 8 & 8 \\ 8 & 8 & 0 \\ 8 & 0 & 8 \end{bmatrix} \begin{bmatrix} \hat{\mu} \\ \hat{\tau}_1 \\ \hat{\tau}_2 \end{bmatrix} = \begin{bmatrix} 399 \\ 182 \\ 217 \end{bmatrix}.$$

One possible solution to these equations is obtained by using the set-to-zero restrictions discussed in Section 6.2, which yields the solution $\hat{\tau}_2 = 0$, $\hat{\tau}_1 = -4.375$, and $\hat{\mu} = 27.125$. (Recall from Chapter 6 that a unique solution does not exist). The residual sum of squares is

$$\begin{aligned} \text{RSS}_2 &= \mathbf{y}'\mathbf{y} - \boldsymbol{\beta}'\mathbf{x}'\mathbf{y} \\ &= 10{,}209 - [(27.125)(399) + (-4.375)(182) + (0)(217)] \\ &= 182.375 \end{aligned}$$

which is based on $16 - 2 = 14$ degrees of freedom. Thus,

$$R(\tau|\mu) = 258.9375 - 182.375 = 76.5625$$

and is based on $15 - 14 = 1$ degree of freedom.

The normal equations for the model in step 3 are:

$$\begin{bmatrix} 16 & 8 & 8 & 5 & 5 & 6 \\ 8 & 8 & 0 & 3 & 2 & 3 \\ 8 & 0 & 8 & 2 & 3 & 3 \\ 5 & 3 & 2 & 5 & 0 & 0 \\ 5 & 2 & 3 & 0 & 5 & 0 \\ 6 & 3 & 3 & 0 & 0 & 6 \end{bmatrix} \begin{bmatrix} \hat{\mu} \\ \hat{\tau}_1 \\ \hat{\tau}_2 \\ \hat{\beta}_1 \\ \hat{\beta}_2 \\ \hat{\beta}_3 \end{bmatrix} = \begin{bmatrix} 399 \\ 182 \\ 217 \\ 112 \\ 119 \\ 168 \end{bmatrix}.$$

To obtain a solution, we let $\hat{\tau}_2 = 0$ and $\hat{\beta}_3 = 0$ (see Chapter 6). This system of equations can then be reduced to an equivalent system by deleting the rows and columns that correspond to $\hat{\tau}_2$ and $\hat{\beta}_3$. After doing this, the system of normal equations reduces to

$$\begin{bmatrix} 16 & 8 & 5 & 5 \\ 8 & 8 & 3 & 2 \\ 5 & 3 & 5 & 0 \\ 5 & 2 & 0 & 5 \end{bmatrix} \begin{bmatrix} \hat{\mu} \\ \hat{\tau}_1 \\ \hat{\beta}_1 \\ \hat{\beta}_2 \end{bmatrix} = \begin{bmatrix} 399 \\ 182 \\ 112 \\ 119 \end{bmatrix}.$$

The solution to this reduced system is

$$\hat{\mu} = 30.154, \quad \hat{\tau}_1 = -4.308, \quad \hat{\beta}_1 = -5.169, \quad \text{and} \quad \hat{\beta}_2 = -4.631.$$

The residual sum of squares for this model is

$$\text{RSS}_3 = \mathbf{y}'\mathbf{y} - \hat{\boldsymbol{\beta}}'\mathbf{x}'\mathbf{y} = 10{,}209 - 10{,}117.369 = 91.631,$$

which is based on 12 degrees of freedom. Thus

$$R(\beta|\mu,\tau) = \text{RSS}_2 - \text{RSS}_3 = 182.375 - 91.631 = 90.744$$

which is based on $14 - 12 = 2$ degrees of freedom.

The normal equations for the model in step 4 are

$$\begin{bmatrix} 16 & 8 & 8 & 5 & 5 & 6 & 3 & 2 & 3 & 2 & 3 & 3 \\ 8 & 8 & 0 & 3 & 2 & 3 & 3 & 2 & 3 & 0 & 0 & 0 \\ 8 & 0 & 0 & 2 & 3 & 3 & 0 & 0 & 0 & 2 & 3 & 3 \\ 5 & 3 & 2 & 5 & 0 & 0 & 3 & 0 & 0 & 2 & 0 & 0 \\ 5 & 2 & 3 & 0 & 5 & 0 & 0 & 2 & 0 & 0 & 3 & 0 \\ 6 & 3 & 3 & 0 & 0 & 6 & 0 & 0 & 3 & 0 & 0 & 3 \\ 3 & 3 & 0 & 3 & 0 & 0 & 3 & 0 & 0 & 0 & 0 & 0 \\ 2 & 2 & 0 & 0 & 2 & 0 & 0 & 2 & 0 & 0 & 0 & 0 \\ 3 & 3 & 0 & 0 & 0 & 3 & 0 & 0 & 3 & 0 & 0 & 0 \\ 2 & 0 & 2 & 2 & 0 & 0 & 0 & 0 & 0 & 2 & 0 & 0 \\ 3 & 0 & 3 & 0 & 3 & 0 & 0 & 0 & 0 & 0 & 3 & 0 \\ 3 & 0 & 3 & 0 & 0 & 3 & 0 & 0 & 0 & 0 & 0 & 3 \end{bmatrix} \begin{bmatrix} \hat{\mu} \\ \hat{\tau}_1 \\ \hat{\tau}_2 \\ \hat{\beta}_1 \\ \hat{\beta}_2 \\ \hat{\beta}_3 \\ \hat{\gamma}_{11} \\ \hat{\gamma}_{12} \\ \hat{\gamma}_{13} \\ \hat{\gamma}_{21} \\ \hat{\gamma}_{22} \\ \hat{\gamma}_{23} \end{bmatrix} = \begin{bmatrix} 399 \\ 182 \\ 217 \\ 112 \\ 119 \\ 168 \\ 60 \\ 50 \\ 72 \\ 52 \\ 69 \\ 96 \end{bmatrix}.$$

To obtain a solution, we let $\hat{\tau}_2 = 0$, $\hat{\beta}_3 = 0$, $\hat{\gamma}_{13} = 0$, $\hat{\gamma}_{21} = 0$, $\hat{\gamma}_{22} = 0$, and $\hat{\gamma}_{23} = 0$ (see Chapter 6). Using the reduction technique, the system reduces to

$$\begin{bmatrix} 16 & 8 & 5 & 5 & 3 & 2 \\ 8 & 8 & 3 & 2 & 3 & 2 \\ 5 & 3 & 5 & 0 & 3 & 0 \\ 5 & 2 & 0 & 5 & 0 & 2 \\ 3 & 3 & 3 & 0 & 3 & 0 \\ 2 & 2 & 0 & 2 & 0 & 2 \end{bmatrix} \begin{bmatrix} \hat{\mu} \\ \hat{\tau}_1 \\ \hat{\beta}_1 \\ \hat{\beta}_2 \\ \hat{\gamma}_{11} \\ \hat{\gamma}_{12} \end{bmatrix} = \begin{bmatrix} 399 \\ 182 \\ 112 \\ 119 \\ 60 \\ 50 \end{bmatrix}.$$

The solution to this reduced system is

$\hat{\mu} = 32$, $\hat{\tau}_1 = -8$, $\hat{\beta}_1 = -6$, $\hat{\beta}_2 = -9$, $\hat{\gamma}_{11} = 2$, and $\hat{\gamma}_{12} = 10$.

The value of $\hat{\sigma}^2$ is 2, since the residual sum of squares is

$$\text{RSS}_4 = \mathbf{y}'\mathbf{y} - \hat{\boldsymbol{\beta}}\mathbf{x}'\mathbf{y} = 10{,}209 - 10{,}189 = 20$$

and is based on 10 degrees of freedom. Also,

$$R(\gamma|\mu, \tau, \beta) = \text{RSS}_3 - \text{RSS}_4 = 91.631 - 20 = 71.631,$$

which is based on 12 − 10 degrees of freedom. The preceding results can be summarized in an analysis of variance table like the one given in Table 10.2.

The sum of squares as well as the test statistic for interaction in the Type I analysis is the same as that obtained using the means model and the matrix procedure in Chapter 9; however, the two procedures give different sum of squares and test statistics for both T and B main effects.

The data in a two-way treatment structure could also be analyzed by first fitting μ, then β, then τ, and finally γ. The only new sums of squares required that are not already given in Table 10.2 are $R(\beta|\mu)$ and

Table 10.2 A Type I Analysis of Variance Table

Source of Variation	*df*	*SS*	*MS*	*F*	α
Total	15	258.9375			
T	1	76.5625	76.5625	38.28	.0001
B	2	90.744	45.372	22.69	.0002
T ∗ *B*	2	71.631	35.815	17.91	.0005
Error	10	20.0	20.0		

$R(\tau|\mu, \beta)$. For unbalanced data cases, the corresponding F-tests for the T and B main effects will usually be different from those obtained by fitting μ, then τ, then β, and finally γ.

It should be recalled from Chapter 6 that the parameter estimates $\hat{\mu}$, $\hat{\tau}_i$, $\hat{\beta}_j$, and $\hat{\gamma}_{ij}$, $i = 1, 2, \ldots, t$; $j = 1, 2, \ldots, b$, are not unbiased estimates of μ, τ_i, β_j, and γ_{ij}. Indeed, the individual parameters are not estimable functions. Under the set to zero restrictions used to solve the normal equations, it is possible to show, for the example, that

$$\left.\begin{array}{lll}
\hat{\mu} & \text{is an unbiased estimate of} & \mu + \tau_2 + \beta_3 + \gamma_{23}, \\
\hat{\tau}_1 & \text{is an unbiased estimate of} & \tau_1 - \tau_2 + \gamma_{13} - \gamma_{23}, \\
\hat{\beta}_1 & \text{is an unbiased estimate of} & \beta_1 - \beta_3 + \gamma_{21} - \gamma_{23}, \\
\hat{\beta}_2 & \text{is an unbiased estimate of} & \beta_2 - \beta_3 + \gamma_{22} - \gamma_{23}, \\
\hat{\gamma}_{11} & \text{is an unbiased estimate of} & \gamma_{11} - \gamma_{13} - \gamma_{21} + \gamma_{23}, \quad \text{and} \\
\hat{\gamma}_{12} & \text{is an unbiased estimate of} & \gamma_{12} - \gamma_{13} - \gamma_{22} + \gamma_{23}.
\end{array}\right\} \qquad (10.2.1)$$

More about estimable functions and their estimates can be found in Section 10.3.

10.3 USING ESTIMABLE FUNCTIONS IN SAS®

In this section we discuss the estimable functions of the model parameters that SAS® makes available. (Readers who do not use SAS® can skip this section.) Since it is easiest to describe estimable functions by using an example, we shall consider again the data in Table 9.1.

The SAS® analysis of this data can be obtained by using the following statements:

```
PROC GLM;
CLASSES T B;
MODEL Y = T B T*B/options;
```

Many options can be used with the last statement. One of the more important of these is the E option, which instructs SAS® to print a general form of the estimable functions of the model parameters.

We recall from Chapter 6 that all linear functions of the parameters in a design model are not necessarily estimable. Using the E option, SAS® prints the information needed to determine which linear combinations of the parameters are estimable and which are not.

The general form of an estimable function given by SAS® GLM is shown in Table 10.3. It means that a linear function of the parameters $\mathbf{l}'\boldsymbol{\beta}$ is estimable if and only if there exist constants L1, L2, L4, L5, L7, and L8 such that

$$\begin{aligned}\mathbf{l}'\boldsymbol{\beta} = {} & (\mathrm{L1})\mu + (\mathrm{L2})\tau_1 + (\mathrm{L1} - \mathrm{L2})\tau_2 + (\mathrm{L4})\beta_1 + (\mathrm{L5})\beta_2 \\ & + (\mathrm{L1} - \mathrm{L4} - \mathrm{L5})\beta_3 + (\mathrm{L7})\gamma_{11} + (\mathrm{L8})\gamma_{12} + (\mathrm{L2} - \mathrm{L7} - \mathrm{L8})\gamma_{13} \\ & + (\mathrm{L4} - \mathrm{L7})\gamma_{21} + (\mathrm{L5} - \mathrm{L8})\gamma_{22} \\ & + (\mathrm{L1} - \mathrm{L2} - \mathrm{L4} - \mathrm{L5} - \mathrm{L7} + \mathrm{L8})\gamma_{23}\end{aligned}$$

where

$$\mathbf{l}' = [l_1, l_2, \ldots, l_{12}] \quad \text{and}$$
$$\boldsymbol{\beta}' = [\mu, \tau_1, \tau_2, \beta_1, \beta_2, \beta_3, \gamma_{11}, \gamma_{12}, \gamma_{13}, \gamma_{21}, \gamma_{22}, \gamma_{23}].$$

For example, from the general form of estimable functions, we can see that:

1. μ is not estimable, since in order for μ to be estimable, we would at least need to have L1 = 1, L2 = 0, and L1 − L2 = 0, and these three equations cannot all be true at the same time.

Table 10.3 General Form of Estimable Functions

EFFECT		COEFFICIENTS
INTERCEPT		L1
T	1	L2
	2	L1 - L2
B	1	L4
	2	L5
	3	L1 - L4 - L5
T*B	1 1	L7
	1 2	L8
	1 3	L2 - L7 - L8
	2 1	L4 - L7
	2 2	L5 - L8
	2 3	L1 - L2 - L4 - L5 + L7 + L8

Table 10.4 A Basis Set of Estimable Functions

L_1	L_2	L_4	L_5	L_7	L_8	*Estimable Function*
1	0	0	0	0	0	$\mu + \tau_2 + \beta_3 + \gamma_{23}$
0	1	0	0	0	0	$\tau_1 - \tau_2 + \gamma_{13} - \gamma_{23}$
0	0	1	0	0	0	$\beta_1 - \beta_3 + \gamma_{21} - \gamma_{23}$
0	0	0	1	0	0	$\beta_2 - \beta_3 + \gamma_{22} - \gamma_{23}$
0	0	0	0	1	0	$\gamma_{11} - \gamma_{13} - \gamma_{21} + \gamma_{23}$
0	0	0	0	0	1	$\gamma_{12} - \gamma_{13} - \gamma_{22} + \gamma_{23}$

2. τ_1 is not estimable, since in order for τ_1 to be estimable, we would at least need to have L1 = 0, L2 = 1, and L1 − L2 = 0, which are contradictory.

3. $\tau_1 - \tau_2$ is not estimable, since in order for $\tau_1 - \tau_2$ to be estimable, we would need to have L1 = 0, L2 = 1, L1 − L2 = −1, L4 = 0, L5 = 0, L1 − L4 − L5 = 0, L7 = 0, L8 = 0, and L2 − L7 − L8 = 0. However, if L2 = 1, L7 = 0, and L8 = 0, then $L_2 - L_7 - L_8 = 1 \neq 0$. Thus, the above equations are inconsistent, and hence $\tau_1 - \tau_2$ is not estimable.

Thus, it is clear that many functions of the parameters are not estimable. But there are also many functions of the parameters which are. A basic set of estimable functions (see Chapter 6) can easily be obtained by successively letting each of the coefficients L1, L2, L4, L5, L7, and L8 be equal to one and all others equal to zero. This *basis set of estimable functions* is given in Table 10.4.

Note that the number of linear functions in the basis set is six, which is equal to the rank of the **X′X** matrix, as discussed in Chapter 6 and also to the number of treatment combinations. Also note that the functions

Table 10.5 Another Basis Set of Estimable Functions

L_1	L_2	L_4	L_5	L_7	L_8	*Estimable Function*
1	$\frac{1}{2}$	$\frac{1}{3}$	$\frac{1}{3}$	$\frac{1}{6}$	$\frac{1}{6}$	$\mu + \bar{\tau}_. + \bar{\beta}_. + \bar{\gamma}_{..}$
0	1	0	0	$\frac{1}{3}$	$\frac{1}{3}$	$\tau_1 - \tau_2 + \bar{\gamma}_{1.} - \bar{\gamma}_{2.}$
0	0	1	0	$\frac{1}{2}$	0	$\beta_1 - \beta_3 + \bar{\gamma}_{.1} - \bar{\gamma}_{.3}$
0	0	0	1	0	$\frac{1}{2}$	$\beta_2 - \beta_3 + \bar{\gamma}_{.2} - \bar{\gamma}_{.3}$
0	0	0	0	1	0	$\gamma_{11} - \gamma_{13} - \gamma_{21} + \gamma_{23}$
0	0	0	0	0	1	$\gamma_{12} - \gamma_{13} - \gamma_{22} + \gamma_{23}$

Table 10.6 Results Obtained with SOLUTION Option in SAS® GLM

PARAMETER		ESTIMATE	T FOR H0: PARAMETER = 0	PR>|T|	STD ERROR OF ESTIMATE
INTERCEPT		32.00000000 B	39.19	0.0001	0.81649658
T	1	−8.00000000 B	−6.93	0.0001	1.15470054
	2	0.00000000 B	.	.	.
B	1	−6.00000000 B	−4.65	0.0009	1.29099445
	2	−9.00000000 B	−7.79	0.0001	1.15470054
	3	0.00000000 B	.	.	.
T*B	1 1	2.00000000 B	1.15	0.2751	1.73205081
	1 2	10.00000000 B	5.77	0.0002	1.73205081
	1 3	0.00000000 B	.	.	.
	2 1	0.00000000 B	.	.	.
	2 2	0.00000000 B	.	.	.
	2 3	0.00000000 B	.	.	.

on the right are those being estimated by $\hat{\mu}$, $\hat{\tau}_1$, $\hat{\beta}_1$, $\hat{\beta}_2$, $\hat{\gamma}_{11}$, and $\hat{\gamma}_{12}$ as given in (10.2.1). There is no unique basis set of estimable functions—another basis set of estimable functions and the values of L1, L2, L4, L5, L7, and L8 needed to obtain them are given in Table 10.5.

When one uses the SOLUTION option with the SAS® GLM model statement, the computer prints out least squares estimates of the model parameters by using the set-to-zero restrictions. The results of the SOLUTION option are shown in Table 10.6.

However, these least squares estimates do not estimate their respective parameters. In fact, as shown above, the individual parameters are not estimable. SAS® GLM indicates this by putting the letter B by the least squares estimate. The functions of the parameters that these estimators are really unbiased estimates of are those given in Table 10.4. That is, $\hat{\mu} = 32$ is the best unbiased estimate of $\mu + \tau_2 + \beta_3 + \gamma_{23}$, $\hat{\tau}_1 = -8$ is the best unbiased estimate of $\tau_1 - \tau_2 + \gamma_{13} - \gamma_{23}$, $\hat{\tau}_2 = 0$ is estimating zero (which it does a good job of doing, too), $\hat{\beta}_1 = -6$ is the best unbiased estimate of $\beta_1 - \beta_3 + \gamma_{21} - \gamma_{23}$, and so on.

The standard errors printed in Table 10.6 are the actual standard errors of the estimator. That is, $\widehat{\text{s.e.}}(\hat{\mu}) = .816$, $\widehat{\text{s.e.}}(\hat{\tau}_1) = 1.155$, and so on. The t-tests which are printed test that the functions being estimated are equal to zero. For example $t = 39.19$, which corresponds to $\hat{\mu}$, tests H_0: $\mu + \tau_2 + \beta_3 + \gamma_{23} = 0$. Such tests are usually not very interesting.

By using the CONTRAST or ESTIMATE statement in the SAS® GLM procedure, we can make inferences about any estimable linear combination of the parameters that we choose. The linear combination need not be a contrast, but it must be estimable. Fortunately, SAS® always checks whether the selected linear combination is an estimable function. If it is, using the ESTIMATE statement gives the best unbiased estimate of the linear combination, the standard error of the estimate, and a t-statistic for testing the hypothesis that the selected linear combination of the model parameters is equal to zero. Using the CONTRAST statement gives a single-degree-of-freedom sum of squares for the linear combination and an F-test for the hypothesis that the linear combination of the parameters is equal to zero. The two tests are equivalent. The proper use of these statements for our example requires the following form:

ESTIMATE '*label*' INTERCEPT c_1 T c_2 c_3 B c_4 c_5 c_6 T∗B c_7 c_8 c_9 c_{10} c_{11} c_{12};

If all of the coefficients of a particular effect are zero, that effect and its coefficients do not need to be included in the statement. To use the CONTRAST statement, one needs only to replace the word ESTIMATE with CONTRAST in the above form.

For example, to obtain the best estimates of the estimable functions given in Table 10.5, we would use:

1. ESTIMATE 'OVERALL MEAN' INTERCEPT 1 T .5 .5 B .33333 .33333 .33333 T∗B .16667 .16667 .16667 .16667 .16667 .16667;
2. ESTIMATE 'T1 − T2' T 1 −1 T∗B .33333 .33333 .33333 −.33333 −.33333 −.33333;
3. ESTIMATE 'B1 − B3' B 1 0 −1 T∗B .5 0 −.5 .5 0 −.5;
4. ESTIMATE 'B2 − B3' B 0 1 −1 T∗B 0 .5 −.5 0 .5 −.5;
5. ESTIMATE 'INT1' T∗B 1 0 −1 −1 0 1;
6. ESTIMATE 'INT2' T∗B 0 1 −1 0 −1 1;

10.4 TYPE I THROUGH TYPE IV HYPOTHESES

Many readers of this book may already be aware that the SAS® GLM procedure gives users the option of selecting one of four types of sums of squares for testing hypotheses. In this section we are mainly concerned with defining and interpreting the corresponding four types of hypotheses that are tested. To illustrate these, we shall use the data in Table 9.1 once more.

As stated in Section 10.2, the Type I sums of squares are obtained by fitting the two-way effects model in a sequential fashion. The sum of squares obtained at each step, which is a measure of the importance of the particular term being considered at that step, is the amount that the residual sum of squares can be reduced by including that term in the model.

Table 10.7 Definitions of Type I and Type II Sums of Squares

		SS	
Source of Variation	*df*	*Type I*	*Type II*
T	$t-1$	$R(\tau\|\mu)$	$R(\tau\|\mu,\beta)$
B	$b-1$	$R(\beta\|\mu,\tau)$	$R(\beta\|\mu,\tau)$
$T*B$	$(t-1)(b-1)$	$R(\gamma\|\mu,\tau,\beta)$	$R(\gamma\|\mu,\tau,\beta)$

The Type II analysis is also obtained by utilizing the model comparison technique. The sums of squares corresponding to each effect are adjusted for every other effect in the model that is at the same or a lower level. Hence, the sum of squares corresponding to the T-effect is $R(\tau|\mu,\beta)$, and the sum of squares corresponding to the B-effect is $R(\beta|\mu,\tau)$. Readers who are slightly confused should see Section 16.3, since the definitions of the Type I and Type II analyses for a three-way treatment structure will help clarify the above discussion. Table 10.7 shows the Type I and Type II sums of squares for two-way effects models.

We believe that most experimenters think in terms of the parameters of the μ_{ij} or the $\mu + \tau_i + \beta_j + \gamma_{ij}$ model. Obviously, then, it is extremely important to know exactly what functions of these model parameters are being tested by the different types of sums of squares. The hypotheses being tested in a Type I analysis are called Type I hypotheses.

Table 10.8 gives the hypotheses tested by a Type I analysis of the data in Table 9.1 in terms of the μ_{ij} model. Later we shall show how to determine these hypotheses from an SAS® output.

Clearly, an experimenter would rarely be interested in the research hypotheses that correspond to T and B in Table 10.8. However, the equations corresponding to $T*B$ are equivalent to testing a no-interac-

Table 10.8 Hypotheses for a Type I Analysis of the μ_{ij} Model for Data in Table 9.1

Source of Variation	*Hypothesis*
T	$3\mu_{11} + 2\mu_{12} + 3\mu_{13} - 2\mu_{21} - 3\mu_{22} - 3\mu_{23} = 0$
B	$37\mu_{11} - 2\mu_{12} - 35\mu_{13} + 28\mu_{21} + 2\mu_{22} - 30\mu_{23} = 0$ and
	$3\mu_{11} + 2\mu_{12} - 5\mu_{13} + 2\mu_{21} + 3\mu_{22} - 5\mu_{23} = 0$
$T*B$	$\mu_{11} - \mu_{12} - \mu_{21} + \mu_{22} = 0$ and
	$\mu_{11} - \mu_{13} - \mu_{21} + \mu_{23} = 0$

Table 10.9 General Forms of Type I Hypotheses in Terms of μ_{ij} Model

Source of Variation	*Type I Hypothesis*
T	$\frac{1}{n_{1\cdot}}\sum_j n_{1j}\mu_{1j} = \frac{1}{n_{2\cdot}}\sum_j n_{2j}\mu_{2j} = \cdots = \frac{1}{n_{t\cdot}}\sum_j n_{tj}\mu_{tj}$
B	$\sum_i \left(n_{ij} - \frac{n_{ij}^2}{n_{i\cdot}}\right)\mu_{ij} = \sum_{j'\neq j}\sum_i \frac{n_{ij}n_{ij'}}{n_{i\cdot}}\mu_{ij'}$ for $j = 1, 2, \ldots, b$
$T * B$	$\mu_{ij} - \mu_{i'j} - \mu_{ij'} + \mu_{i'j'} = 0$ for all i, j, i', and j'

tion hypothesis. That is, the contrasts $\mu_{11} - \mu_{12} - \mu_{21} + \mu_{22}$ and $\mu_{11} - \mu_{13} - \mu_{21} + \mu_{23}$ span the interaction space for this particular example.

Table 10.9 gives the Type I hypotheses for the general two-factor model

$$y_{ijk} = \mu_{ij} + \varepsilon_{ijk} \qquad i = 1, 2, \ldots, t, \quad j = 1, 2, \ldots, b, \quad k = 1, 2, \ldots, n_{ij}.$$

For the data in Table 9.1, the hypotheses tested by the Type II analysis in terms of the μ_{ij} model are given in Table 10.10.

For the general model, the hypothesis tested by the row corresponding to the T-effect for the Type II analysis is

$$\sum_j \left(n_{ij} - \frac{n_{ij}^2}{n_{\cdot j}}\right)\mu_{ij} = \sum_{i'\neq i}\sum_j \frac{n_{ij}n_{i'j}}{n_{\cdot j}}\mu_{i'j} \qquad \text{for} \quad i = 1, 2, \ldots, t. \tag{10.4.1}$$

The other two rows are the same as they were for the Type I analysis.

The Type I and Type II hypotheses in terms of the $\mu + \tau_i + \beta_j + \gamma_{ij}$ model are given in Tables 10.11 and 10.12.

Examination of Tables 10.9 through 10.12 reveals that the hypotheses tested by the Type I and Type II analyses may not be very interesting. In addition, the rejection or acceptance of these hypotheses may not be easy to interpret.

Table 10.10 Type II Hypotheses for μ_{ij} Model for Data in Table 9.1

Source of Variation	*Type II Hypothesis*
T	$4\mu_{11} + 4\mu_{12} + 5\mu_{13} - 4\mu_{21} - 4\mu_{22} - 5\mu_{23} = 0$
B	Same as for Type I analysis
$T * B$	Same as for Type I analysis

Table 10.11 Type I Hypotheses for $\mu + \tau_i + \beta_j + \gamma_{ij}$ Model for Data in Table 9.1

Source of Variation	*Type I Hypothesis*
T	$\tau_1 - \tau_2 + \frac{1}{8}(\beta_1 - \beta_2) + \frac{1}{8}(3\gamma_{11} + 2\gamma_{12} + 3\gamma_{13} - 2\gamma_{21} - 3\gamma_{22} - 3\gamma_{23}) = 0$
B	$\beta_1 - \beta_2 + \frac{1}{65}(37\gamma_{11} - 2\gamma_{12} - 35\gamma_{13} + 28\gamma_{21} + 2\gamma_{22} - 30\gamma_{23}) = 0$ and $\beta_1 + \beta_2 - 2\beta_3 + \frac{1}{5}(3\gamma_{11} + 2\gamma_{12} - 5\gamma_{13} + 2\gamma_{21} + 3\gamma_{22} - 5\gamma_{23}) = 0$
$T * B$	$\gamma_{11} - \gamma_{12} - \gamma_{21} + \gamma_{22} = 0$ and $\gamma_{11} - \gamma_{13} - \gamma_{21} + \gamma_{23}$

A third way to compute sums of squares from an effects model is as follows:

1. For the t levels of treatment T, generate $t - 1$ dummy variables, and for the b levels of treatment B, generate $b - 1$ dummy variables.
2. The interaction between T and B is represented by the products of their corresponding dummy variables.
3. The model with all of the dummy variables for the treatment variables and their interactions is fitted, and the residual sum of squares is obtained. This is equivalent to the residual sum of squares from the full-effects model.
4. Next, the model is fitted that contains all of the dummy variables except those corresponding to the main effect or interaction being tested. The difference between the residual sum of squares of this reduced model and that of the model in (3) is the sum of squares corresponding to that effect.

The resulting analysis is called a *Type III analysis*. It is also known as Yates' weighted squares of means technique. When all treatment

Table 10.12 Type II Hypotheses for $\mu + \tau_i + \beta_j + \gamma_{ij}$ Model for Data in Table 9.1

Source of Variation	*Type II Hypothesis*
T	$\tau_1 - \tau_2 + \frac{1}{13}(4\gamma_{11} + 4\gamma_{12} + 5\gamma_{13} - 4\gamma_{21} - 4\gamma_{22} - 5\gamma_{23}) = 0$
B	Same as for Type I analysis
$T * B$	Same as for Type I analysis

Table 10.13 Type III Hypotheses for μ_{ij} Model

Source of Variation	*Hypothesis*	
T	$\bar{\mu}_{1\cdot} = \bar{\mu}_{2\cdot} = \cdots = \bar{\mu}_{t\cdot}$	
B	$\bar{\mu}_{\cdot 1} = \bar{\mu}_{\cdot 2} = \cdots = \bar{\mu}_{\cdot b}$	
$T * B$	$\mu_{ij} - \mu_{i'j} - \mu_{ij'} + \mu_{i'j'} = 0$	for all i, j, i', and j'

combinations are observed, the hypotheses tested by a Type III analysis are the same as those tested for balanced data sets. These Type III hypotheses for a means model are given in Table 10.13 and for an effects model in Table 10.14. These hypotheses are usually the ones desired by experimenters.

SAS® (1982) introduced a fourth way of generating sums of squares corresponding to the main effects and their interactions. When all treatment combinations are observed, the hypotheses tested by this Type IV analysis are the same as those tested by the Type III analysis; however, when some treatment combinations are not observed, the Type III and Type IV analyses do not agree. We shall discuss the construction of Type IV hypotheses in Chapter 14.

To conclude this section, we make the following recommendations for analyzing data in a two-way treatment structure model with no missing treatment combinations:

1. If the experimenter wants to compare the effects of the two treatments, he or she should look at hypotheses tested by a Type III analysis. These hypotheses are equivalent to the hypotheses tested in the balanced or equal-subclass-numbers case.
2. If the experimenter is interested in building a model with which to predict the effects of particular treatment combinations, then he or she should use Type I and/or Type II analyses.
3. In survey experiments, the number of observations per treatment combination is often proportional to the frequency with which those

Table 10.14 Type III Hypotheses for $\mu + \tau_i + \beta_j + \gamma_{ij}$ Model

Source of Variation	*Hypothesis*	
T	$\tau_1 + \bar{\gamma}_{1\cdot} = \tau_2 + \bar{\gamma}_{2\cdot} = \cdots = \tau_t + \bar{\gamma}_{t\cdot}$	
B	$\beta_1 + \bar{\gamma}_{\cdot 1} = \beta_2 + \bar{\gamma}_{\cdot 2} = \cdots = \beta_b + \bar{\gamma}_{\cdot b}$	
$T * B$	$\gamma_{ij} - \bar{\gamma}_{i\cdot} - \bar{\gamma}_{\cdot j} + \bar{\gamma}_{\cdot\cdot} = 0$	for all i and j

combinations actually occur in the population. In this case, the experimenter may be most interested in the hypotheses based on $R(\tau|\mu)$ and $R(\beta|\mu)$, since these sums of squares test hypotheses about the weighted averages of the row means and the column means with the weights proportional to the observed sample sizes. This may require two Type I analyses, one with T first in the model and another with B first.

10.5 USING TYPE I THROUGH TYPE IV ESTIMABLE FUNCTIONS IN SAS®

In this section, we show how one can use the information provided by the GLM procedure in SAS® to determine the hypotheses being tested by the different types of sums of squares. (Readers who do not use SAS® can skip this section.) In Section 10.3 we discussed the general form of estimable functions obtained by using the E option on the MODEL statement. If the E1 option is chosen, SAS® will print the general form of the Type I estimable functions for each effect. The results of this for the data in Table 9.1 is shown in Table 10.15.

From Table 10.15, we see that $\mathbf{l}'\boldsymbol{\beta}$, a linear combination of the parameter vector $\boldsymbol{\beta}' = [\mu, \tau_1, \tau_2, \beta_1, \beta_2, \beta_3, \gamma_{11}, \gamma_{12}, \gamma_{13}, \gamma_{21}, \gamma_{22}, \gamma_{23}]$ is a Type I estimable function for T if and only if there exists a constant L2 such that

$$\begin{aligned}\mathbf{l}'\boldsymbol{\beta} = {} & (\mathrm{L2})\tau_1 - (\mathrm{L2})\tau_2 + (.125 \cdot \mathrm{L}_2)\beta_1 - (.125 \cdot \mathrm{L2})\beta_2 + (.375 \cdot \mathrm{L2})\gamma_{11} \\ & + (.25 \cdot \mathrm{L2})\gamma_{12} + (.375 \cdot \mathrm{L2})\gamma_{13} - (.25 \cdot \mathrm{L2})\gamma_{21} \\ & - (.375 \cdot \mathrm{L2})\gamma_{22} - (.375 \cdot \mathrm{L2})\gamma_{23}.\end{aligned}$$

A basis set for the Type I estimable functions for T can be obtained by choosing a specific value for L2, say L2 = 1 or L2 = 8. We are free to choose only one value of the L's, which corresponds to the 1 degree of freedom associated with the test; all other values are determined for us. Taking L2 = 1, a basis set for the Type I estimable functions is

$$\left\{\tau_1 - \tau_2 + \frac{1}{8}\beta_1 - \frac{1}{8}\beta_2 + \frac{3}{8}\gamma_{11} + \frac{1}{4}\gamma_{12} + \frac{3}{8}\gamma_{13} - \frac{1}{4}\gamma_{21} - \frac{3}{8}\gamma_{22} - \frac{3}{8}\gamma_{23}\right\}.$$

This function of the parameters is the one that is compared to zero by a Type I analysis. See Table 10.11.

Another basis set can be constructed by taking L2 = 8. This set is given by

$$\{8\tau_1 - 8\tau_2 + \beta_1 - \beta_2 + 3\gamma_{11} + 2\gamma_{12} + 3\gamma_{13} - 2\gamma_{21} - 3\gamma_{22} - 3\gamma_{23}\}.$$

Since $\mu_{ij} = \mu + \tau_i + \beta_j + \gamma_{ij}$, the hypotheses being tested in the μ_{ij} model are determined by assigning the coefficients of the γ_{ij} terms in the $\mu + \tau_i + \beta_j + \gamma_{ij}$ model to the corresponding μ_{ij}'s. Thus, the Type I

Table 10.15 Type I Estimable Functions from SAS® for Data in Table 9.1

TYPE I ESTIMABLE FUNCTIONS FOR: T			TYPE I ESTIMABLE FUNCTIONS FOR: T*B		
EFFECT		COEFFICIENTS	EFFECT		COEFFICIENTS
INTERCEPT		0	INTERCEPT		0
T	1	L2	T	1	0
	2	−L2		2	0
B	1	0.125*L2	B	1	0
	2	−0.125*L2		2	0
	3	0		3	0
T*B	1 1	0.375*L2	T*B	1 1	L7
	1 2	0.25*L2		1 2	L8
	1 3	0.375*L2		1 3	−L7 − L8
	2 1	−0.25*L2		2 1	−L7
	2 2	−0.375*L2		2 2	−L8
	2 3	−0.375*L2		2 3	7 + L8

TYPE I ESTIMABLE FUNCTIONS FOR: B		
EFFECT		COEFFICIENTS
INTERCEPT		0
T	1	0
	2	0
B	1	L4
	2	L5
	3	−L4 − L5
T*B	1 1	0.5692*L4 + 0.0308*L5
	1 2	−0.0308*L4 + 0.4308*L5
	1 3	−0.5385*L4 − 0.4615*L5
	2 1	0.4308*L4 − 0.0308*L5
	2 2	0.0308*L4 + 0.5692*L5
	2 3	−0.4615*L4 − 0.5385*L5

hypotheses for T in terms of the μ_{ij} model is

$$\frac{3}{8}\mu_{11} + \frac{1}{4}\mu_{12} + \frac{3}{8}\mu_{13} - \frac{1}{4}\mu_{21} - \frac{3}{8}\mu_{22} - \frac{3}{8}\mu_{23} = 0$$

or equivalently,

$$3\mu_{11} + 2\mu_{12} + 3\mu_{13} - 2\mu_{21} - 3\mu_{22} - 3\mu_{23} = 0$$

which is the hypothesis given in Table 10.8.

If the E2 option on the MODEL statement is chosen, SAS® will print the general form of the Type II estimable functions for each effect. The results for the data in Table 9.1 are shown in Table 10.16.

Table 10.16 Type II Estimable Functions from SAS® for Data in Table 9.1

TYPE II ESTIMABLE FUNCTIONS FOR: T

EFFECT		COEFFICIENTS
INTERCEPT		0
T	1	L2
	2	−L2
B	1	0
	2	0
	3	0
T*B	1 1	0.3077*L2
	1 2	0.3077*L2
	1 3	0.3846*L2
	2 1	−0.3077*L2
	2 2	−0.3077*L2
	2 3	−0.3846*L2

TYPE II ESTIMABLE FUNCTIONS FOR: T*B

EFFECT		COEFFICIENTS
INTERCEPT		0
T	1	0
	2	0
B	1	0
	2	0
	3	0
T*B	1 1	L7
	1 2	L8
	1 3	−L7 − L8
	2 1	−L7
	2 2	−L8
	2 3	L7 + L8

TYPE II ESTIMABLE FUNCTIONS FOR: B

EFFECT		COEFFICIENTS
INTERCEPT		0
T	1	0
	2	0
B	1	L4
	2	L5
	3	−L4 − L5
T*B	1 1	0.5692*L4 + 0.0308*L5
	1 2	−0.0308*L4 + 0.4308*L5
	1 3	−0.5385*L4 − 0.4615*L5
	2 1	0.4308*L4 − 0.0308*L5
	2 2	0.0308*L4 + 0.5692*L5
	2 3	−0.4615*L4 − 0.5385*L5

From Table 10.16, we see that $\mathbf{l}'\boldsymbol{\beta}$ is a Type II estimable function for B if and only if there exist constants L4 and L5 such that

$$\begin{aligned}\mathbf{l}'\boldsymbol{\beta} = (\text{L4})\beta_1 &+(\text{L5})\beta_2 - (\text{L4} + \text{L5})\beta_3 + (0.5692 \cdot \text{L4} + 0.0308 \cdot \text{L5})\gamma_{11}\\ &+(-0.0308 \cdot \text{L4} + 0.4308 \cdot \text{L5})\gamma_{12}\\ &+(-0.5385 \cdot \text{L4} - 0.4615 \cdot \text{L5})\gamma_{13}\\ &+(0.4308 \cdot \text{L4} - 0.0308 \cdot \text{L5})\gamma_{21}\\ &+(0.0308 \cdot \text{L4} + 0.5692 \cdot \text{L5})\gamma_{22}\\ &+(-0.4615 \cdot \text{L4} - 0.5385 \cdot \text{L5})\gamma_{23}.\end{aligned}$$

With a little luck or by using the general forms given in Table 10.9, one can determine that the decimal numbers given in the above expression have a lowest common denominator of 65. Choosing L4 = 1 and L5 = 0 and then L4 = 0 and L5 = 1 provides one basis set for the Type II estimable functions for B. This set is

$$\left\{\beta_1 - \beta_3 + \frac{1}{65}(37\gamma_{11} - 2\gamma_{12} - 35\gamma_{13} + 28\gamma_{21} + 2\gamma_{22} - 30\gamma_{23}),\right.$$
$$\left.\beta_2 - \beta_3 + \frac{1}{65}(2\gamma_{11} + 28\gamma_{12} - 30\gamma_{13} - 2\gamma_{21} + 37\gamma_{22} - 35\gamma_{23})\right\}$$

for the $\mu + \tau_i + \beta_j + \gamma_{ij}$ model and

$$\left\{\frac{1}{65}(37\mu_{11} - 2\mu_{12} - 35\mu_{13} + 28\mu_{21} + 2\mu_{22} - 30\mu_{23}),\right.$$
$$\left.\frac{1}{65}(2\mu_{11} + 28\mu_{12} - 30\mu_{13} - 2\mu_{21} + 37\mu_{22} - 35\mu_{23})\right\}$$

for the μ_{ij} model. This latter basis set is equivalent to

$$\{37\mu_{11} - 2\mu_{12} - 35\mu_{13} + 28\mu_{21} + 2\mu_{22} - 30\mu_{23},$$
$$2\mu_{11} + 28\mu_{12} - 30\mu_{13} - 2\mu_{21} + 37\mu_{22} - 35\mu_{23}\}.$$

The hypotheses for B given in Table 10.12 can be obtained by letting L4 = 1 and L5 = 0 and then L4 = 1 and L5 = 1. In this case we can choose values for two of the L's, namely L4 and L5. Thus, there are 2 degrees of freedom corresponding to the Type II sum of squares for B.

From the general form of the Type I estimable functions for $T * B$ in Table 10.15, we see that we have two L's for which we can choose values. A basis set for the Type I estimable functions for $T * B$ can be obtained by letting L7 = 1 and L8 = 0 and then L7 = 0 and L8 = 1. These two choices yield

$$\{\gamma_{11} - \gamma_{13} - \gamma_{21} + \gamma_{23}, \gamma_{12} - \gamma_{13} - \gamma_{22} - \gamma_{23}\}$$

for the $\mu + \tau_i + \beta_j + \gamma_{ij}$ model and hence,

$$\{\mu_{11} - \mu_{13} - \mu_{21} + \mu_{23}, \mu_{12} - \mu_{13} - \mu_{22} + \mu_{23}\}$$

for the μ_{ij} model. The reader might wish to verify that every 2×2 table difference $\mu_{ij} - \mu_{i'j} - \mu_{ij'} + \mu_{i'j'}$ can be obtained as some linear combination of these two functions in the basis set.

From Table 10.16, the general form of Type II estimable functions for T is given by

$$(\text{L2})\tau_1 - (\text{L2})\tau_2 + (0.3077 \cdot \text{L2})\gamma_{11} + (0.3077 \cdot \text{L2})\gamma_{12} + (0.3846 \cdot \text{L2})\gamma_{13}$$
$$- (0.3077 \cdot \text{L2})\gamma_{21} - (0.3077 \cdot \text{L2})\gamma_{22} - (0.3846 \cdot \text{L2})\gamma_{23}.$$

The lowest common denominator of these decimal fractions is 13; thus, by taking L2 = 1, a basis set of Type II estimable functions for T is

$$\left\{\tau_1 - \tau_2 + \frac{1}{13}(4\gamma_{11} + 4\gamma_{12} + 5\gamma_{13} - 4\gamma_{21} - 4\gamma_{22} - 5\gamma_{23})\right\}$$

for the $\mu + \tau_i + \beta_j + \gamma_{ij}$ model, and taking L2 = 13, we obtain a basis set for the μ_{ij} model of

$$\{4\mu_{11} + 4\mu_{12} + 5\mu_{13} - 4\mu_{21} - 4\mu_{22} - 5\mu_{23}\}.$$

Finally, we consider the Type III and Type IV estimable functions. For the data in Table 9.1, these are the same; they are shown in Table 10.17. A basis set of Type III estimable functions for T in the $\mu + \tau_i + \beta_j + \gamma_{ij}$ model (taking L2 = 1) is

$$\left\{\tau_1 - \tau_2 + \frac{1}{3}(\gamma_{11} + \gamma_{12} + \gamma_{13} - \gamma_{21} - \gamma_{22} - \gamma_{23})\right\},$$

which is equivalent to $\{\tau_1 - \tau_2 + \bar{\gamma}_{1\cdot} - \bar{\gamma}_{2\cdot}\}$. For the μ_{ij} model, a basis set is $\{\bar{\mu}_{1\cdot} - \bar{\mu}_{2\cdot}\}$. In a similar manner, we see that basis sets for the Type

Table 10.17 Type III Estimable Functions from SAS® for Data in Table 9.1

TYPE III ESTIMABLE FUNCTIONS FOR: T

EFFECT		COEFFICIENTS
INTERCEPT		0
T	1	L2
	2	-L2
B	1	0
	2	0
	3	0
T*B	1 1	0.3333*L2
	1 2	0.3333*L2
	1 3	0.3333*L2
	2 1	-0.3333*L2
	2 2	-0.3333*L2
	2 3	-0.3333*L2

TYPE III ESTIMABLE FUNCTIONS FOR: T*B

EFFECT		COEFFICIENTS
INTERCEPT		0
T	1	0
	2	0
B	1	0
	2	0
	3	0
T*B	1 1	L7
	1 2	L8
	1 3	-L7 - L8
	2 1	-L7
	2 2	-L8
	2 3	L7 + L8

TYPE III ESTIMABLE FUNCTIONS FOR: B

EFFECT		COEFFICIENTS
INTERCEPT		0
T	1	0
	2	0
B	1	L4
	2	L5
	3	-L4 - L5
T*B	1 1	0.5*L4
	1 2	0.5*L5
	1 3	-0.5*L4 - 0.5*L5
	2 1	0.5*L4
	2 2	0.5*L5
	2 3	-0.5*L4 - 0.5*L5

III estimable functions for B are

$$\{\beta_1 - \beta_3 + \bar{\gamma}_{\cdot 1} - \bar{\gamma}_{\cdot 3}, \beta_2 - \beta_3 + \bar{\gamma}_{\cdot 2} - \bar{\gamma}_{\cdot 3}\}$$

for the $\mu + \tau_i + \beta_j + \gamma_{ij}$ model and

$$\{\bar{\mu}_{\cdot 1} - \bar{\mu}_{\cdot 3}, \bar{\mu}_{\cdot 2} - \bar{\mu}_{\cdot 3}\}$$

for the μ_{ij} model. Note that H_0: $\bar{\mu}_{\cdot 1} - \bar{\mu}_{\cdot 3} = 0$ and $\bar{\mu}_{\cdot 2} - \bar{\mu}_{\cdot 3} = 0$ is equivalent to H_0: $\bar{\mu}_{\cdot 1} = \bar{\mu}_{\cdot 2} = \bar{\mu}_{\cdot 3}$. The Type IV estimable functions are identical to the Type III estimable functions in this case.

10.6 POPULATION MARGINAL MEANS AND LEAST SQUARES MEANS

The population marginal means for the two-way effects model are defined by

$$\mu + \tau_i + \bar{\beta}_{\cdot} + \bar{\gamma}_{i\cdot} \qquad i = 1, 2, \ldots, t,$$

for the T-treatments and

$$\mu + \bar{\tau}_{\cdot} + \beta_j + \bar{\gamma}_{\cdot j} \qquad j = 1, 2, \ldots, b,$$

for the B-treatments.

The best estimates of these marginal means are $\hat{\mu} + \hat{\tau}_i + \hat{\bar{\beta}}_{\cdot} + \hat{\bar{\gamma}}_{i\cdot} = \hat{\bar{\mu}}_{i\cdot}$, $i = 1, 2, \ldots, t$, and $\hat{\mu} + \hat{\bar{\tau}}_{\cdot} + \hat{\beta}_j + \hat{\bar{\gamma}}_{\cdot j} = \hat{\bar{\mu}}_{\cdot j}$, $j = 1, 2, \ldots, b$, respectively, and are often called *least squares means*. Their respective standard errors are given by equations (9.5.2) and (9.5.4). To make inferences about linear combinations of the population marginal means, use (9.5.5) and (9.5.6).

10.7 COMPUTER ANALYSES

Nearly all the statistical computing packages have been developed in order to deal with effects models rather than means models. Since three major types of hypotheses can be tested for the balanced two-way treatment structure with unequal subclass numbers, one must be careful to determine which type is tested by the statistical package being used. Although the computing packages have been developed with effects models in mind, the means model can easily be implemented as well. Using the means model allows the user to specify meaningful contrasts among the treatment means.

The names we have used in our discussion, that is, Types I, II, III, and IV, correspond to the names used by SAS®. Table 10.18 compares available BMDP and SPSS analyses with definitions of the Type I through Type IV hypotheses.

Readers who use other computing packages are encouraged to analyze the data in Table 9.1 and compare their analyses with those given in this chapter and Chapter 9.

Table 10.18 Analyses of Two-Way Effects Models When All Treatment Combinations Are Observed

	STATISTICAL COMPUTING PACKAGE			
	BMD		*SPSS*	
Hypothesis Type	*P2V*	*P4V*	*ANOVA*	*MANOVA*
Type I	Not Possible	BETWEEN = SIZES*	OPTION 10	Default
Type II	Not Possible	BETWEEN = SIZES*	Default	Not Given†
Type III, Type IV	Default	BETWEEN = EQUAL	OPTION 9	SSTYPE(UNIQUE)

*These two types are both given simultaneously.
†These can be obtained by using the default option with the effects in a different order.

Estimates of the population marginal means can be obtained within the SAS® GLM procedure by adding the following statement after the model card:

LSMEANS T B T∗B/*options*;

This card causes estimates of the T marginal means, the B marginal means, and the $T*B$ marginal means, which are the $\hat{\mu}_{ij}$'s, to be printed. Two useful options that are also available are STDERR and PDIFF. The STDERR option instructs SAS® to print the standard errors of the estimated population marginal means; the PDIFF option tells SAS® to give the significance levels of the appropriate t-tests for comparing each population marginal mean to all other population marginal means within the same effect.

SAS® GLM also allows means to be calculated by using a MEANS/options; statement. However, with unbalanced data the MEANS statement does not give unbiased estimates of the population marginal means, instead it causes estimates of weighted averages of the row means and of the column means to be printed. That is,

MEANS T B/options;

gives the best unbiased estimates of the weighted means

$$\tilde{\mu}_{i\cdot} = \frac{\sum_{j=1}^{b} n_{ij}\mu_{ij}}{n_{i\cdot}}, \qquad i = 1, 2, \ldots, t, \quad \text{and}$$

$$\tilde{\mu}_{\cdot j} = \frac{\sum_{i=1}^{t} n_{ij}\mu_{ij}}{n_{\cdot j}}, \qquad j = 1, 2, \ldots, b.$$

Probably the only case in which an experimenter may be interested in these weighted means is in survey type experiments (see the third recommendation at the end of Section 10.4).

For the data in Table 9.1, the best estimates of these weighted means are given by $\hat{\tilde{\mu}}_{1.} = 22.75$, $\hat{\tilde{\mu}}_{2.} = 27.125$, $\hat{\tilde{\mu}}_{.1} = 22.4$, $\hat{\tilde{\mu}}_{.2} = 23.8$, and $\hat{\tilde{\mu}}_{.3} = 28.0$.

BMD-P4V prints both types of marginal means. Those labeled

MEAN

are the best estimates of the weighted marginal means, $\tilde{\mu}_{i.}$ and $\tilde{\mu}_{.j}$, while those labeled

WTD · MEAN

are the best estimates of the population marginal means, $\bar{\mu}_{i.}$ and $\bar{\mu}_{.j}$.

SPSS ANOVA outputs the means as deviations from the grand mean. These means depend on the ANOVA option selected. SPSS MANOVA uses a notation opposite to that of BMD-P4V but similar to that used in our discussion. The estimates of the population marginal means, $\bar{\mu}_{i.}$ and $\bar{\mu}_{.j}$, are labeled

UNWGT.,

and estimates of the weighted means, $\tilde{\mu}_{i.}$ and $\tilde{\mu}_{.j}$, are labeled

WGT.

CONCLUDING REMARKS

In this chapter, we considered the analysis of two-way treatment structures with unequal subclass numbers using the effects model, given the assumption that all treatment combinations were observed. The effects model was used, even though the means model provides answers to all questions that can be raised, because much of the existing statistical computing software utilizes the effects model.

Type I through Type III analyses and the conditions for using each analysis type were discussed. In almost all cases, the Type III analysis will be preferred. The Type III analysis is the same as the analysis given in Chapter 9. Population marginal means were contrasted with weighted marginal means, and the appropriateness of each kind of mean was considered.

The statistical analyses obtained from several popular statistical computing packages were compared, and the analyses given by SAS® GLM were discussed in detail.

11

Analyzing Large Balanced Two-Way Experiments with Unequal Subclass Numbers

CHAPTER OUTLINE

In this chapter we present a method for obtaining an approximate analysis of balanced-treatment-structure experiments that have an unequal number of observations in each subclass.

11.1 FEASIBILITY PROBLEMS

We generally recommend using a general computing package like one of those discussed in Section 10.7 to analyze unbalanced data sets in which every treatment combination is observed at least once. However, many situations arise in practice where it is not feasible to do so, especially when there are several treatment factors with each factor having several different levels. In these cases, the general procedures require several matrix inversions, and the size of the matrices to be inverted often exceeds the capabilities of the computer. At the very least, the matrices require an extraordinary amount of time (and hence money) to be inverted.

For example, consider an experiment with four factors, each factor occurring at five levels. The size of the largest matrix requiring inversion in this situation is 624×624. It is almost impossible for even a mighty computer to invert such a large matrix. Obviously, such an experiment is not too unusual.

An alternative way to analyze these types of messy data situations is called the *method of unweighted means*. In Section 11.2 the method is described for two-way treatment structures. While one rarely encounters two-way experiments that cannot be analyzed by general computing procedures, the method of unweighted means is easily discussed and understood for the two-way experiment, and the discussion readily generalizes to larger experimental situations. Before continuing, we want to point out that it is not the number of observations measured that makes general procedures unfeasible, but the number of treatment combinations being studied.

11.2 METHOD OF UNWEIGHTED MEANS

Basically, the method of unweighted means determines the various sums of squares corresponding to each of the effects by using the means of the various treatment combinations. The formulas given here are the ones used for equal sample sizes; however, if the sample sizes are only slightly unequal, they still give quite accurate results. The correct sums of squares are obtained by using the Type III analysis, which the formulas presented here approximate.

For testing H_{01}: $\bar{\mu}_{1\cdot} = \bar{\mu}_{2\cdot} = \cdots = \bar{\mu}_{t\cdot}$, we use

$$SS_{H_{01}} = b\sum \hat{\bar{\mu}}_{i\cdot}^2 - bt\hat{\bar{\mu}}_{\cdot\cdot}^2,$$

which is based on $t-1$ degrees of freedom. For testing H_{02}: $\bar{\mu}_{\cdot 1} = \bar{\mu}_{\cdot 2} = \cdots = \bar{\mu}_{\cdot b}$, we use

$$SS_{H_{02}} = t\sum \hat{\bar{\mu}}_{\cdot j}^2 - bt\hat{\bar{\mu}}_{\cdot\cdot}^2,$$

which is based on $b - 1$ degrees of freedom. For testing H_{03}: $\mu_{ij} - \mu_{i'j} - \mu_{ij'} + \mu_{i'j'} = 0$ for all i, i', j, and j', we use

$$SS_{H_{03}} = \sum_i \sum_j \hat{\mu}_{ij}^2 - b\sum_i \hat{\bar{\mu}}_{i\cdot}^2 - t\sum_j \hat{\bar{\mu}}_{\cdot j}^2 + bt\hat{\bar{\mu}}_{\cdot\cdot}^2,$$

which is based on $(b - 1)(t - 1)$ degrees of freedom.

Since the above sums of squares are computed on the basis of means, the usual sum of squares for error,

$$\text{ESS} = \sum_{i,j,k} (y_{ijk} - \bar{y}_{ij\cdot})^2,$$

must be adjusted. This is because the $\hat{\mu}_{ij}$'s have variances given by σ^2/n_{ij} rather than σ^2, the variance of the y_{ijk}'s. The adjustment is made by multiplying the sum of squares for error by $1/\tilde{n}$ where

$$\frac{1}{\tilde{n}} = \frac{1}{bt}\left[\sum_{ij}\left(\frac{1}{n_{ij}}\right)\right].$$

Since the quantity $\tilde{n}$ is the harmonic mean of the sample sizes, it is one possible average of the n_{ij}'s. The degrees of freedom for error are still $N - bt$.

The analysis of variance table for an unweighted means analysis is given in Table 11.1. This analysis yields reasonable approximations to the F-distribution only when the cell sample sizes are not too unequal. The usual recommendation is to use this analysis if the sample sizes vary by no more than a factor of 2.

Table 11.1 Unweighted Means Analysis

Source of Variation	*df*	*SS*
T	$t - 1$	$b\sum_i (\hat{\bar{\mu}}_{i\cdot} - \hat{\bar{\mu}}_{\cdot\cdot})^2$
B	$b - 1$	$t\sum_j (\hat{\bar{\mu}}_{\cdot j} - \hat{\bar{\mu}}_{\cdot\cdot})^2$
$T * B$	$(t - 1)(b - 1)$	$\sum_{ij} (\hat{\mu}_{ij} - \hat{\bar{\mu}}_{i\cdot} - \hat{\bar{\mu}}_{\cdot j} + \hat{\bar{\mu}}_{\cdot\cdot})^2$
Error	$N - bt$	$\frac{1}{\tilde{n}} \cdot \text{ESS}$

11.3 SIMULTANEOUS INFERENCE AND MULTIPLE COMPARISONS

Before considering an example, we should make some remarks about comparing population marginal means. The best estimate of the T_i marginal mean is $\hat{\bar{\mu}}_{i\cdot}$, and its estimated standard error is

$$\frac{\hat{\sigma}}{b}\sqrt{\sum_j\left(\frac{1}{n_{ij}}\right)}.$$

The best estimate of the B_j marginal mean is

$$\hat{\bar{\mu}}_{\cdot j}=\frac{1}{t}\sum_i \hat{\mu}_{ij},$$

and its estimated standard error is

$$\frac{\hat{\sigma}}{t}\sqrt{\sum_i\frac{1}{n_{ij}}}.$$

One should note that, in the above formulas, $\hat{\sigma}$ is replaced by the usual estimator and not the adjusted estimator.

The experimenter might consider using an unweighted means analysis to determine where there are significant effects, if any. If there are significant effects, contrasts of the significant effects can be compared by using a Fisher's LSD (t-test) multiple comparison procedure.

The appropriate t-statistic for testing $\bar{\mu}_{i\cdot}=\bar{\mu}_{i'\cdot}$ is

$$t_c=\frac{\hat{\bar{\mu}}_{i\cdot}-\hat{\bar{\mu}}_{i'\cdot}}{(\hat{\sigma}/b)\left[\Sigma_j(1/n_{ij})+\Sigma_j(1/n_{i'j})\right]^{1/2}},$$

and the appropriate t-statistic for testing $\bar{\mu}_{\cdot j}=\bar{\mu}_{\cdot j'}$ is

$$t_c=\frac{\hat{\bar{\mu}}_{\cdot j}-\hat{\bar{\mu}}_{\cdot j'}}{(\hat{\sigma}/t)\sqrt{\Sigma_i(1/n_{ij})+\Sigma_i(1/n_{ij'})}}.$$

11.4 AN EXAMPLE OF THE METHOD OF UNWEIGHTED MEANS

As an example, we again use the data in Table 9.1. Recall that the error sum of squares was 20 with 10 degrees of freedom, and the table of means was as below:

$\hat{\mu}_{ij}$	B_1	B_2	B_3	$\hat{\bar{\mu}}_{i\cdot}$
T_1	20	25	24	23
T_2	26	23	32	27
$\hat{\bar{\mu}}_{\cdot j}$	23	24	28	$25=\hat{\bar{\mu}}_{\cdot\cdot}$

Table 11.2 Unweighted Means Analysis of Data in Table 9.1

Source of Variation	*df*	*SS*	*MS*	*F*	$\hat{\alpha}$
T	1	24	24	30.85	.0005
B	2	28	14	17.99	.0005
$T * B$	2	28	14	17.99	.0005
Error (adj)	10	$\frac{7}{18}(20) = 7.78$	.778		

The value of $\tilde{n}$ is obtained from

$$\frac{1}{\tilde{n}} = \frac{1}{3 \cdot 2}\left(\frac{1}{3} + \frac{1}{2} + \frac{1}{3} + \frac{1}{2} + \frac{1}{3} + \frac{1}{3}\right)$$
$$= \frac{1}{6} \cdot \frac{14}{6} = \frac{7}{18},$$

so that ESS(adj) $= \frac{7}{18}(20) = 7.78$. We also obtain:

$$\sum_{ij} \hat{\mu}_{ij}^2 = 20^2 + 25^2 + \cdots + 32^2 = 3{,}830,$$

$$\sum_{i} \hat{\bar{\mu}}_{i\cdot}^2 = 23^2 + 27^2 = 1{,}258,$$

$$\sum_{j} \hat{\bar{\mu}}_{\cdot j}^2 = 23^2 + 24^2 + 28^2 = 1{,}889, \quad \text{and}$$

$$\hat{\bar{\mu}}_{\cdot\cdot}^2 = 625.$$

Hence,

$$SS_{H_{01}} = 3(1{,}258) - 6(625) = 24,$$

$$SS_{H_{02}} = 2(1{,}889) - 6(625) = 28, \quad \text{and}$$

$$SS_{H_{03}} = 3{,}830 - 3(1{,}258) - 2(1{,}889) + 6(625) = 28.$$

The analysis of variance table is given in Table 11.2.

11.5 COMPUTER ANALYSES

The sums of squares necessary for Table 11.2 can easily be obtained by combining several computing procedures. The unadjusted error sum of squares, $(N - bt)\hat{\sigma}^2$, can be obtained most efficiently by considering the experiment as a one-way experiment and utilizing the means model. To obtain the T, B, and $T * B$ sum of squares required for the unweighted means analysis, one must do the following:

1. Obtain the cell means for each treatment combination.

2. Put these means in a data set.
3. Analyze the means using a two-way model and any computing procedure developed for equal subclass experiments.

After obtaining all the appropriate sums of squares, one must construct the analysis of variance table by hand.

As an example, consider the data in Table 9.1. The required statements to get an unweighted means analysis using SAS® are:

```
DATA OLD; INPUT T B Y; TRT =
          10*T + B; CARDS;
                    [Data goes here.]

PROC ANOVA; CLASSES TRT; }  Get ESS from here.
MODEL Y = TRT;           }

PROC SORT; BY T B;                      }
PROC MEANS NOPRINT; BY T B;             }  Computes
OUTPUT OUT = NEW MEAN = MHAT;           }  means.
VARIABLES Y;                            }

PROC ANOVA DATA = NEW;   }  Analyzes the means; get
CLASSES T B;             }  T, B, and T * B sums of
MODEL MHAT = T B T*B;    }  squares from here.
```

CONCLUDING REMARKS

In this chapter, we introduced a method for obtaining satisfactory statistical analyses of experiments involving large numbers of different treatment combinations, each observed at least once. The techniques are useful whenever the number of different treatment combinations is greater than 75 and nearly mandatory whenever there are more than 150. They can also be used by persons who have software available for analyzing balanced but not unbalanced experiments.

12

Case Study: Balanced Two-Way Treatment Structure with Unequal Subclass Numbers

CHAPTER OUTLINE

In this chapter, we analyze a set of data arising from a two-way treatment structure conducted in a randomized complete block design. The experiment was intended to have been balanced, but due to unforeseen circumstances some treatment combinations were missing from some blocks. We still assume, however, that every treatment combination is observed at least once. The cases where some treatment combinations are never observed are discussed in Chapters 13 through 15.

12.1 FAT–SURFACTANT EXAMPLE

A bakery scientist wanted to study the effects of combining three different fats with each of three different surfactants on the specific volume of bread loaves baked from doughs mixed from each of the nine treatment combinations. Four flours of the same type but from different sources were used as blocking factors. That is, loaves were made using all nine treatment combinations for each of the four flours. Unfortunately, one container of yeast turned out to be ineffective, and the data from the ten loaves made with that yeast had to be removed from the analysis. Fortunately, all nine Fat∗Surfactant treatment combinations were observed at least once. The data are given in Table 12.1.

The data in Table 12.1 were analyzed using the SAS® GLM procedure. Since all treatment combinations are observed at least once, and since some of the same treatment combinations are observed in each block (flour), the Type III sums of squares test hypotheses that are interesting and easy to interpret. Thus, the estimable functions need not be requested in the SAS® GLM analysis. In addition, all marginal means

Table 12.1 Specific Volumes from Baking Experiment

		FLOUR			
Fat	*Surfactant*	*1*	*2*	*3*	*4*
	1	6.7	4.3	5.7	—
1	2	7.1	—	5.9	5.6
	3	—	5.5	6.4	5.8
	1	—	5.9	7.4	7.1
2	2	—	5.6	—	6.8
	3	6.4	5.1	6.2	6.3
	1	7.1	5.9	—	—
3	2	7.3	6.6	8.1	6.8
	3	—	7.5	9.1	—

are estimable, and their estimates have been obtained by using the LSMEANS option. The resulting analysis is shown below.

```
          OPTIONS LS=74 NODATE;
          DATA;
          INPUT FAT SURF FLOUR1 - FLOUR4;
          CARDS;

NOTE: DATA SET WORK.DATA1 HAS 9 OBSERVATIONS AND 6 VARIABLES. 366
OBS/TRK.
NOTE: THE DATA STATEMENT USED 0.10 SECONDS AND 210K.

          PROC PRINT;
          TITLE CHAPTER 12 - FAT*SURFACTANT EXAMPLE;

NOTE: THE PROCEDURE PRINT USED 0.14 SECONDS AND 202K AND PRINTED
PAGE 1.
          DATA; SET; DROP FLOUR1 - FLOUR4;
          FLOUR=1; SPVOL=FLOUR1;OUTPUT;
          FLOUR=2; SPVOL=FLOUR2;OUTPUT;
          FLOUR=3; SPVOL=FLOUR3;OUTPUT;
          FLOUR=4; SPVOL=FLOUR4;OUTPUT;

NOTE: DATA SET WORK.DATA2 HAS 36 OBSERVATIONS AND 4 VARIABLES.
529 OBS/TRK.
NOTE: THE DATA STATEMENT USED 0.09 SECONDS AND 202K.

          PROC GLM;
          CLASSES FLOUR FAT SURF;
          MODEL SPVOL=FLOUR FAT|SURF;
          LSMEANS FAT|SURF/STDERR PDIFF;

NOTE: THE PROCEDURE GLM USED 0.79 SECONDS AND 268K AND PRINTED
PAGES 2 TO 5.
NOTE: SAS USED 268K MEMORY.

NOTE: SAS INSTITUTE INC.
      SAS CIRCLE
      PO BOX 8000
      CARY, N.C. 27511-8000
```

CHAPTER 12 - FAT*SURFACTANT EXAMPLE

OBS	FAT	SURF	FLOUR1	FLOUR2	FLOUR3	FLOUR4
1	1	1	6.7	4.3	5.7	.
2	1	2	7.1	.	5.9	5.6
3	1	3	.	5.5	6.4	5.8
4	2	1	.	5.9	7.4	7.1
5	2	2	.	5.6	.	6.8
6	2	3	6.4	5.1	6.2	6.3
7	3	1	7.1	5.9	.	.
8	3	2	7.3	6.6	8.1	6.8
9	3	3	.	7.5	9.1	.

CHAPTER 12 - FAT*SURFACTANT EXAMPLE

GENERAL LINEAR MODELS PROCEDURE

CLASS LEVEL INFORMATION

CLASS	LEVELS	VALUES
FLOUR	4	1 2 3 4
FAT	3	1 2 3
SURF	3	1 2 3

NUMBER OF OBSERVATIONS IN DATA SET = 36

NOTE: ALL DEPENDENT VARIABLES ARE CONSISTENT WITH RESPECT TO THE PRESENCE OR ABSENCE OF MISSING VALUES. HOWEVER, ONLY 26 OBSERVATIONS IN DATA SET CAN BE USED IN THIS ANALYSIS.

CHAPTER 12 - FAT*SURFACTANT EXAMPLE

GENERAL LINEAR MODELS PROCEDURE

DEPENDENT VARIABLE: SPVOL

SOURCE	DF	SUM OF SQUARES	MEAN SQUARE
MODEL	11	22.51952891	2.04722990
ERROR	14	2.31585570	0.16541826
CORRECTED TOTAL	25	24.83538462	
MODEL F =	12.38		PR > F = 0.0001

R - SQUARE	C.V.	ROOT MSE	SPVOL MEAN
0.906752	6.2869	0.40671644	6.46923077

SOURCE	DF	TYPE I SS	F VALUE	PR > F
FLOUR	3	6.39309890	12.88	0.0003
FAT	2	10.33041605	31.23	0.0001
SURF	2	0.15724944	0.48	0.6314
FAT*SURF	4	5.63876453	8.52	0.0011

SOURCE	DF	TYPE III SS	F VALUE	PR > F
FLOUR	3	8.69081097	17.51	0.0001
FAT	2	10.11784983	30.58	0.0001
SURF	2	0.99720998	3.01	0.0815
FAT*SURF	4	5.63876453	8.52	0.0011

CHAPTER 12 - FAT*SURFACTANT EXAMPLE

GENERAL LINEAR MODELS PROCEDURE

LEAST SQUARES MEANS

FAT	SPVOL LSMEAN	STD ERR LSMEAN	PROB > |T| HO:LSMEAN=0	LSMEAN NUMBER
1	5.85019684	0.13648939	0.0001	1
2	6.57713120	0.14771267	0.0001	2
3	7.47251449	0.15648429	0.0001	3

PROB > |T| HO: LSMEAN(I)=LSMEAN(J)

I/J	1	2	3
1	.	0.0029	0.0001
2	0.0029	.	0.0010
3	0.0001	0.0010	.

NOTE: TO ENSURE OVERALL PROTECTION LEVEL, ONLY PROBABILITIES ASSOCIATED WITH PRE-PLANNED COMPARISONS SHOULD BE USED.

SURF	SPVOL LSMEAN	STD ERR LSMEAN	PROB > |T| HO:LSMEAN=0	LSMEAN NUMBER
1	6.39597022	0.15018107	0.0001	1
2	6.59993558	0.14324083	0.0001	2
3	6.90393673	0.14732474	0.0001	3

PROB > |T| HO: LSMEAN(I)=LSMEAN(J)

I/J	1	2	3
1	.	0.3480	0.0291
2	0.3480	.	0.1608
3	0.0291	0.1608	.

NOTE: TO ENSURE OVERALL PROTECTION LEVEL, ONLY PROBABILITIES ASSOCIATED WITH PRE-PLANNED COMPARISONS SHOULD BE USED.

FAT	SURF	SPVOL LSMEAN	STD ERR LSMEAN	PROB > |T| HO: LSMEAN=0	LSMEAN NUMBER
1	1	5.53635388	0.24036653	0.0001	1
1	2	5.89132489	0.23921852	0.0001	2
1	3	6.12291175	0.24137422	0.0001	3
2	1	7.02291175	0.24137422	0.0001	4
2	2	6.70848186	0.30057982	0.0001	5
2	3	6.00000000	0.20335822	0.0001	6
3	1	6.62864505	0.30066843	0.0001	7
3	2	7.20000000	0.20335822	0.0001	8
3	3	8.58889843	0.30013634	0.0001	9

CHAPTER 12 - FAT*SURFACTANT EXAMPLE

GENERAL LINEAR MODELS PROCEDURE

LEAST SQUARES MEANS FOR EFFECT FAT*SURF
PROB > |T| HO: LSMEAN(I)=LSMEAN(J)

DEPENDENT VARIABLE: SPVOL

I/J	1	2	3	4	5	6	7	8
1	.	0.3156	0.1105	0.0007	0.0098	0.1630	0.0118	0.0001
2	0.3156	.	0.5099	0.0052	0.0546	0.7344	0.0788	0.0009
3	0.1105	0.5099	.	0.0169	0.1428	0.7028	0.2203	0.0042
4	0.0007	0.0052	0.0169	.	0.4184	0.0059	0.3341	0.5836
5	0.0098	0.0546	0.1428	0.4184	.	0.0712	0.8550	0.1971
6	0.1630	0.7344	0.7028	0.0059	0.0712	.	0.1053	0.0009
7	0.0118	0.0788	0.2203	0.3341	0.8550	0.1053	.	0.1378
8	0.0001	0.0009	0.0042	0.5836	0.1971	0.0009	0.1378	.
9	0.0001	0.0001	0.0001	0.0010	0.0006	0.0001	0.0004	0.0018

PROB > |T| HO: LSMEAN(I)=LSMEAN(J)

I/J	9
1	0.0001
2	0.0001
3	0.0001
4	0.0010
5	0.0006
6	0.0001
7	0.0004
8	0.0018
9	.

NOTE: TO ENSURE OVERALL PROTECTION LEVEL, ONLY PROBABILITIES ASSOCIATED WITH PRE-PLANNED COMPARISONS SHOULD BE USED.

The Type III F-value for the Fat $*$ Surfactant interaction is $F = 8.52$, which is significant at the $p = .0011$ level. Thus, the surfactants should be compared within each Fat level, and the fats should be compared within each Surfactant level. Figure 12.1 gives a plot of the two-way least squares means; sample means located within the same circle are not significantly different. The p-values used are those given in the LSMEANS table.

From Figure 12.1, we can make the following observations:

1. The combination of fat 3 with surfactant 3 gives a response that is significantly higher than those given by all other treatment combinations.
2. Fat 3 generally gives a response that is significantly higher than that given by fat 1.
3. There is no difference in the surfactant levels used with fat 1.

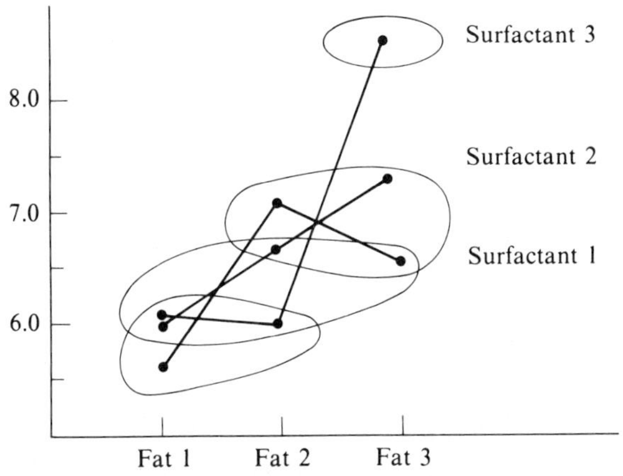

Figure 12.1 Plot of least squares means. Means located within the same circle are not significantly different.

In the next chapter, the case where some treatment combinations are never observed is discussed. Such cases require additional care in selecting a proper analysis.

CONCLUDING REMARKS

In this chapter, we considered the analysis of a balanced two-way treatment structure in a randomized complete block design when some treatment combinations are missing in some blocks. The analysis described is appropriate only when each treatment combination is observed at least once. SAS® GLM was used to obtain the analysis.

13

Using the Means Model to Analyze Two-Way Treatment Structures With Missing Treatment Combinations

CHAPTER OUTLINE

In this and the next two chapters, we discuss the analysis of two-way treatment structures when some treatment combinations are never observed. These kinds of experimental situations often occur in practice, mostly by chance but sometimes by design. When the experimenter does have control over the experiment, extreme care should be taken to insure that all treatment combinations are observed.

Many statistical packages contain routines that calculate test statistics for experiments with missing treatment combinations, but as we shall show in this chapter, the observed values of those test statistics often have little, if any, meaning. Thus, the statistical packages available give the experimenter a false sense of security about the analysis when, in fact, the analysis automatically provided is not generally an analysis of interest. The following sections point out some of the problems and provide methods to obtain a correct, meaningful analysis.

13.1 PARAMETER ESTIMATION

As in Chapter 9, the effect of missing treatment combinations complicates the analysis enough that we again use a very simple set of hypothetical data to aid our discussion. A realistic example is discussed in Chapter 15.

Consider the hypothetical data in Table 13.1 from a two-way treatment structure in a completely randomized design with treatment T and treatment B each having three levels.

Let μ_{ij} represent the response expected when treatments T_i and B_j are applied to a randomly selected experimental unit. A general model for this experiment is

$$y_{ijk} = \mu_{ij} + \varepsilon_{ijk} \qquad i = 1, 2, \ldots, t, \quad j = 1, 2, \ldots, b, \quad k = 0, 1, \ldots, n_{ij} \tag{13.1.1}$$

where $\varepsilon_{ijk} \sim$ i.i.d. $N(0, \sigma^2)$. If $n_{ij} = 0$, then the treatment combination of T_i and B_j is not observed.

Whenever treatment combinations are missing, certain hypotheses cannot be tested without making some additional assumptions about the parameters in the model. Hypotheses involving parameters corresponding to the missing cells generally cannot be tested. For example, for the data

Table 13.1 A Two-Way Experiment with Missing Treatment Combinations

	B_1	B_2	B_3
T_1	2, 6		8, 6
T_2	3	14	12, 9
T_3	6	9	

in Table 13.1, it is not possible to estimate any linear combinations (or test any hypotheses) that involve parameters μ_{12} and μ_{33} unless one is willing to make some assumptions about them.

One common assumption is that there is no interaction between the levels of T and the levels of B. In our opinion, this assumption should not be made without some supporting experimental evidence. All too often we find experimenters willing to assume no interaction exists between the treatment levels in any of their experiments mainly because they do not understand how to deal with such an interaction or because they believe they are not interested in it.

These are definitely not justifiable reasons for assuming no interaction between the two sets of treatments. If interaction exists, the experimenter must deal with it and be interested in making inferences about it. In Chapter 8 we discussed methods for dealing with interaction when all treatment combinations are observed. The kinds of questions considered there can also be considered here.

As previously stated, it is not possible to make inferences about functions of parameters involving missing treatment combinations. For example, it is not possible to test $\bar{\mu}_{1\cdot} = \bar{\mu}_{2\cdot} = \bar{\mu}_{3\cdot}$ or $\bar{\mu}_{\cdot 1} = \bar{\mu}_{\cdot 2} = \bar{\mu}_{\cdot 3}$, since these hypotheses involve parameters about which we have no information. Indeed, it is not possible to estimate all of the expected marginal means. For the above data, one cannot estimate $\bar{\mu}_{1\cdot}$, $\bar{\mu}_{3\cdot}$, $\bar{\mu}_{\cdot 2}$, or $\bar{\mu}_{\cdot 3}$. However, one can estimate $\bar{\mu}_{2\cdot}$ and $\bar{\mu}_{\cdot 1}$, since these functions of the parameters do not involve parameters corresponding to missing cells.

As one would expect, the best estimates of the parameters of model (13.1) are

$$\hat{\mu}_{ij} = \bar{y}_{ij\cdot} \qquad i = 1, 2, \ldots, t, \quad j = 1, 2, \ldots, b \qquad \text{if} \quad n_{ij} > 0,$$

and

$$\hat{\sigma}^2 = \frac{\Sigma_{ijk}\left(y_{ijk} - \bar{y}_{ij\cdot}\right)^2}{N - C}$$

where $N = n_{\cdot\cdot}$ and $C =$ the total number of treatment combinations observed. If $n_{ij} = 0$, μ_{ij} is not estimable since there is no information about μ_{ij}. When $n_{ij} > 0$, the sampling distribution of $\hat{\mu}_{ij}$ is $N(\mu_{ij}, \sigma^2/n_{ij})$, $i = 1, 2, \ldots, t$, $j = 1, 2, \ldots, b$, and the sampling distribution of $(N - C)\hat{\sigma}^2/\sigma^2$ is $\chi^2(N - C)$. In addition, $\hat{\mu}_{ij}$, $i = 1, 2, \ldots, t$, $j = 1, 2, \ldots, b$, $n_{ij} > 0$, and $\hat{\sigma}^2$ are statistically independent.

13.2 HYPOTHESIS TESTING AND CONFIDENCE INTERVALS

Clearly, one method of analyzing experiments with missing treatment combinations is to use the procedures discussed in Chapter 1; in fact, this is often the best method. That is, the procedures in Chapter 1 can be used to test hypotheses about any linear combinations of the μ_{ij}'s corresponding to observed treatment combinations.

EXAMPLE 13.1

Suppose we wish to obtain a 95% confidence interval for $\bar{\mu}_{2\cdot}$ in Table 13.1. We first note that $\hat{\mu}_{11} = 4$, $\hat{\mu}_{13} = 7$, $\hat{\mu}_{21} = 3$, $\hat{\mu}_{22} = 14$, $\hat{\mu}_{23} = 10.5$, $\hat{\mu}_{31} = 6$, $\hat{\mu}_{32} = 9$, and

$$\hat{\sigma}^2 = \frac{\begin{array}{c}(2-4)^2 + (6-4)^2 + (8-7)^2 + (6-7)^2 \\ + (12-10.5)^2 + (9-10.5)^2\end{array}}{(10-7)}$$

$$= \frac{14.5}{3} = 4.833$$

and is based on 3 degrees of freedom. The best estimate of $\bar{\mu}_{2\cdot}$ is

$$\hat{\bar{\mu}}_{2\cdot} = \frac{\hat{\mu}_{21} + \hat{\mu}_{22} + \hat{\mu}_{23}}{3} = \frac{3 + 14 + 10.5}{3} = 9.167,$$

and its estimated standard error is

$$\widehat{\text{s.e.}}(\hat{\bar{\mu}}_{2\cdot}) = \frac{\hat{\sigma}}{3}\sqrt{\frac{1}{n_{21}} + \frac{1}{n_{22}} + \frac{1}{n_{23}}}$$

$$= \frac{2.198}{3}\sqrt{1 + 1 + \frac{1}{2}} = 1.158.$$

Thus, a 95% confidence interval for $\bar{\mu}_{2\cdot}$ is

$$9.167 - (3.182)(1.158) < \bar{\mu}_{2\cdot} < 9.167 + (3.182)(1.158)$$

or $5.48 < \bar{\mu}_{2\cdot} < 12.85$.

EXAMPLE 13.2

Suppose we wish to determine whether there is interaction in the above set of data. The interaction space would have 4 degrees of freedom if all treatment combinations were observed, but in this case only two linearly independent contrasts measure interaction. Two possibilities are

$$\mu_{11} - \mu_{13} - \mu_{21} + \mu_{23} \quad \text{and} \quad \mu_{21} - \mu_{22} - \mu_{31} + \mu_{32}.$$

Thus, we test

$$H_0\colon \mu_{11} - \mu_{13} - \mu_{21} + \mu_{23} = 0 \quad \text{and} \quad \mu_{21} - \mu_{22} - \mu_{31} + \mu_{32} = 0.$$

Equivalently, we can test the hypothesis $\mathbf{C}\boldsymbol{\mu} = \mathbf{0}$ where

$$\mathbf{C} = \begin{bmatrix} 1 & -1 & -1 & 0 & 1 & 0 & 0 \\ 0 & 0 & 1 & -1 & 0 & -1 & 1 \end{bmatrix} \text{ and } \boldsymbol{\mu} = \begin{bmatrix} \mu_{11} \\ \mu_{13} \\ \mu_{21} \\ \mu_{22} \\ \mu_{23} \\ \mu_{31} \\ \mu_{32} \end{bmatrix}.$$

The sum of squares due to H_0 is

$$SS_{H_0} = (\mathbf{C}\hat{\boldsymbol{\mu}})'(\mathbf{CDC}')^{-1}(\mathbf{C}\hat{\boldsymbol{\mu}})$$

where

$$\mathbf{D} = \text{Diag}\left(\frac{1}{2}, \frac{1}{2}, 1, 1, \frac{1}{2}, 1, 1\right).$$

We get

$$\begin{aligned} SS_{H_0} &= [4.5, \ -8]\begin{bmatrix} \frac{5}{2} & -1 \\ -1 & 4 \end{bmatrix}^{-1}\begin{bmatrix} 4.5 \\ -8 \end{bmatrix} \\ &= \frac{1}{9}[4.5, \ -8]\begin{bmatrix} 4 & 1 \\ 1 & \frac{5}{2} \end{bmatrix}\begin{bmatrix} 4.5 \\ -8 \end{bmatrix} \\ &= 18.778 \end{aligned}$$

and is based on 2 degrees of freedom, which comes from the rank of $\mathbf{C}$. The appropriate F-statistic is

$$F = \frac{18.778/2}{4.833} = 1.94$$

with two and three degrees of freedom; this F is not significant at the 10% level.

We note that if we had observed the two missing treatment combinations, there may have been significant interaction. Our interpretation is basically that treatments T_1 and T_2 do not interact with B_1 and B_3, and treatments T_2 and T_3 do not interact with B_1 and B_2. We cannot make any general statements about interaction between the T treatments and the B treatments.

EXAMPLE 13.3

It is not possible to test $\bar{\mu}_{1\cdot} = \bar{\mu}_{2\cdot} = \bar{\mu}_{3\cdot}$ in the data for Table 13.1 because of the two missing cells. However, it is possible to test

$$H_0: \frac{\mu_{11} + \mu_{13}}{2} = \frac{\mu_{21} + \mu_{23}}{2} \quad \text{and}$$
$$\frac{\mu_{21} + \mu_{22}}{2} = \frac{\mu_{31} + \mu_{32}}{2}.$$

In a very broad sense, these are T main-effect type hypotheses. The experimenter must determine if either of these hypotheses are of interest; we use them to illustrate this method.

To test the hypotheses given by H_0, we can take

$$\mathbf{C} = \begin{bmatrix} 1 & 1 & -1 & 0 & -1 & 0 & 0 \\ 0 & 0 & 1 & 1 & 0 & -1 & -1 \end{bmatrix}.$$

Then the sum of squares due to H_0 is

$$\begin{aligned} SS_{H_0} &= [-2.5,\ \ 2]\begin{bmatrix} \frac{5}{2} & -1 \\ -1 & 4 \end{bmatrix}^{-1}\begin{bmatrix} -2.5 \\ 2 \end{bmatrix} \\ &= \frac{1}{9}[-2.5,\ \ 2]\begin{bmatrix} 4 & 1 \\ 1 & \frac{5}{2} \end{bmatrix}\begin{bmatrix} -2.5 \\ 2 \end{bmatrix} \\ &= 2.778 \end{aligned}$$

with 2 degrees of freedom. The appropriate F-statistic is

$$F = \frac{2.778/2}{4.833} = .291$$

with 2 and 3 degrees of freedom and hence is not significant.

The reader might have noticed that $\mathbf{CDC'}$ is the same for both of these last two examples; this is coincidental and is not generally true.

13.3 COMPUTER ANALYSES

When some treatment combinations are not observed, it is often best to consider the experiment as a one-way experiment and use computing routines similar to those described in Section 1.7. However, since many statistical packages provide certain tests automatically when an effects model is used, many experimenters have preferred them. Such analyses are described in Chapter 14.

CONCLUDING REMARKS

In this chapter we discussed some of the complications that result whenever some treatment combinations are not observed. In this chapter we used the means model (the effects model is used in Chapter 14). The important thing to remember when some treatment combinations are not observed is that some hypotheses of interest may not be testable unless some additional assumptions can be made about the parameters in the model. However, such assumptions should not be made without the evidence to support them.

14

Using the Effects Model to Analyze Two-Way Treatment Structures with Missing Treatment Combinations

CHAPTER OUTLINE

In Chapter 13 we discussed the use of the means model in analyzing two-way treatment structures when some treatment combinations are not observed. In this chapter we consider the use of the effects model for analyzing the same problem. Using the effects model does not enable one to answer any questions that cannot be answered by using the means model, and vice versa. While the means model is very simple and easy to understand, the effects model appears to be much more complex than it really is. We prefer to use the means model and are discussing the effects model here only because many statistical packages seem to recommend and encourage its use.

14.1 TYPE I AND TYPE II HYPOTHESES

Type I and Type II analyses for two-way treatment structures with missing treatment combinations can be defined like those for such structures where all combinations are observed. That is, successive models can be fit, and the resulting reductions in the residual sum of squares are determined as different effects are added to the model.

14.1.1 Type I Analysis

To illustrate, we shall construct a Type I analysis of the data in Table 13.1 (see Chapter 6 for the general formula). Readers interested only in the results should see Tables 14.1 through 14.4 and skip the derivations.

First, we fit the model $y_{ijk} = \mu + \varepsilon_{ijk}$. The normal equation for this model is $10\hat{\mu} = 75$. A solution is $\hat{\mu} = \frac{75}{10} = 7.5$, and $\text{RSS}_1 = \mathbf{y}'\mathbf{y} - \hat{\boldsymbol{\beta}}'\mathbf{X}'\mathbf{y} = 687 - (7.5)(75) = 124.5$, which is based on $10 - 1 = 9$ degrees of freedom.

Next, we fit the model $y_{ijk} = \mu + \tau_i + \varepsilon_{ijk}$. The normal equations corresponding to this model are

$$\begin{bmatrix} 10 & 4 & 4 & 2 \\ 4 & 4 & 0 & 0 \\ 4 & 0 & 4 & 0 \\ 2 & 0 & 0 & 2 \end{bmatrix} \begin{bmatrix} \hat{\mu} \\ \hat{\tau}_1 \\ \hat{\tau}_2 \\ \hat{\tau}_3 \end{bmatrix} = \begin{bmatrix} 75 \\ 22 \\ 38 \\ 15 \end{bmatrix}.$$

One possible solution to these equations is obtained by using the set-to-zero restrictions discussed in Section 6.2. The solution is $\hat{\mu} = 7.5$, $\hat{\tau}_1 = -2$, $\hat{\tau}_2 = 2$, and $\hat{\tau}_3 = 0$. The residual sum of squares is

$$\text{RSS}_2 = \mathbf{y}'\mathbf{y} - \hat{\boldsymbol{\beta}}'\mathbf{X}'\mathbf{y} = 687$$
$$-[(7.5)(75) + (-2)(22) + (2)(38) + (0)(15)] = 92.5$$

and is based on $10 - 3 = 7$ degrees of freedom. Thus $R(\tau|\mu) = 124.5 - 92.5 = 32$ and is based on $9 - 7 = 2$ degrees of freedom.

Next, we fit the model $y_{ijk} = \mu + \tau_i + \beta_j + \varepsilon_{ijk}$. The normal equations corresponding to this model are

$$\begin{bmatrix} 10 & 4 & 4 & 2 & 4 & 2 & 4 \\ 4 & 4 & 0 & 0 & 2 & 0 & 2 \\ 4 & 0 & 4 & 0 & 1 & 1 & 2 \\ 2 & 0 & 0 & 2 & 1 & 1 & 0 \\ 4 & 2 & 1 & 1 & 4 & 0 & 0 \\ 2 & 0 & 1 & 1 & 0 & 2 & 0 \\ 4 & 2 & 2 & 0 & 0 & 0 & 4 \end{bmatrix} \begin{bmatrix} \hat{\mu} \\ \hat{\tau}_1 \\ \hat{\tau}_2 \\ \hat{\tau}_3 \\ \hat{\beta}_1 \\ \hat{\beta}_2 \\ \hat{\beta}_3 \end{bmatrix} = \begin{bmatrix} 75 \\ 22 \\ 38 \\ 15 \\ 17 \\ 23 \\ 35 \end{bmatrix}.$$

To obtain a solution, we let $\hat{\tau}_3 = 0$ and $\hat{\beta}_3 = 0$. Then this system can be reduced to an equivalent system by deleting the rows and columns corresponding to $\hat{\tau}_3$ and $\hat{\beta}_3$. Thus, the system reduces to

$$\begin{bmatrix} 10 & 4 & 4 & 4 & 2 \\ 4 & 4 & 0 & 2 & 0 \\ 4 & 0 & 4 & 1 & 1 \\ 4 & 2 & 1 & 4 & 0 \\ 2 & 0 & 1 & 0 & 2 \end{bmatrix} \begin{bmatrix} \hat{\mu} \\ \hat{\tau}_1 \\ \hat{\tau}_2 \\ \hat{\beta}_1 \\ \hat{\beta}_2 \end{bmatrix} = \begin{bmatrix} 75 \\ 22 \\ 38 \\ 17 \\ 23 \end{bmatrix}.$$

The solution to this reduced system is

$$\hat{\mu} = 8.3889, \quad \hat{\tau}_1 = -.8333, \quad \hat{\tau}_2 = 1.5556, \quad \hat{\beta}_1 = -4.1111,$$
$$\text{and} \quad \hat{\beta}_2 = 2.3333.$$

The residual sum of squares is

$$\begin{aligned} \text{RSS}_3 &= \mathbf{y}'\mathbf{y} - \hat{\boldsymbol{\beta}}'\mathbf{X}\mathbf{y} = 687 - 653.7222 \\ &= 33.2778 \end{aligned}$$

and is based on $10 - 5 = 5$ degrees of freedom. Thus

$$\begin{aligned} R(\beta|\mu, \tau) &= \text{RSS}_2 - \text{RSS}_3 \\ &= 92.5 - 33.2778 \\ &= 59.2222 \end{aligned}$$

with $7 - 5 = 2$ degrees of freedom.

Finally, we fit the full model

$$y_{ijk} = \mu + \tau_i + \beta_j + \gamma_{ij} + \varepsilon_{ijk}.$$

The normal equations corresponding to this model are

$$\begin{bmatrix} 10&4&4&2&4&2&4&2&0&2&1&1&2&1&1&0\\ 4&4&0&0&2&0&2&2&0&2&0&0&0&0&0&0\\ 4&0&4&0&1&1&2&0&0&0&1&1&2&0&0&0\\ 2&0&0&2&1&1&0&0&0&0&0&0&0&1&1&0\\ 4&2&1&1&4&0&0&2&0&0&1&0&0&1&0&0\\ 2&0&1&1&0&2&0&0&0&0&0&1&0&0&1&0\\ 4&2&2&0&0&0&4&0&0&2&0&0&2&0&0&0\\ 2&2&0&0&2&0&0&2&0&0&0&0&0&0&0&0\\ 0&0&0&0&0&0&0&0&0&0&0&0&0&0&0&0\\ 2&2&0&0&0&0&2&0&0&2&0&0&0&0&0&0\\ 1&0&1&0&1&0&0&0&0&0&1&0&0&0&0&0\\ 1&0&1&0&0&1&0&0&0&0&0&1&0&0&0&0\\ 2&0&2&0&0&0&2&0&0&0&0&0&2&0&0&0\\ 1&0&0&1&1&0&0&0&0&0&0&0&0&1&0&0\\ 1&0&0&1&0&1&0&0&0&0&0&0&0&0&1&0\\ 0&0&0&0&0&0&0&0&0&0&0&0&0&0&0&0 \end{bmatrix} \begin{bmatrix} \hat{\mu}\\ \hat{\tau}_1\\ \hat{\tau}_2\\ \hat{\tau}_3\\ \hat{\beta}_1\\ \hat{\beta}_2\\ \hat{\beta}_3\\ \hat{\gamma}_{11}\\ \hat{\gamma}_{12}\\ \hat{\gamma}_{13}\\ \hat{\gamma}_{21}\\ \hat{\gamma}_{22}\\ \hat{\gamma}_{23}\\ \hat{\gamma}_{31}\\ \hat{\gamma}_{32}\\ \hat{\gamma}_{33} \end{bmatrix} = \begin{bmatrix} 75\\ 22\\ 38\\ 15\\ 17\\ 23\\ 35\\ 8\\ 0\\ 14\\ 3\\ 14\\ 21\\ 6\\ 9\\ 0 \end{bmatrix}.$$

In order to solve this system, we shall again use the set-to-zero restrictions. However, some minor adjustments are required because of the two missing cells. Had all cells been filled, the set-to-zero restrictions would have us taking

$$\hat{\tau}_3 = 0, \quad \hat{\beta}_3 = 0, \quad \hat{\gamma}_{13} = 0, \quad \hat{\gamma}_{23} = 0, \quad \hat{\gamma}_{31} = 0, \quad \hat{\gamma}_{32} = 0, \quad \text{and} \quad \hat{\gamma}_{33} = 0.$$

Since the cell corresponding to $\hat{\gamma}_{33}$ is missing, we need to set another estimator, either $\hat{\gamma}_{21}$ or $\hat{\gamma}_{22}$, equal to zero. We choose $\hat{\gamma}_{22} = 0$. Then the rows and columns of $\mathbf{X}'\mathbf{X}$ and the rows of $\mathbf{X}'\mathbf{y}$ corresponding to $\hat{\tau}_3$, $\hat{\beta}_3$, $\hat{\gamma}_{13}$, $\hat{\gamma}_{22}$, $\hat{\gamma}_{23}$, $\hat{\gamma}_{31}$, and $\hat{\gamma}_{32}$ can be deleted from the system to get a reduced system. Also, the rows and columns corresponding to $\hat{\gamma}_{12}$ and $\hat{\gamma}_{33}$ (the missing cells) can be deleted. Thus, the reduced system is

$$\begin{bmatrix} 10&4&4&4&2&2&1\\ 4&4&0&2&0&2&0\\ 4&0&4&1&1&0&1\\ 4&2&1&4&0&2&1\\ 2&0&1&0&2&0&0\\ 2&2&0&2&0&2&0\\ 1&0&1&1&0&0&1 \end{bmatrix} \begin{bmatrix} \hat{\mu}\\ \hat{\tau}_1\\ \hat{\tau}_2\\ \hat{\beta}_1\\ \hat{\beta}_2\\ \hat{\gamma}_{11}\\ \hat{\gamma}_{21} \end{bmatrix} = \begin{bmatrix} 75\\ 22\\ 38\\ 17\\ 23\\ 8\\ 3 \end{bmatrix}.$$

The solution to this reduced system is $\hat{\mu} = 5.5$, $\hat{\tau}_1 = 1.5$, $\hat{\tau}_2 = 5.0$, $\hat{\beta}_1 = .5$,

Table 14.1 Type I Analysis for Data in Table 13.1

Source of Variation	*df*	*SS*	*MS*	*F*
Total	9	124.5		
T	2	32.0	16.0	3.31
B	2	59.22	29.61	6.13
T * *B*	2	18.78	9.39	1.94
Error	3	14.5	4.833	

$\hat{\beta}_2 = 3.5$, $\hat{\gamma}_{11} = -3.5$, $\hat{\gamma}_{21} = -8.0$, and

$$\begin{aligned} \text{RSS}_4 &= \mathbf{y}'\mathbf{y} - \hat{\boldsymbol{\beta}}'\mathbf{X}\mathbf{y} \\ &= 687 - 672.5 = 14.5 \end{aligned}$$

with $10 - 7 = 3$ degrees of freedom. Thus,

$$\begin{aligned} R(\gamma|\mu, \tau, \beta) &= \text{RSS}_3 - \text{RSS}_4 \\ &= 33.27778 - 14.5 = 18.7778 \end{aligned}$$

with $5 - 3 = 2$ degrees of freedom.

The Type I analysis of variance table for this problem is given in Table 14.1.

14.1.2 Type II Analysis

To construct a Type II analysis, we still need $R(\tau|\mu, \beta)$, which would involve comparing the two models

$$y_{ijk} = \mu + \beta_j + \varepsilon_{ijk} \quad \text{and} \quad y_{ijk} = \mu + \tau_i + \beta_j + \varepsilon_{ijk}.$$

The resulting Type II analysis is given in Table 14.2.

Table 14.2 Type II Analysis for Data in Table 13.1

Source of Variation	*df*	*SS*	*MS*	*F*
Total	9	124.5		
T	2	10.72	5.36	1.11
B	2	59.22	29.61	6.13
T * *B*	2	18.78	9.39	1.94
Error	3	14.5	4.833	

Table 14.3 Type I Hypotheses in $\mu + \tau_i + \beta_j + \gamma_{ij}$ Model for Data in Table 13.1

Source of Variation	*Hypothesis*
T	$\tau_1 - \tau_3 - \frac{1}{2}\beta_2 + \frac{1}{2}\beta_3 + \frac{1}{2}\gamma_{11} + \frac{1}{2}\gamma_{13} - \frac{1}{2}\gamma_{31} - \frac{1}{2}\gamma_{32} = 0$, and
	$\tau_2 - \tau_3 - \frac{1}{4}\beta_1 - \frac{1}{4}\beta_2 + \frac{1}{2}\beta_3 + \frac{1}{4}\gamma_{21} + \frac{1}{4}\gamma_{22} + \frac{1}{2}\gamma_{23} - \frac{1}{2}\gamma_{31} - \frac{1}{2}\gamma_{32} = 0$
B	$\beta_1 - \beta_3 + \frac{5}{9}\gamma_{11} - \frac{5}{9}\gamma_{13} + \frac{1}{3}\gamma_{21} + \frac{1}{9}\gamma_{22} - \frac{4}{9}\gamma_{23} + \frac{1}{9}\gamma_{31} - \frac{1}{9}\gamma_{32} = 0$, and
	$\beta_2 - \beta_3 + \frac{1}{3}\gamma_{11} - \frac{1}{3}\gamma_{13} + \frac{2}{3}\gamma_{22} - \frac{2}{3}\gamma_{23} - \frac{1}{3}\gamma_{31} + \frac{1}{3}\gamma_{32} = 0$
$T * B$	$\gamma_{11} - \gamma_{13} - \gamma_{22} + \gamma_{23} - \gamma_{31} + \gamma_{32} = 0$, and
	$\gamma_{21} - \gamma_{22} - \gamma_{31} + \gamma_{32} = 0$

Tables 14.3 through 14.6 show the Type I and Type II hypotheses for the data in Table 13.1. The entries in these tables can be determined from an SAS® analysis of Table 13.1 or by using the general formulas given in Table 10.9 and (10.4.1), which are also correct for the missing-cells problem.

Next, we discuss possible interpretations of the Type I and Type II main-effect hypotheses. The Type I and Type II hypotheses will generally not make much sense unless one's objective is to build a simple model for making predictions rather than to test hypotheses about the effects of the different treatment combinations. For model building, the interpretations are exactly the same as they were in Chapter 10, where we discussed the case with no missing treatment combinations; as in Chapter 10, if the numbers of observations in each cell are proportional to the actual numbers of each treatment combination existing in the population, then the experimenter might be interested in $R(\tau|\mu)$ and $R(\beta|\mu)$. Both of

Table 14.4 Type II Hypotheses in $\mu + \tau_i + \beta_j + \gamma_{ij}$ Model for Data in Table 13.1

Source of Variation	*Hypothesis*
T	$\tau_1 - \tau_3 + \frac{2}{3}\gamma_{11} + \frac{1}{3}\gamma_{13} + \frac{1}{3}\gamma_{22} - \frac{1}{3}\gamma_{23} - \frac{2}{3}\gamma_{31} - \frac{1}{3}\gamma_{32} = 0$, and
	$\tau_2 - \tau_3 + \frac{2}{9}\gamma_{11} - \frac{2}{9}\gamma_{13} + \frac{1}{3}\gamma_{21} + \frac{4}{9}\gamma_{22} + \frac{2}{9}\gamma_{23} - \frac{5}{9}\gamma_{31} - \frac{4}{9}\gamma_{32} = 0$
B	Same as for Type I analysis
$T * B$	Same as for Type I analysis

Table 14.5 Type I Hypotheses in μ_{ij} Model for Data in Table 13.1

Source of Variation	*Hypothesis*
T	$\mu_{11} + \mu_{13} - \mu_{31} - \mu_{32} = 0$ and $\mu_{21} + \mu_{22} + 2\mu_{23} - 2\mu_{31} - 2\mu_{32} = 0$
B	$5\mu_{11} - 5\mu_{13} + 3\mu_{21} + \mu_{22} - 4\mu_{23} + \mu_{31} - \mu_{32} = 0$ and $\mu_{11} - \mu_{13} + 2\mu_{22} - 2\mu_{23} - \mu_{31} + \mu_{32} = 0$
$T * B$	$\mu_{11} - \mu_{13} - \mu_{22} + \mu_{23} - \mu_{31} + \mu_{32} = 0$ and $\mu_{21} - \mu_{22} - \mu_{31} + \mu_{32} = 0$

Table 14.6 Type II Hypotheses in μ_{ij} Model for Data in Table 13.1

Source of Variation	*Hypothesis*
T	$2\mu_{11} + \mu_{13} + \mu_{22} - \mu_{23} - 2\mu_{31} - \mu_{32} = 0$ and $2\mu_{11} - 2\mu_{13} + 3\mu_{21} + 4\mu_{22} + 2\mu_{23} - 5\mu_{31} - 4\mu_{32} = 0$
B	Same as for Type I analysis
$T * B$	Same as for Type I analysis

these terms can be obtained by conducting two Type I analyses, each with a different model statement. One model statement would have T as the first effect in the model; the other, B.

14.2 TYPE III HYPOTHESES

When all treatment combinations are observed, the Type III hypotheses are the same as those tested when there are equal subclass numbers. When some treatment combinations are missing, however, such hypotheses cannot be tested since they involve parameters about which we have no information. For the data in Table 13.1, we cannot estimate $\bar{\mu}_{1\cdot}$ and $\bar{\mu}_{3\cdot}$ since we cannot estimate μ_{12} and μ_{33}. Likewise, we cannot estimate $\bar{\mu}_{\cdot 2}$ and $\bar{\mu}_{\cdot 3}$. Hence, it is not possible to test $\bar{\mu}_{1\cdot} = \bar{\mu}_{2\cdot} = \bar{\mu}_{3\cdot}$ or $\bar{\mu}_{\cdot 1} = \bar{\mu}_{\cdot 2} = \bar{\mu}_{\cdot 3}$.

Both the Type I and Type II hypotheses for the main effects depend on the numbers of observations in each cell. As long as there is at least one observation in a cell, then that cell mean is estimable. Thus functions of the parameters that are estimable depend only on which treatment combinations are observed and not on how many times they are observed.

Table 14.7 Type III Analysis of Data in Table 13.1

Source of Variation	*df*	*SS*	*MS*	*F*
Total	9	124.5		
T	2	8.4135	4.2067	.87
B	2	61.9615	30.9807	6.41
$T * B$	2	18.7778	9.3889	1.94
Error	3	14.5	4.8333	

Type III hypotheses are developed so that they do not depend on the cell sizes, but only on which cells are observed. This is consistent with the definition of Type III hypotheses for two-way experiments where all treatment combinations are observed. That is, the hypotheses $\bar{\mu}_{1\cdot} = \bar{\mu}_{2\cdot} = \cdots = \bar{\mu}_{t\cdot}$ and $\bar{\mu}_{\cdot 1} = \bar{\mu}_{\cdot 2} = \cdots = \bar{\mu}_{\cdot b}$ do not depend on the cell sizes.

We are not going to discuss the construction of Type III hypotheses for the missing-data case. Even though the objectives are reasonable, we think that the Type III hypotheses are the worst hypotheses to consider in this situation because there seems to be no reasonable way to interpret them. To illustrate, a Type III analysis of the data in Table 13.1 is given in Table 14.7. The hypotheses being tested in the Type III analysis are given in Tables 14.8 and 14.9.

Examination of Tables 14.8 and 14.9 reveals that the Type III hypotheses are not meaningful except for the $T * B$ interaction. The results in Tables 14.7 through 14.9 were taken from an SAS® GLM analysis of the data.

Table 14.8 Type III Hypotheses in $\mu + \tau_i + \beta_j + \gamma_{ij}$ Model for Data in Table 13.1

Source of Variation	*Hypothesis*
T	$\tau_1 - \tau_3 + \frac{2}{3}\gamma_{11} + \frac{1}{3}\gamma_{13} + \frac{1}{3}\gamma_{22} - \frac{1}{3}\gamma_{23} - \frac{2}{3}\gamma_{31} - \frac{1}{3}\gamma_{32} = 0$ and $\tau_2 - \tau_3 + \frac{2}{15}\gamma_{11} - \frac{2}{15}\gamma_{13} + \frac{2}{5}\gamma_{21} + \frac{7}{15}\gamma_{22} + \frac{2}{15}\gamma_{23} - \frac{8}{15}\gamma_{31} - \frac{7}{15}\gamma_{32} = 0$
B	$\beta_1 - \beta_3 + \frac{7}{15}\gamma_{11} - \frac{7}{15}\gamma_{13} + \frac{2}{5}\gamma_{21} + \frac{2}{15}\gamma_{22} - \frac{8}{15}\gamma_{23} + \frac{2}{15}\gamma_{31} - \frac{2}{15}\gamma_{32} = 0$ and $\beta_2 - \beta_3 + \frac{1}{3}\gamma_{11} - \frac{1}{3}\gamma_{13} + \frac{2}{3}\gamma_{22} - \frac{2}{3}\gamma_{23} - \frac{1}{3}\gamma_{31} + \frac{1}{3}\gamma_{32} = 0$
$T * B$	Same as for Type I analysis

Table 14.9 Type III Hypotheses in μ_{ij} Model for Data in Table 13.1

Source of Variation	*Hypothesis*
T	$2\mu_{11} + \mu_{13} + \mu_{22} - \mu_{23} - 2\mu_{31} - \mu_{32} = 0$ and $2\mu_{11} - 2\mu_{13} + 6\mu_{21} + 7\mu_{22} + 2\mu_{23} - 8\mu_{31} - 7\mu_{32} = 0$
B	$7\mu_{11} - 7\mu_{13} + 6\mu_{21} + 2\mu_{22} - 8\mu_{23} + 2\mu_{31} - 2\mu_{32} = 0$ and $\mu_{11} - \mu_{13} + 2\mu_{22} - 2\mu_{23} - \mu_{31} + \mu_{32} = 0$
$T * B$	Same as for Type I analysis

14.3 TYPE IV HYPOTHESES

In the previous two sections, we attempted to show that none of the main-effect hypotheses tested by Type I, Type II, or Type III analyses are entirely satisfactory when there are missing treatment combinations, since they rarely have any reasonable interpretations. Such hypotheses are extremely difficult to interpret because the coefficients of cell means occurring in the same row or column are rarely the same. Type IV hypotheses are constructed so that the cell mean coefficients are balanced; hence, the resulting hypotheses are interpretable.

To illustrate, let us look at all possible Type IV hypotheses that are testable in the data set of Table 13.1. Basically, for a two-way treatment structure, a hypothesis is a Type IV type hypothesis if it compares the levels of one treatment averaged over one or more common levels of the other treatment. Thus, the hypotheses are expected marginal means hypotheses except that when treatment combinations are missing, one cannot average across all levels of the other treatment but only across some of the levels of the other factor.

Figure 14.1 gives the pattern of the cells that were observed in Table 13.1. All possible Type IV hypotheses for T for the cell means in Figure 14.1 are given in Figure 14.2.

Similar results can be obtained for the Type IV hypotheses for B. All possible Type IV hypotheses for B are given in Figure 14.3.

	B_1	B_2	B_3
T_1	μ_{11}	—	μ_{13}
T_2	μ_{21}	μ_{22}	μ_{23}
T_3	μ_{31}	μ_{32}	—

Figure 14.1 Cell means that can be estimated.

$$\left(\frac{\mu_{11}+\mu_{13}}{2}\right)=\left(\frac{\mu_{21}+\mu_{23}}{2}\right)$$
$$\left(\frac{\mu_{21}+\mu_{22}}{2}\right)=\left(\frac{\mu_{31}+\mu_{32}}{2}\right)^*$$
$$\mu_{11}=\mu_{21}$$
$$\mu_{11}=\mu_{31}{}^*$$
$$\mu_{21}=\mu_{31}$$
$$\mu_{22}=\mu_{32}$$
$$\mu_{13}=\mu_{23}$$

Figure 14.2 All possible Type IV hypotheses for *T*. The two hypotheses marked with asterisks were tested simultaneously by an SAS® GLM Type IV analysis.

SAS® GLM automatically generates some Type IV hypotheses that can usually be interpreted, but an appropriate interpretation cannot be made without first examining the Type IV estimable functions to see what hypotheses are generated and tested. That is, there is no unique interpretation appropriate for all data sets as there was in the case of no missing treatment combinations. In fact, renumbering or reordering the treatments before doing the analysis may result in different Type IV hypotheses and hence different sums of squares and *F*-values. Thus, the Type IV analysis obtained probably depends on what the treatments are called or how they are numbered. Obviously, this is not very desirable, but it is unavoidable. SAS® GLM indicates this situation by placing an asterisk on the printed degrees of freedom and noting that "OTHER TYPE IV TESTABLE HYPOTHESES EXIST WHICH MAY YIELD DIFFERENT SS."

The Type IV hypothesis for *T* that was tested by an SAS® GLM analysis of the data in Table 13.1 is equivalent to simultaneously testing

$$\left(\frac{\mu_{11}+\mu_{21}}{2}\right)=\left(\frac{\mu_{13}+\mu_{23}}{2}\right)^*$$
$$\left(\frac{\mu_{21}+\mu_{31}}{2}\right)=\left(\frac{\mu_{22}+\mu_{32}}{2}\right)$$
$$\mu_{11}=\mu_{13}$$
$$\mu_{21}=\mu_{22}$$
$$\mu_{21}=\mu_{23}$$
$$\mu_{22}=\mu_{23}{}^*$$
$$\mu_{31}=\mu_{32}$$

Figure 14.3 All possible Type IV hypotheses for *B*. The two hypotheses marked with asterisks were tested simultaneously by an SAS® GLM Type IV analysis.

$\mu_{11} = \mu_{31}$ and $(\mu_{21} + \mu_{22})/2 = (\mu_{31} + \mu_{32})/2$. Thus, the Type IV hypothesis for T simultaneously compares the effect of T_1 and T_3 at level 1 of B and the effect of T_2 and T_3 averaged over levels 1 and 2 of B. We note that level 3 of B is not involved at all in this set.

Type IV hypotheses not tested by SAS® GLM but probably just as interesting to the experimenter as those automatically tested are given in Figure 14.2.

In order to test such interesting Type IV hypotheses, one can (and should) use the ESTIMATE or CONTRAST option. For example, to test all of the Type IV hypotheses in Figure 14.2 when using the effects model, we would use the following statements after the MODEL statement:

```
ESTIMATE 'T1 VS T2 AVE OVER B1 AND B3' T 1
-1 0 T*B .5 .5 -.5 0 -.5 0 0;

ESTIMATE 'T2 VS T3 AVE OVER B1 AND B2' T 0
1 -1 T*B
0 0 .5 .5 0 -.5 -.5;

ESTIMATE 'T1 VS T2 AT B1' T 1 -1 0 T*B
1 0 -1 0 0 0 0;

ESTIMATE 'T1 VS T3 AT B1' T 1 0 -1 T*B
1 0 0 0 0 -1 0;

ESTIMATE 'T2 VS T3 AT B1' T 0 1 -1 T*B
0 0 1 0 0 -1 0;

ESTIMATE 'T2 VS T3 AT B2' T 0 1 -1 T*B
0 0 0 1 0 0 -1;

ESTIMATE 'T1 VS T2 AT B3' T 1 -1 0 T*B
0 1 0 0 -1 0 0;
```

If the reader does not have SAS® available, all of the hypotheses in Figures 14.2 and 14.3 can be tested by using the means model and procedures similar to those given in Section 13.2.

14.4 POPULATION MARGINAL MEANS AND LEAST SQUARES MEANS

Population marginal means and least squares means are defined here just as they were in Sections 9.5 and 10.6. However, if a particular treatment is not observed in combination with all levels of the other treatment, then the corresponding population marginal mean is not estimable. In this case the table of two-way cell means (for example, the $\hat{\mu}_{ij}$'s) can be used to compare each observed treatment combination to all other observed treatment combinations. If a data set is quite sparse, few if any population marginal means will be estimable.

For the data in Table 13.1, $\bar{\mu}_{2\cdot}$ and $\bar{\mu}_{\cdot 1}$ are the only population marginal means that are estimable. Their best estimates are $\hat{\bar{\mu}}_{2\cdot} = 9.167$

and $\hat{\mu}_{\cdot 1} = 4.333$, respectively. In general, the best estimate of $\Sigma c_{ij}\mu_{ij}$ is $\Sigma c_{ij}\hat{\mu}_{ij}$, and its estimated standard error is

$$\widehat{\text{s.e.}}\left(\sum c_{ij}\hat{\mu}_{ij}\right) = \hat{\sigma}\sqrt{\sum \frac{c_{ij}^2}{n_{ij}}},$$

where the sums are taken over all nonempty cells. A $(1 - \alpha)$ 100% confidence interval for $\Sigma c_{ij}\mu_{ij}$ is given by

$$\sum_{ij} c_{ij}\hat{\mu}_{ij} \pm t_{\alpha/2,\nu}\hat{\sigma}\sqrt{\sum \frac{c_{ij}^2}{n_{ij}}}.$$

A t-statistic with ν degrees of freedom for testing $\Sigma c_{ij}\mu_{ij} = 0$ is given by

$$t_c = \frac{\Sigma_{ij} c_{ij}\hat{\mu}_{ij}}{\hat{\sigma}\sqrt{\Sigma\left(c_{ij}^2/n_{ij}\right)}}.$$

In both instances, ν is the degrees of freedom of $\hat{\sigma}^2$, the error mean square.

For persons wishing to make multiple comparisons, we recommend using the $\hat{\alpha}$-values given by the above t-tests whenever the F-value for comparing all treatment combinations is significant. If this F-value is not significant, then one should use Bonferroni's method on all comparisons of interest. That is, declare two linear combinations significantly different if the $\hat{\alpha}$-value obtained is less than α/p where p is the total number of planned comparisons. For data snooping and unplanned comparisons, one should use a Scheffé procedure. See Johnson (1973).

14.5 COMPUTER ANALYSES

The reader should use his or her statistical package to analyze the examples given in this chapter and in Chapters 15 and 17. Comparing the results of the analyses so obtained with those given in this book will give the reader valuable insight into the kinds of hypotheses tested by the packages he or she is accustomed to using. We know of no package that handles the analysis of data with missing treatment combinations adequately or completely. Several do a good job with unbalanced data provided that there are no missing treatment combinations.

Anyone who does many statistical analyses on data with missing combinations should learn how to use a package that allows a specified set of hypotheses to be tested. Then, and only then, can one be sure that the hypotheses tested are reasonable, meaningful, and interpretable. Some statistical computing packages that allow the user to specify his or her own hypotheses include SAS® GLM, SPSS MANOVA, and BMD-P4V. If one uses a means model, then BMD-P1V and SPSS ONEWAY, as well as SAS® GLM, enable the user to specify his own hypotheses.

Table 14.10 Possible Statistical Analyses of Two-Way Effects Models with Missing Treatment Combinations Derived by Using BMDP and SPSS

	STATISTICAL COMPUTING PACKAGE		
Hypothesis	*BMD-P4V*	*SPSS ANOVA*	*SPSS MANOVA*
Type I	BETWEEN = SIZES*	OPTION 10	Default
Type II	BETWEEN = SIZES*	Default	No automatic option
Type III	Not Possible	OPTION 9†	SSTYPE (UNIQUE)
Type IV	Not Possible	Not Possible	Not Possible

*These two types are given simultaneously.
†SPSS ANOVA option 9 will function only if at least one level of every factor has data in every cell. This level must be specified as the first level of that factor.

The definitions of the Type I through Type IV hypotheses that we have used are consistent with those used by SAS® GLM. Table 14.10 compares available BMDP and SPSS analyses to these definitions of Type I through Type IV hypotheses. We note that BMD-P2V will not work for data with missing treatment combinations. Also, BMD-P4V apparently does not include interaction terms when there are missing cells.

CONCLUDING REMARKS

In summary, a good analysis of data with missing treatment combinations requires a great deal of thought. An experimenter or statistician cannot simply run a computer program on the data and then select numbers from that program to report in a paper. Unfortunately, this has been done and is being done by an extremely large number of experimenters and data analysts. We hope that anyone who has studied this chapter will never do it again. Those willing to exert the necessary effort to analyze their data correctly are advised to use the means model discussed in Chapter 13.

In Chapter 15, a more realistic example is discussed.

15

Case Study: Two-Way Treatment Structure with Missing Treatment Combinations

CHAPTER OUTLINE

In Chapters 13 and 14 we discussed the analysis of two-way treatment structures in a completely randomized design structure when there were missing treatment combinations. In this chapter we illustrate how to analyze a two-way treatment structure in a randomized complete block design when some treatment combinations are not observed.

Consider the data in Table 15.1, which is obtained from the experiment described in Chapter 12 but with several more values missing.

Figure 15.1 shows the treatment combinations observed at least once. Any hypothesis that involves treatment combinations (fat 1, surfactant 3) or (fat 2, surfactant 2), cannot be tested unless additional assumptions are made. In this discussion, we let FS_{ij} represent the response expected when fat i and surfactant j are assigned to a randomly selected experimental unit.

To get the error sum of sum of squares for this experiment, we can fit either an effects model or a means model in a randomized block design structure by using any of the available statistical packages. The model to fit is

SPVOL = BLK FAT SURF FAT*SURF

or

SPVOL = BLK TRTCOMB

where TRTCOMB takes on a different value for each of the different observed treatment combinations. After fitting one of these models, one finds that the error sum of squares is equal to 2.0941, with 11 degrees of

Table 15.1 Specific Volumes from Baking Experiment in Chapter 12

		FLOUR			
Fat	*Surfactant*	*1*	*2*	*3*	*4*
	1	6.7	4.3	5.7	—
1	2	7.1	—	5.9	5.6
	3	—	—	—	—
	1	—	5.9	7.4	7.1
2	2	—	—	—	—
	3	6.4	5.1	6.2	6.3
	1	7.1	5.9	—	—
3	2	7.3	6.6	8.1	6.8
	3	—	7.5	9.1	—

		Surfactant 1	2	3
Fat	1	X	X	
	2	X		X
	3	X	X	X

Figure 15.1 Treatment combinations observed in baking experiment.

freedom. Thus

$$\hat{\sigma}^2 = \frac{2.0941}{11} = .1904.$$

If all treatment combinations were observed, there would be 4 degrees of freedom for interaction hypotheses. Since two treatment combinations are not observed, only two degrees of freedom remain for interaction hypotheses. Two independent contrasts in the interaction space are

$$FS_{11} - FS_{12} - FS_{31} + FS_{32} \quad \text{and} \quad FS_{21} - FS_{23} - FS_{31} + FS_{33}.$$

An SAS® GLM Fat*Surfactant Type IV analysis tests these two contrasts equal to zero simultaneously. The value of the F-statistic for testing the two contrasts equal to zero simultaneously is

$$F = \frac{5.4002/2}{.1904} = 14.18$$

with 2 and 11 degrees of freedom.

All possible Type IV hypotheses for Fat are specified in Table 15.2.

Hypotheses 4 and 5 were automatically tested simultaneously in an SAS® GLM Type IV analysis for Fat. Hypotheses 1 through 3 can be

Table 15.2 Type IV Hypotheses for Fat

Hypothesis		*df*	*F*	$\hat{\alpha}$
1.	$FS_{11} = FS_{21} = FS_{31}$	2	8.69	.006
2.	$FS_{12} = FS_{32}$	1	15.18	.003
3.	$FS_{23} = FS_{33}$	1	43.89	.000
4.	$FS_{11} + FS_{12} = FS_{31} + FS_{32}$	2	13.25	.001
5.	$FS_{21} + FS_{23} = FS_{31} + FS_{32}$			

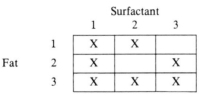

Table 15.3 Type IV Hypotheses for Surfactant

Hypothesis		*df*	*F*	$\hat{\alpha}$
1.	$FS_{11} = FS_{12}$	1	.85	.376
2.	$FS_{21} = FS_{23}$	1	9.11	.012
3.	$FS_{31} = FS_{32} = FS_{33}$	2	9.95	.003
4.	$FS_{11} + FS_{31} = FS_{12} + FS_{32}$	1	2.64	.132
5.	$FS_{21} + FS_{31} = FS_{23} + FS_{33}$	1	2.79	.123

tested by using the CONTRAST option, while hypotheses 2 and 3 can also be examined by using the ESTIMATE option. The testing results are also shown in Table 15.2.

All possible Type IV hypotheses for Surfactant are specified in Table 15.3.

The last equality in hypothesis 3 and hypothesis 5 were automatically tested simultaneously in a SAS® GLM Type IV analysis for Surfactant. The F-value given was $F = 6.34$ with 2 and 11 degrees of freedom. All five hypotheses can be tested by using the CONTRAST option, and all but hypothesis 3 by using the ESTIMATE option. All of the single-degree-of-freedom hypotheses in Tables 15.2 and 15.3 can be specified as options in BMD-P2V, BMD-P4V, and SPSS MANOVA.

Since there is significant interaction in these data, it is probably best to compare all observed treatment combinations by examining the least

Table 15.4 Least Squares Means and p-Values for Pairwise t-Tests for Data in Table 15.1

		p-VALUE						
Treatment Combination	*Least Squares Mean*	FS_{11}	FS_{12}	FS_{21}	FS_{23}	FS_{31}	FS_{32}	FS_{33}
FS_{11}	5.541		.376	.002	.204	.020	.001	.000
FS_{12}	5.883			.010	.736	.101	.003	.000
FS_{21}	7.022				.012	.390	.610	.003
FS_{23}	6.000					.130	.003	.000
FS_{31}	6.642						.182	.001
FS_{32}	7.200							.004
FS_{33}	8.595							

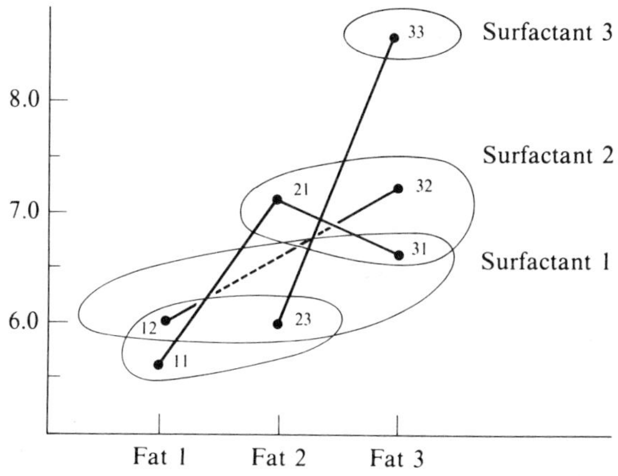

Figure 15.2 Fat * Surfactant least square means. Means located within the same circle are not significantly different.

squares means of the seven treatment combinations observed. The least squares means and P-values for comparing them pairwise by using t-tests are given in Table 15.4. We note that when the design structure is completely randomized, the population cell means are "best" estimated by taking the mean of the observations for each cell. (This is not true in a randomized block design.) These best estimates can easily be obtained with a computing package or by using the methods of Chapter 6. From these estimates, one can construct Figure 15.2, where means not significantly different have been enclosed in the same circle.

CONCLUDING REMARKS

This chapter illustrated the analysis of a two-way treatment structure experiment in a randomized complete block design when some treatment combinations are not observed.

16

Analyzing Three-Way and Higher-Order Treatment Structures

CHAPTER OUTLINE

In Chapters 7 through 15, we discussed the analysis of two-way treatment structures. The methods and results given in those nine chapters can be generalized to more complex treatment structures; such analyses become only slightly more complicated as the complexity of the treatment structure increases. We illustrate the method of generalization by specifically addressing the analysis of three-factor treatment structures.

In Section 16.1, we give a general strategy to follow when analyzing higher-order treatment structures. Section 16.2 discusses the analysis of balanced and unbalanced treatment structures. The discussion of unbalanced experiments includes the case where each treatment combination is observed at least once and the case where some treatment combinations are missing.

16.1 GENERAL STRATEGY

Suppose that treatments T_i, B_j, and C_k are applied simultaneously to the same experimental unit. Let μ_{ijk} represent the expected response to the treatment combination (T_i, B_j, C_k) for $i = 1, 2, \ldots, t$, $j = 1, 2, \ldots, b$, and $k = 1, 2, \ldots, c$. There is no three-factor interaction among these treatment combinations provided that

$$(\mu_{ijk} - \mu_{i'jk} - \mu_{ij'k} + \mu_{i'j'k}) - (\mu_{ijk'} - \mu_{i'jk'} - \mu_{ij'k'} + \mu_{i'j'k'}) = 0$$

for all i, i', j, j', k, and k'. This implies that the $T * B$ interaction at level k of factor C is the same as the $T * B$ interaction at level k' of factor C for all values of k and k'. Similarly, the $T * C$ interaction is the same at all levels of factor B, and the $B * C$ interaction is the same at all levels of factor T. Equivalent expressions of the no-interaction statements are:

1. $$\mu_{ijk} - \bar{\mu}_{ij\cdot} - \bar{\mu}_{i\cdot k} - \bar{\mu}_{\cdot jk} + \bar{\mu}_{i\cdot\cdot} + \bar{\mu}_{\cdot j\cdot} + \bar{\mu}_{\cdot\cdot k} - \bar{\mu}_{\cdot\cdot\cdot} = 0$$

   for all i, j, and k, and
2. there exist parameters

   $$\mu, \tau_1, \tau_2, \ldots, \tau_t, \quad \beta_1, \beta_2, \ldots, \beta_b, \quad \xi_1, \xi_2, \ldots, \xi_c, \quad \gamma_{11}, \gamma_{12}, \ldots, \gamma_{tb}, \quad \eta_{11}, \eta_{12}, \ldots, \eta_{tc}, \quad \text{and} \quad \theta_{11}, \theta_{12}, \ldots, \theta_{bc}$$

   such that

   $$\mu_{ijk} = \mu + \tau_i + \beta_j + \xi_k + \gamma_{ij} + \eta_{ik} + \theta_{jk} \qquad \text{for all } i, j, \text{ and } k,$$

   that is, the μ_{ijk}'s can be described by main effects and two-factor interaction effects.

When analyzing three-way treatment structures, the first and most important step is to determine whether there is a three-factor interaction,

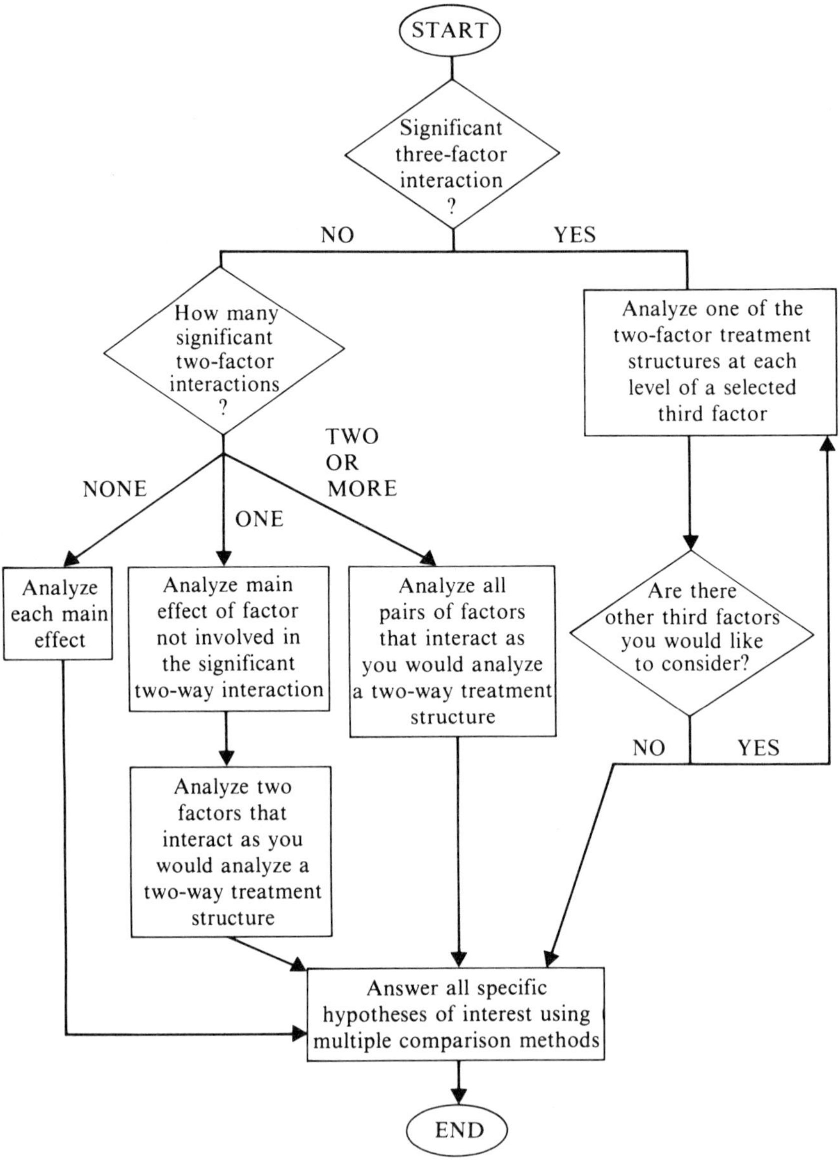

Figure 16.1 Strategy for analyzing three-factor experiments.

even though the experimenter may not be interested in it. If there is no three-factor interaction, then the second step is to determine whether there are any two-factor interactions. If there are also no two-factor interactions, then each of the main effects can be analyzed.

If a three-factor interaction exists, the experimenter should analyze the two-way treatment structures of two treatment factors at each level of a selected third treatment factor, usually the factor of least interest. Obviously, these two-way analyses could be done by letting each treatment be the selected one.

The types of analyses that can be obtained with statistical computing packages are similar to those available for two-way treatment structures.

Figure 16.1 presents a general strategy for analyzing three-way treatment structures. This strategy can also be applied to four-way and higher-order treatment structures.

16.2 BALANCED AND UNBALANCED EXPERIMENTS

If all treatment combinations are observed an equal number of times, the resulting data can be analyzed by using SAS® ANOVA, BMD-P2V, or SPSS ANOVA. A model should first be used that includes all interaction terms.

16.2.1 Balanced Treatment Structures

If all treatment combinations are observed but an unequal number of times, then SAS® GLM (Type III), BMD-P2V, BMD-P4V (BETWEEN = EQUAL option), SPSS ANOVA (option 9), or SPSS MANOVA (SSTYPE = UNIQUE) should be used. If every treatment combination is observed at least once, all main-effect and interaction hypotheses can still be tested, and the questions answered are the same as those in balanced treatment structures.

16.2.2 Unbalanced Treatment Structures

If some treatment combinations are missing, then, as was the case in Chapter 13, no hypothesis involving the missing treatment combinations can be tested. The experimenter should specify Type IV hypotheses of interest, which can be tested by using the matrix procedure described in Chapter 1 or by the contrast options available in most statistical computing packages.

The example in Chapter 17 demonstrates some of the steps that may be required in order to obtain a complete analysis of data with missing treatment combinations.

CONCLUDING REMARKS

This chapter discussed the analysis of three-way and higher-way treatment structure experiments. A flow chart was given that provides a general strategy for analyzing such experiments.

It is important to examine the highest-order interaction effects first. Many experimenters avoid considering higher-order interactions because they are often not quite sure how to deal with these interactions. Such temptations should be avoided. Even though experience has shown us that very high order interactions are seldom significant, they must be dealt with whenever they are. The techniques discussed in Chapter 8 can be generalized to three-way and higher-way treatment structures and may help determine which treatment combinations are causing the interaction.

17

Case Study: Three-Way Treatment Structure with Many Missing Treatment Combinations

CHAPTER OUTLINE

In this chapter we show a complete analysis of a three-way treatment structure when many treatment combinations are missing.

17.1 NUTRITION SCORES EXAMPLE

Some home economists conducted a survey experiment to study how much lower-socioeconomic-level mothers knew about nutrition and to judge the effect of a training program designed to increase their knowledge of nutrition. A test was administered to the mothers both before and after the training program, and the changes in their test scores were measured. These changes are reported in Table 17.1. The mothers tested were classified according to three factors: age, race, and whether they were receiving food stamps.

First, the effects of the 15 treatment combinations that were observed are compared. The resulting estimate of σ^2 is $\hat{\sigma}^2 = 28.56$ with 92 degrees of freedom, and the F-statistic for comparing the 15 treatment

Table 17.1 Change in Scores on Nutrition Test from Pretraining to Posttraining

	GROUP					
	NO FOOD STAMPS RECEIVED			FOOD STAMPS RECEIVED		
Age Classification	*Black*	*Hispanic*	*White*	*Black*	*Hispanic*	*White*
1			5, −2, −10	4, 4		−8, 9
2	−4, −2, 0, 0, 5, −6, 2		7, 2, −13, 2, 3, 3, −4, −5			5, 0, 10, 3 3, 7, 7, 4
3	3, −14, −14, −1, 3, 1	−1, 6	−20, 6, 9, −5, 3, −1, 3, 0, 4, −3, 2, 3, −5, 2, −1, −1, 6, −8, 0, 2	1, 5, 15, 9	0	4, 5, 0, 5, 2, 8, 1, −2, 6, 6, 4, −5, 6, 3, 7, 4, 5, 12, 3, 8, 3, 8, 13, 4, 7, 9, 3, 12, 11, 4, 12
4	0			−3		−6, −5, 5, 8, 5, 6, 7, 6, 2, 7, 5

combination means is $F = 2.67$ with 14 and 92 degrees of freedom. Its corresponding p-value is $p = .0026$. This indicates there are significant differences in the means of the 15 groups studied.

17.2 AN SAS® GLM ANALYSIS

The data in Table 17.1 were also analyzed with SAS® GLM by using the following control statements:

```
PROC GLM; CLASSES GROUP AGE RACE;
MODEL GAIN = GROUP|AGE|RACE/SOLUTION
E4 SS1 SS4;
LSMEANS GROUP|AGE|RACE/PDIFF STDERR;
```

The resulting Type IV analysis of variance table is summarized in Table 17.2.

Table 17.2 shows that there are zero degrees of freedom for the three-factor interaction hypothesis, which indicates that there are no contrasts in the observations that can be used for estimating a three-factor interaction. This does not imply that there is not any three-factor interaction, only that there are no testable hypotheses in the three-factor interaction effects.

The Type IV F-values seem to suggest that there are no significant differences between any of the factors used to classify the data. This seems a bit strange, particularly since the overall means comparison test gives $F = 2.67$ and is highly significant. Since this F-statistic is significant, some of the groups should have significantly different means. Why, then, are none of the Type IV F-values significant? The answer lies in the

Table 17.2 Type IV Analysis of Variance Table for Data in Table 17.1

Source of Variation	*df*	*SS*	*MS*	*F*	*p*
Group	1*	75.7378	75.7378	2.65	.1068
Age	3*	41.5258	13.8419	.48	.6977
Group * Age	3	91.5762	30.5254	1.07	.3670
Race	2*	11.6770	5.8385	.20	.8155
Group * Race	2	113.7034	56.8517	1.99	.1424
Age * Race	3	87.3014	29.1005	1.02	.3891
Group * Age * Race	0	0.0	0.0		
Error	92	2,627.4724	28.5595		

*Other testable Type IV hypotheses may yield different F-statistics.

Type IV estimable functions that SAS® GLM used to generate its Type IV hypotheses.

Next, we examine the Type IV hypotheses for these data generated by SAS® in order to determine what the Type IV F-values in the analysis of variance table are actually testing. For this purpose, let μ_{ijk} represent the expected response from the treatment combination (group i, age j, and race k) where $i = 0, 1$ (0 indicating "does not receive food stamps" and 1 indicating "receives food stamps"), $j = 1, 2, 3, 4$ (indicating the age classification), and $k = B, H, W$ (B indicating "black," H "Hispanic," and W "white"). The hypotheses tested by the Type IV analysis are given in Table 17.3.

It seems likely that few, if any, of the hypotheses tested by the Type IV analysis of SAS® GLM will be of particular interest to the experimenter. We do not even consider the Type I through Type III hypotheses, since they usually make little sense in cases where there are missing cells.

Table 17.3 Type IV Hypotheses Tested by SAS® GLM

Source of Variation	*Hypothesis*
Group	$\mu_{01W} + \mu_{02W} + \mu_{03B} + \mu_{03H} + \mu_{03W} + \mu_{04B}$ $= \mu_{11W} + \mu_{12W} + \mu_{13B} + \mu_{13H} + \mu_{13W} + \mu_{14B}$
Age	$\mu_{11B} + \mu_{11W} = \mu_{14B} + \mu_{14W}$, $\mu_{02B} + \mu_{12W} = \mu_{04B} + \mu_{14W}$, and $\mu_{03B} + \mu_{13B} + \mu_{13W} = \mu_{04B} + \mu_{14B} + \mu_{14W}$
*Group * Age*	$(\mu_{01W} - \mu_{03W} - \mu_{11W} + \mu_{13W}) + (\mu_{03B} - \mu_{04B} - \mu_{13B} + \mu_{14B}) = 0$, $(\mu_{02W} - \mu_{03W} - \mu_{12W} + \mu_{13W}) + (\mu_{03B} - \mu_{04B} - \mu_{13B} + \mu_{14B}) = 0$, and $\mu_{03B} - \mu_{04B} - \mu_{13B} + \mu_{14B} = 0$
Race	$\mu_{02B} + \mu_{03B} + \mu_{11B} + \mu_{13B} + \mu_{14B} = \mu_{02W} + \mu_{03W} + \mu_{11W} + \mu_{13W} + \mu_{14W}$ and $\mu_{03H} + \mu_{13H} = \mu_{03W} + \mu_{13W}$
*Group * Race*	$\mu_{03B} - \mu_{03W} - \mu_{13B} - \mu_{13W} = 0$ and $\mu_{03H} - \mu_{03W} - \mu_{13H} + \mu_{13W} = 0$
*Age * Race*	$\mu_{11B} - \mu_{11W} - \mu_{14B} + \mu_{14W} = 0$, $\mu_{02B} - \mu_{02W} - \mu_{03B} + \mu_{03W} = 0$, and $\mu_{13B} - \mu_{13W} - \mu_{14B} + \mu_{14W} = 0$
*Group * Age * Race*	None

17.3 A COMPLETE ANALYSIS

Since we cannot test for three-factor interaction and hence do not know whether there is any, we examine the two-way analyses at each level of a third treatment factor. Let us suppose that the experimenter is most interested in the effects of, or differences between, the Race*Group combinations. We thus examine these two-way treatment combinations at each level of the Age factor, as shown in Table 17.4.

For Age = 1 in Table 17.4, we observe the following: (1) It is not possible to test for Group*Race interaction in this age group, since no contrast exists that measures a two-factor interaction; (2) the only Type IV hypothesis comparing groups that can be tested is $\mu_{01W} = \mu_{11W}$; (3) the only Type IV hypothesis concerning races that can be tested is $\mu_{11B} = \mu_{11W}$; and (4) the hypothesis $\mu_{01W} = \mu_{11B}$ can also be tested, but it is probably of secondary interest since it involves different levels in both factors.

A t-statistic for testing the hypothesis in (2) is

$$t = \frac{\hat{\mu}_{01W} - \hat{\mu}_{11W}}{\widehat{\text{s.e.}}(\hat{\mu}_{01W} - \hat{\mu}_{11W})} = \frac{-\frac{7}{3} - \frac{1}{2}}{\sqrt{28.56(\frac{1}{3} + \frac{1}{2})}} = -.581$$

which is significant at the $\hat{\alpha} = .563$ level. The t-tests for the other two hypotheses can be obtained in a similar fashion.

As the hypotheses in (2), (3), and (4) involve only pairs of means, the significance levels of tests of these hypotheses can also be obtained from the three-way least squares means table produced by SAS® GLM. This table is reproduced in Table 17.5, where the three appropriate p-values

Table 17.4 Group*Race Table for Each Age Level for Data in Table 17.1

		Race		
	Group	*B*	*H*	*W*
AGE = 1	0			X
	1	X		X
AGE = 2	0	X		X
	1			X
AGE = 3	0	X	X	X
	1	X	X	X
AGE = 4	0	X		
	1	X		X

Note: "X" indicates that the cell was observed at least once in the experiment.

Table 17.5 Three-Way Least Squares Means, Standard Errors, and Pairwise Comparisons for Data in Table 17.1

			POPULATION CELL MEAN														
Population Cell Mean	*Least Squares Means*	*Standard Errors*	μ_{01W}	μ_{02B}	μ_{02W}	μ_{03B}	μ_{03H}	μ_{03W}	μ_{04B}	μ_{11B}	μ_{11W}	μ_{12W}	μ_{13B}	μ_{13H}	μ_{13W}	μ_{14B}	μ_{14W}
μ_{01W}	−2.333	3.085		.662	.638	.725	.324	.521	.706	.198	.563	.049	.108	.706	.108	.914	.090
μ_{02B}	−.714	2.020			.974	.323	.455	.827	.901	.274	.778	.046	.016	.901	.007	.690	.096
μ_{02W}	−.625	1.889				.295	.461	.850	.912	.277	.791	.042	.015	.912	.005	.676	.090
μ_{03B}	−3.667	2.182					.161	.167	.527	.082	.342	.004	.002	.527	.000	.908	.008
μ_{03H}	2.500	3.779						.497	.703	.780	.709	.575	.283	.703	.456	.403	.783
μ_{03W}	−.200	1.195							.971	.292	.860	.026	.010	.971	.000	.610	.059
μ_{04B}	0.000	5.344								.543	.939	.392	.212	1.000	.321	.692	.516
μ_{11B}	4.000	3.779									.514	.836	.451	.543	.717	.288	.930
μ_{11W}	.500	3.779										.303	.133	.939	.210	.594	.477
μ_{12W}	4.875	1.889											.425	.392	.798	.168	.619
μ_{13B}	7.500	2.672												.213	.466	.082	.219
μ_{13H}	0.000	5.344													.321	.692	.516
μ_{13W}	5.419	.960														.124	.344
μ_{14B}	−3.000	5.344															.238
μ_{14W}	3.636	1.611															

Table 17.6 Testable Hypotheses for Age = 2 for Data in Table 17.1

Hypothesis	*p-Value*	*Importance*
$\mu_{02B} = \mu_{02W}$	.974	Primary
$\mu_{02W} = \mu_{12W}$	.042	Primary
$\mu_{02B} = \mu_{12W}$	.046	Secondary

Note: The *p*-values can be obtained from the three-way least squares means table or from pairwise *t*-tests.

have been circled. The *p*-value for the hypothesis in (2) is .563; in (3), .514; and in (4), .198. Thus, none of these three hypotheses can be rejected.

Tables 17.6 and 17.7 give the testable hypotheses, the *p*-values of their respective test statistics, and a (subjective) importance rating for the age groups, Age = 2 and Age = 4, respectively.

Finally, we examine the Age = 3 group. In this group all Group * Race combinations are observed; hence, it is possible to test for Group * Race interaction. If this interaction is not significant, we can examine the main-effect means in this group. These three hypotheses are specified by

$$\text{H}_{01}: \quad \mu_{03B} - \mu_{03W} - \mu_{13B} + \mu_{13W} = 0 \quad \text{and}$$

$$\mu_{03H} - \tfrac{1}{2}(\mu_{03B} + \mu_{03W}) - \mu_{13H} + \tfrac{1}{2}(\mu_{13B} + \mu_{13W}) = 0,$$

$$\text{H}_{02}: \quad \bar{\mu}_{03\cdot} = \bar{\mu}_{13\cdot}, \quad \text{and}$$

$$\text{H}_{03}: \quad \bar{\mu}_{\cdot 3B} = \bar{\mu}_{\cdot 3H} = \bar{\mu}_{\cdot 3W},$$

respectively.

Table 17.7 Testable Hypotheses for Age = 4 for Data in Table 17.1

Hypothesis	*p-Value*	*Importance*
$\mu_{04B} = \mu_{14B}$	.692	Primary
$\mu_{14B} = \mu_{14W}$	.238	Primary
$\mu_{04B} = \mu_{14W}$	.516	Secondary

Note: The *p*-values can be obtained from the three-way least squares means table or from pairwise *t*-tests.

The test statistic for H_{01} was given by the SAS® GLM analysis, while the tests of the other two were not. Hypothesis H_{01} is equivalent to the one tested by the Group * Race Type IV *F*-value, which can be seen by examining the estimable functions for Group * Race in Table 17.3. The tests for these three hypotheses can be obtained by using the CONTRAST option in GLM, as illustrated below, or by hand. In this case, it's probably easier to do the testing by hand. When using SAS®, it's also easier to use a MEANS model. With this model, only the coefficients corresponding to the Group * Age * Race effect need to be entered.

The SAS® GLM CONTRAST statements needed for an effects model are shown below.

For H_{01}: CONTRAST 'H01' GROUP 0 0 AGE 0 0 0 0
GROUP * RACE 0 0 0 0 0 0 0 0 RACE 0 0
0 GROUP * RACE 1 0 −1 −1 0 1
AGE * RACE 0 0 0 0 0 0 0 0 0
GROUP * AGE * RACE 0 0 0 1 0 −1 0 0 0
0 −1 0 1 0 0, GROUP 0 0 AGE 0 0 0 0
GROUP * AGE 0 0 0 0 0 0 0 0 RACE 0 0
0 GROUP * RACE −.5 1 −.5 .5 −1 .5
AGE * RACE 0 0 0 0 0 0 0 0 0
GROUP * AGE * RACE 0 0 0 −.5 1 −.5 0 0
0 0 .5 −1 .5 0 0;

For H_{02}: CONTRAST 'H02' GROUP 1 −1 AGE 0 0 0
0 GROUP * AGE 0 0 1 0 0 0 −1 0 RACE
0 0 0 GROUP * RACE .333333
.333333 .333333 −.333333 −.333333 −.333333
AGE * RACE 0 0 0 0 0 0 0 0 0
GROUP * AGE * RACE 0 0 0 .333333 .333333
.333333 0 0 0 0 −.33333 −.33333 −.33333 0 0;

For H_{03}: CONTRAST 'H03' GROUP 0 0 AGE 0 0 0 0
GROUP * AGE 0 0 0 0 0 0 0 0 RACE 1
−1 0 GROUP * RACE .5 −.5 0 .5 −.5 0
AGE * RACE 0 0 0 0 1 −1 0 0 0
GROUP * AGE * RACE 0 0 0 .5 −.5 0 0 0 0
0 .5 −.5 0 0 0, GROUP 0 0 AGE 0 0 0 0

GROUP*AGE 0 0 0 0 0 0 0 0 RACE 1 0

−1 GROUP*RACE .5 0 −.5 .5 0 −.5

AGE*RACE 0 0 0 0 1 0 −1 0 0

GROUP*AGE*RACE 0 0 0 .5 0 −.5 0 0 0 0

.5 0 −.5 0 0;

The SAS® results obtained by using these contrast statements are given in Table 17.8.

Next, we use the results in Section 1.4 to obtain the values of the test statistics for the above three hypotheses. Using the notation of Section 1.4, we take

$$\boldsymbol{\mu}' = [\mu_{03B} \quad \mu_{03H} \quad \mu_{03W} \quad \mu_{13B} \quad \mu_{13H} \quad \mu_{13W}],$$

$$\mathbf{D} = \text{Diag}[\tfrac{1}{6} \quad \tfrac{1}{2} \quad \tfrac{1}{20} \quad \tfrac{1}{4} \quad 1 \quad \tfrac{1}{31}], \text{ and}$$

$$\hat{\boldsymbol{\mu}}' = [-3.667 \quad 2.500 \quad -.200 \quad 7.500 \quad 0 \quad 5.419].$$

For H_{01}, we take

$$\mathbf{C} = \begin{bmatrix} 1 & 0 & -1 & -1 & 0 & 1 \\ -\frac{1}{2} & 1 & -\frac{1}{2} & \frac{1}{2} & -1 & \frac{1}{2} \end{bmatrix}$$

which provides

$$SS_{H_{01}} = (\mathbf{C}\hat{\boldsymbol{\mu}})'(\mathbf{CDC}')^{-1}(\mathbf{C}\hat{\boldsymbol{\mu}})$$

$$= \begin{bmatrix} -5.548 \\ 10.893 \end{bmatrix}' \begin{bmatrix} .4989 & -.1672 \\ -.1672 & 1.6247 \end{bmatrix}^{-1} \begin{bmatrix} -5.548 \\ 10.893 \end{bmatrix} = 113.7.$$

For H_{02}, we take

$$\mathbf{C} = [1 \quad 1 \quad 1 \quad -1 \quad -1 \quad -1],$$

which provides

$$SS_{H_{02}} = (-14.286)(1.999)^{-1}(-14.286) = 102.1.$$

Table 17.8 Results from SAS® CONTRAST Statements

Contrast	*df*	*SS*	*F*	*p*
Group * Race	2	113.70344186	1.99	0.1424
Group, Age = 3	1	102.10009991	3.57	0.0618
Race, Age = 3	2	7.80545113	0.14	0.8724

For H_{03}, we take

$$\mathbf{C} = \begin{bmatrix} 1 & -1 & 0 & 1 & -1 & 0 \\ 1 & 0 & -1 & 1 & 0 & -1 \end{bmatrix},$$

and calculate

$$SS_{H_{03}} = (\mathbf{C}\hat{\boldsymbol{\mu}})'(\mathbf{CDC}')^{-1}(\mathbf{C}\hat{\boldsymbol{\mu}}).$$

We get

$$SS_{H_{03}} = \begin{bmatrix} 1.333 \\ -1.386 \end{bmatrix}' \begin{bmatrix} 1.917 & .417 \\ .417 & .4989 \end{bmatrix}^{-1} \begin{bmatrix} 1.333 \\ -1.386 \end{bmatrix} = 7.805$$

with 2 degrees of freedom. The resulting F-statistics and their corresponding significance probabilities are $F = 1.99$ and $\hat{\alpha} = .142$ for H_{01}, $F = 3.57$ and $\hat{\alpha} = .062$ for H_{02}, and $F = .14$ and $\hat{\alpha} = .872$ for H_{03}.

If the experimenter is also interested in the effects of the Race * Age combinations, these combinations can be analyzed at each value of the Group factor. A similar situation exists if the experimenter wanted to examine the effects of the Group * Age combinations for each race. Both of these analyses can be done either by hand, by using CONTRAST options, or by using three-way least squares means when possible, as illustrated above for the different levels of the Age factor.

The analysis of higher-order cross-classified treatment structures can be carried out in ways similar to those illustrated in this chapter.

CONCLUDING REMARKS

This chapter presented the analysis of a three-way treatment structure having a large number of missing treatment combinations. A SAS® GLM analysis was obtained and interpreted. Questions not answered by the SAS® analysis were also raised, and techniques for answering these questions were illustrated.

18

Random Models and Variance Components

CHAPTER OUTLINE

Models with more than one random component are applied to several situations, including random and mixed models where some or all of the factors in the treatment structure are random and split-plot and repeated measures designs. The parameters of interest for such models include the variances associated with the distributions of the random components (usually called *variance components*). It is important to be able to identify the random components of a model and be able to utilize them in the analysis of the model. When carrying out an analysis of variance for a given model, the expected values of the mean squares (which are functions of the variance components) are needed in order to construct proper test statistics and determine standard errors for comparisons between treatment means. It is also important to be able to obtain estimates of the variance components, test hypotheses, and construct confidence intervals about functions of the variance components.

The discussion of random models and methods of analyzing them is divided into four chapters. This chapter defines the random effects model and describes a general procedure for computing expectations of sums of squares. The procedure can easily be used by a computer program to evaluate the expectations of sums of squares. The problem of estimation is discussed in Chapter 19, methods for testing hypotheses and constructing confidence intervals are presented in Chapter 20, and a detailed analysis of an example is presented in Chapter 21.

18.1 INTRODUCTION

The philosophy behind the use of random models is quite different from that behind the use of the fixed-effects models (discussed in the previous chapters) in both the sampling scheme and the parameters of interest. Before these differences are discussed, *random effect* and *fixed effect* should be defined.

Definition 18.1 A factor is *random* if its levels consist of a random sample of levels from a population of possible levels.

Definition 18.2 A factor is *fixed* if its levels are selected by a nonrandom process or if its levels consist of the entire population of possible levels.

Thus, in order to determine whether a factor is a fixed effect or a random effect, one needs to know how the experimenter selected the levels of that factor. If all possible levels of the factor or a set of selected levels of the factor are included in the experiment, the factor is considered as a fixed effect. If some form of randomization is used to select the levels included in the experiment, then the factor is random.

For example, suppose a plant breeder wants to study a characteristic (say, yield) of wheat varieties. There is a large number of possible varieties (a population of varieties), but if he wants to study a certain set of varieties, then he would select just those varieties for his experiment. In this case, the factor Variety is called a fixed effect, since the levels of varieties are chosen or fixed. However, if the plant breeder is interested in how a characteristic is distributed among the varieties in the population, then he is not interested in which set of varieties is included in the experiment. In this case, the plant breeder can randomly select the varieties to be included in the experiment from the population of varieties. Therefore the factor Variety in this experiment is a random effect.

When constructing a model to describe a given experimental situation, it must be stated whether a factor is random or fixed. The models considered in the previous chapters were constructed under the assumption that all factors in the treatment structure were fixed. However, the idea of a random effect was alluded to when blocking was introduced in Chapter 4, where it was assumed that the factor Blocks was a random effect.

Three basic types of models can be constructed, depending on the assumptions about the factors in the treatment structure. These types of models are defined below.

Definition 18.3 A model is called a *fixed* or *fixed effects model* if all of the factors in the treatment structure are fixed effects.

Definition 18.4 A model is called a *random* or *random effects model* if all of the factors in the treatment structure are random effects.

Definition 18.5 A model is called a *mixed* or *mixed effects model* if some of the factors in the treatment structure are fixed effects and some are random effects.

The models discussed in this chapter are all random effects models. The discussion of mixed models is presented in Chapters 22 and 23. To help motivate the application and analysis of random models, the example below is presented.

EXAMPLE 18.1

A consumer group studied the variation in coffee prices in U.S. cities with populations of at least 20,000. The three factors the group investigated were states, cities within states, and stores within cities within states. The treatment structure is a three-way, two-level, nested system involving State × City × Store, where City is nested within State and Store is nested within City. The sampling procedure used was to select r states at random ($r < 50$) from the population of all possible states; then from the ith randomly selected state, select t_i cities at random from the

c_i cities in state i ($t_i < c_i$) with populations of at least 20,000; and finally, randomly select n_{ij} stores ($n_{ij} < s_{ij}$) from the s_{ij} stores in the jth city selected from the ith state. For each randomly selected store, the price of a particular grade of coffee was determined.

A model describing the collected data is

$$y_{ijk} = \mu + s_i + c_{ij} + a_{ijk} \qquad i = 1, 2, \ldots, r \quad j = 1, 2, \ldots, t_i$$
$$k = 1, 2, \ldots, n_{ij}$$

where μ denotes the average price of coffee in the United States, s_i denotes the effect of the ith randomly selected state, c_{ij} denotes the effect of the jth randomly selected city from state i, and a_{ijk} denotes the effect of the kth randomly selected store from the jth city in the ith state. It is assumed that (1) s_i is distributed i.i.d. $N(0, \sigma_s^2)$, (2) c_{ij} is distributed i.i.d. $N(0, \sigma_c^2)$, and (3) a_{ijk} is distributed i.i.d. $N(0, \sigma_a^2)$. It is also assumed that the random factors are all distributed independently of each other. Thus, the parameters of this random model are μ, σ_s^2, σ_c^2, and σ_a^2. The terms s_i, c_{ij}, and a_{ijk} are random variables and are not parameters of the model.

18.2 GENERAL RANDOM MODEL IN MATRIX FORM

In order to describe methods used to evaluate the expectations of sums of squares, it is necessary to have some general notation. This section presents a matrix representation of a random model, which is used in later sections. To help visualize the general model and its expression in terms of matrices, a random model for a one-way treatment structure in a completely randomized design structure is examined.

EXAMPLE 18.2: One-Way Random Model

A model describing a one-way treatment structure where treatment is a random effect is

$$y_{ij} = \mu + a_i + b_{ij} \qquad i = 1, 2, \ldots, t, \quad j = 1, 2, \ldots, n_i \tag{18.2.1}$$

where μ is the population mean of the response, a_i denotes the effect of treatment i and is assumed to be distributed i.i.d. $N(0, \sigma_a^2)$, and b_{ij} denotes the random error of the jth observation of the ith treatment, which is assumed to be distributed i.i.d. $N(0, \sigma_b^2)$. It is also assumed that a_i and b_{ij} are independent random variables. These assumptions allow the variances and covariances of the observations to be evaluated. The variance of an observation is

$$\begin{aligned}\text{Var}(y_{ij}) &= \text{Var}(\mu + a_i + b_{ij}) = \text{Var}(a_i) + \text{Var}(b_{ij}) \\ &= \sigma_a^2 + \sigma_b^2.\end{aligned}$$

Thus there are two components of the variance of y_{ij}, the variance of the population of treatments and the variance of the experimental units (hence the name *variance components* or *components of variance*).

The covariance of two observations obtained from the same treatment is

$$\begin{aligned}\operatorname{Cov}(y_{ij}, y'_{ij}) &= \operatorname{Cov}(\mu + a_i + b_{ij}, \mu + a_i + b'_{ij}) \\ &= \operatorname{Cov}(a_i, a_i) = \operatorname{Var}(a_i) = \sigma_a^2.\end{aligned}$$

The covariance between two observations obtained from different treatments is zero. Hence, observations obtained from the same treatment are correlated, whereas observations obtained from different treatments are uncorrelated.

The above model can be described in matrix notation. The matrix model corresponding to model (18.2.1) is

$$\mathbf{y} = \mathbf{j}\mu + \mathbf{X}_1\mathbf{a} + \mathbf{b} \tag{18.2.2}$$

or

$$\begin{bmatrix} y_{11} \\ y_{12} \\ \vdots \\ y_{1n_1} \\ y_{21} \\ y_{22} \\ \vdots \\ y_{2n_2} \\ \vdots \\ y_{t1} \\ y_{t2} \\ \vdots \\ y_{tn_t} \end{bmatrix} = \begin{bmatrix} 1 \\ 1 \\ \vdots \\ 1 \\ 1 \\ 1 \\ \vdots \\ 1 \\ \vdots \\ 1 \\ 1 \\ \vdots \\ 1 \end{bmatrix} \mu + \begin{bmatrix} 1 & 0 & \cdots & 0 \\ 1 & 0 & \cdots & 0 \\ \vdots & \vdots & & \vdots \\ 1 & 0 & \cdots & 0 \\ 0 & 1 & \cdots & 0 \\ 0 & 1 & \cdots & 0 \\ \vdots & \vdots & & \vdots \\ 0 & 1 & \cdots & 0 \\ \vdots & \vdots & & \vdots \\ 0 & 0 & \cdots & 1 \\ 0 & 0 & \cdots & 1 \\ \vdots & \vdots & & \vdots \\ 0 & 0 & \cdots & 1 \end{bmatrix} \begin{bmatrix} a_1 \\ a_2 \\ \vdots \\ a_t \end{bmatrix} + \begin{bmatrix} b_{11} \\ b_{12} \\ \vdots \\ b_{1n_1} \\ b_{21} \\ b_{22} \\ \vdots \\ b_{2n_2} \\ \vdots \\ b_{t1} \\ b_{t2} \\ \vdots \\ b_{tn_t} \end{bmatrix},$$

where $\mathbf{j}$ is an $N \times 1$ vector of ones ($N = \sum_{i=1}^{t} n_i$), $\mathbf{X}_1$ is an $N \times t$ design matrix, $\mathbf{a}$ is the $t \times 1$ vector random variable assumed to be distributed as the multivariate normal distribution $N_t(\mathbf{0}, \sigma_a^2 \mathbf{I}_t)$, and $\mathbf{b}$ is the $N \times 1$ vector random variable assumed to be distributed $N_N(\mathbf{0}, \sigma_b^2 \mathbf{I}_N)$. The covariance matrix of $\mathbf{y}$ is

$$\begin{aligned}\boldsymbol{\Sigma} &= \operatorname{Var}(\mathbf{y}) = \operatorname{Var}(\mathbf{j}\mu + \mathbf{X}_1\mathbf{a} + \mathbf{b}) \\ &= \mathbf{X}_1 \operatorname{Var}(\mathbf{a}) \mathbf{X}'_1 + \operatorname{Var}(\mathbf{b}) \\ &= \sigma_a^2 \mathbf{X}_1 \mathbf{X}'_1 + \sigma_b^2 \mathbf{I}_N.\end{aligned}$$

The variances of the y_{ij}'s and the covariances between pairs of y_{ij}'s can be obtained from the diagonal and off-diagonal elements of $\boldsymbol{\Sigma}$, respectively.

The general random model will have k random components representing the main effects and interactions between the factors of the treatment structure, the overall mean μ, and the experimental unit error vector. The general random model, written in matrix notation, is

$$\mathbf{y} = \mathbf{j}_n\mu + \mathbf{X}_1\mathbf{a}_1 + \mathbf{X}_2\mathbf{a}_2 + \cdots + \mathbf{X}_k\mathbf{a}_k + \boldsymbol{\varepsilon} \tag{18.2.3}$$

where $\mathbf{a}_i$, $i = 1, 2, \ldots, k$, denotes the effects of the treatment structure and is assumed to be distributed as

$$\mathbf{a}_1 \sim N_{t_1}\left(\mathbf{0}, \sigma_1^2\mathbf{I}_{t_1}\right), \ldots, \mathbf{a}_k \sim N_{t_k}\left(\mathbf{0}, \sigma_k^2\mathbf{I}_{t_k}\right), \boldsymbol{\varepsilon} \sim N_N\left(\mathbf{0}, \sigma_\varepsilon^2\mathbf{I}_N\right),$$

and the $\mathbf{X}_i$'s are the partitions of the design matrix associated with the respective components of the model.

The covariance matrix of $\mathbf{y}$ is

$$\boldsymbol{\Sigma} = \sigma_1^2\mathbf{X}_1\mathbf{X}_1' + \cdots + \sigma_k^2\mathbf{X}_k\mathbf{X}_k' + \sigma_\varepsilon^2\mathbf{I}_n.$$

This matrix form of the general random model is used to describe the method for evaluating the expectations of sums of squares involving the observations.

18.3 COMPUTING EXPECTED MEAN SQUARES

The expected values of the sums of squares from an analysis of variance of a random model involve the variance components. For a given model, two methods can be used to evaluate the expected mean squares (remember that a mean square is the sum of squares divided by its degrees of freedom). The first method is to evaluate algebraically the expected values by using the model assumptions; the second method is to evaluate the expected values by means of a computer. The algebraic method is presented by applying it to the sum of squares obtained from the analysis of a one-way random model. The second method is discussed in general terms and demonstrated by examples.

18.3.1 Algebraic Method

There are two variance components in the one-way random model in equation (18.2.1). Two sums of squares are used in the analysis of variance—the sum of squares within, designated Q_1, and the sum of squares between, designated Q_2, where

$$Q_1 = \sum_{i=1}^{t}\sum_{j=1}^{n_i}(y_{ij} - \bar{y}_{i\cdot})^2 = \sum_{i=1}^{t}\sum_{j=1}^{n_i} y_{ij}^2 - \sum_{i=1}^{t} n_i\bar{y}_{i\cdot}^2$$

and

$$Q_2 = \sum_{i=1}^{t} n_i(\bar{y}_{i\cdot} - \bar{y}_{\cdot\cdot})^2 = \sum_{i=1}^{t} n_i\bar{y}_{i\cdot}^2 - N\bar{y}_{\cdot\cdot}^2 \qquad \text{where } N = \sum_{i=1}^{t} n_i.$$

In terms of the random variables of model (18.2.1), the quantities in Q_1 and Q_2 are expressed as

$$\left.\begin{aligned} y_{ij} &= \mu + a_i + b_{ij}, \\ \bar{y}_{i\cdot} &= \mu + a_i + \bar{b}_{i\cdot}, \qquad \text{and} \\ \bar{y}_{\cdot\cdot} &= \mu + \tilde{a}_{\cdot} + \bar{b}_{\cdot\cdot} \qquad \text{where} \quad \tilde{a} = \frac{\sum_{i=1}^{t} n_i a_i}{N}. \end{aligned}\right\} \tag{18.3.1}$$

Substituting the terms in (18.3.1) into Q_1, we get

$$Q_1 = \sum_{i=1}^{t} \sum_{j=1}^{n_1} \left(\mu + a_i + b_{ij} - \mu - a_i - \bar{b}_{1\cdot}\right)^2 = \sum_{i=1}^{t} \sum_{j=1}^{n_i} \left(b_{ij} - \bar{b}_{i\cdot}\right)^2.$$

The expectation of Q_1 can be evaluated as

$$\begin{aligned} E(Q_1) &= \sum_{i=1}^{t} E \sum_{j=1}^{n_i} \left(b_{ij} - \bar{b}_{i\cdot}\right)^2 \\ &= \sum_{i=1}^{t} \sum_{j=1}^{n_i} \left[E\left(b_{ij}^2\right) + E\left(\bar{b}_{i\cdot}^2\right) - 2E\left(b_{ij}\bar{b}_{i\cdot}\right)\right] \qquad \text{(by squaring)} \\ &= \sum_{i=1}^{t} \sum_{j=1}^{n_i} \left(\sigma_b^2 + \frac{\sigma_b^2}{n_i} - 2\frac{\sigma_b^2}{n_i}\right) \\ &\quad \left(\text{using } E\left(b_{ij}^2\right) = \sigma_b^2,\ E\left(\bar{b}_{i\cdot}^2\right) = \frac{\sigma_b^2}{n_i}\right) \\ &= \sum_{i=1}^{t} \sum_{j=1}^{n_i} \frac{(n_i - 1)\sigma_b^2}{n_i} = \sigma_b^2 \sum_{i=1}^{t} (n_i - 1) = (N - t)\sigma_b^2. \end{aligned}$$

Substituting the expressions in (18.3.1) into the expression for Q_2 provides

$$\begin{aligned} Q_2 &= \sum_{i=1}^{t} n_i \left(\mu + a_i + \bar{b}_{1\cdot} - \mu - \tilde{a}_{\cdot} - \bar{b}_{\cdot\cdot}\right)^2 \\ &= \sum_{i=1}^{t} n_i \left[(a_i - \tilde{a}_{\cdot}) + \left(\bar{b}_{i\cdot} - \bar{b}_{\cdot\cdot}\right)\right]^2. \end{aligned}$$

The expectation of Q_2 is then evaluated as

$$\begin{aligned} E(Q_2) &= \sum_{i=1}^{t} n_i \left[E(a_i - \tilde{a}_{\cdot})^2 + E\left(\bar{b}_{i\cdot} - \bar{b}_{\cdot\cdot}\right)^2\right] \\ &= \sum_{i=1}^{t} n_i \left[E(a_i)^2 + E(\tilde{a}_{\cdot})^2 - 2E(a_i \tilde{a}_{\cdot}) + E\left(\bar{b}_{i\cdot}\right)^2 \right. \\ &\qquad \left. + E\left(\bar{b}_{\cdot\cdot}\right)^2 - 2E\left(\bar{b}_{i\cdot}\bar{b}_{\cdot\cdot}\right)\right]. \end{aligned}$$

This expectation is evaluated in two parts. The first part evaluates the expectation involving the b's. The distributions associated with the means $\bar{b}_{i.}$ and $\bar{b}_{..}$ are

$$\bar{b}_{i.} \sim N\left(0, \frac{\sigma_b^2}{n_i}\right) \quad \text{and} \quad \bar{b}_{..} \sim N\left(0, \frac{\sigma_b^2}{N}\right).$$

Then the expectation of the part of Q_2 involving the b's is

$$\sum_{i=1}^{t} n_i\left[E(\bar{b}_{i.})^2 + E(\bar{b}_{..})^2 - 2E(\bar{b}_{i.}\bar{b}_{..})\right]$$

$$= \sum_{i=1}^{t} n_i\left\{\sigma_b^2\left(\frac{1}{n_i}\right) + \frac{\sigma_b^2}{N} - 2E\left[\bar{b}_{i.}\frac{\sum_{i'=1}^{t} n_{i'}b_{i'.}}{N}\right]\right\}$$

$$= \sum_{i=1}^{t} n_i\left[\sigma_b^2\left(\frac{1}{n_i}\right) + \frac{\sigma_b^2}{N} - 2\frac{n_i}{N}E(\bar{b}_{i.})^2\right]$$

$$= \sum_{i=1}^{t} n_i\left[\sigma_b^2\left(\frac{1}{n_i}\right) + \sigma_b^2\left(\frac{1}{N}\right) - 2\left(\frac{n_i}{N}\right)\left(\frac{\sigma_b^2}{n_i}\right)\right]$$

$$= \sum_{i=1}^{t} n_i\left(\frac{1}{n_i} - \frac{1}{N}\right)\sigma_b^2$$

$$= \sum_{i=1}^{t}\left(1 - \frac{n_i}{N}\right)\sigma_b^2 = (t-1)\sigma_b^2.$$

To evaluate the second part of the expectation of Q_2, involving the a's, use the fact that the weighted mean $\tilde{a}_.$ satisfies

$$\tilde{a}_. = \sum_{i=1}^{t} \frac{n_i a_i}{N}.$$

Since the a_i's are independent,

$$\operatorname{Var}(\tilde{a}_.) = E(\tilde{a}_.)^2 = \sum_{i=1}^{t}\left(\frac{n_i^2}{N^2}\right)\sigma_a^2 = \frac{1}{N^2}\left(\sum_{i=1}^{t} n_i^2\right)\sigma_a^2.$$

Also,

$$\operatorname{Cov}(\tilde{a}_., a_i) = E(\tilde{a}_. a_i) = \left(\frac{n_i}{N}\right)E(a_i)^2 = \left(\frac{n_i}{N}\right)\sigma_a^2,$$

since $E(a_{i'}a_i) = 0$ for $i \neq i'$.

Thus, the expectation of the part of Q_2 involving the a's is

$$\sum_{i=1}^{t} n_i\left[E(a_i)^2 + E(\tilde{a}_.)^2 - 2E(\tilde{a}_. a_i)\right]$$

$$= \sum_{i=1}^{t} n_i\sigma_a^2 + \sum_{i=1}^{t} \frac{n_i\sigma_a^2}{N^2}\left(\sum_{i=1}^{t} n_i^2\right) - 2\sum_{i=1}^{t}\left(\frac{n_i^2}{N}\right)\sigma_a^2$$

$$= \left(\sum_{i=1}^{t} n_i - \frac{\sum_{i=1}^{t} n_i^2}{N}\right)\sigma_a^2 = \left(N - \frac{\sum_{i=1}^{t} n_i^2}{N}\right)\sigma_a^2.$$

Finally, putting the two parts together, we get for the expectation of Q_2

$$E(Q_2) = (t-1)\sigma_b^2 + \left(N - \frac{\sum n_i^2}{N}\right)\sigma_a^2.$$

The above discussion evaluates the expectations of the two sums of squares. The two mean squares are $Q_1/(N-t)$ and $Q_2/(t-1)$, and their respective expectations are

$$E\left(\frac{Q_1}{N-t}\right) = \sigma_b^2$$

and

$$E\left(\frac{Q_2}{t-1}\right) = \sigma_b^2 + \left(N - \frac{\sum n_i^2}{N}\right)\frac{\sigma_a^2}{t-1}.$$

18.3.2 Computing Method

Hartley (1967) described a computer technique for computing expectations of mean squares that he called *synthesis*. In order to describe and later apply the technique, the expectation of a sum of squares computed from the general random model (18.2.3) is expressed in matrix notation. A sum of squares can always be represented in a quadratic form as

$$Q = \mathbf{y}'\mathbf{A}\mathbf{y} \qquad (18.3.2)$$

where $\mathbf{y}$ is the vector of observations and $\mathbf{A}$ is the appropriately chosen symmetric matrix of constants (Graybill, 1976). As an example, the sample variance of a vector of n observations is

$$s^2 = \sum_{i=1}^{n} \frac{(y_i - \bar{y})^2}{n-1}$$

$$= \mathbf{y}'\frac{(\mathbf{I}_n - (1/n)\mathbf{J}_n)\mathbf{y}}{n-1}$$

where $\mathbf{I}_n$ is an $n \times n$ identity matrix, $\mathbf{J}_n$ is an $n \times n$ matrix of ones, and the matrix of the quadratic form is

$$\mathbf{A} = \frac{\mathbf{I}_n - (1/n)\mathbf{J}_n}{n-1}.$$

For different models, certain choices of $\mathbf{A}$ yield the desired sums of squares, but fortunately, as will be seen shortly, it is not necessary to know—or know how to determine—the matrix $\mathbf{A}$. You only need to know that it exists.

For the general random model (18.2.3) and its corresponding covariance matrix $\boldsymbol{\Sigma}$, the expectation of a quadratic form (Graybill, 1976) $\mathbf{y}'\mathbf{A}\mathbf{y}$ is

$$E(\mathbf{y}'\mathbf{A}\mathbf{y}) = \text{Tr}[\mathbf{A}\boldsymbol{\Sigma}] + \mu^2\mathbf{j}_n'\mathbf{A}\mathbf{j}_n \tag{18.3.3}$$

where $\text{Tr}[\mathbf{B}] = \sum_{i=1}^{n} b_{ii}$ where b_{ii}, $i = 1, 2, \ldots, n$, denotes the diagonal elements of the square matrix $\mathbf{B}$. The sums of squares in the analysis of variance are constructed such that $\mu^2\mathbf{j}_n'\mathbf{A}\mathbf{j}_n = 0$. In this way the expectations of the sums of squares do not depend on μ. Thus, the expectations of sums of squares that satisfy $\mathbf{j}_n'\mathbf{A}\mathbf{j}_n = 0$ are given by

$$E(\mathbf{y}'\mathbf{A}\mathbf{y}) = \text{Tr}[\mathbf{A}\boldsymbol{\Sigma}]. \tag{18.3.4}$$

The matrix $\boldsymbol{\Sigma}$ is

$$\boldsymbol{\Sigma} = \sigma_1^2\mathbf{X}_1\mathbf{X}_1' + \cdots + \sigma_k^2\mathbf{X}_k\mathbf{X}_k' + \sigma_\varepsilon^2\mathbf{I}_n.$$

Thus the expectation of the quadratic form $\mathbf{y}'\mathbf{A}\mathbf{y}$ is

$$\begin{aligned} E(\mathbf{y}'\mathbf{A}\mathbf{y}) &= \text{Tr}\left[\mathbf{A}\left(\sigma_1^2\mathbf{X}_1\mathbf{X}_1' + \cdots + \sigma_k^2\mathbf{X}_k\mathbf{X}_k' + \sigma_\varepsilon^2\mathbf{I}\right)\right] \\ &= \text{Tr}\left[\sigma_1^2\mathbf{A}\mathbf{X}_1\mathbf{X}_1' + \cdots + \sigma_k^2\mathbf{A}\mathbf{X}_k\mathbf{X}_k' + \sigma_\varepsilon^2\mathbf{A}\right] \\ &= \sigma_1^2\text{Tr}\left[\mathbf{A}\mathbf{X}_1\mathbf{X}_1'\right] + \cdots + \sigma_k^2\text{Tr}\left[\mathbf{A}\mathbf{X}_k\mathbf{X}_k'\right] + \sigma_\varepsilon^2\text{Tr}[\mathbf{A}]. \end{aligned}$$

The coefficient of σ_ε^2 is $\text{Tr}[\mathbf{A}]$, which is equal to the degrees of freedom associated with the sum of squares $\mathbf{y}'\mathbf{A}\mathbf{y}$. The coefficient of σ_s^2 is $\text{Tr}[\mathbf{A}\mathbf{X}_s\mathbf{X}_s']$ for $s = 1, 2, \ldots, k$. One property of the trace operator is that $\text{Tr}[\mathbf{A}\mathbf{X}_s\mathbf{X}_s'] = \text{Tr}[\mathbf{X}_s'\mathbf{A}\mathbf{X}_s]$, where $\text{Tr}[\mathbf{X}_s'\mathbf{A}\mathbf{X}_s]$ is the sum of the diagonal elements of $\mathbf{X}_s'\mathbf{A}\mathbf{X}_s$ or

$$\text{Tr}\left[\mathbf{X}_s'\mathbf{A}\mathbf{X}_s\right] = \sum_{j=1}^{t_s} \mathbf{x}_{sj}'\mathbf{A}\mathbf{x}_{sj}.$$

But $\mathbf{x}_{sj}'\mathbf{A}\mathbf{x}_{sj}$ is the same sum of squares as $\mathbf{y}'\mathbf{A}\mathbf{y}$ except that the column vector $\mathbf{x}_{sj}$ is used as data in place of $\mathbf{y}$. Hence, a program that calculates $\mathbf{y}'\mathbf{A}\mathbf{y}$ can also be used to calculate $\mathbf{x}_{sj}'\mathbf{A}\mathbf{x}_{sj}$. Thus, the coefficient of σ_s^2 in the expectation of $\mathbf{y}'\mathbf{A}\mathbf{y}$ is the sum

$$\mathbf{x}_{s1}'\mathbf{A}\mathbf{x}_{s1} + \cdots + \mathbf{x}_{st_s}'\mathbf{A}\mathbf{x}_{st_s}.$$

If $\mathbf{A}$ is known, then the above sums of squares can be evaluated explicitly. If $\mathbf{A}$ is not known, which is likely since a computer code is probably used to calculate $\mathbf{y}'\mathbf{A}\mathbf{y}$, each $\mathbf{x}'_{sj}\mathbf{A}\mathbf{x}_{sj}$ can be computed by letting the computer calculate the sum of squares where the column $\mathbf{x}_{sj}$ is used as the data (instead of $\mathbf{y}$). Thus, the sum of squares must be computed for each column of $[\mathbf{X}_1, \ldots, \mathbf{X}_k]$ as if it were data, and then the expectation of the sum of squares is evaluated as

$$E(\mathbf{y}'\mathbf{A}\mathbf{y}) = \sigma_1^2\left(\sum_{j=1}^{t_1} \mathbf{x}'_{1j}\mathbf{A}\mathbf{x}_{1j}\right) + \cdots + \sigma_k^2\left(\sum_{j=1}^{t_k} \mathbf{x}'_{kj}\mathbf{A}\mathbf{x}_{kj}\right) + \sigma_\varepsilon^2 v$$

where v is the degrees of freedom associated with $\mathbf{y}'\mathbf{A}\mathbf{y}$.

To help demonstrate the idea of synthesis, the expectations of the sums of squares, Q_1 and Q_2, are recomputed for the one-way random effects model. First, a specific model is used to show how to compute the expectations and then the expectations are computed for the general one-way random model.

The matrix form for a model to describe the yield of four varieties of wheat in an experiment with four replications is

$$\begin{bmatrix} y_{11}\\ y_{12}\\ y_{13}\\ y_{14}\\ y_{21}\\ y_{22}\\ y_{23}\\ y_{24}\\ y_{31}\\ y_{32}\\ y_{33}\\ y_{34}\\ y_{41}\\ y_{42}\\ y_{43}\\ y_{44} \end{bmatrix} = \begin{bmatrix} 1\\1\\1\\1\\1\\1\\1\\1\\1\\1\\1\\1\\1\\1\\1\\1 \end{bmatrix}\mu + \begin{bmatrix} 1&0&0&0\\ 1&0&0&0\\ 1&0&0&0\\ 1&0&0&0\\ 0&1&0&0\\ 0&1&0&0\\ 0&1&0&0\\ 0&1&0&0\\ 0&0&1&0\\ 0&0&1&0\\ 0&0&1&0\\ 0&0&1&0\\ 0&0&0&1\\ 0&0&0&1\\ 0&0&0&1\\ 0&0&0&1 \end{bmatrix}\begin{bmatrix} a_1\\ a_2\\ a_3\\ a_4 \end{bmatrix} + \mathbf{b}$$

or

$$\mathbf{y} = \mathbf{j}_{16}\mu + [\mathbf{x}_1, \mathbf{x}_2, \mathbf{x}_3, \mathbf{x}_4]\mathbf{a} + \mathbf{b}.$$

The expectation of Q_1, the within sum of squares, is

$$E(Q_1) = E(\mathbf{y}'\mathbf{A}_W\mathbf{y}) = \sigma_a^2\left[\sum_{i=1}^{4} \mathbf{x}'_i\mathbf{A}_W\mathbf{x}_i\right] + 12\sigma_b^2$$

where 12 is the degrees of freedom associated with Q_1 and $\mathbf{A}_W$ denotes the matrix of the sum of squares Q_1. To obtain the coefficient of σ_a^2, we compute the within sum of squares by using $\mathbf{x}_1$ as data, then by using $\mathbf{x}_2$ as data, then $\mathbf{x}_3$, and finally $\mathbf{x}_4$. The within sum of squares for column $\mathbf{x}_1$

is

$$W_1 = \sum_{i=1}^{4} \sum_{j=1}^{4} x_{1ij}^2 - 4 \sum_{i=1}^{4} \bar{x}_{1i\cdot}^2$$
$$= 4 - 4(1) = 0.$$

Likewise, the within sums of squares for columns $\mathbf{x}_2$, $\mathbf{x}_3$, and $\mathbf{x}_4$ are also zero, meaning that the coefficient of σ_a^2 in $E(Q_1)$ is zero; that is, $E(Q_1) = 12\sigma_b^2$.

The expectation of Q_2, the between sum of squares, is

$$E(Q_2) = E(\mathbf{y}'\mathbf{A}_B\mathbf{y}) = \sigma_a^2\left(\sum_{i=1}^{4} \mathbf{x}_i'\mathbf{A}_B\mathbf{x}_i\right) + 3\sigma_b^2$$

where 3 is the degrees of freedom associated with Q_2 and $\mathbf{A}_B$ denotes the matrix of the sum of squares Q_2. To compute the coefficient of σ_a^2, compute the between sum of squares for each column $\mathbf{x}_1$, $\mathbf{x}_2$, $\mathbf{x}_3$, and $\mathbf{x}_4$ and add them together. The between sum of squares computed by using $\mathbf{x}_1$ as the data is

$$b_1 = 4\sum_{i=1}^{4} \bar{x}_{1i\cdot}^2 - 16\bar{x}_{1\cdot\cdot}^2$$
$$= 4(1^2 + 0^2 + 0^2 + 0^2) - 16(.25)^2$$
$$= 3.$$

The between sum of squares for each of the columns $\mathbf{x}_2$, $\mathbf{x}_3$, and $\mathbf{x}_4$ is also 3; thus, the coefficient of σ_a^2 is 3 + 3 + 3 + 3 = 12. Putting this in the expression for $E(Q_2)$,

$$E(Q_2) = 12\sigma_a^2 + 3\sigma_b^2.$$

Next, use the synthesis technique to compute the expectations of the between and within sums of squares for the general one-way random effects model in equation (18.2.1). The matrix form of the model is

$$\begin{bmatrix} y_{11} \\ \vdots \\ y_{1n_1} \\ y_{21} \\ \vdots \\ y_{2n_2} \\ \vdots \\ y_{t1} \\ \vdots \\ y_{tn_t} \end{bmatrix} = \begin{bmatrix} 1 \\ \vdots \\ 1 \\ 1 \\ \vdots \\ 1 \\ \vdots \\ 1 \\ \vdots \\ 1 \end{bmatrix} \mu + \overset{\begin{matrix} \mathbf{x}_1 & \mathbf{x}_2 & \cdots & \mathbf{x}_t \end{matrix}}{\begin{bmatrix} 1 & 0 & \cdots & 0 \\ \vdots & \vdots & & \vdots \\ 1 & 0 & \cdots & 0 \\ 0 & 1 & \cdots & 0 \\ \vdots & \vdots & & \vdots \\ 0 & 1 & \cdots & 0 \\ \vdots & \vdots & & \vdots \\ 0 & 0 & \cdots & 1 \\ \vdots & \vdots & & \vdots \\ 0 & 0 & \cdots & 1 \end{bmatrix}} \begin{bmatrix} a_1 \\ a_2 \\ \vdots \\ a_t \end{bmatrix} + \mathbf{b}$$

where $\mathbf{x}_i$ is superimposed over the respective column. The within sum of squares is

$$Q_1 = \sum_{i=1}^{t} \sum_{j=1}^{n_i} y_{ij}^2 - \sum_{i=1}^{t} n_i \bar{y}_{i\cdot}^2 = \mathbf{y}'\mathbf{A}_W\mathbf{y}$$

and its expectation is

$$E(Q_1) = \sigma_a^2 \sum_{i=1}^{t} \mathbf{x}_i'\mathbf{A}_W\mathbf{x}_i + (N - t)\sigma_b^2$$

where $N = \sum_{i=1}^{t} n_i$ and $\mathbf{A}_W$ is the matrix for the within sum of squares. The within sum of squares for the first column, $\mathbf{x}_1$, is

$$\begin{aligned} u_1 &= \sum_{i=1}^{t} \sum_{j=1}^{n_i} x_{1ij}^2 - \sum_{i=1}^{t} n_i \bar{x}_{1i\cdot}^2 \\ &= n_1 - \left[n_1(1) + n_2(0) + \cdots + n_t(0)\right] \\ &= 0. \end{aligned}$$

Likewise, the within sum of squares for each of the other columns, $\mathbf{x}_2, \ldots, \mathbf{x}_t$, is also zero, and the coefficient of σ_a^2 in $E(Q_1)$ is zero. Thus,

$$E(Q_1) = (N - t)\sigma_b^2.$$

The between sum of squares is

$$Q_2 = \sum_{i=1}^{t} n_i \bar{y}_{i\cdot}^2 - N\bar{y}_{\cdot\cdot}^2 = \mathbf{y}'\mathbf{A}_B\mathbf{y},$$

and its expectation is

$$E(Q_2) = \sigma_a^2 \sum_{i=1}^{t} \mathbf{x}_i'\mathbf{A}_B\mathbf{x}_i + (t - 1)\sigma_b^2$$

where $\mathbf{A}_B$ is the matrix for the between sum of squares. The between sum of squares for the first column, $\mathbf{x}_1$, is

$$b_1 = \sum_{i=1}^{t} n_i \bar{x}_{1i\cdot}^2 - N\bar{x}_{1\cdot\cdot}^2.$$

For column $\mathbf{x}_1$, $\bar{x}_{11\cdot} = 1, \bar{x}_{12\cdot} = \cdots = \bar{x}_{1t\cdot} = 0$ and $\bar{x}_{\cdot\cdot} = n_1/N$. Thus,

$$\begin{aligned} b_1 &= n_1(1)^2 + n_2(0)^2 + \cdots + n_t(0)^2 - N\left(\frac{n_1}{N}\right)^2 \\ &= \frac{n_1 - n_1^2}{N}. \end{aligned}$$

The between sum of squares for the other columns is similarly obtained. Thus the coefficient of σ_a^2 in $E(Q_2)$ is

$$\sum_{i=1}^{t}\left(\frac{n_i - n_i^2}{N}\right) = N - \frac{\sum_{i=1}^{t} n_i^2}{N}.$$

Combining these results, the expectation of the between sum of squares is

$$E(Q_2) = (t-1)\sigma_b^2 + \left(N - \frac{\sum_{i=1}^{t} n_i^2}{N}\right)\sigma_a^2.$$

The expectations of Q_1 and Q_2 obtained via synthesis are equivalent to those obtained algebraically.

Next, we shall apply the method of synthesis to a two-way random model. The two-way treatment structure in a completely randomized design structure where both row treatments and column treatments are random effects can be modeled by

$$y_{ijk} = \mu + a_i + b_j + c_{ij} + \varepsilon_{ijk} \qquad i = 1,2,\ldots,b,$$
$$j = 1,2,\ldots,t, \quad k = 1,2,\ldots,n_{ij},$$

where a_i denotes the random effects of each row treatment and is assumed to be distributed i.i.d. $N(0,\sigma_a^2)$, b_j denotes the random effect of each column treatment and is assumed to be distributed i.i.d. $N(0,\sigma_b^2)$, c_{ij} denotes the random effect of each combination of row treatment i with column treatment j and is assumed to be distributed i.i.d. $N(0,\sigma_c^2)$, and ε_{ijk} denotes the experimental unit error and is assumed to be distributed i.i.d. $N(0,\sigma_\varepsilon^2)$. The schematic in Figure 18.1 represents data from an unbalanced two-way treatment structure in a completely randomized design structure.

The matrix model corresponding to the data in Figure 18.1 is

$$\begin{bmatrix} y_{111}\\ y_{112}\\ y_{113}\\ y_{121}\\ y_{122}\\ y_{131}\\ y_{132}\\ y_{211}\\ y_{212}\\ y_{221}\\ y_{222}\\ y_{223}\\ y_{231}\\ y_{232} \end{bmatrix} = \begin{bmatrix} 1\\1\\1\\1\\1\\1\\1\\1\\1\\1\\1\\1\\1\\1 \end{bmatrix}\mu + \begin{bmatrix} 1&0\\1&0\\1&0\\1&0\\1&0\\1&0\\1&0\\0&1\\0&1\\0&1\\0&1\\0&1\\0&1\\0&1 \end{bmatrix}\begin{bmatrix} a_1\\ a_2 \end{bmatrix} + \begin{bmatrix} 1&0&0\\1&0&0\\1&0&0\\0&1&0\\0&1&0\\0&0&1\\0&0&1\\1&0&0\\1&0&0\\0&1&0\\0&1&0\\0&1&0\\0&0&1\\0&0&1 \end{bmatrix}\begin{bmatrix} b_1\\ b_2\\ b_3 \end{bmatrix} + \begin{bmatrix} 1&0&0&0&0&0\\1&0&0&0&0&0\\1&0&0&0&0&0\\0&1&0&0&0&0\\0&1&0&0&0&0\\0&0&1&0&0&0\\0&0&1&0&0&0\\0&0&0&1&0&0\\0&0&0&1&0&0\\0&0&0&0&1&0\\0&0&0&0&1&0\\0&0&0&0&1&0\\0&0&0&0&0&1\\0&0&0&0&0&1 \end{bmatrix}\begin{bmatrix} c_{11}\\ c_{12}\\ c_{13}\\ c_{21}\\ c_{22}\\ c_{23} \end{bmatrix} + \boldsymbol{\varepsilon}$$

Column Treatment

		1	2	3
Row Treatment	1	y_{111} y_{112} y_{113}	y_{121} y_{122}	y_{131} y_{132}
	2	y_{211} y_{212}	y_{221} y_{222} y_{223}	y_{231} y_{232}

Figure 18.1 Example of two-way data.

or $\mathbf{y} = \mathbf{j}_{14}\mu + \mathbf{X}_1\mathbf{a} + \mathbf{X}_2\mathbf{b} + \mathbf{X}_3\mathbf{c} + \boldsymbol{\varepsilon}$. The parameters of the model are μ, σ_a^2, σ_b^2, σ_c^2, and σ_ε^2.

The sums of squares for this unbalanced two-way experiment can be computed in several different ways (see Chapters 9 and 10). In the analysis for this model, there are four different sums of squares. To demonstrate the method of synthesis, select four sums of squares that would correspond to the balanced case of SSROWS, SSCOLUMNS, SSINTERACTION, and SSERROR but have been modified for the unequal sample sizes. The four sums of squares are

$$Q_1 = \sum_{i=1}^{b} \frac{y_{i\cdot\cdot}^2}{n_{i\cdot}} - \frac{y_{\cdot\cdot\cdot}^2}{N} \qquad \text{(SSROWS)},$$

$$Q_2 = \sum_{j=1}^{t} \frac{y_{\cdot j\cdot}^2}{n_{\cdot j}} - \frac{y_{\cdot\cdot\cdot}^2}{N} \qquad \text{(SSCOLUMNS)},$$

$$Q_3 = \sum_{i=1}^{b}\sum_{j=1}^{t} \frac{y_{ij\cdot}^2}{n_{ij}} - Q_1 - Q_2 - \frac{y_{\cdot\cdot\cdot}^2}{N} \qquad \text{(SSINTERACTION)},$$

$$Q_4 = \sum_{i=1}^{b}\sum_{j=1}^{t}\sum_{k=1}^{n_{ij}} (y_{ijk} - \bar{y}_{ij\cdot})^2 \qquad \text{(SSERROR)}.$$

For the data in Figure 18.1,

$$Q_2 = \frac{y_{\cdot 1\cdot}^2}{5} + \frac{y_{\cdot 2\cdot}^2}{5} + \frac{y_{\cdot 3\cdot}^2}{4} - \frac{y_{\cdot\cdot\cdot}^2}{14}$$

and

$$E(Q_2) = k_1\sigma_a^2 + k_2\sigma_b^2 + k_3\sigma_c^2 + 2\sigma_\varepsilon^2$$

where 2 is the degrees of freedom associated with Q_2.

Now, use synthesis to determine k_1, k_2, and k_3. To determine k_1, compute Q_2 for each column of $\mathbf{X}_1$. The value of Q_2 using the first column of $\mathbf{X}_1$ as data is

$$\frac{3^2}{5} + \frac{2^2}{5} + \frac{2^2}{4} - \frac{7^2}{14} = 0.1,$$

and the value of Q_2 using the second column of $\mathbf{X}_1$ as data is

$$\frac{2^2}{5} + \frac{3^2}{5} + \frac{2^2}{4} - \frac{7^2}{14} = 0.1;$$

thus, $k_1 = 0.1 + 0.1 = 0.2$. To determine k_2, compute Q_2 for each column of $\mathbf{X}_2$. The value of Q_2 for column 1 of $\mathbf{X}_2$ is

$$\frac{5^2}{5} + \frac{0^2}{5} + \frac{0^2}{4} - \frac{5^2}{14} = 3.214,$$

for column 2 of $\mathbf{X}_2$ is

$$\frac{0^2}{5} + \frac{5^2}{5} + \frac{0^2}{4} - \frac{5^2}{14} = 3.214,$$

and for column 3 of $\mathbf{X}_2$ is

$$\frac{0^2}{5} + \frac{0^2}{5} + \frac{4^2}{4} - \frac{4^2}{14} = 2.857.$$

Thus the value of k_2 is

$$\begin{aligned} k_2 &= 3.214 + 3.214 + 2.857 \\ &= 9.285. \end{aligned}$$

To determine k_3, compute Q_2 for each column of $\mathbf{X}_3$. The value of Q_2 for column 1 of $\mathbf{X}_3$ is

$$\frac{3^2}{5} + \frac{0^2}{5} + \frac{0^2}{5} - \frac{3^2}{14} = 1.157,$$

for column 2 of $\mathbf{X}_3$ is

$$\frac{0^2}{5} + \frac{2^2}{5} + \frac{0^2}{4} - \frac{2^2}{14} = 0.514,$$

for column 3 of $\mathbf{X}_3$ is

$$\frac{0^2}{5} + \frac{0^2}{5} + \frac{2^2}{4} - \frac{2^2}{14} = 0.714,$$

for column 4 of $\mathbf{X}_3$ is

$$\frac{2^2}{5} + \frac{0^2}{5} + \frac{0^2}{4} - \frac{2^2}{14} = 0.514,$$

for column 5 of $\mathbf{X}_3$ is

$$\frac{0^2}{5} + \frac{3^2}{5} + \frac{0^2}{4} - \frac{3^2}{14} = 1.157,$$

and for column 6 of $\mathbf{X}_3$ is

$$\frac{0^2}{5} + \frac{0^2}{5} + \frac{2^2}{4} - \frac{2^2}{14} = 0.714.$$

Thus, the value of k_3 is

$$k_3 = 1.157 + 0.514 + 0.714 + 0.514 + 1.157 + 0.714$$
$$= 4.770$$

By putting these values together, the expectation of Q_2 is

$$E(Q_2) = 0.2\sigma_a^2 + 9.285\sigma_b^2 + 4.770\sigma_c^2 + 2\sigma_\varepsilon^2.$$

By using the same technique, the expectations of Q_1 and Q_3 are determined to be

$$E(Q_1) = 7\sigma_a^2 + .143\sigma_b^2 + 2.429\sigma_c^2 + \sigma_\varepsilon^2$$

and

$$E(Q_3) = 4.371\sigma_c^2 + 2\sigma_\varepsilon^2.$$

In general, the expectation of the SSERROR is the degrees of freedom associated with SSERROR times σ_ε^2. In this case

$$E(\text{SSERROR}) = E(Q_4) = 8\sigma_\varepsilon^2.$$

There are various ways to compute sums of squares for unbalanced treatment structures, including the Type I through Type IV sums of squares of SAS® GLM. The computations of the expectations of the Type I and Type IV sums of squares from SAS® GLM are shown below.

Table 18.1 contains the results of computing the two types of sums of squares for the data (denoted by y in the table), each column of $\mathbf{X}_1$ (denoted by a_1 and a_2), each column of $\mathbf{X}_2$ (denoted by b_1, b_2, and b_3), and each column of $\mathbf{X}_3$ (denoted by c_{11}, c_{12}, c_{13}, c_{21}, c_{22}, and c_{23}). By using the sums of squares in Table 18.1, the expectation of the Type I SSROWS is

$$E[\text{SSROWS(I)}] = \sigma_a^2(3.5 + 3.5) + \sigma_b^2(.0714 + .0714 + 0)$$
$$+\sigma_c^2(.6428 + .2857 + .2857 + .2857 + .6428 + .2857) + (1)\sigma_\varepsilon^2$$
$$= 7\sigma_a^2 + .1428\sigma_b^2 + 2.4284\sigma_c^2 + \sigma_\varepsilon^2.$$

By using similar methods, the expectations of all Type I and Type IV sums of squares can be obtained. The results are listed below. Type I sums of squares:

$$E(\text{SSROWS}) = \sigma_\varepsilon^2 + 7.0\sigma_a^2 + .1428\sigma_b^2 + 2.4284\sigma_c^2$$
$$E(\text{SSCOLUMNS}) = 2\sigma_\varepsilon^2 + 9.1429\sigma_b^2 + 4.6252\sigma_c^2$$
$$E(\text{SSINTERACTION}) = 2\sigma_\varepsilon^2 + 4.5178\sigma_c^2$$
$$E(\text{SSERROR}) = 8\sigma_\varepsilon^2$$

Table 18.1 Type I and Type IV Sums of Squares for y and Each Column of $[X_1, X_2, X_3]$ Data in Figure 18.1

	SAS® SUMS OF SQUARES					
	TYPE I			TYPE IV		
Effect	*Row*	*Column*	*Interaction*	*Row*	*Column*	*Interaction*
y	.64286	14.75966	109.14509	.1667	8.9098	109.1451
a_1	3.5	0	0	3.375	0	0
a_2	3.5	0	0	3.375	0	0
b_1	.0714	3.1429	0	0	3.1059	0
b_2	.0714	3.1429	0	0	3.1059	0
b_3	0	2.8571	0	0	2.8235	0
c_{11}	.6428	.9378	.7765	.375	.7765	.7765
c_{12}	.2857	.6521	.7765	.375	.7765	.7765
c_{13}	.2857	.7227	.7059	.375	.7059	.7059
c_{21}	.2857	.6521	.7765	.375	.7765	.7765
c_{22}	.6428	.9378	.7765	.375	.7765	.7765
c_{23}	.2857	.7227	.7059	.375	.7059	.7059

Note: SSERROR = 30.67 with 8 degrees of freedom.

Type IV sums of squares:

$$E(\text{SSROWS}) = \sigma_\varepsilon^2 + 6.75\sigma_a^2 + 2.25\sigma_c^2$$

$$E(\text{SSCOLUMNS}) = 2\sigma_\varepsilon^2 + 9.0353\sigma_b^2 + 4.5178\sigma_c^2$$

$$E(\text{SSINTERACTION}) = 2\sigma_\varepsilon^2 + 4.5178\sigma_c^2$$

$$E(\text{SSERROR}) = 8\sigma_\varepsilon^2$$

There are several alternative methods for computing sums of squares (with names attached) in the statistical literature that should be discussed here. Those techniques that can be applied to mixed models as well as random models are so indicated.

Henderson (1953) introduced four methods of computing sums of squares, called Henderson's Methods I, II, III, and IV (also see Searle, 1971). Methods I and III are discussed here. The Analysis of Variance Method, or Henderson's Method I, which is appropriate only for random models, is a technique that consists of computing sums of squares analogous to those computed for balanced data except that they are altered to account for unequal numbers of observations per treatment combination. The two-way classification model is used below to demonstrate this method.

For the model

$$y_{ijk} = \mu + a_i + b_j + c_{ij} + \varepsilon_{ijk} \qquad i = 1, 2, \ldots, b,$$
$$j = 1, 2, \ldots, t, \quad k = 1, 2, \ldots, n,$$

the sums of squares are

$$\mathrm{SSA} = nt \sum_{i=1}^{b} (\bar{y}_{i\cdot\cdot} - \bar{y}_{\cdot\cdot\cdot})^2 = \sum_{i=1}^{b} \frac{y_{i\cdot\cdot}^2}{nt} - \frac{y_{\cdot\cdot\cdot}^2}{nbt}$$

$$\mathrm{SSB} = nb \sum_{j=1}^{t} (\bar{y}_{\cdot j\cdot} - \bar{y}_{\cdot\cdot\cdot})^2 = \sum_{j=1}^{t} \frac{y_{\cdot j\cdot}^2}{nb} - \frac{y_{\cdot\cdot\cdot}^2}{nbt}$$

$$\mathrm{SSA}*\mathrm{B} = n \sum_{i=1}^{b} \sum_{j=1}^{t} (\bar{y}_{ij\cdot} - \bar{y}_{i\cdot\cdot} - \bar{y}_{\cdot j\cdot} + \bar{y}_{\cdot\cdot\cdot})^2$$

$$= \sum_{i-1}^{b} \sum_{j=1}^{t} \frac{y_{ij\cdot}^2}{n} - \sum_{i=1}^{b} \frac{y_{i\cdot\cdot}^2}{nt} - \sum_{j=1}^{t} \frac{y_{\cdot i\cdot}^2}{nb} + \frac{y_{\cdot\cdot\cdot}^2}{nbt}.$$

For unequal sample sizes, replace nbt with $n_{\cdot\cdot}$, nb with $n_{\cdot j}$, and nt with $n_{i\cdot}$. Thus, the sums of squares become

$$\mathrm{SSA} = \sum_{i=1}^{b} \frac{y_{i\cdot\cdot}^2}{n_{i\cdot}} - \frac{y_{\cdot\cdot\cdot}^2}{n_{\cdot\cdot}}$$

$$\mathrm{SSB} = \sum_{j=1}^{t} \frac{y_{\cdot j\cdot}^2}{n_{\cdot j}} - \frac{y_{\cdot\cdot\cdot}^2}{n_{\cdot\cdot}}$$

$$\mathrm{SSA}*\mathrm{B} = \sum_{i=1}^{b} \sum_{j=1}^{t} \frac{y_{ij\cdot}^2}{n_{ij}} - \sum_{i=1}^{t} \frac{y_{i\cdot\cdot}^2}{n_{\cdot\cdot}} - \sum_{j=1}^{b} \frac{y_{\cdot j\cdot}^2}{n_{\cdot j}} + \frac{y_{\cdot\cdot\cdot}^2}{n_{\cdot\cdot}}.$$

These are the sums of squares used at the beginning of the discussion of the data in Figure 18.1.

The fitting-constants method, or Henderson's Method III, involves fitting various linear models to the data and then computing the corresponding sums of squares. The method uses what is called the reduction in the sums of squares due to fitting the whole model and those due to fitting various submodels (see Chapter 10).

To set the notation, consider the model

$$\mathbf{y} = \mathbf{X}_1\mathbf{b}_1 + \mathbf{X}_2\mathbf{b}_2 + \mathbf{X}_3\mathbf{b}_3 + \boldsymbol{\varepsilon}.$$

The reduction in the total sum of squares due to fitting the whole model is

$$R(\mathbf{b}_1, \mathbf{b}_2, \mathbf{b}_3) = \mathbf{y}'\mathbf{y} - \text{SSERROR}(\mathbf{b}_1, \mathbf{b}_2, \mathbf{b}_3)$$

where SSERROR($\mathbf{b}_1, \mathbf{b}_2, \mathbf{b}_3$) is the residual sum of squares after fitting the whole model. The reduction due to fitting $\mathbf{b}_1$ and $\mathbf{b}_2$ is

$$R(\mathbf{b}_1, \mathbf{b}_2) = \mathbf{y}'\mathbf{y} - \text{SSERROR}(\mathbf{b}_1, \mathbf{b}_2)$$

where SSERROR($\mathbf{b}_1, \mathbf{b}_2$) is the residual sum of squares for the model

$$\mathbf{y} = \mathbf{b}_1\mathbf{X}_1 + \mathbf{b}_2\mathbf{X}_2 + \boldsymbol{\varepsilon}.$$

The reduction due to $\mathbf{b}_3$ after fitting $\mathbf{b}_1$ and $\mathbf{b}_2$ is denoted by $R(\mathbf{b}_3|\mathbf{b}_1, \mathbf{b}_2)$ and is given by

$$\begin{aligned} R(\mathbf{b}_3|\mathbf{b}_1, \mathbf{b}_2) &= R(\mathbf{b}_1, \mathbf{b}_2, \mathbf{b}_3) - R(\mathbf{b}_1, \mathbf{b}_2) \\ &= \text{SSERROR}(\mathbf{b}_1, \mathbf{b}_2) - \text{SSERROR}(\mathbf{b}_1, \mathbf{b}_2, \mathbf{b}_3). \end{aligned}$$

Likewise, the reduction due to $\mathbf{b}_1$ after fitting $\mathbf{b}_2$ is

$$\begin{aligned} R(\mathbf{b}_1|\mathbf{b}_2) &= R(\mathbf{b}_1, \mathbf{b}_2) - R(\mathbf{b}_2) \\ &= \text{SSERROR}(\mathbf{b}_2) - \text{SSERROR}(\mathbf{b}_1, \mathbf{b}_2), \end{aligned}$$

and the reduction due to $\mathbf{b}_1, \mathbf{b}_2$ after fitting $\mathbf{b}_3$ is

$$\begin{aligned} R(\mathbf{b}_1, \mathbf{b}_2|\mathbf{b}_3) &= R(\mathbf{b}_1, \mathbf{b}_2, \mathbf{b}_3) - R(\mathbf{b}_3) \\ &= \text{SSERROR}(\mathbf{b}_3) - \text{SSERROR}(\mathbf{b}_1, \mathbf{b}_2, \mathbf{b}_3). \end{aligned}$$

One of the advantages of using this technique is that $E[R(\mathbf{b}_1, \mathbf{b}_2|\mathbf{b}_3)]$ does not depend on $\mathbf{b}_3$ if it is a fixed effect or σ_3^2 if it is a random effect unless $\mathbf{b}_3$ denotes an interaction between $\mathbf{b}_1$ and $\mathbf{b}_2$ or with $\mathbf{b}_1$ or $\mathbf{b}_2$.

For the model

$$\mathbf{y} = \mu\mathbf{j} + \mathbf{X}_1\mathbf{b} + \mathbf{X}_2\mathbf{t} + \mathbf{X}_3\mathbf{g} + \boldsymbol{\varepsilon},$$

one set of possible sums of squares (which are SAS® GLM Type I sums of squares) is

$$\begin{aligned} R(\mathbf{b}|\mu) &= R(\mathbf{b}, \mu) - R(\mu) \\ R(\mathbf{t}|\mu, \mathbf{b}) &= R(\mu, \mathbf{b}, \mathbf{t}) - R(\mu, \mathbf{b}) \\ R(\mathbf{g}|\mu, \mathbf{b}, \mathbf{t}) &= R(\mu, \mathbf{b}, \mathbf{t}, \mathbf{g}) - R(\mu, \mathbf{b}, \mathbf{t}) \\ \text{SSERROR}(\mu, \mathbf{g}, \mathbf{b}, \mathbf{t}) &= \mathbf{y}'\mathbf{y} - R(\mu, \mathbf{b}, \mathbf{t}, \mathbf{g}). \end{aligned}$$

The expectations of these sums of squares are

$$\begin{aligned} E[\text{SSERROR}(\mu, \mathbf{g}, \mathbf{b}, \mathbf{t})] &= (n-p)\sigma_\varepsilon^2, \\ E[R(\mathbf{g}|\mu, \mathbf{b}, \mathbf{t})] &= k_1\sigma_\varepsilon^2 + k_2\sigma_g^2, \\ E[R(\mathbf{t}|\mu, \mathbf{b})] &= k_3\sigma_\varepsilon^2 + k_4\sigma_g^2 + k_5\sigma_t^2, \quad \text{and} \\ E[R(\mathbf{b}|\mu)] &= k_6\sigma_\varepsilon^2 + k_7\sigma_g^2 + k_8\sigma_t^2 + k_9\sigma_b^2. \end{aligned}$$

The Type I sums of squares given by $R(\mathbf{t}|\mu)$, $R(\mathbf{b}|\mu, \mathbf{t})$, and $R(\mathbf{g}|\mu, \mathbf{b}, \mathbf{t})$ could also be used, and the method of fitting constants can be used for both random and mixed models.

The method of synthesis can be used to evaluate the expectations of sums of squares for any model involving random effects or multiple error terms. These expectations can be used to estimate the variance components, test hypotheses, and construct confidence intervals about functions of the variance components. Methods for estimation are discussed in Chapter 19, and inference techniques are presented in Chapter 20.

CONCLUDING REMARKS

In this chapter, the concepts of random and fixed effects were defined as well as the concept of a random model. The model is expressed in matrix form in order to describe methods for computing the expected mean squares. An unbalanced one-way treatment structure in a completely randomized design structure was used to demonstrate the algebraic and the synthesis methods of computing expected mean squares. Different methods for computing sums of squares were described, and an unbalanced two-way treatment structure in a completely randomized design structure was used to demonstrate the computations of the respective expected mean squares.

19

Methods for Estimating Variance Components

CHAPTER OUTLINE

There are several ways to estimate variance components for the general random model. All of the procedures yield the same estimators when the design is balanced (equal n's and no missing cells) but different estimators when it is not. The three techniques discussed in this chapter are the method of moments, maximum likelihood methods, and the MINQUE method. The method of moments produces unbiased estimates, maximum likelihood estimators are consistent and have the usual large-sample-size properties of maximum likelihood estimates, and the MINQUE method produces estimates having minimum variance within the class of quadratic unbiased estimates.

When the design is unbalanced, method-of-moments estimates are easiest to compute (with existing computer programs), whereas the other two methods require iterative procedures. On the other hand, the maximum likelihood and MINQUE methods provide estimators with better properties than does the method of moments.

19.1 METHOD OF MOMENTS

The method of moments has been used to obtain estimates of variance components since Eisenhart (1947) gave the name MODEL II to the random model. Many researchers worked on the method of moments during the next 20 years and derived estimators, tests of hypotheses, and confidence intervals about variance components (see Searle, 1971, and Graybill, 1976, for good lists of references). In this section, a generalized version of the method of moments is discussed.

19.1.1 Description

The method-of-moments technique for estimating the variance components of the general random model of equation (18.2.3) involves the following:

1. obtaining a set of sums of squares,
2. evaluating the expectations of each sum of squares,
3. equating the expectations of the sums of squares to the observed value of each sum of squares, and
4. solving the resulting system of equations for the estimates of the variance components.

When $\mathbf{a}_i$, $i = 1, 2, \ldots, k$, and $\boldsymbol{\varepsilon}$ of model (18.2.3) are normally distributed and the sums of squares are distributed independently of each other, the resulting estimators are minimum variance unbiased. If $\mathbf{a}_i$, $i = 1, 2, \ldots, k$, and $\boldsymbol{\varepsilon}$ have the same first four moments as those of a normal distribution, the estimators are minimum variance quadratic unbiased (Graybill, 1976, p. 632).

The method-of-moments technique does not require the assumption of normality in order to obtain estimators. The only known property these estimators possess without the assumption of normality or the assumption that the distributions of the random vectors have the same first four moments of a normal distribution is that they are unbiased.

To apply the method-of-moments technique, one must do the following:

1. Compute as many sums of squares and then corresponding mean squares as there are variance components in the model.
2. Evaluate the expectations of these mean squares in terms of the variance components. (The resulting expectations do not involve μ, and each variance component must be included in the expectation of at least one sum of squares, which can be evaluated by using the techniques of Chapter 18.)
3. Equate the observed mean squares to their respective expected mean squares, thus generating a system of linear equations in the variance components.
4. Solve these equations for the estimates of the variance components.

The key to the method of moments is determining how to compute sums of squares and then evaluating the expectations of the resulting mean squares. These topics were discussed in Chapter 18.

If the model has $k + 1$ variance components, $\sigma_\varepsilon^2, \sigma_1^2, \ldots, \sigma_k^2$, then $k + 1$ sums of squares are required. Let $Q_0 = \mathbf{y}'\mathbf{A}_0\mathbf{y}, Q_1 = \mathbf{y}'\mathbf{A}_1\mathbf{y}, \ldots, Q_k = \mathbf{y}'\mathbf{A}_k\mathbf{y}$, denote the sums of squares with expectations

$$E(Q_i) = a_{i0}\sigma_\varepsilon^2 + a_{i1}\sigma_1^2 + \cdots + a_{ik}\sigma_k^2 \qquad i = 0, 1, \ldots, k.$$

Equate the sums of squares to their expectations as

$$Q_i = a_{i0}\hat{\sigma}_\varepsilon^2 + a_{i1}\hat{\sigma}_1^2 + \cdots + a_{ik}\hat{\sigma}_k^2 \qquad i = 0, 1, \ldots, k,$$

or, in matrix notation,

$$\begin{bmatrix} Q_0 \\ Q_1 \\ \vdots \\ Q_k \end{bmatrix} = \begin{bmatrix} a_{00} & a_{01} & \cdots & a_{0k} \\ a_{10} & a_{11} & \cdots & a_{1k} \\ \vdots & \vdots & & \vdots \\ a_{k0} & a_{k1} & \cdots & a_{kk} \end{bmatrix} \begin{bmatrix} \hat{\sigma}_\varepsilon^2 \\ \hat{\sigma}_1^2 \\ \vdots \\ \hat{\sigma}_k^2 \end{bmatrix}$$

or $\mathbf{Q} = \mathbf{A}\hat{\boldsymbol{\sigma}}^2$.

The method-of-moments estimators are the solutions to the system of equations, or $\hat{\boldsymbol{\sigma}}^2 = \mathbf{A}^{-1}\mathbf{Q}$. The estimators are obtained without restricting them to the parameter space. In many models and methods for computing sums of squares, the **A** matrix is triangular and thus the estimators can be obtained without inverting **A**.

Each estimator is a linear combination of the observed sums of squares, $Q_0, Q_1, \ldots, Q_k$, as

$$\hat{\sigma}_i^2 = b_{i0}Q_0 + b_{i1}Q_1 + \cdots + b_{ik}Q_k \qquad i = 0, 1, \ldots, k$$

where $\mathbf{b}' = (b_{i0}, b_{i1}, \ldots, b_{ik})$ is the ith row of $\mathbf{A}^{-1}$. The variance of $\hat{\sigma}_i^2$ is

$$\operatorname{Var}(\hat{\sigma}_i^2) = \operatorname{Var}(b_{i0}Q_0 + \cdots + b_{ik}Q_k);$$

when Q_i, $i = 0, 1, \ldots, k$, are independent, the variance is

$$\operatorname{Var}(\hat{\sigma}_i^2) = b_{i0}^2\operatorname{Var}(Q_0) + \cdots + b_{ik}^2\operatorname{Var}(Q_k).$$

A summary of method-of-moment estimators and their variances for several models is presented in Searle (1971, chap. 11).

For most balanced models, the method-of-moments estimators are uniformly minimum variance unbiased estimators of the variance components (Graybill, 1976). Thus, for nearly balanced models, the method-of-moments estimators should have fairly good properties.

19.1.2 Applications

EXAMPLE 19.1: Unbalanced One-Way Model

The unbalanced one-way random model of Example 18.2 is

$$y_{ij} = \mu + a_i + b_{ij} \qquad i = 1, 2, \ldots, t \quad j = 1, 2, \ldots, n_i.$$

Two sums of squares are the sum of squares within, Q_1, and the sums of squares between, Q_2, where

$$Q_1 = \sum_{i=1}^{t} \sum_{j=1}^{n_i} y_{ij}^2 - \sum_{i=1}^{t} n_i \bar{y}_{i\cdot}^2$$

and

$$Q_2 = \sum_{i=1}^{t} n_i \bar{y}_{i\cdot}^2 - \left(\sum n_i\right)\bar{y}_{\cdot\cdot}^2.$$

The expectations of Q_1 and Q_2 were evaluated in Chapter 18 as

$$E(Q_1) = (N - t)\sigma_b^2$$

and

$$E(Q_2) = (t - 1)\sigma_b^2 + \left(N - \frac{\Sigma n_i^2}{N}\right)\sigma_a^2$$

where $N = \Sigma n_i$.

The equations obtained by equating the sums of squares to their expectations are

$$Q_1 = (N - t)\hat{\sigma}_b^2$$

and

$$Q_2 = (t-1)\hat{\sigma}_b^2 + \left(N - \frac{\Sigma n_i^2}{N}\right)\hat{\sigma}_a^2,$$

or, in matrix notation,

$$\begin{bmatrix} Q_1 \\ Q_2 \end{bmatrix} = \begin{bmatrix} N-t & 0 \\ t-1 & N - \dfrac{\Sigma n_i^2}{N} \end{bmatrix} \begin{bmatrix} \hat{\sigma}_b^2 \\ \hat{\sigma}_a^2 \end{bmatrix}.$$

The equation can also be generated by equating the mean squares to their expectations, which amounts to dividing each equation by the degrees of freedom. The solutions are the same.

The resulting system of equations to be solved involving the mean squares is

$$\frac{Q_1}{N-t} = \hat{\sigma}_b^2$$

and

$$\frac{Q_2}{t-1} = \hat{\sigma}_b^2 + \frac{\left[N - \left(\Sigma n_i^2/N\right)\right]\hat{\sigma}_a^2}{t-1}.$$

The solution to this system of equations is

$$\hat{\sigma}_b^2 = \frac{Q_1}{N-t}$$

and

$$\hat{\sigma}_a^2 = \frac{Q_2 - (t-1)\hat{\sigma}_b^2}{N - \left(\Sigma n_i^2/N\right)}.$$

EXAMPLE 19.2: Wheat Varieties—One-Way Random Model

An experimenter randomly selected four varieties of wheat from a population of varieties of wheat and conducted an experiment. The experiment had a completely randomized design structure with a maximum of four plots per variety (the plot is the experimental unit). The experimenter randomly selected 20 plants from each plot and rated the amount of insect damage done to each plant on a scale from 0 to 10. Thus, the response measured on each plot is the mean of the 20 ratings. The data are shown in Figure 19.1.

The computations necessary to compute the sums of squares, Q_1 and Q_2, are given in Figure 19.1 along with the resulting sums of squares, their expected mean squares, the system of equations, and the resulting

Variety

A	B	C	D
3.90	3.60	4.15	3.35
4.05	4.20	4.60	
4.25	4.05	4.15	3.80
	3.85	4.40	

$$\bar{y}_{..} = \frac{52.35}{13} = 4.0269 \qquad Q_1 = .50792$$

$$\sum_{i=1}^{4} \sum_{j=1}^{n_i} y_{ij}^2 = 212.1275 \qquad Q_2 = .81016$$

$$\sum_{i=1}^{4} n_i \bar{y}_{i\cdot}^2 = 211.61958$$

Expected Mean Squares

$$E\left(\frac{Q_1}{9}\right) = \sigma_b^2 \qquad E\left(\frac{Q_2}{3}\right) = \sigma_b^2 + 3.1795\sigma_a^2$$

System of Equations

$$\frac{Q_1}{9} = .05644 = \hat{\sigma}_b^2$$

$$\frac{Q_2}{3} = .27005 = \hat{\sigma}_b^2 + 3.1795\hat{\sigma}_a^2$$

Solution to System of Equations

$$\hat{\sigma}_b^2 = .05644$$

$$\hat{\sigma}_a^2 = .06719$$

Figure 19.1 Insect damage on four randomly selected varieties of wheat.

solution. The information obtained from the estimators is that the plot-to-plot variance within a variety is about .056, while the variance of the population of varieties is about .067. These two components of variance give the estimate of the variance of a randomly selected plot planted to a randomly selected variety as

$$\hat{\sigma}_a^2 + \hat{\sigma}_b^2 = .056 + .067 = .123.$$

When we analyze experiments with treatment structures that are other than one-way, there is no universally accepted technique for obtaining sums of squares from which to derive estimates of the variance components. The methods presented in Chapter 18 for computing sums of squares are used to estimate the variance components for a two-way random model.

EXAMPLE 19.3: Data for Two-Way Design in Figure 18.1

The data in Figure 19.2 are observations for Figure 18.1, where the expectations of several types of sums of squares were evaluated via synthesis. The values Q_1, Q_2, Q_3, and Q_4 correspond to Henderson's Method I sums of squares.

By equating the values of the sums of squares to their respective expected values, the following system of equations is obtained:

$$\begin{aligned}
.6428 &= \hat{\sigma}_\varepsilon^2 + .1429\hat{\sigma}_b^2 + 2.4286\hat{\sigma}_c^2 + 7\hat{\sigma}_a^2, \\
15.2143 &= 2\hat{\sigma}_\varepsilon^2 + .2\hat{\sigma}_a^2 + 4.77\hat{\sigma}_c^2 + 9.286\hat{\sigma}_b^2, \\
108.6905 &= 2\hat{\sigma}_\varepsilon^2 + 6.37\hat{\sigma}_c^2, \quad \text{and} \\
30.6666 &= 8\hat{\sigma}_\varepsilon^2.
\end{aligned}$$

The solution to the system of equations, which provides estimates of the variance components, is

$$\begin{aligned}
\hat{\sigma}_\varepsilon^2 &= 3.83325, \\
\hat{\sigma}_c^2 &= 15.8593, \\
\hat{\sigma}_b^2 &= -10.8817, \quad \text{and} \\
\hat{\sigma}_a^2 &= -8.2523.
\end{aligned}$$

Column Treatments

Row Treatment	1	2	3
1	10 12 11	13 15	21 19
2	16 18	13 19 14	11 13

$Q_1 = .6428$ $\quad Q_2 = 15.2143$

$Q_3 = 108.6905$ $\quad Q_4 = \text{SSERROR} = 30.666$

Figure 19.2 Data for Example 19.3.

Example 19.3 points out one of the problems often encountered when using the method-of-moments technique: It can yield negative estimates of the variance components, which are not admissible. This problem is addressed in more detail in Section 20.3, where a modified method-of-moments procedure is proposed. Listed below are the method-of-moments estimates for Example 19.3 obtained by solving the systems of equations generated by the method of fitting constants or Henderson's Method III sums of squares (Type I) and the Type IV sums of squares. (The expectations of these sums of squares were evaluated by synthesis and are listed in Section 18.3.2.)

Type I estimates:

$$\hat{\sigma}_\varepsilon^2 = 3.8333$$

$$\hat{\sigma}_c^2 = \frac{109.14509 - 2(3.8333)}{4.5178} = 22.4620$$

$$\hat{\sigma}_b^2 = \frac{14.75966 - 2(3.8333) - 4.6252(22.4620)}{9.1429} = -10.5877$$

$$\hat{\sigma}_a^2 = \frac{.64286 - 3.8333 - .1428(-10.71322) - 2.4284(22.4620)}{7.0}$$

$$= -8.0329$$

Type IV estimates:

$$\hat{\sigma}_\varepsilon^2 = 3.8333$$

$$\hat{\sigma}_c^2 = 22.4620$$

$$\hat{\sigma}_b^2 = \frac{8.9098 - 2(3.8333) - 4.5178(22.4620)}{9.0353} = -11.108$$

$$\hat{\sigma}_a^2 = \frac{.1667 - 3.8333 - 2.25(22.4620)}{6.75} = -8.0305.$$

Confidence intervals and tests of hypotheses constructed from methods-of-moments estimators are discussed in Chapter 20.

19.2 MAXIMUM LIKELIHOOD ESTIMATORS

In statistics, the most common technique of estimation is the method of maximum likelihood. The process uses the assumed distribution of the observations and constructs a likelihood function, which is a function of the model parameters. The maximum likelihood estimators are those values of the parameters from the parameter space that maximize the value of the likelihood function. In practice, the $\log_e$ of the likelihood function is maximized.

The parameter space for the general random model described in Section 18.2 is

$$\left\{-\infty < \mu < \infty; 0 < \sigma_i^2 < \infty, i = 1, 2, \ldots, k; 0 < \sigma_\varepsilon^2 < \infty\right\}.$$

For the general random model, the distribution of the vector of observations is

$$\mathbf{y} \sim N\left(\mathbf{j}_n\mu, \sigma_\varepsilon^2\mathbf{I}_n + \sigma_1^2\mathbf{X}_1\mathbf{X}_1' + \sigma_2^2\mathbf{X}_2\mathbf{X}_2' + \cdots + \sigma_k^2\mathbf{X}_k\mathbf{X}_k'\right).$$

The likelihood function of the observations is

$$L\left(\mu, \sigma_\varepsilon^2, \sigma_1^2, \sigma_2^2, \ldots, \sigma_k^2\right) = (2\Pi)^{-n/2}|\Sigma|^{-1/2} \times \exp\left[-\frac{1}{2}(\mathbf{y} - \mathbf{j}_n\mu)'\Sigma^{-1}(\mathbf{y} - \mathbf{j}_n\mu)\right].$$

The process of maximizing the likelihood function over the parameter space generally requires an iterative procedure utilizing likelihood equations generated by taking either the first derivatives or the first and second derivatives of the likelihood function with respect to each parameter of the model.

When the data are from a balanced design (equal n's and no missing cells), the set of likelihood equations generated by equating the first derivatives of the likelihood function to zero can often be solved explicitly. The solutions obtained are not restricted to the parameter space, since some of the estimated variance components can have negative values. For some balanced models, it can be shown that the maximum likelihood estimate of σ_i^2 is $\hat{\sigma}_i^2 = 0$ when the solution for σ_i^2 to the likelihood equations is negative (Searle, 1971). For unbalanced designs, an iterative technique is required where estimation is restricted to values belonging to the parameter space.

Several computational algorithms have been developed for maximizing the likelihood function and thus providing maximum likelihood estimates of the model parameters (Hemmerle and Hartley, 1973) and (Corbeil and Searle, 1976). The large sample-size variances of the maximum likelihood estimates can be obtained by inverting the matrix of second derivatives where the derivatives are evaluated at the values of the maximum likelihood estimates. Maximum likelihood estimators and their variances have been obtained for several experimental designs and are reported in Searle (1971).

Maximum likelihood techniques are being incorporated into some of the popular statistical packages. The program BMD-P3V seems to work quite satisfactorily, but the program SAS® VARCOMP does not have an adequate iteration procedure and often diverges when it should not or says it converges when it does not. (In addition, SAS® VARCOMP does not maximize over the parameter space.) However, both programs work quite satisfactorily for balanced designs.

In using programs, the experimenter should thoroughly investigate the algorithm being used and determine its properties, that is, whether it always yields meaningful estimates and whether it maximizes the likelihood function over the parameter space.

The maximum likelihood estimates for the variance components in Examples 19.2 and 19.3 were obtained using BMD-P3V. The maximum likelihood estimates for σ_ε^2 and σ_a^2 in Example 19.2 are $\hat{\sigma}_\varepsilon^2 = .05749$ and $\hat{\sigma}_a^2 = .04855$. The maximum likelihood estimates for the parameters of the model in Example 19.3 are $\hat{\sigma}_\varepsilon^2 = 3.8428$, $\hat{\sigma}_c^2 = 7.40708$, $\hat{\sigma}_b^2 = 0$, and $\hat{\sigma}_a^2 = 0$. The algorithm in BMD-P3V does the maximization over the parameter space, as is shown by the values of $\hat{\sigma}_a^2$ and $\hat{\sigma}_b^2$ being assigned the value of zero.

19.3 MINQUE METHOD

In general, the only property that the method-of-moments estimators of variance components possess is that they are unbiased. Rao (1971) described a general procedure for obtaining *m*inimum *n*orm *q*uadratic *u*nbiased *e*stimators (MINQUE) for variance components. For the general random model of Section 18.2, the MINQUE of a linear combination of variances

19.3.1 Description

$$\theta = C_0\sigma_\varepsilon^2 + C_1\sigma_1^2 + \cdots + C_k\sigma_k^2$$

is a quadratic function of the observations that is unbiased for θ and has minimum variance within the class of quadratic unbiased estimators of θ. Thus, MINQUE estimators of variance components possess the minimum variance property, whereas the method-of-moments estimators generally do not.

The estimate of θ is a quadratic function of $\mathbf{y}$; thus, for some matrix $\mathbf{A}$, the estimator of θ has the form $\mathbf{y}'\mathbf{A}\mathbf{y}$. The expectation of $\mathbf{y}'\mathbf{A}\mathbf{y}$ is

$$E(\mathbf{y}'\mathbf{A}\mathbf{y}) = \mathrm{Tr}(\boldsymbol{\Sigma}\mathbf{A}) + \mu^2\mathbf{j}_n'\mathbf{A}\mathbf{j}_n$$

and by assumption, $E(\mathbf{y}'\mathbf{A}\mathbf{y}) = \theta$. Since the expectation does not depend on μ, $\mathbf{A}$ must be chosen to satisfy $\mathbf{j}_n'\mathbf{A}\mathbf{j}_n = 0$. Under the conditions of normality, the variance of $\mathbf{y}'\mathbf{A}\mathbf{y}$ when $\mathbf{j}_n'\mathbf{A}\mathbf{j}_n = 0$ is

$$\mathrm{Var}(\mathbf{y}'\mathbf{A}\mathbf{y}) = 2\,\mathrm{Tr}[\boldsymbol{\Sigma}\mathbf{A}]^2.$$

Thus the MINQUE of θ is $\mathbf{y}'\mathbf{A}\mathbf{y}$ where $\mathbf{A}$ is chosen such that $\mathrm{Tr}[\mathbf{A}\boldsymbol{\Sigma}] = \theta$ and $\mathrm{Tr}[\mathbf{A}\boldsymbol{\Sigma}]^2$ is minimized over the parameter space

$$\{0 < \sigma_\varepsilon^2 < \infty; 0 < \sigma_i^2 < \infty, i = 1, 2, \ldots, k\}.$$

Rao (1971) shows that the MINQUE of $\boldsymbol{\sigma}^{2\prime} = (\sigma_\varepsilon^2, \sigma_1^2, \ldots, \sigma_k^2)$ is $\hat{\boldsymbol{\sigma}}^2 = \mathbf{S}^{-1}\mathbf{u}$ where $\mathbf{S}$ is a $(k+1) \times (k+1)$ matrix with elements

$$S_{ii'} = \mathrm{Tr}[\mathbf{X}_i\mathbf{X}_i'\mathbf{R}\mathbf{X}_{i'}\mathbf{X}_{i'}'\mathbf{R}] \qquad i, i' = 0, 1, \ldots, k,$$

$\mathbf{u}$ is a $(k+1) \times 1$ vector with elements

$$u_i = \mathbf{y}'\mathbf{R}\mathbf{X}_i\mathbf{X}_i'\mathbf{R}\mathbf{y} \qquad i = 0, 1, \ldots, k$$

and

$$\mathbf{R} = \Sigma^{-1}\left[\mathbf{I} - \mathbf{j}(\mathbf{j}'\Sigma^{-1}\mathbf{j})^{-1}\mathbf{j}'\Sigma^{-1}\right].$$

The solution for $\hat{\boldsymbol{\sigma}}^2$ depends on the elements of Σ that are functions of the unknown variance components. In order to compute the MINQUE of $\boldsymbol{\sigma}^2$, some constants must be substituted into Σ for $\sigma_\varepsilon^2, \sigma_i^2, \ldots, \sigma_k^2$. For that set of constants, the estimate of $\boldsymbol{\sigma}^2$ is MINQUE (and is a quadratic function of $\mathbf{y}$).

Usually it is best to substitute values for $\sigma_\varepsilon^2, \ldots, \sigma_k^2$ into Σ that are close to the true values. One possible procedure is to obtain values from other experiments. Another method is to use an iterative process constructed by using some initial values of the variance components, say $\sigma_{\varepsilon 0}^2, \sigma_{10}^2, \ldots, \sigma_{k0}^2$, to start the process. Use those initial values to evaluate Σ and obtain $\hat{\boldsymbol{\sigma}}_{(1)}^2$. Here $\hat{\boldsymbol{\sigma}}_{(1)}^2$ depends on the values chosen for $\boldsymbol{\sigma}_0^2$. Then use $\hat{\boldsymbol{\sigma}}_{(1)}^2$ to evaluate Σ to obtain the second iteration estimate, $\hat{\boldsymbol{\sigma}}_{(2)}^2$. The process is terminated when there is very little change from one iteration to the next. The resulting iterative MINQUE estimators are no longer quadratic functions of $\mathbf{y}$, since the elements of Σ are functions of $\mathbf{y}$. The final estimator of $\boldsymbol{\sigma}^2$, say at step $m + 1$, can be called MINQUE given the previous value of $\hat{\boldsymbol{\sigma}}_{(m)}^2$.

For balanced models, Swallow and Searle (1978) have shown that the equations simplify so that an explicit solution can be obtained. The solutions are identical to those provided by the method of moments.

When there are unequal sample sizes and/or empty cells, the iterative procedure seems appropriate. In using the iterative procedure, the inverse of Σ is required at each step. The matrix Σ is $n \times n$ and for most experiments is far too large to be practically inverted by a computer. For some specific models, the task of inverting Σ can be reduced by taking into account properties of Σ and obtaining an explicit expression for Σ^{-1} (Swallow and Searle, 1978). Other computational algorithms are being developed, which will hopefully make MINQUE estimators easier to obtain.

The MINQUE estimators (either evaluated at constants for σ_i^2 or at a given previous step) are linear combinations of quadratic forms of $\mathbf{y}$. Thus, the variance can be evaluated since the variance of a quadratic form $\mathbf{y}'\mathbf{B}\mathbf{y}$ is $2\,\mathrm{Tr}(\mathbf{B}\Sigma)^2$. Swallow and Searle (1978) show how to use the expressions to obtain variances for the estimators in an unbalanced one-way model. They computed and compared the variances of the MINQUE and method-of-moments estimators for the unbalanced one-way model with varying numbers of populations, sample sizes, and values of the variances. The variances of the MINQUE estimators were evaluated as if the true values of σ_a^2 and σ_ε^2 were used in the estimation process. The estimate of σ_ε^2 obtained from the method of moments was quite comparable to that from the MINQUE method where the variance of the MINQUE was no more than 4% smaller than the method-of-moments estimator.

For fairly balanced models (n_i's not too different), the variance of the MINQUE of σ_a^2 was no more than 10% smaller than the variance of the method-of-moments estimator of σ_a^2. For many unbalanced sample sizes, the variance of the MINQUE was as much as 60% smaller than the corresponding method-of-moments estimator. The comparisons should be much more alike if values other than the true σ_a^2 and σ_ε^2 are used in the estimation process.

MINQUE is a very viable method for estimating variance components and can be used for situations where numerical techniques allow the computations to be done easily by a computer. The researcher should watch the literature for new algorithms for computing MINQUEs and for their implementation into statistical software. SAS® VARCOMP has a MINQUE option, but it only does one iteration, which does not seem to be sufficient for some unbalanced designs.

19.3.2 Application

The conclusion of this section presents an example of the unbalanced one-way design with the MINQUE estimators.

EXAMPLE 19.4: Unbalanced One-Way Design

The equations from Swallow and Searle (1978) are presented for a general one-way model; they are then applied to the data in Example 19.3. The model is

$$y_{ij} = \mu + a_i + \varepsilon_{ij} \qquad i = 1, 2, \ldots, t \quad j = 1, 2, \ldots, n_i$$

where $\mathbf{a} \sim N(\mathbf{0}, \sigma_a^2 \mathbf{I}_t)$ and $\boldsymbol{\varepsilon} \sim N(\mathbf{0}, \sigma_\varepsilon^2 \mathbf{I}_N)$. Define

$$K_i = \frac{n_i}{\sigma_{\varepsilon 0}^2 + n_i \sigma_{a0}^2} \quad \text{and} \quad K = \frac{1}{\sum_{i=1}^t K_i}.$$

The elements of the matrix $\mathbf{S}$ are

$$S_{11} = \sum_{i=1}^t K_i^2 - 2K \sum_{i=1}^t K_i^3 + K^2 \left(\sum_{i=1}^t K_i^2 \right)^2,$$

$$S_{12} = \sum_{i=1}^t \frac{K_i^2}{n_i} - 2K \sum_{i=1}^t \frac{K_i^3}{n_i} + K^2 \left(\sum_{i=1}^t K_i^2 \right) \left(\sum_{i=1}^t \frac{K_i^2}{n_i} \right),$$

and

$$S_{22} = \frac{N - t}{\sigma_{\varepsilon 0}^4} + \sum_{i=1}^t \frac{K_i^2}{n_i^2} - 2K \sum_{i=1}^t \frac{K_i^3}{n_i^2} + K^2 \left(\sum_{i=1}^t \frac{K_i^2}{n_i} \right)^2.$$

The elements of the vector **U** are

$$U_1 = \sum_{i=1}^{t} K_i^2 \left(\bar{y}_{i\cdot} - K \sum_{i=1}^{t} K_i \bar{y}_{i\cdot} \right)^2$$

and

$$U_2 = \frac{\sum_{i=1}^{t} \sum_{j=1}^{n_i} y_{ij}^2 - \sum_{i=1}^{t} n_i \bar{y}_{i\cdot}^2}{\sigma_{\varepsilon 0}^4} + \sum_{i=1}^{t} \frac{K_i^2 \left(\bar{y}_{i\cdot} - K \sum_{i=1}^{t} K_i \bar{y}_{i\cdot} \right)^2}{n_i}.$$

For a given set of values $\sigma_{\varepsilon 0}^2$ and σ_{a0}^2, the MINQUE estimators of σ_ε^2 and σ_a^2 are $\hat{\boldsymbol{\sigma}}^2 = \mathbf{S}^{-1}\mathbf{U}$, or

$$\hat{\sigma}_\varepsilon^2 = \frac{S_{11}U_2 - S_{12}U_1}{C}$$

and

$$\hat{\sigma}_a^2 = \frac{S_{22}U_1 - S_{12}U_2}{C}$$

where $C = S_{11}S_{22} - S_{12}^2$. The variances of the estimators are

$$\text{Var}(\hat{\sigma}_\varepsilon^2) = \frac{2S_{11}}{C}$$

and

$$\text{Var}(\hat{\sigma}_a^2) = \frac{2S_{22}}{C},$$

and

$$\text{Cov}(\hat{\sigma}_\varepsilon^2, \hat{\sigma}_a^2) = \frac{-2S_{12}}{C}.$$

The estimators are not very sensitive to the choice of $\sigma_{\varepsilon 0}^2$ and σ_{a0}^2, but the variances are. Table 19.1 contains the estimators and their variances as a function of the starting values listed for the data in Figure 19.1.

The last line in Table 19.1 uses the method-of-moments estimators as values for $\sigma_{\varepsilon 0}^2$ and σ_{a0}^2. The variances of the estimators are quite wild for $\sigma_{\varepsilon 0}^2$ and σ_{a0}^2 that are far from the estimators of σ_ε^2 and σ_a^2. If an iterative procedure is used, the solution converges to $\hat{\sigma}_\varepsilon^2 = .057003$ and $\hat{\sigma}_a^2 = .073155$ with $\text{Var}(\hat{\sigma}_\varepsilon^2) = .000721$, $\text{Var}(\hat{\sigma}_a^2) = .005694$, and $\text{Cov}(\hat{\sigma}_\varepsilon^2, \hat{\sigma}_a^2) = -.000235$. The iterative procedure was started at ($\sigma_{\varepsilon 0}^2 = 1$, $\sigma_{a0}^2 = 2$) and ($\sigma_{\varepsilon 0}^2 = 1000$, $\sigma_{a0}^2 = 50$) among others, and converged each time to the above values in four iterations. After two iterations, the estimators were quite stable but the variances were still changing. The estimates of σ_ε^2 and σ_a^2 obtained from SAS® VARCOMP with the MINQUE method are $\hat{\sigma}_\varepsilon^2 = .06503$ and $\hat{\sigma}_a^2 = .05638$.

Table 19.1 MINQUEs for Data in Figure 19.1

$\sigma^2_{\varepsilon 0}$	σ^2_{a0}	$\hat{\sigma}^2_\varepsilon$	$\hat{\sigma}^2_a$	$\text{Var}(\hat{\sigma}^2_\varepsilon)$	$\text{Var}(\hat{\sigma}^2_a)$	$\text{Cov}(\hat{\sigma}^2_\varepsilon, \hat{\sigma}^2_a)$
1	2	.05671	.07489	.222	3.639	−0.728
10	20	.05671	.07489	22.2	363.9	−7.276
5	1	.0606	.06306	5.517	4.963	−1.717
.05644	.06719	.05707	.07282	.000707	.00496	−.000229

The SAS® VARCOMP was also used to obtain estimates for the parameters of the model in Example 19.3. The estimates are $\hat{\sigma}^2_\varepsilon = 3.9132$, $\hat{\sigma}^2_c = 22.3196$, $\hat{\sigma}^2_b = -10.5495$, and $\hat{\sigma}^2_a = -7.9954$. Again, the SAS® technique cannot do the optimization over the parameter space because it can produce negative estimates.

CONCLUDING REMARKS

This chapter presented three methods, the method of moments, the maximum likelihood method, and the MINQUE method, for obtaining estimates of the variance components of a random model. Examples were described for each method. The method-of-moments estimators, which are the easiest to compute, were discussed in detail. The maximum likelihood and MINQUE methods are difficult to compute and generally require a good computer algorithm. The MINQUE method was described in detail for the unbalanced one-way model.

20

Methods for Making Inferences About Variance Components

CHAPTER OUTLINE

When an experimenter designs an experiment involving factors that are random effects, he generally wishes to make inferences about specific variance components. In particular, if σ_a^2 is the variance component for the levels of factor A, the experimenter may wish to determine if there is enough evidence to conclude that $\sigma_a^2 > 0$. That can be accomplished (1) by testing the hypothesis H_0: $\sigma_a^2 = 0$ versus H_a: $\sigma_a^2 > 0$, (2) by constructing a confidence interval about σ_a^2, or (3) by constructing a lower confidence limit for σ_a^2. This chapter addresses these problems for random models. Several cases are discussed, including using the method of moments for balanced designs and using the methods of moments, maximum likelihood, and MINQUE for unbalanced designs.

20.1 TESTING HYPOTHESES

There are two basic techniques for testing hypotheses about variance components. The first technique uses method-of-moments estimators and sums of squares from the analysis of variance table to construct F-statistics. For most balanced models, the F-statistics are distributed exactly as F-distributions, whereas for unbalanced models the distributions are approximate F-distributions with the approximations becoming poorer as the designs become more unbalanced. The second technique is the likelihood-ratio test, which is asymptotically distributed as a chi-square distribution.

For balanced designs, the F-statistic approach is probably better than the likelihood-ratio test, while for unbalanced designs there is no clear-cut choice. For very unbalanced designs, with a large amount of data, the likelihood ratio is likely to be better if it is based on a good maximum likelihood algorithm.

20.1.1 Balanced Models

If the model is balanced, then the sums of squares obtained by the usual analysis of variance are independently distributed as scalar multiples of chi-square random variables. Let u denote a sum of squares based on v degrees of freedom with the expected mean square $\sigma_\varepsilon^2 + k_1\sigma_1^2 + k_2\sigma_2^2$; then

$$\frac{u}{\sigma_\varepsilon^2 + k_1\sigma_1^2 + k_2\sigma_2^2}$$

is distributed as a chi-square random variable with v degrees of freedom. For many hypotheses of the form H_0: $\sigma_1^2 = 0$ versus H_a: $\sigma_1^2 > 0$, there are two independent sums of squares, denoted by u_1 and u_2, based on v_1 and v_2 degrees of freedom, respectively, with expectations

$$E\left(\frac{u_1}{v_1}\right) = \sigma_\varepsilon^2 + k_1\sigma_1^2 + k_2\sigma_2^2 + k_3\sigma_3^2$$

and

$$E\left(\frac{u_2}{v_2}\right) = \sigma_\varepsilon^2 + k_2\sigma_2^2 + k_3\sigma_3^2.$$

The hypothesis H_0: $\sigma_1^2 = 0$ versus H_a: $\sigma_1^2 > 0$ is equivalent to H_0: $E(u_1/v_1) = E(u_2/v_2)$ versus H_a: $E(u_1/v_1) > E(u_2/v_2)$. The test statistic is $F = (u_1/v_1)/(u_2/v_2)$, which, under the conditions of H_0, is distributed as an F-distribution with v_1 and v_2 degrees of freedom. The hypothesis is rejected for large values of F. The following two examples demonstrate the above procedure.

EXAMPLE 20.1: Two-Way Treatment Structure with Both Factors Random in a Completely Randomized Design Structure

A model for the two-way treatment structure with both factors random in a completely randomized design structure is

$$y_{ijk} = \mu + a_i + b_j + c_{ij} + \varepsilon_{ijk}$$

for $i = 1, 2, \ldots, a$; $j = 1, 2, \ldots, b$; and $k = 1, 2, \ldots, n$, where $a_i \sim$ i.i.d. $N(0, \sigma_a^2)$, $b_j \sim$ i.i.d. $N(0, \sigma_b^2)$, $c_{ij} \sim$ i.i.d. $N(0, \sigma_c^2)$, and $\varepsilon_{ijk} \sim$ i.i.d. $N(0, \sigma_\varepsilon^2)$. The analysis of variance table for the model is shown in Table

Table 20.1 Analysis of Variance Table for Example 20.1

Source of Variation	*df*	*SS*	*EMS*
A	$a-1$	$nb\sum_{i=1}^{a}(\bar{y}_{i\cdot\cdot} - \bar{y}_{\cdot\cdot\cdot})^2$	$\sigma_\varepsilon^2 + n\sigma_c^2 + nb\sigma_a^2$
B	$b-1$	$na\sum_{i=1}^{b}(\bar{y}_{\cdot j\cdot} - \bar{y}_{\cdot\cdot\cdot})^2$	$\sigma_\varepsilon^2 + n\sigma_c^2 + na\sigma_b^2$
AB	$(a-1)(b-1)$	$n\sum_{i=1}^{a}\sum_{j=1}^{b}(\bar{y}_{ij\cdot} - \bar{y}_{i\cdot\cdot} - \bar{y}_{\cdot j\cdot} + \bar{y}_{\cdot\cdot\cdot})^2$	$\sigma_\varepsilon^2 + n\sigma_c^2$
Error	$ab(n-1)$	$\sum_{i=1}^{a}\sum_{j=1}^{b}\sum_{k=1}^{n}(y_{ijk} - \bar{y}_{ij\cdot})^2$	σ_ε^2

Hypothesis	*Test Statistic*
H_0: $\sigma_a^2 = 0$ vs. H_a: $\sigma_a^2 > 0$	$F = \dfrac{\text{MSA}}{\text{MSAB}}$
H_0: $\sigma_b^2 = 0$ vs. H_a: $\sigma_b^2 > 0$	$F = \dfrac{\text{MSB}}{\text{MSAB}}$
H_0: $\sigma_{ab}^2 = 0$ vs. H_a: $\sigma_{ab}^2 > 0$	$F = \dfrac{\text{MSAB}}{\text{MSERROR}}$

20.1. Also listed in Table 20.1 are the hypotheses to be tested about the variance components and the corresponding test statistics. The test statistics are constructed by examining the expected mean squares to select the proper numerators and denominators.

EXAMPLE 20.2: Three-Way Treatment Structure with Two Factors Crossed and One Factor Having Levels Nested within Levels of One Crossed Factor

The data in Figure 20.1 are from a design where the three factors in the treatment structure are random, the levels of A are crossed with the levels of B, and the levels of C are nested within the levels of B, all in a completely randomized design structure. A model to describe the data is

$$y_{ijkm} = \mu + a_i + b_j + (ab)_{ij} + c_{k(j)} + (ac)_{ik(j)} + \varepsilon_{ijkm}$$

for $i = 1, 2, \ldots, a$; $j = 1, 2, \ldots, b$; $k = 1, 2, \ldots, c$; and $m = 1, 2, \ldots, n$. The terms in the model are μ, which denotes the overall mean; $a_i \sim$ i.i.d. $N(0, \sigma_a^2)$, which denotes the effect of level i of factor A; $b_j \sim$ i.i.d. $N(0, \sigma_b^2)$, which denotes the effect of level j of factor B; $(ab)_{ij} \sim$ i.i.d. $N(0, \sigma_{ab}^2)$, which denotes the interaction between treatments A and B; $c_{k(j)} \sim$ i.i.d. $N(0, \sigma_{c(b)}^2$, which denotes the effect of level k of factor C nested within the jth level of treatment B; $(ac)_{ik(j)} \sim$ i.i.d. $N(0, \sigma_{ac(b)}^2$, which denotes the interaction between treatments A and C nested within the levels of B; and $\varepsilon_{ijkm} \sim$ i.i.d. $N(0, \sigma_\varepsilon^2)$, which denotes the sampling error.

The analysis of variance table with expected mean squares is shown in Table 20.2. By examining the expected mean squares, F-statistics can be constructed as: $F =$ MSA/MSAB tests H_0: $\sigma_a^2 = 0$ versus H_a: $\sigma_a^2 > 0$; $F =$ MSAB/MSAC(B) tests H_0: $\sigma_{ab}^2 = 0$ versus H_a: $\sigma_{ab}^2 > 0$; $F =$ MSC(B)/MSAC(B) tests H_0: $\sigma_{c(b)}^2 = 0$ versus H_a: $\sigma_{c(b)}^2 > 0$; and $F =$ MSAC(B)/MSERROR tests H_0: $\sigma_{ac(b)}^2 = 0$ versus H_a: $\sigma_{ac(b)}^2 > 0$.

			Factor A			
			A_1		A_2	
	B_1	C_1	20	20	25	26
		C_2	23	22	26	27
Factor B	B_2	C_3	36	34	38	36
		C_4	39	38	40	39

Figure 20.1 Data for Example 20.2

Table 20.2 Analysis of Variance Table for Example 20.2

Source of Variation	*df*	*SS*	*EMS*
A	$a-1$	$nbc\sum_i(\bar{y}_{i\cdots}-\bar{y}_{\cdots\cdot})^2$	$\sigma_\varepsilon^2+n\sigma_{ac(b)}^2+nc\sigma_{ab}^2+nbc\sigma_a^2$
B	$b-1$	$na\sum_j(\bar{y}_{\cdot j\cdot\cdot}-\bar{y}_{\cdots\cdot})^2$	$\sigma_\varepsilon^2+n\sigma_{ac(b)}^2+na\sigma_{c(b)}^2+nc\sigma_{ab}^2+nac\sigma_b^2$
AB	$(a-1)(b-1)$	$nc\sum_i\sum_j(\bar{y}_{ij\cdot\cdot}-\bar{y}_{i\cdots}-\bar{y}_{\cdot j\cdot\cdot}+\bar{y}_{\cdots\cdot})^2$	$\sigma_\varepsilon^2+n\sigma_{ac(b)}^2+nc\sigma_{ab}^2$
$C(B)$	$b(c-1)$	$na\sum_j\sum_k(\bar{y}_{\cdot jk\cdot}-\bar{y}_{\cdot j\cdot\cdot})^2$	$\sigma_\varepsilon^2+n\sigma_{ac(b)}^2+na\sigma_{c(b)}^2$
$AC(B)$	$(a-1)b(c-1)$	$n\sum_i\sum_j\sum_k(\bar{y}_{ijk\cdot}-\bar{y}_{ij\cdot\cdot}-\bar{y}_{\cdot jk\cdot}+\bar{y}_{\cdot j\cdot\cdot})^2$	$\sigma_\varepsilon^2+n\sigma_{ac(b)}^2$
Error	$abc(n-1)$	$\sum_i\sum_j\sum_k\sum_m(y_{ijkm}-\bar{y}_{ijk\cdot})^2$	σ_ε^2

However, there is no F-statistic to test H_0: $\sigma_b^2=0$ versus H_a: $\sigma_b^2>0$ since none of the mean squares have the expected value $\sigma_\varepsilon^2+n\sigma_{ac(b)}^2+na\sigma_{c(b)}^2+nc\sigma_{ab}^2$, the expected value of MSB when $\sigma_b^2=0$. But there is a linear combination of mean squares (not including MSB) that has the desired expectation, that is, E[MSC(B) + MSAB − MSAC(B)] $=\sigma_\varepsilon^2+n\sigma_{ac(b)}^2+na\sigma_{c(b)}^2+nc\sigma_{ac}^2$. Let Q = MSC(B) + MSAB − MSAC(B); then the test statistic is $F=\text{MSB}/Q$, which is approximately distributed as an F-distribution with $b-1$ and r degrees of freedom. To determine r, the degrees of freedom, the distribution of $rQ/E(Q)$ is approximated by a chi-square distribution. The technique, called a Satterthwaite (1946) approximation, uses $Q=q_1\text{MS}_1+\cdots+q_k\text{MS}_k$ where MS_i is a mean square based on f_i degrees of freedom, the mean squares are independent, and q_i is a known constant. Then $rQ/E(Q)$ is approximately distributed as a central chi-square random variable based on r degrees of freedom where

$$r=\frac{(Q)^2}{\sum_{i=1}^k(q_i\text{MS}_i)^2/f_i}.$$

Assume U is a mean square based on f degrees of freedom, independent of $\text{MS}_1,\text{MS}_2,\ldots,\text{MS}_k$, with expectation $E(U)=E(Q)+k\sigma_0^2$. The statistic to test H_0: $\sigma_0^2=0$ versus H_a: $\sigma_0^2>0$ is $F=U/Q$, which is approximately distributed as an F-distribution with f and r degrees of freedom. The statistic to test H_0: $\sigma_b^2=0$ versus H_a: $\sigma_b^2>0$ is

Table 20.3 Analysis of Variance Table for Data in Figure 20.1

Source of Variation	*df*	*SS*	*MS*	*F*
A	1	39.0625	39.0625	3.698
B	1	770.0625	770.0625	
AB	1	10.5625	10.5625	13.000
$C(B)$	2	24.1250	12.0625	14.846
$AC(B)$	2	1.6250	.8125	1.000
Error	8	6.5000	.8125	

$F = \text{MSB}/Q$, which is approximately distributed as an F-distribution with $b - 1$ and r degrees of freedom where

$$r = \frac{Q^2}{\dfrac{[\text{MSC(B)}]^2}{b(c-1)} + \dfrac{(\text{MSAB})^2}{(a-1)(b-1)} + \dfrac{[\text{MSAC(B)}]^2}{b(a-1)(c-1)}}.$$

Table 20.3 contains the analysis of variance table for the data in Figure 20.1. Included in the table are the F-statistics for testing the respective hypotheses. To test H_0: $\sigma_b^2 = 0$ versus H_a: $\sigma_b^2 > 0$, let

$$\begin{aligned} Q &= \text{MSC(B)} + \text{MSAB} - \text{MSAC(B)} \\ &= 12.0625 + 10.5625 - .8125 = 21.8125 \end{aligned}$$

and

$$\begin{aligned} r &= \frac{(21.8125)^2}{(12.0625)^2/2 + (10.5625)^2/1 + (.8125)^2/2} \\ &= \frac{475.7852}{184.9785} = 2.57. \end{aligned}$$

The test statistic is $F = 770.0625/21.8125 = 35.304$, which is based on 1 and 2.57 degrees of freedom. For small degrees of freedom, one should obtain the percentage points by using interpolation. To obtain the value of $F_{.05}(1, 2.57)$, use $F_{.05}(1,2) = 4.3027$ and $F_{.05}(1,3) = 3.1825$; then

$$F_{.05}(1, 2.57) = 4.3027 - .57(4.3027 - 3.1825) = 3.6642.$$

To test hypotheses about variance components in balanced designs, the F-test constructed from a ratio of two mean squares should be used whenever possible. When the ratio of two mean squares cannot be used, the Satterthwaite approximation seems to be effective.

20.1.2 Unbalanced Models

For unbalanced models, the sums of squares in the analysis of variance table are no longer necessarily independent, although sets of sums of squares may be independent for some special cases. The residual or error sum of squares is always independent of the other sums of squares in the analysis of variance table. Thus, for any mean square U with expectation $\sigma_\varepsilon^2 + k\sigma_0^2$, the statistic $F = U/\text{MSERROR}$ tests the hypothesis H_0: $\sigma_0^2 = 0$ versus H_a: $\sigma_0^2 > 0$. Under the conditions of H_0, F is distributed as a central F-distribution with u and v degrees of freedom where u are the degrees of freedom associated with U and v are the degrees of freedom associated with MSERROR.

Mean squares with expectations involving more variance components than σ_ε^2 and σ_0^2 cannot be used to obtain test statistics about single variance components having exact F-distributions. The reason the ratios are not exactly distributed as F is that the respective mean squares are not independently distributed. If the design is not too unbalanced, then the F-ratios may be adequate.

For unbalanced designs, to test $\sigma_a^2 = 0$, there will be one mean square U_1 with expectation

$$E(U_1) = \sigma_\varepsilon^2 + k_1\sigma_a^2 + k_2\sigma_b^2 + k_3\sigma_c^2.$$

No other mean square has expectation $\sigma_\varepsilon^2 + k_2\sigma_b^2 + k_3\sigma_c^2$. One method is to take a linear combination of other mean squares, say $Q = \Sigma_{i=2}^k q_i \text{MS}_i$ where $E(Q) = \sigma_\varepsilon^2 + k_2\sigma_b^2 + k_3\sigma_c^2$. The Satterthwaite approximation can be used to approximate the distribution of Q, that is, to find r such that $rQ/E(Q)$ is approximately distributed as a chi-square random variable with r degrees of freedom. The approximation is twofold, since (1) the degrees of freedom are approximated and (2) the mean squares making up Q are not necessarily independently distributed as required by the approximation.

The second method employs the likelihood ratio, which involves evaluating the value of the likelihood function for the complete model and evaluating the value of the likelihood function under the conditions of H_0. For the general random model

$$\mathbf{y} = \mathbf{j}_n\mu + \mathbf{X}_1\mathbf{a}_1 + \mathbf{X}_2\mathbf{a}_2 + \cdots + \mathbf{X}_k\mathbf{a}_k + \boldsymbol{\varepsilon},$$

the likelihood function is

$$L(\mu, \sigma_1^2, \sigma_2^2, \ldots, \sigma_k^2) = (2\Pi)^{-n/2}|\boldsymbol{\Sigma}|^{-1/2}$$
$$\exp\left[-\frac{1}{2}(\mathbf{y} - \mathbf{j}_n\mu)'\boldsymbol{\Sigma}^{-1}(\mathbf{y} - \mathbf{j}_n\mu)\right]$$

where $\boldsymbol{\Sigma} = \sigma_\varepsilon^2\mathbf{I} + \sigma_1^2\mathbf{X}_1\mathbf{X}_1' + \sigma_2^2\mathbf{X}_2\mathbf{X}_2' + \cdots + \sigma_k^2\mathbf{X}_k\mathbf{X}_k'$. The likelihood function subject to the conditions of H_0: $\sigma_1^2 = 0$ is

$$L_0\left(\mu, 0, \sigma_2^2, \sigma_2^2, \ldots, \sigma_k^2\right) = (2\Pi)^{-n/2}|\Sigma_0|^{-1/2}$$
$$\exp\left[-\frac{1}{2}(\mathbf{y} - \mathbf{j}_n\mu)'\Sigma_0^{-1}(\mathbf{y} - \mathbf{j}_n\mu)\right]$$

where $\Sigma_0 = \sigma_\varepsilon^2\mathbf{I} + \sigma_2^2\mathbf{X}_2\mathbf{X}_2' + \sigma_3^2\mathbf{X}_3\mathbf{X}_3' + \cdots + \sigma_k^2\mathbf{X}_k\mathbf{X}_k'$.

Next we must obtain maximum likelihood estimators for the parameters of both likelihood functions and evaluate the likelihood functions at those estimators. The likelihood ratio is

$$LR\left(\sigma_1^2 = 0\right) = \frac{L_0\left(\hat{\mu}_0, 0, \hat{\sigma}_{20}^2, \hat{\sigma}_{30}^2, \ldots, \hat{\sigma}_{k0}^2\right)}{L\left(\hat{\mu}, \hat{\sigma}_1^2, \hat{\sigma}_2^2, \ldots, \hat{\sigma}_k^2\right)}$$

where $\hat{\sigma}_{i0}^2$ denotes the maximum likelihood estimate from the likelihood function under the conditions of H_0. Under these conditions, the asymptotic distribution of

$$-2\log_e\left[\text{LR}\left(\sigma_1^2 = 0\right)\right] = -2\log_e\left[L_0\left(\hat{\mu}_0, 0, \hat{\sigma}_{20}^2, \hat{\sigma}_{30}^2, \ldots, \hat{\sigma}_{k0}^2\right)\right]$$
$$+2\log_e\left[L\left(\hat{\mu}, \hat{\sigma}_1^2, \hat{\sigma}_2^2, \ldots, \hat{\sigma}_k^2\right)\right]$$

is central chi-square with 1 degree of freedom. The reason there is one degree of freedom is that there is one less parameter in $L_0(\cdot)$ than in $L(\cdot)$. The decision rule is to reject H_0 if $-2\log_e[\text{LR}(\sigma_1^2 = 0)] > \chi_\alpha^2(1)$. The evaluation of the likelihood functions, which require maximum likelihood estimates, is not an easy task computationally and generally requires a good computer algorithm. The likelihood-ratio test statistics for the next two examples were computed by using BMD-P3V.

EXAMPLE 20.3: Wheat Varieties—One-Way Random Model

The maximum likelihood estimates of the parameters for the model describing the data in Example 19.2 are $\hat{\mu} = 3.991$, $\hat{\sigma}_b^2 = .057$, and $\hat{\sigma}_a^2 = .049$. The value of $-2\log_e L(\hat{\mu}, \hat{\sigma}_b^2, \hat{\sigma}_a^2)$ equals 4.967. Under the conditions of H_0: $\sigma_a^2 = 0$, the maximum likelihood estimates of the parameters are $\hat{\mu}_0 = 4.027$, $\hat{\sigma}_{b0}^2 = .101$, and $\hat{\sigma}_{a0}^2 = 0$. The value of $-2\log_e L_0(\hat{\mu}_0, \hat{\sigma}_{b0}^2, 0)$ equals 7.138. The value of $-2\log_e$ of the likelihood ratio is $7.138 - 4.967 = 2.171$. The value 2.171 is compared with percentage points of a central chi-square distribution with one degree of freedom.

EXAMPLE 20.4: Unbalanced Two-Way Design

Example 19.3 provided an analysis of the data in Figure 19.2. The maximum likelihood estimates of the model parameters are $\hat{\mu} = 18.846$, $\hat{\sigma}_a^2 = 0$, $\hat{\sigma}_b^2 = 0$, $\hat{\sigma}_c^2 = 7.407$, and $\hat{\sigma}_\varepsilon^2 = 3.843$. The value of

$-2\log_e L(\hat{\mu}, \hat{\sigma}_\varepsilon^2, \hat{\sigma}_a^2, \hat{\sigma}_b^2, \hat{\sigma}_c^2) = 68.726$. To test $\sigma_a^2 = 0$ or $\sigma_b^2 = 0$, the value of the $-2\log_e$ likelihood function is unchanged, since zero is the maximum likelihood estimator of σ_a^2 and σ_b^2. Thus, one would conclude that $\sigma_a^2 = 0$ and $\sigma_b^2 = 0$ since the values of the test statistics are both zero. The estimates of the parameters under H_0: $\sigma_c^2 = 0$ are $\hat{\mu}_0 = 14.643$, $\hat{\sigma}_{a0}^2 = 0$, $\hat{\sigma}_{b0}^2 = 0$, $\hat{\sigma}_{c0}^2 = 0$, and $\hat{\sigma}_{\varepsilon 0}^2 = 11.087$. The value of $-2\log_e[L(\hat{\mu}_0, \hat{\sigma}_{a0}^2, \hat{\sigma}_{b0}^2, \hat{\sigma}_{c0}^2, \hat{\sigma}_{\varepsilon 0}^2)] = 73.411$. The test statistic is $-2\log_e \text{LR}(\sigma_c^2 = 0) = 73.411 - 68.726 = 4.685$, which is to be compared against a percentage point of a chi-square distribution with one degree of freedom.

The method of moments using the Satterthwaite approximation can be used to obtain approximate tests of $\sigma_a^2 = 0$ and $\sigma_b^2 = 0$ and an exact test of $\sigma_c^2 = 0$. To test H_0: $\sigma_c^2 = 0$ versus H_a: $\sigma_c^2 > 0$, the test statistic is

$$F = \frac{\text{MSINTERACTION}}{\text{MSERROR}} = \frac{54.345}{3.833} = 14.178,$$

which is compared with a percentage point of an F-distribution with 2 and 8 degrees of freedom. The expectations of the Type I sums of squares are in Section 18.3.2. The expectation of the MSCOLUMN is

$$E(\text{MSCOLUMN}) = \sigma_\varepsilon^2 + 4.5714\sigma_b^2 + 2.313\sigma_c^2.$$

To test H_0: $\sigma_b^2 = 0$ versus H_a: $\sigma_b^2 > 0$, a mean square needs to be constructed from MSERROR and MSINTERACTION, which have expected mean squares σ_ε^2 and $\sigma_\varepsilon^2 + 2.259\sigma_c^2$, respectively. The mean square is

$$\begin{aligned} Q &= \left(\frac{2.313}{2.259}\right)\text{MSINTERACTION} + \left(\frac{1 - 2.313}{2.259}\right)\text{MSERROR} \\ &= (1.024)\text{MSINTERACTION} - (.024)\text{MSERROR} \\ &= (1.024)(54.573) - (.024)(3.833) = 55.793. \end{aligned}$$

The approximate distribution of $rQ/E(Q)$ is that of a chi-square distribution with r degrees of freedom where

$$\begin{aligned} r &= \frac{(55.793)^2}{[(1.024)(54.573)]^2/2 + [(.024)(3.833)]^2/8} \\ &= 1.994, \end{aligned}$$

which can be round to 2 degrees of freedom. The test statistic is $F = 7.379/55.793 = .133$, which must be compared with a percentage point from an F-distribution with 2 and 1.994 degrees of freedom.

When the design is not too unbalanced and the sample size is small, the tests of hypotheses employing the Satterthwaite approximation are quite appropriate. When the design is very unbalanced and the sample size is large, the likelihood ratio test should be used. For nonextreme designs, neither method is clearly preferable. In these cases, a small simulation may need to be run to determine the properties of the procedures in order to make a good choice.

20.2 CONSTRUCTING CONFIDENCE INTERVALS

A few procedures provide exact confidence intervals about some of the variance components of some models, but most confidence intervals are approximate, relying on the Satterthwaite approximation.

For the general random model, a $(1 - \alpha)100\%$ confidence interval about σ_ε^2 is

$$\frac{v\hat{\sigma}_\varepsilon^2}{\chi^2_{\alpha/2,v}} \leq \sigma_\varepsilon^2 \leq \frac{v\hat{\sigma}_\varepsilon^2}{\chi^2_{1-\alpha/2,v}}$$

where $\hat{\sigma}_\varepsilon^2 = \text{SSRESIDUAL}/v$, v is the degrees of freedom associated with $\hat{\sigma}_\varepsilon^2$ and $\chi^2_{1-\alpha/2,v}$ and $\chi^2_{\alpha/2,v}$ denote percentage points from a chi-square distribution with $1 - \alpha/2$ area and $\alpha/2$ area to the right of the points under the curve, respectively.

20.2.1 Balanced Models

Since all sums of squares (not including SSRESIDUAL) in the analysis of variance table for a balanced model are independent of the residual sum of squares, the next result can be used to construct a confidence interval about a variance component, say σ_1^2, with a confidence coefficient of at least size $1 - \alpha$ if there is a mean square in the model with expectation $\sigma_\varepsilon^2 + a\sigma_1^2$. Let Q_1 = MSRESIDUAL based on u_1 degrees of freedom and Q_2 be a mean square based on u_2 degrees of freedom with expectation $\sigma_\varepsilon^2 + a\sigma_1^2$. A set of exact simultaneous $(1 - \alpha)100\%$ confidence intervals about σ_ε^2 and $\sigma_\varepsilon^2 + a\sigma_1^2$ is

$$\frac{u_1 Q_1}{\chi^2_{\rho/2,u_1}} \leq \sigma_\varepsilon^2 \leq \frac{u_1 Q_1}{\chi^2_{1-\rho/2,u_1}}$$

$$\frac{u_2 Q_2}{\chi^2_{\rho/2,u_2}} \leq \sigma_\varepsilon^2 + a\sigma_1^2 \leq \frac{u_2 Q_2}{\chi^2_{1-\rho/2,u_2}}$$

where $\rho = 1 - \sqrt{1 - \alpha}$.

The intersection of these two regions on the $(\sigma_\varepsilon^2, \sigma_1^2)$ plane provides an $(1 - \alpha)100\%$ simultaneous confidence region for $(\sigma_\varepsilon^2, \sigma_1^2)$. A graph of the confidence region is shown in Figure 20.2. A $(1 - \alpha)100\%$ confidence interval about σ_1^2 is obtained by determining the maximum and minimum of σ_1^2 over the confidence region, which are shown by c and d, respectively. The values of c and d can be determined by solving for the value of σ_1^2 where the respective lines intersect. Those lines intersect at

$$c = \frac{u_2 Q_2/\chi^2_{\rho/2,u_2} - u_1 Q_1/\chi^2_{1-\rho/2,u_1}}{a}$$

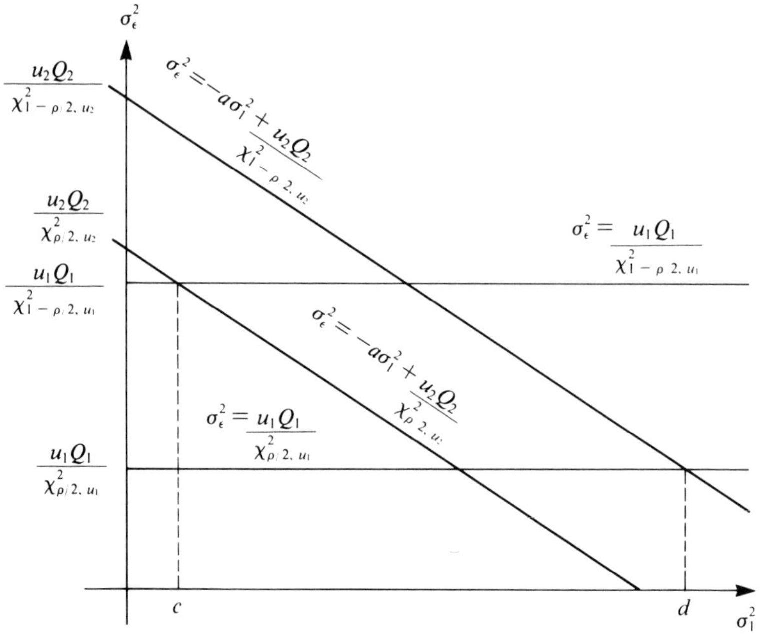

Figure 20.2 Graph of confidence region for $(\sigma_\epsilon^2, \sigma_1^2)$.

and

$$d = \frac{u_2Q_2/\chi^2_{1-\rho/2,u_2} - u_1Q_1/\chi^2_{\rho/2,u_1}}{a},$$

which provides $c \leq \sigma_1^2 \leq d$ as the $(1 - \alpha)100\%$ confidence interval about σ_1^2.

Another confidence interval, proposed by Williams (1962), has limits

$$c_1 = \frac{u_2\left[Q_2 - Q_1F_{\alpha/2,u_2,u_1}\right]}{a\chi^2_{\alpha/2,u_2}}$$

and

$$d_1 = \frac{u_2\left[Q_2 - Q_1F_{1-\alpha/2,u_2,u_1}\right]}{a\chi^2_{1-\alpha/2,u_2}}$$

and has a confidence coefficient of at least $1 - 2\alpha$. Simulation studies (Boardman, 1974, and an unpublished study by the authors) have shown that if ρ is chosen to be α, the two confidence intervals have confidence coefficients of at least $1 - \alpha$ (instead of $1 - 2\alpha$ and $(1 - \alpha)^2$, as presented). Thus the confidence intervals are used with $\rho = \alpha$ to construct $(1 - \alpha)100\%$ confidence intervals. The interval of Williams is a little shorter than the one described by $c \le \sigma_1^2 \le d$. Thus the (c, d) interval is a little more conservative than Williams's.

EXAMPLE 20.5

The data in Figure 20.3 are from a one-way treatment structure where the levels of the treatment were randomly selected in a completely randomized design structure with five treatments and three observations per treatment. An analysis of variance table including the expected mean squares is provided in Table 20.4. A 95% confidence interval about σ_ε^2 is

$$\frac{16}{20.483} \le \sigma_\varepsilon^2 \le \frac{16}{3.247}$$

or

$$.781 \le \sigma_\varepsilon^2 \le 4.928.$$

A 95% confidence interval about $\sigma_\varepsilon^2 + 3\sigma_1^2$ is

$$\frac{128}{11.143} \le \sigma_\varepsilon^2 + 3\sigma_1^2 \le \frac{128}{.484}$$

or

$$11.487 \le \sigma_\varepsilon^2 + 3\sigma_1^2 \le 264.463.$$

By using the above confidence region, a 95% confidence interval about σ_1^2 is $c \le \sigma_1^2 \le d$ where

$$c = \frac{11.487 - 4.928}{3} = 2.186$$

and

$$d = \frac{264.463 - .781}{3} = 87.894$$

		Worker 1	2	3	4	5
Station	1	2	6	7	3	12
	2	5	6	9	5	11
	3	4	8	8	6	13

Figure 20.3 Number of units assembled in Example 20.5.

Table 20.4 Analysis of Variance Table for Data in Example 20.5

Source of Variation	*df*	*SS*	*MS*	*E(MS)*
Workers	4	128	32	$\sigma_\varepsilon^2 + 3\sigma_\varepsilon^2$
Error	10	16	1.6	σ_ε^2

or

$$2.186 \le \sigma_1^2 \le 87.894.$$

The above procedure can be used to construct a confidence interval about any variance component, say σ_1^2 (in the balanced model), where there are two independent mean squares Q_1 and Q_2 based on u_1 and u_2 degrees of freedom, respectively, such that $E(Q_2) = E(Q_1) + a\sigma_1^2$. Let $\sigma_0^2 = E(Q_1)$ and replace σ_ε^2 by σ_0^2 in the previous development; then the $(1 - \alpha)100\%$ confidence interval about σ_1^2 is $c \le \sigma_1^2 \le d$. This result implies that if the method-of-moments technique can be used to test a hypothesis about a variance component, the above technique can be used to construct a confidence interval about that variance component. The result of Williams (1962) has also been applied to this case (Graybill, 1976, theorem 15.3.5).

EXAMPLE 20.6

To demonstrate the above more general procedure, a 90% confidence interval about σ_a^2 is constructed for the data in Example 20.2 (including the expected mean squares in Table 20.2 and the analysis of variance information in Table 20.3). Let $Q_2 =$ MSA and $Q_1 =$ MSAB, both of which are based on 1 degree of freedom. The expectations are

$$E(Q_1) = \sigma_\varepsilon^2 + 2\sigma_{ac(b)}^2 + 4\sigma_{ab}^2$$

and

$$E(Q_2) = \sigma_\varepsilon^2 + 2\sigma_{ac(b)}^2 + 4\sigma_{ab}^2 + 8\sigma_a^2.$$

Let $\sigma_0^2 = \sigma_\varepsilon^2 + 2\sigma_{ac(b)}^2 + 4\sigma_{ab}^2$, $\sigma_1^2 = \sigma_a^2$, and $a = 8$. Then the limits of the confidence interval are

$$c = \frac{(1 \cdot 39.0625/3.8415) - (1 \cdot 10.5625/.00393)}{8}$$

$$= -334.69$$

and

$$d = \frac{(1 \cdot 39.0625/.00393) - (1 \cdot 10.5625/3.8415)}{8}$$
$$= 1242.10$$

or $-334.69 \le \sigma_a^2 \le 1242.10$, which can be truncated to $0 \le \sigma_a^2 \le 1242.10$.

The large negative limit occurs when, say, σ_a^2 is not significantly different from zero. To eliminate the possibility of a negative limit on σ_a^2, first test the hypothesis that $\sigma_a^2 = 0$. If you fail to reject, then conclude that $\sigma_a^2 = 0$; and if you do reject, construct the confidence interval. The same technique can be used to construct confidence intervals about σ_{ab}^2, $\sigma_{ac(b)}^2$, and $\sigma_{c(b)}^2$, but it cannot be used for σ_b^2. For cases like σ_b^2, a Satterthwaite approximation can be used to construct a confidence interval about σ_b^2. The result says that $r\hat{\sigma}_b^2/\sigma_b^2$ is distributed as a chi-square random variable with r degrees of freedom where

$$\hat{\sigma}_b^2 = \sum_{i=1}^{k} q_i \mathrm{MS}_i,$$

MS_i, $i = 1, 2, \ldots, k$, is a set of independent mean squares based on f_i, $i = 1, 2, \ldots, k$, degrees of freedom, q_i, $i = 1, 2, \ldots, k$, is a set of known constants, some values of which may be zero, and

$$r = \frac{(\hat{\sigma}_b^2)^2}{\sum_{i=1}^{k}(q_i \mathrm{MS}_i)^2/f_i}.$$

An approximate $(1 - \alpha)100\%$ confidence interval about σ_b^2 is

$$\frac{r\hat{\sigma}_b^2}{\chi_{1-\alpha/2}^2(r)} \le \sigma_b^2 \le \frac{r\hat{\sigma}_b^2}{\chi_{\alpha/2}^2(r)}.$$

EXAMPLE 20.7

To demonstrate the use of the Satterthwaite approximation to construct a confidence interval about a variance component, we shall use the data from Example 20.2. The estimate of σ_b^2 is

$$\hat{\sigma}_b^2 = \frac{\mathrm{MSB} - [\mathrm{MSC(B)} + \mathrm{MSAB} - \mathrm{MSAC(B)}]}{8}$$
$$= \frac{1}{8}\mathrm{MSB} - \frac{1}{8}\mathrm{MSC(B)} - \frac{1}{8}\mathrm{MSAB} + \frac{1}{8}\mathrm{MSAC(B)}$$
$$= \frac{1}{8}(770.0625 - 12.0625 - 10.5625 + .8125)$$
$$= 93.531$$

and

$$r = \frac{(93.631)^2}{\left[\frac{(770.0625/8)^2}{1} + \frac{(12.0625/8)^2}{2} + \frac{(10.5625/8)^2}{1} + \frac{(.8125/8)^2}{2}\right]}$$

$$= .944$$

(which is rounded to 1 so that chi-square tables can be used). An approximate 90% confidence interval about σ_b^2 is

$$\frac{93.531}{3.8415} \leq \sigma_b^2 \leq \frac{93.531}{.00393}$$

or

$$24.348 \leq \sigma_b^2 \leq 23{,}799.237.$$

20.2.2 Unbalanced Models

The sums of squares from the analysis of variance table for an unbalanced model are not necessarily independently distributed chi-square random variables. Thus, the techniques used for balanced models cannot be applied to unbalanced models without violating the assumptions. If the model is not too unbalanced, the sums of squares will nearly be independent chi-square random variables, and the balanced-model techniques should provide effective confidence intervals.

For unbalanced models and large sample sizes, the maximum likelihood estimates and their asymptotic properties can be used to construct confidence intervals about the variance components. A good maximum likelihood estimation algorithm should provide estimators as well as the asymptotic variances. If $\hat{\sigma}_1^2$ is the maximum likelihood estimate of σ_1^2 with variance $\text{Var}(\hat{\sigma}_1^2)$, then a $(1-\alpha)100\%$ asymptotic confidence interval about σ_1^2 is

$$\hat{\sigma}_1^2 - Z_{\alpha/2}\sqrt{\text{Var}(\hat{\sigma}_1^2)} \leq \sigma_1^2 \leq \hat{\sigma}_1^2 + Z_{\alpha/2}\sqrt{\text{Var}(\hat{\sigma}_1^2)}$$

where $Z_{\alpha/2}$ denotes the upper $\alpha/2$ percentage point of a $N(0,1)$ distribution.

EXAMPLE 20.8

For the data in Example 19.2, the output from BMD-P3V provided $\hat{\sigma}_a^2 = .04855$ with variance $(.048164)^2$. A 95% asymptotic confidence interval about σ_a^2 is

$$.04855 - 1.96 \cdot .048164 \leq \sigma_a^2 \leq .04855 + 1.96 \cdot .048164$$

or

$$-.0458 \leq \sigma_a^2 \leq .1430,$$

which can be truncated as $0 \leq \sigma_a^2 \leq .1430$. The problem of negative lower limits can be eliminated if one first tests $\sigma_a^2 = 0$ and computes the confidence interval only if the hypothesis is rejected.

The MINQUE technique can also provide an estimate of the variance of the variance component estimators. Again, the confidence interval is constructed by using the normal distribution, as shown in the next example.

EXAMPLE 20.9

The data in Example 19.2 were used in Example 19.4 to obtain MINQUE estimators. The estimate of σ_a^2 obtained from the iterative procedure is $\hat{\sigma}_a^2 = .073155$ with $\text{Var}(\hat{\sigma}_a^2) = .005684$. A 95% approximate confidence interval about σ_a^2 is

$$.073155 - 1.96 \cdot \sqrt{.005684} \leq \sigma_a^2 \leq .073155 + 1.96 \cdot \sqrt{.005684}$$

or

$$-.07461 \leq \sigma_a^2 \leq .2209,$$

which can be truncated to $0 \leq \sigma_a^2 \leq .2209$.

EXAMPLE 20.10

We shall now apply the method of Section 20.2.1 to the data of Example 19.2. Let $Q_1 = .05644$, $Q_2 = .27005$, $u_1 = 9$, $u_2 = 3$, and $a = 3.1795$. A 95% confidence interval about σ_a^2 is $c \leq \sigma_a^2 \leq d$ where

$$c = \frac{(3 \cdot .27005)/9.3484 - (12 \cdot .05644)/2.7004}{3.1795}$$

$$= -.05162$$

and

$$d = \frac{(3 \cdot .27005)/.2158 - (12 \cdot .05644)/19.023}{3.1795}$$

$$= 1.16900$$

or $-.05162 \leq \sigma_a^2 \leq 1.16900$, which can be truncated to $0 \leq \sigma_a^2 \leq 1.16900$.

The above three examples show how different the confidence intervals can be for a given problem. When the design is unbalanced and the sample size is not large, one suggestion is to use the widest confidence interval. The next section incorporates hypothesis testing into the method-of-moments estimation.

20.3 MODIFIED METHOD-OF-MOMENTS ESTIMATION PROCEDURE

Chapter 19 explained that using the method of moments as an estimation technique can yield negative estimates of variances. The following is a proposal for eliminating these negative estimates.

The proposal is to build models and try to determine which variance components to include in the final model. Start with the complete model, fit it to the data, and use the procedures in Section 20.1 to test hypotheses about the variance components of the model. If all variance components are determined to be statistically significant at some chosen α (say, $\alpha < .3$) level, then estimate the components of variance (which will be nonnegative); confidence intervals can be constructed using the methods of Section 20.2. If some of the variance components are not statistically significant at the chosen α level, eliminate the least important variance component (the one with the largest estimated significance level as determined by the test statistic) and fit the reduced model.

Continue to eliminate variance components from the model until all the remaining ones are statistically significant. The estimates of the variance components that are eliminated are all zero, and the remainder are estimated from the reduced model. By using this model-building method, the method-of-moments estimators will be nonnegative. This method is used in the example of Chapter 21.

CONCLUDING REMARKS

This chapter described methods to test hypotheses and construct confidence intervals about variance components. Included were exact methods that can be used in making inferences for most balanced models. Approximate methods evolving around the Satterthwaite approximation were described for unbalanced models and for some problems with balanced models. Several examples were included to demonstrate the techniques, and a method to build a variance components model was presented, which can be used in conjunction with the method of moments to eliminate negative estimates of variance components.

21

Case Study: Analysis of a Random Model

CHAPTER OUTLINE

The previous three chapters described the methods for analyzing random models and provided some examples to demonstrate the various techniques of analysis. This chapter presents the analysis of a more complex experimental situation, which includes model building, testing hypotheses, and constructing confidence intervals.

21.1 DATA

In this experiment, the efficiency of workers in assembly lines at several plants was studied. Three plants were randomly selected. Within each plant, three assembly sites were randomly selected, and four workers were randomly selected from the pool of workers at each plant. Each worker was to work four times at each assembly site in his or her plant, but because of scheduling problems and other priorities, that was not accomplished.

The efficiency scores as well as the plant number, site number, and worker number are listed in Table 21.1. The model used to describe the data is

$$y_{ijkl} = \mu + p_i + s_{j(i)} + w_{k(i)} + sw_{jk(i)} + \varepsilon_{ijkl} \qquad (21.1.1)$$

where p_i is the ith plant effect, $s_{j(i)}$ is the jth site effect within plant i, $w_{k(i)}$ is the kth Worker effect within plant i, $sw_{jk(i)}$ is the interaction effect of Site and Worker in plant i, and ε_{ijkl} is the error term. It is assumed that $p_i \sim N(0, \sigma_p^2)$, $s_{j(i)} \sim N(0, \sigma_s^2)$, $w_{k(i)} \sim N(0, \sigma_w^2)$, $sw_{jk(i)} \sim N(0, \sigma_{sw}^2)$, and $\varepsilon_{ijkl} \sim N(0, \sigma_\varepsilon^2)$.

21.2 ESTIMATION

The method of moments employing the Type I sums of squares, maximum likelihood, and MINQUE(0) estimates (the one-step estimates provided by SAS® VARCOMP) of the variance components are in Table 21.2. The expected mean squares of the Type I sums of squares are given in Table 21.3. Since the Type I method of moments and MINQUE(0) estimates include negative estimates, the method described in Section 20.3 is used to build an adequate variance component model.

21.3 MODEL BUILDING

The method of moments based on the Type I sums of squares will be used to build a model. The process starts by investigating the variance component in the last sum of squares and then working up the line to the variance component in the first sum of squares, where each component is tested to see if it is equal to zero. The observed and expected mean squares are listed in Table 21.3.

The first step is to test H_0: $\sigma_{sw}^2 = 0$ versus H_a: $\sigma_{sw}^2 > 0$. The appropriate test statistic is

$$F_{c_{sw}} = \frac{\text{MSSITE} * \text{WORKER(PLANT)}}{\text{MSERROR}} = 21.42.$$

Table 21.1 Data for Case Study

Worker	SITE			
Plant 1	*1*	*2*	*3*	*4*
1	100.6	110.0	100.0	98.2
	106.8	105.8	102.5	99.5
	100.6		97.6	
			98.7	
			98.7	
2	92.3	103.2	96.4	108.0
	92.0	100.5		108.9
	97.2	100.2		107.9
	93.9	97.7		
	93.0			
3	96.9	92.5	86.8	94.4
	96.1	85.9		93.0
	100.8	85.2		91.0
		89.4		
		88.7		
Plant 2	*5*	*6*	*7*	*8*
4	82.6	96.5	87.9	83.6
		100.1	93.5	82.7
		101.9	88.9	87.7
		97.9	92.8	88.0
		95.9		82.5
5	72.7	71.7	78.4	82.1
		72.1	80.4	79.9
		72.4	83.8	81.9
		71.4	77.7	82.6
			81.2	78.6
6	82.5	80.9	96.3	77.7
	82.1	84.0	92.4	78.6
	82.0	82.2	92.0	77.2
		83.4	95.8	78.8
		81.5		80.5

Table 21.1 (continued)

	Site			
Plant 3	*9*	*10*	*11*	*12*
7	107.6	96.1	101.1	109.1
	108.8	98.5		
	107.2	97.3		
	104.2	93.5		
	105.4			
8	97.1	91.9	88.0	89.6
	94.2		91.4	86.0
	91.5		90.3	91.2
	99.2		91.5	87.4
			85.7	
9	87.1	97.8	95.9	101.4
		95.9	89.7	100.1
				102.1
				98.4

Table 21.2 Estimates of Variance Components for Data in Table 21.1

	ESTIMATE		
Parameter	*Method of Moments (Type I)*	*Maximum Likelihood*	*MINQUE(0)*
σ_p^2	48.806	29.599	54.206
σ_s^2	−4.878	0	−4.416
σ_w^2	24.255	28.945	24.230
σ_{sw}^2	35.617	28.759	29.839
σ_ϵ^2	4.983	4.982	6.616

Table 21.3 Observed Mean Squares for Type I Sums of Squares for Data in Table 21.1

Source of Variation	*df*	MEAN SQUARE *Observed*	*Expected*
Plant	2	2,313.759	$\sigma_\varepsilon^2 + 3.928\sigma_{sw}^2 + 10.307\sigma_s^2 + 13.136\sigma_w^2 + 38.941\sigma_p^2$
Worker(Plant)	6	456.942	$\sigma_\varepsilon^2 + 3.902\sigma_{sw}^2 + .656\sigma_s^2 + 13.037\sigma_w^2$
Site(Plant)	9	84.049	$\sigma_\varepsilon^2 + 3.479\sigma_{sw}^2 + 9.193\sigma_s^2$
Site * Worker(Plant)	18	106.738	$\sigma_\varepsilon^2 + 2.857\sigma_{sw}^2$
Error	82	4.983	σ_ε^2

The sampling distribution of $F_{c_{sw}}$ is $F_{(18,82)}$, for which the significance level is less than .0001.

The second step is to test H_0: $\sigma_s^2 = 0$ versus H_s: $\sigma_s^2 > 0$. The E[MSSITE(PLANT)] equals $\sigma_\varepsilon^2 + 3.479\sigma_{sw}^2 + 9.193\sigma_s^2$, and no other mean square exists with expectation $\sigma_\varepsilon^2 + 3.479\sigma_{sw}^2$, which can be used as a divisor for the test statistic. Thus, a mean square Q_s^* needs to be constructed from MSERROR and MSSITE * WORKER(PLANT) such that $E(Q_s^*) = \sigma_\varepsilon^2 + 3.479\sigma_{sw}^2$. Such a Q_s^* is

$$\begin{aligned} Q_s^* &= 3.479\left[\frac{\text{MSSITE} * \text{WORKER(PLANT)}}{2.857}\right] \\ &\quad + \left(1 - \frac{3.479}{2.857}\right)\text{MSERROR} \\ &= 1.218[\text{MSSITE} * \text{WORKER(PLANT)}] - (.218)\text{MSERROR} \\ &= 128.921 \end{aligned}$$

By using the Satterthwaite approximation, $r_s Q_s^*/(\sigma_\varepsilon^2 + 3.479\sigma_{sw}^2)$ is approximately distributed as a chi-square random variable with r_s degrees of freedom where

$$\begin{aligned} r_s &= \frac{(Q^*)^2}{[1.217\ \text{MSSITE} * \text{WORKER(PLANT)}]^2/18 + [(.217)\ \text{MSERROR}]^2/82} \\ &= \frac{(128.921)^2}{[1.217(106.738)]^2/18 + [.217(4.983)]^2/82} \\ &= 17.7 \end{aligned}$$

The test statistic is

$$F_{c_s} = \frac{\text{MSSITE(PLANT)}}{Q_s^*} = .65,$$

which is distributed as $F(9, 17.7)$. The test statistic is nonsignificant; therefore H_0: $\sigma_s^2 = 0$ is not rejected. Since σ_s^2 is negligible, eliminate it from the model and fit a reduced model to the data. The reduced model is

$$y_{ijkl} = \mu + p_i + w_{k(i)} + sw_{jk(i)} + \varepsilon_{ijkl}. \tag{21.3.1}$$

The degrees of freedom, Type I mean squares, and expected mean squares for the reduced model are listed in Table 21.4.

The next step is to check and see if any of the terms in the reduced model are negligible. The statistic to test H_0: $\sigma_{sw}^2 = 0$ is

$$F_{c_{sw}} = \frac{\text{MSSITE} * \text{WORKER(PLANT)}}{\text{MSERROR}} = 19.90$$

which is distributed as $F_{(27,82)}$. The significance level is less than .0001.

A Q_w^* needs to be computed to test H_0: $\sigma_w^2 = 0$ versus H_a: $\sigma_w^2 > 0$. The mean square Q_w^* needs to be constructed from MSSITE * WORKER(PLANT) and MSERROR such that

$$E(Q_w^*) = \sigma_\varepsilon^2 + 3.902\sigma_{sw}^2.$$

Then

$$\begin{aligned} Q_w^* &= 3.902\left[\frac{\text{MSSITE} * \text{WORKER(PLANT)}}{3.064}\right] \\ &\quad + \left(1 - \frac{3.902}{3.064}\right)\text{MSERROR} \\ &= 1.2732[\text{MSSITE} * \text{WORKER(PLANT)}] - .2732(\text{MSERROR}) \\ &= 1.2732(99.175) - .2732(4.983) = 124.908. \end{aligned}$$

Table 21.4 Observed Mean Squares of Type I Sums of Squares for Reduced Model for Data in Table 21.1

Source of Variation	*df*	MEAN SQUARE *Observed*	MEAN SQUARE *Expected*
Plant	2	2,313.759	$\sigma_\varepsilon^2 + 3.928\sigma_{sw}^2 + 13.136\sigma_w^2 + 38.941\sigma_p^2$
Worker(Plant)	6	456.942	$\sigma_\varepsilon^2 + 3.902\sigma_{sw}^2 + 13.037\sigma_w^2$
Site * Worker(Plant)	27	99.175	$\sigma_\varepsilon^2 + 3.064\sigma_{sw}^2$
Error	82	4.983	σ_ε^2

The test statistic is

$$F_{c_w} = \frac{456.942}{124.908} = 3.658$$

which is distributed as $F_{(6, r_w)}$ where

$$r_w = \frac{(Q_w^*)^2}{[(1.2732)\text{MSSITE}*\text{WORKER(PLANT)}]^2/27 + ((.2732)\text{MSERROR})^2/82}$$

$$= \frac{(124.908)^2}{(126.270)^2/27 + (1.361)^2/82}$$

$$= 26.4$$

By interpolation, $F_{.10(6,26.4)} = 1.66$; thus, the null hypothesis $\sigma_w^2 = 0$ is rejected at $\alpha < .10$.

Finally, another Q_p^* needs to be constructed from MSWORKER (PLANT), MSSITE ∗ WORKER(PLANT), and MSERROR such that

$$E(Q_p^*) = \sigma_\varepsilon^2 + 3.928\sigma_{sw}^2 + 13.136\sigma_w^2$$

so that a statistic can be constructed to test H_0: $\sigma_p^2 = 0$ versus H_a: $\sigma_p^2 > 0$. The required Q_p^* is

$$Q_p^* = \frac{13.136}{13.037}[\text{MSWORKER(PLANT)}]$$

$$+ \left[\frac{3.928 - \frac{13.136}{13.037}(3.902)}{3.064}\right]\text{MSSITE}*\text{WORKER(PLANT)}$$

$$+ \left[1 - \frac{13.136}{13.037} - \frac{3.928 - \frac{13.136}{13.037}(3.902)}{3.064}\right]\text{MSERROR}$$

$$= (1.00759)\text{MSWORKER(PLANT)}$$

$$- (.00118)\text{MSSITE}*\text{WORKER(PLANT)} - (.00641)\text{MSERROR}$$

$$= (1.00759)(456.942) - (.00118)(99.175) - (.00641)(4.983)$$

$$= 460.412 - .117 - .032 = 460.26.$$

The test statistic is

$$F_{c_p} = \frac{\text{MSPLANT}}{Q_p^*} = 5.027$$

Table 21.5 Estimates of Variance Components for Reduced Model for Data in Table 21.1

	ESTIMATE		
Parameter	*Method of Moments (Type I)*	*Maximum Likelihood*	*MINQUE(0)*
σ_p^2	47.597	29.599	53.103
σ_w^2	25.469	28.945	25.345
σ_{sw}^2	30.739	28.759	25.420
σ_ε^2	4.983	4.982	6.612

which is distributed $F_{(2,r_p)}$ where $r_p = (Q_p^*)^2/D$ and

$$D = \frac{[(1.00759)\text{MSWORKER(PLANT)}]^2}{6} + \frac{[(.00118)\text{MSSITE}*\text{WORKER(PLANT)}]^2}{27} + \frac{[(.00641)\text{MSERROR}]^2}{82}$$
$$= 76.74$$

The $r_p = 5.996$ and $F_{.10}(2, 5.996) = 3.46$, and the hypothesis is rejected at $\alpha < .10$.

All of the variance components in model (21.3.1) are significantly different from zero at $\alpha < .10$. The estimates of the variance components for the reduced model are shown in Table 21.5.

21.4 CONFIDENCE INTERVALS

The next step in the analysis is to construct confidence intervals about the variance components of the reduced model in equation (21.3.1). The confidence intervals are computed by the Satterthwaite approximation and the method described in Section 20.2.

Both techniques have the same confidence interval for σ_ε^2. A 90% confidence interval about σ_ε^2 is

$$\frac{(82)\text{MSERROR}}{\chi^2_{.05,82}} \le \sigma_\varepsilon^2 \le \frac{(82)\text{MSERROR}}{\chi^2_{.95,82}}$$

or

$$\frac{(82)(4.983)}{104.2} \le \sigma_\varepsilon^2 \le \frac{(82)(4.983)}{62.1}$$

or

$$3.921 \le \sigma_\varepsilon^2 \le 6.580.$$

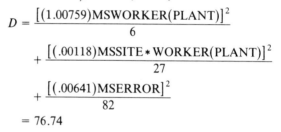

For the Satterthwaite approximation to the confidence interval about σ^2_{sw}, use

$$\hat{\sigma}^2_{sw} = \frac{\text{MSSITE} * \text{WORKER(PLANT)}}{3.064} - \frac{\text{MSERROR}}{3.064} = 30.739.$$

Then $r_{sw}(\hat{\sigma}^2_{sw}/\sigma^2_{sw})$ is approximately distributed as a chi-square random variable with r_{sw} degrees of freedom where

$$r_{sw} = \frac{\left(\hat{\sigma}^2_{sw}\right)^2}{\left[\frac{\text{MSSITE} * \text{WORKER(PLANT)}}{3.064}\right]^2 /27 + \left(\frac{\text{MSERROR}}{3.064}\right)^2 /82}$$
$$= 24.33.$$

A 90% approximate confidence interval about σ^2_{sw} is

$$\frac{(24.33)\hat{\sigma}^2_{sw}}{\chi^2_{.05,24.33}} \le \sigma^2_{sw} \le \frac{(24.33)\hat{\sigma}^2}{\chi^2_{.95,24.33}}$$

or

$$20.31 \le \sigma^2_{sw} \le 53.04$$

where $\chi^2_{.95,24.33} = 14.10$ and $\chi^2_{.05,24.33} = 36.83$ were obtained by interpolation.

To construct an approximate confidence interval about σ^2_w, use

$$\hat{\sigma}^2_w = \frac{\text{MSWORKER(PLANT)}}{13.037} - \left(\frac{1.2738}{13.037}\right)\text{MSSITE} * \text{WORKER(PLANT)} + \left(\frac{.2732}{13.037}\right)\text{MSERROR}$$
$$= 25.469.$$

The approximate degrees of freedom associated with $\hat{\sigma}^2_w$ are $r_w = 3.12$. A 90% approximate confidence interval about σ^2_w is

$$\frac{(3.12)\hat{\sigma}^2_w}{\chi^2_{.05,3.12}} \le \sigma^2_w \le \frac{(3.12)\hat{\sigma}^2_w}{\chi^2_{.95,3.12}}$$

or

$$9.92 \le \sigma^2_w \le 65.30.$$

To construct an approximate confidence interval about σ^2_p, use

$$\hat{\sigma}^2_p = \frac{\text{MSPLANT}}{38.941} - \left(\frac{1.0076}{38.941}\right)\text{MSWORKER(PLANT)} + \left(\frac{.00118}{38.941}\right)\text{MSSITE} * \text{WORKER(PLANT)} + \left(\frac{.00641}{38.941}\right)\text{MSERROR}$$
$$= 47.597.$$

The approximate degrees of freedom associated with $\hat{\sigma}_p^2$ are $r_p = 1.27$. Thus a 90% approximate confidence interval about $\hat{\sigma}_p^2$ is

$$13.675 \le \sigma_p^2 \le 1{,}970.35.$$

The final stage in the analysis is to use the method in Section 20.2 to construct confidence intervals about the variance components. A 90% confidence interval about σ_{sw}^2 is c to d where

$$c = \frac{(27)\text{MSSITE}*\text{WORKER(PLANT)}/\chi^2_{.05,27} - (82)\text{MSERROR}/\chi^2_{.95,82}}{3.064}$$

and

$$d = \frac{(27)\text{MSSITE}*\text{WORKER(PLANT)}/\chi^2_{.95,27} - (82)\text{MSERROR}/\chi^2_{.05,82}}{3.064}$$

or, since $c = 16.53$ and $d = 52.83$,

$$16.53 \le \sigma_{sw}^2 \le 52.83.$$

A 90% confidence interval about σ_w^2 is c to d where

$$c = \frac{(6)\text{MSWORKER(PLANT)}/\chi^2_{.05,6} - r_w Q_w^*/\chi^2_{.95,r_w}}{14.037}$$

and

$$d = \frac{(6)\text{MSWORKER(PLANT)}/\chi^2_{.95,6} - r_w Q_2^*/\chi^2_{.05,r_w}}{13.037}$$

where $r_w = 26.4$ and $Q_w^* = 124.908$. The 90% confidence interval about σ_w^2 is

$$.571 \le \sigma_w^2 \le 121.807.$$

A 90% confidence interval about σ_p^2 is c to d where

$$c = \frac{(2)\text{MSPLANT}/\chi^2_{.05,2} - r_p Q_p^*/\chi^2_{.95,r_p}}{38.941}$$

and

$$d = \frac{(2)\text{MSPLANT}/\chi^2_{.95,2} - r_p Q_p^*/\chi^2_{.05,r_p}}{38.941}$$

where $r_p = 5.996$ and $Q_p^* = 460.26$. The 90% confidence interval is $-23.091 \le \sigma_p^2 \le 1{,}182.7$, or $0 \le \sigma_p^2 \le 1{,}182.7$.

The confidence intervals obtained from the two methods are somewhat different, with larger discrepancies for smaller degrees of freedom. The Satterthwaite approximation confidence intervals are somewhat shorter than the confidence intervals obtained by the method described in Section 20.2.

CONCLUDING REMARKS

In this chapter, a complex unbalanced random model was analyzed in detail, with all three methods of estimation described in Chapter 19 used to estimate the variance components. The methods for testing hypotheses and constructing confidence intervals described in Chapter 20 were also demonstrated. One of the variance components in the original model was found to be nonsignificant. The model-building approach described in Chapter 20 was used to provide an adequate model with meaningful estimates of the remaining variance components.

22
Analysis of Mixed Models

CHAPTER OUTLINE

Mixed models are used to describe data from experiments whose treatment structures involve some factors that are fixed and some that are random. Thus, models for describing such experiments involve two parts, one part describing the random effects and the other describing the fixed effects. Consequently, the analysis of a mixed model consists of two types of analyses, a random analysis and a fixed analysis. This chapter discusses the construction of mixed models and the necessary steps for doing the random analysis and the fixed analysis. Both balanced and unbalanced treatment structures are considered.

22.1 INTRODUCTION TO MIXED MODELS

When the treatment structures of experiments involve both random factors and fixed factors, the models for describing such experiments are called *mixed models*. Since there are two types of factors in the treatment structure, the model has two parts, a random part and a fixed part. In order to construct such models, the following rule is used to determine whether a specific interaction is a random or a fixed effect.

Rule: If a main effect is a random effect, then any interaction involving that main effect is also a random effect. The only interactions that are fixed effects are those whose main effects are all fixed effects.

For example, a model for a three-way treatment structure where A and B are fixed effects and C is a random effect is

$$y_{ijkm} = \mu + \alpha_i + \beta_j + \gamma_{ij} + c_k + d_{ik} + f_{jk} + g_{ijk} + \varepsilon_{ijkm}$$

where

$$\mu + \alpha_i + \beta_j + \gamma_{ij}$$

is the fixed effects part of the model and

$$c_k + d_{ik} + f_{jk} + g_{ijk} + \varepsilon_{ijkm}$$

is the random effects part of the model with distributions $c_k \sim$ i.i.d. $N(0, \sigma_c^2)$; $d_{ik} \sim$ i.i.d. $N(0, \sigma_d^2)$; $f_{jk} \sim$ i.i.d. $N(0, \sigma_f^2)$; $g_{ijk} \sim$ i.i.d. $N(0, \sigma_g^2)$; $\varepsilon_{ijkm} \sim$ i.i.d. $N(0, \sigma_\varepsilon^2)$; and c_k, d_{ik}, f_{jk}, g_{ijk}, and ε_{ijkm} are all independent random variables. The interaction between A and B, denoted by γ_{ij}, is a fixed effect, while all other interactions are random effects since they all involve the random factor C, denoted by the index k.

The general mixed model in matrix notation is

$$\mathbf{y} = \mathbf{X}_1\boldsymbol{\beta} + \mathbf{X}_2\mathbf{a}_2 + \cdots + \mathbf{X}_k\mathbf{a}_k + \boldsymbol{\varepsilon}$$

where $\mathbf{X}_1\boldsymbol{\beta}$ is the fixed part of the model and $\mathbf{X}_2\mathbf{a}_2 + \cdots + \mathbf{X}_k\mathbf{a}_k + \boldsymbol{\varepsilon}$ is the random part of the model with $\mathbf{a}_i \sim$ i.i.d. $N(0, \sigma_i^2\mathbf{I})$, $i = 2, 3, \ldots, k$, and $\boldsymbol{\varepsilon} \sim N(\mathbf{0}, \sigma_\varepsilon^2 \mathbf{I})$. The population parameters of the mixed model are $\boldsymbol{\beta}$, $\sigma_2^2, \sigma_3^2, \ldots, \sigma_k^2$, and σ_ε^2. The analysis of the random part consists of estimating, testing hypotheses, and constructing confidence intervals about the variance components, $\sigma_2^2, \sigma_3^2, \ldots, \sigma_k^2$, and σ_ε^2. The analysis of the fixed part consists of estimating, testing hypotheses, and constructing confidence intervals about estimable functions of $\boldsymbol{\beta}$. The analysis of the mixed model is presented in the next two sections, one section for each part of the analysis.

22.2 ANALYSIS OF RANDOM PART OF MIXED MODEL

The analysis of the mixed model starts with the random part. A random model can be constructed from the general mixed model by fitting the fixed effects part of the model and then computing the residuals of that model. The resulting model does not depend on the fixed effects part of the model. The general mixed model can be expressed as

$$\mathbf{y} = \mathbf{X}_1\boldsymbol{\beta} + \boldsymbol{\varepsilon}^*$$

where

$$\operatorname{Var}(\boldsymbol{\varepsilon}^*) = \boldsymbol{\Sigma} = \sigma_2^2 \mathbf{X}_2\mathbf{X}_2' + \sigma_3^2 \mathbf{X}_3\mathbf{X}_3' + \cdots + \sigma_k^2 \mathbf{X}_k\mathbf{X}_k' + \sigma_\varepsilon^2 \mathbf{I}_n.$$

The least squares estimator of $\boldsymbol{\beta}$ is $\hat{\boldsymbol{\beta}} = (\mathbf{X}'\mathbf{X})^{-}\mathbf{X}'\mathbf{y}$, and the vector of residuals is

$$\mathbf{r} = \mathbf{y} - \mathbf{X}_1\hat{\boldsymbol{\beta}} = (\mathbf{I} - \mathbf{X}_1\mathbf{X}_1^{-})\mathbf{y}.$$

A model for the vector of residuals is

$$\mathbf{r} = (\mathbf{I} - \mathbf{X}_1\mathbf{X}_1^{-})\mathbf{X}_2\mathbf{a}_2 + (\mathbf{I} - \mathbf{X}_1\mathbf{X}_1^{-})\mathbf{X}_3\mathbf{a}_3 + \cdots + (\mathbf{I} - \mathbf{X}_1\mathbf{X}_1^{-})\mathbf{X}_k\mathbf{a}_k + (\mathbf{I} - \mathbf{X}_1\mathbf{X}_1^{-})\boldsymbol{\varepsilon}.$$

The residual model does not depend on the fixed effects of the model, $\mathbf{X}_1\boldsymbol{\beta}$, and thus is called a random model. The methods for analyzing random models discussed in Chapters 18, 19, and 20 can be used to analyze the residual random model. The three techniques, the method of moments, the maximum likelihood method, and the MINQUE method, are the topics of the next three subsections.

22.2.1 Method of Moments

As described in Chapter 19 for the analysis of a random model, the method-of-moments technique requires computing sums of squares, determining their expectations, and then estimating the variance components from the system of equations obtained by equating the observed mean squares to their expected values. In order to estimate the variance components, sums of squares must be obtained whose expectations do not depend on the fixed effects part of the model. The method of fitting constants discussed in Section 19.1 provides such sums of squares when the fixed effects part of the model is fit first and then the random effects part is fit.

To demonstrate the analysis of the random part of the mixed model, consider a two-way treatment structure in a completely randomized design structure with one fixed factor, denoted by β, and the other random factor, denoted by t. The resulting two-way mixed model is

$$y_{ijk} = \mu + \beta_i + t_j + g_{ij} + \varepsilon_{ijk} \qquad i = 1, 2, \ldots, b, \quad j = 1, 2, \ldots, t,$$
$$k = 0, 1, \ldots, n_{ij},$$

where it is assumed that t_j is distributed i.i.d. $N(0, \sigma_t^2)$, g_{ij} is distributed i.i.d. $N(0, \sigma_g^2)$, and ε_{ijk} is distributed i.i.d. $N(0, \sigma_\varepsilon^2)$.

The sums of squares obtained from the method of fitting constants are $R(\beta|\mu)$, $R(t|\mu, \beta)$, $R(g|\mu, \beta, t)$, and SSERROR. The expectations of the last three mean squares are

$$E[\text{MSR}(t|\mu, \beta)] = \sigma_\varepsilon^2 + k_1\sigma_g^2 + k_2\sigma_t^2$$
$$E[\text{MSR}(g|\mu, \beta, t)] = \sigma_\varepsilon^2 + k_3\sigma_g^2$$

and $E(\text{MSERROR}) = \sigma_\varepsilon^2$. The expectations of these mean squares do not involve the fixed effects parameters and thus can be used to estimate the variance components and to test hypotheses about them. The techniques used in Chapters 19 and 20 can also be used to analyze the random part of the mixed model.

Because estimates of the variance components are required for the fixed part of the analysis, it is important that they be nonnegative. The technique described in Section 20.3 can be used to build a suitable model. The method is demonstrated for two examples, one balanced and one unbalanced, in Chapter 23.

22.2.2 Maximum Likelihood Method

The method of maximum likelihood can be applied to the complete likelihood function, denoted by

$$L(\boldsymbol{\beta}, \sigma_2^2, \ldots, \sigma_k^2, \sigma_\varepsilon^2) = (2\pi)^{-n/2}|\boldsymbol{\Sigma}|^{-1/2}$$
$$\times \exp\left[-\tfrac{1}{2}(\mathbf{y} - \mathbf{X}_1\boldsymbol{\beta})\boldsymbol{\Sigma}^{-1}(\mathbf{y} - \mathbf{X}_1\boldsymbol{\beta})\right],$$

which provides equations for simultaneously estimating the fixed and random effects (Hartley and Rao, 1967). Corbeil and Searle (1976) expressed the likelihood function in two parts, one involving the fixed effects and one free of the fixed effects. They then obtained maximum likelihood estimators of the variance components from that part of the model free of the fixed effects, which they call *re*stricted *m*aximum *l*ikelihood estimators, or REML estimators.

To construct the restricted likelihood function, let $\mathbf{H}$ be an $nx(n - p)$ matrix of rank $n - p$ such that $\mathbf{HH}' = \mathbf{I} - \mathbf{X}_1\mathbf{X}_1^-$. Define the transformation

$$\mathbf{Z} = \begin{bmatrix} \mathbf{Z}_1 \\ \mathbf{Z}_2 \end{bmatrix} = \begin{bmatrix} \mathbf{X}_1' \\ \mathbf{H}' \end{bmatrix} \mathbf{y} \qquad \text{where} \quad \begin{bmatrix} \mathbf{X}_1' \\ \mathbf{H}' \end{bmatrix}$$

is an $n \times n$ nonsingular matrix. Thus the transformation from $\mathbf{y}$ to $\mathbf{Z}$ is one to one. The distribution of $\mathbf{Z}$ is

$$\begin{bmatrix} \mathbf{Z}_1 \\ \mathbf{Z}_2 \end{bmatrix} \sim N\left[\begin{bmatrix} \mathbf{X}_1'\boldsymbol{\beta} \\ \mathbf{0} \end{bmatrix}, \begin{bmatrix} \mathbf{X}_1'\boldsymbol{\Sigma}\mathbf{X}_1 & \mathbf{X}_1'\boldsymbol{\Sigma}\mathbf{H} \\ \mathbf{H}'\boldsymbol{\Sigma}\mathbf{X}_1 & \mathbf{H}'\boldsymbol{\Sigma}\mathbf{H} \end{bmatrix}\right].$$

The joint likelihood function of $(\mathbf{Z}_1, \mathbf{Z}_2)$ can be partitioned into the likelihood of $\mathbf{Z}_1$ given $\mathbf{Z}_2$ times the marginal likelihood of $\mathbf{Z}_2$. The marginal likelihood of $\mathbf{Z}_2$ does not depend on the fixed effects of the model and is

$$L(\sigma_2^2, \sigma_3^2, \ldots, \sigma_k^2, \sigma_\varepsilon^2) = (2\pi)^{-n-p/2} |\mathbf{H}'\boldsymbol{\Sigma}\mathbf{H}|^{-1/2} \times \exp\left[-\tfrac{1}{2}\mathbf{y}'\mathbf{H}(\mathbf{H}'\boldsymbol{\Sigma}\mathbf{H})^{-1}\mathbf{H}'\mathbf{y}\right].$$

The REML estimators of the variance components are those that maximize $L(\sigma_2^2, \sigma_3^2, \ldots, \sigma_\varepsilon^2)$. This restricted likelihood function is for the random variable $\mathbf{Z}_2$. Since the residuals for the residual model in Section 22.2. are $\boldsymbol{\gamma} = \mathbf{H}\mathbf{Z}_2$, the restricted likelihood function utilizes the information from the residual model that does not depend on the fixed effects. Again, it is important to obtain nonnegative estimates of the variance components since these estimates are used in the analysis of the fixed effects part of the model. The computer algorithm BMD-P3V provides REML estimators of the variance components for a mixed model. The REML estimates are obtained for the two examples in Chapter 23.

22.2.3 MINQUE Method

The MINQUE method for a random model is described in Section 19.3. In that application, the mean of the model is $\mathbf{j}_n\mu$, while here the mean is $\mathbf{X}_1\boldsymbol{\beta}$. The estimator can be generalized to the general mixed model (Swallow and Searle, 1978), where the MINQUE estimator of

$$\boldsymbol{\sigma}^2 = (\sigma_2^2, \sigma_3^2, \ldots, \sigma_k^2, \sigma_\varepsilon^2)'$$

is $\hat{\boldsymbol{\sigma}}^2 = \mathbf{S}^{-1}\mathbf{u}$ with matrix $\mathbf{S}$ having elements

$$S_{i-1,i'-1} = \operatorname{Tr}[\mathbf{X}_i\mathbf{X}_i'\mathbf{R}\mathbf{X}_i\mathbf{X}_i'\mathbf{R}] \qquad i, i' = 2, 3, \ldots, k+1$$

where $i = k + 1$ corresponds to ε and $\mathbf{X}_{k+1} = \mathbf{I}$,

$$R = \boldsymbol{\Sigma}^{-1}(\mathbf{I} - \mathbf{X}_1)(\mathbf{X}_1'\boldsymbol{\Sigma}^{-1}\mathbf{X}_1)^{-}\mathbf{X}_1'\boldsymbol{\Sigma}^{-1}),$$

and $\mathbf{u}$ has elements $u_i = \mathbf{y}'\mathbf{R}\mathbf{X}_i\mathbf{X}_{i'}\mathbf{R}\mathbf{y}$.

The MINQUE estimators are difficult to compute, although efficient algorithms for computing them are being developed. Be sure to use one that provides nonnegative estimators, since the estimators need to be used in analyzing the fixed part of the model. One-iteration MINQUE estimators are computed for the two examples discussed in Chapter 23.

22.3 ANALYSIS OF FIXED PART OF MIXED MODEL

The analysis of the fixed part of a mixed model consists of all aspects of the analysis of a fixed model, as described below.

22.3.1 Estimation

There are several methods for estimating estimable functions of $\boldsymbol{\beta}$ in the mixed model. The general mixed model can be expressed as

$$\mathbf{y} = \mathbf{X}_1\boldsymbol{\beta} + \boldsymbol{\varepsilon}^* \quad \text{where} \quad \operatorname{Var}(\boldsymbol{\varepsilon}^*) = \sigma_2^2\mathbf{X}_2\mathbf{X}_2' + \sigma_3^2\mathbf{X}_3\mathbf{X}_3' + \cdots + \sigma_k^2\mathbf{X}_k\mathbf{X}_k' + \sigma_\varepsilon^2\mathbf{I}.$$

A linear combination, $\mathbf{a}'\boldsymbol{\beta}$ is estimable for this mixed model if and only if there exists a vector $\mathbf{c}$ such that $E(\mathbf{c}'\mathbf{y}) = \mathbf{a}'\boldsymbol{\beta}$. This definition is the same as that for a linear model (see Chapter 6). The estimate of $\mathbf{a}'\boldsymbol{\beta}$, an estimable function of $\boldsymbol{\beta}$, is $\mathbf{a}'\hat{\boldsymbol{\beta}}$ where $\hat{\boldsymbol{\beta}}$ is a solution to a set of normal equations or to the likelihood equations.

The least squares estimator of $\mathbf{a}'\boldsymbol{\beta}$ is $\mathbf{a}'\hat{\boldsymbol{\beta}}_{\mathrm{LS}}$ where $\hat{\boldsymbol{\beta}}_{\mathrm{LS}} = (\mathbf{X}_1'\mathbf{X}_1)^-\mathbf{X}_1'\mathbf{y}$ or some other solution for $\hat{\boldsymbol{\beta}}$ in $\mathbf{X}_1'\mathbf{X}_1\hat{\boldsymbol{\beta}} = \mathbf{X}_1'\mathbf{y}$. The least squares estimator of $\boldsymbol{\beta}$ does not use the covariance matrix $\boldsymbol{\Sigma}$, the covariance matrix of $\mathbf{y}$.

If the elements of $\boldsymbol{\Sigma}$ are known (that is, $\sigma_2^2, \sigma_3^2, \ldots, \sigma_k^2$ and σ_ε^2 are known), the best linear unbiased estimator (BLUE) of $\mathbf{a}'\boldsymbol{\beta}$ is $\mathbf{a}'\hat{\boldsymbol{\beta}}_{\mathrm{B}}$ where

$$\hat{\boldsymbol{\beta}}_{\mathrm{B}} = \left(\mathbf{X}_1'\boldsymbol{\Sigma}^{-1}\mathbf{X}_1\right)^-\mathbf{X}_1'\boldsymbol{\Sigma}^{-1}\mathbf{y}$$

or some other solution for $\hat{\boldsymbol{\beta}}$ in

$$\mathbf{X}_1'\boldsymbol{\Sigma}^{-1}\mathbf{X}_1\hat{\boldsymbol{\beta}} = \mathbf{X}_1'\boldsymbol{\Sigma}^{-1}\mathbf{y}.$$

For most balanced designs and some simple unbalanced designs,

$$\mathbf{X}_1'\boldsymbol{\Sigma}^{-1}\mathbf{X}_1 = \left(g(\boldsymbol{\sigma}^2)\right)^{-1}\mathbf{X}_1'\mathbf{X}_1 \quad \text{and} \quad \mathbf{X}_1'\boldsymbol{\Sigma}^{-1}\mathbf{y} = \left(g(\boldsymbol{\sigma}^2)\right)^{-1}\mathbf{X}_1'\mathbf{y}$$

where $g(\boldsymbol{\sigma}^2)$ is a known linear function of the variance components. Thus the BLUE of $\mathbf{a}'\boldsymbol{\beta}$ is $\mathbf{a}'\hat{\boldsymbol{\beta}}_{\mathrm{B}}$ where $\hat{\boldsymbol{\beta}}_{\mathrm{B}} = (\mathbf{X}_1'\mathbf{X}_1)^-\mathbf{X}_1'\mathbf{y}$, which does not depend on the variance components. Thus for most balanced designs, the BLUE of an estimable function of $\boldsymbol{\beta}$ can be computed even though the variance components are not known. The variance of $\mathbf{a}'\hat{\boldsymbol{\beta}}_{\mathrm{B}}$ in the balanced case is $g(\boldsymbol{\sigma}^2)\mathbf{a}'(\mathbf{X}_1'\mathbf{X}_1)^-\mathbf{a}$.

When the designs are unbalanced and the variance components are unknown, life is not so easy. Because the BLUE does not exist (since it depends on the unknown variances), a weighted least squares estimator can be obtained where $\hat{\boldsymbol{\Sigma}}$ is used as the weighting matrix. The estimated covariance matrix is

$$\hat{\boldsymbol{\Sigma}} = \hat{\sigma}_2^2 \mathbf{X}_2\mathbf{X}_2' + \hat{\sigma}_3^2 \mathbf{X}_3\mathbf{X}_3' + \cdots + \hat{\sigma}_k^2 \mathbf{X}_k\mathbf{X}_k' + \hat{\sigma}_\varepsilon^2 \mathbf{I}_n$$

where $\hat{\sigma}_2^2, \hat{\sigma}_3^2, \ldots, \hat{\sigma}_k^2$ and $\hat{\sigma}_\varepsilon^2$ are the estimators of the variance components obtained from one of the methods discussed in Section 22.2. The weighted least squares estimator of $\mathbf{a}'\boldsymbol{\beta}$ is $\mathbf{a}'\hat{\boldsymbol{\beta}}_\omega$ where

$$\hat{\boldsymbol{\beta}}_\omega = \left(\mathbf{X}_1'\hat{\boldsymbol{\Sigma}}^{-1}\mathbf{X}_1\right)^{-}\mathbf{X}_1'\hat{\boldsymbol{\Sigma}}^{-1}\mathbf{y}$$

or some other solution for $\hat{\boldsymbol{\beta}}$ in

$$\mathbf{X}_1'\hat{\boldsymbol{\Sigma}}^{-1}\mathbf{X}_1\hat{\boldsymbol{\beta}} = \mathbf{X}_1'\hat{\boldsymbol{\Sigma}}^{-1}\mathbf{y}.$$

For most designs, $\mathbf{a}'\hat{\boldsymbol{\beta}}_\omega$ converges to $\mathbf{a}'\boldsymbol{\beta}$ as the sample size increases. For the convergence, some care must be taken so that as the sample size increases, the number of parameters does not go to infinity. The large sample variance of $\mathbf{a}'\hat{\boldsymbol{\beta}}_\omega$ is $\mathbf{a}'(\mathbf{X}_1'\hat{\boldsymbol{\Sigma}}^{-1}\mathbf{X}_1)^{-}\mathbf{a}$.

The method of maximum likelihood can also be used to obtain an estimator of $\mathbf{a}'\boldsymbol{\beta}$. The maximum likelihood estimate of $\mathbf{a}'\boldsymbol{\beta}$ is $\mathbf{a}'\hat{\boldsymbol{\beta}}_M$ where $\hat{\boldsymbol{\beta}}_M$ is a solution to the likelihood equations. The variance of $\mathbf{a}'\hat{\boldsymbol{\beta}}_M$ is $\mathbf{a}'\mathbf{W}\mathbf{a}$ where $\mathbf{W}$ is the partition of the generalized inverse of the information matrix corresponding to $\hat{\boldsymbol{\beta}}_M$.

The examples in Chapter 23 demonstrate the above estimators for both a balanced and an unbalanced design.

22.3.2 Constructing Confidence Intervals

For most balanced designs, the BLUE of an estimable function, $\mathbf{a}'\boldsymbol{\beta}$, is

$$\mathbf{a}'\hat{\boldsymbol{\beta}} = \mathbf{a}'\left(\mathbf{X}_1'\mathbf{X}_1\right)^{-}\mathbf{X}_1'\mathbf{y}$$

and has variance

$$g(\boldsymbol{\sigma}^2)\mathbf{a}'\left(\mathbf{X}_1'\mathbf{X}_1\right)^{-}\mathbf{a} = g(\boldsymbol{\sigma}^2)c$$

where $g(\boldsymbol{\sigma}^2)$ is a linear combination of the variance components. The estimator of $g(\boldsymbol{\sigma}^2)$ may be a single mean square, say $\widehat{g(\boldsymbol{\sigma}^2)} = \mathrm{MS}_i$, from the analysis of variance, or it may be a linear combination of mean squares, say

$$\widehat{g(\boldsymbol{\sigma}^2)} = q_1\mathrm{MS}_1 + q_2\mathrm{MS}_2 + \cdots + q_{k+1}\mathrm{MS}_{k+1}$$

for constants $q_1, q_2, \ldots, q_{k+1}$. When $g(\boldsymbol{\sigma}^2)$ is estimated by a single mean square, then a $(1-\alpha)100\%$ confidence interval about $\mathbf{a}'\boldsymbol{\beta}$ is

$$\mathbf{a}'\hat{\boldsymbol{\beta}} \pm t_{\alpha/2,\nu_i}\sqrt{\mathrm{MS}_i \times c}$$

where ν_i is the degrees of freedom associated with MS_i. When $g(\sigma^2)$ is estimated by a linear combination of mean squares, each mean square has its own degrees of freedom.

Since the degrees of freedom for each mean square can be different, the percentage point used in computing the confidence interval must be approximated. There are two methods for obtaining the approximate percentage point. The first method takes a weighted average of the values of $t_{\alpha/2,\nu_i}$, $i = 1, 2, \ldots, k + 1$, as

$$t^*_{\alpha/2} = \frac{q_1\text{MS}_1 t_{\alpha/2,\nu_1} + q_2\text{MS}_2 t_{\alpha/2,\nu_2} + \cdots + q_{k+1}\text{MS}_{k+1} t_{\alpha/2,\nu_{k+1}}}{q_1\text{MS}_1 + q_2\text{MS}_2 + \cdots + q_{k+1}\text{MS}_{k+1}}.$$

The resulting confidence interval is

$$\mathbf{a}'\hat{\boldsymbol{\beta}} \pm t^*_{\alpha/2}\sqrt{\widehat{g(\sigma^2)}c}\,.$$

The second method approximates the degrees of freedom used in selecting the t-value by using the Satterthwaite approximation (see Chapter 20). The degrees of freedom for the t-value are approximated by

$$r = \frac{\left(\widehat{g(\sigma^2)}\right)^2}{(q_1\text{MS}_1)^2/\nu_1 + (q_2\text{MS}_2)^2/\nu_2 + \cdots + (q_{k+1}\text{MS}_{k+1})^2/\nu_k}$$

The resulting confidence interval is

$$\mathbf{a}'\hat{\boldsymbol{\beta}} \pm t_{\alpha/2,r}\sqrt{\widehat{g(\sigma^2)}c}\,.$$

Applications of these approximations are shown in Chapter 23 as well as in Chapters 25, 26, 27, and 32.

For unbalanced designs, the weighted least squares estimator of an estimable function is (from Section 22.3.1) $\mathbf{a}'\hat{\boldsymbol{\beta}}_\omega$ where

$$\hat{\boldsymbol{\beta}}_\omega = \left(\mathbf{X}_1'\hat{\boldsymbol{\Sigma}}^{-1}\mathbf{X}_1\right)^{-}\mathbf{X}_1'\hat{\boldsymbol{\Sigma}}^{-1}\mathbf{y}$$

and the estimate of the variance of $\mathbf{a}'\hat{\boldsymbol{\beta}}_\omega$ is approximated by

$$\mathbf{a}'(\mathbf{X}_1'\hat{\boldsymbol{\Sigma}}^{-1}\mathbf{X}_1)\mathbf{a}.$$

For a large sample size,

$$\frac{\mathbf{a}'\hat{\boldsymbol{\beta}}_\omega - \mathbf{a}'\boldsymbol{\beta}}{\sqrt{\mathbf{a}'(\mathbf{X}_1'\hat{\boldsymbol{\Sigma}}^{-1}\mathbf{X}_1)\mathbf{a}}} \sim N(0,1).$$

Thus a $(1 - \alpha)100\%$ asymptotic confidence interval about $\mathbf{a}'\boldsymbol{\beta}$ is

$$\mathbf{a}'\hat{\boldsymbol{\beta}}_\omega \pm Z_{\alpha/2}\sqrt{\mathbf{a}'(\mathbf{X}_1'\hat{\boldsymbol{\Sigma}}^{-1}\mathbf{X}_1)\mathbf{a}}$$

where $Z_{\alpha/2}$ is the upper $\alpha/2$ percentage point of the standard normal distribution, $N(0,1)$.

For small to moderate sample sizes, the above interval will be too short. An approximate t-value should be constructed as was done with t^* or the Satterthwaite approximation. A conservative confidence interval can be obtained by replacing $Z_{\alpha/2}$ with $t_{\alpha/2,\nu_m}$ where ν_m is the smallest number of degrees of freedom associated with a sum of squares used to estimate the variance components. A final suggestion is to compute t^* for the unbalanced design by combining the t-values as is done for balanced designs. For designs that are not too unbalanced, this method should provide adequate confidence intervals. An example in Chapter 23 demonstrates the above ideas.

If $\boldsymbol{\beta}$ is estimated by using the method of maximum likelihood, then the asymptotic distribution of

$$\frac{\mathbf{a}'\hat{\boldsymbol{\beta}}_M - \mathbf{a}'\boldsymbol{\beta}}{\sqrt{\mathbf{a}'\mathbf{W}\mathbf{a}}}$$

is $N(0, 1)$ where $\mathbf{W}$ is the covariance matrix of $\hat{\boldsymbol{\beta}}_M$. A $(1 - \alpha)100\%$ confidence interval about $\mathbf{a}'\boldsymbol{\beta}$ is

$$\mathbf{a}'\hat{\boldsymbol{\beta}}_M \pm Z_{\alpha/2}\sqrt{\mathbf{a}'\mathbf{W}\mathbf{a}}\,.$$

For large sample sizes, the above confidence interval based on the asymptotic distribution of the maximum likelihood estimate should provide adequate coverage. However, for small to moderate sample sizes, the confidence interval will generally be too narrow.

22.3.3 Testing Hypotheses

The confidence intervals can be used to test hypotheses for either a single estimable function or a set of estimable functions by using a simultaneous confidence interval method. However, the fixed effects part of the model generally involves main effects and interactions between fixed factors. Most researchers want to be able to test hypotheses about those main effects and interactions in order to determine which confidence intervals need to be constructed.

For balanced models, most mean squares for the fixed effects have expectations of the form

$$E(\text{MS}_F) = \sigma_\varepsilon^2 + C_1\sigma_1^2 + C_2\sigma_2^2 + \cdots + C_k\sigma_k^2 + Q^2$$

where Q^2 is zero if and only if the null hypothesis is true. There also exists a mean square in the model with expectation of the form

$$E(\text{MS}_i) = \sigma_\varepsilon^2 + C_1\sigma_1^2 + C_2\sigma_2^2 + \cdots + C_k\sigma_k^2.$$

The test statistic is $F = \text{MS}_F/\text{MS}_i$, which under the conditions of the null hypothesis is distributed as an F-distribution with υ_F and υ_i degrees of freedom corresponding to MS_F and MS_i, respectively. By looking at the expectations of all of the mean squares, the proper denominators can be selected for testing hypotheses about the fixed effects.

In unbalanced designs and some balanced designs, there does not exist a mean square with expectation

$$\sigma_\varepsilon^2 + C_1\sigma_1^2 + C_2\sigma_2^2 + \cdots + C_k\sigma_k^2.$$

In that case, a mean square must be constructed that has the proper expectation where the constructed mean square is a linear combination of the mean squares used to estimate the variance components. Let

$$u = q_1\text{MS}_1 + q_2\text{MS}_2 + \cdots + q_k\text{MS}_k + q_{k+1}\text{MS}_{k+1}$$

be such that

$$E(u) = \sigma_\varepsilon^2 + C_1\sigma_1^2 + C_2\sigma_2^2 + \cdots + C_k\sigma_k^2.$$

The test statistic is $F = \text{MS}_\text{F}/u$, which under the conditions of the null hypothesis is approximately distributed as an F-distribution with υ_F and r degrees of freedom where

$$r = \frac{u^2}{(q_1\text{MS}_1)^2/\upsilon_1 + (q_2\text{MS}_2)^2/\upsilon_2 + \cdots + (q_{k+1}\text{MS}_{k+1})^2/\upsilon_{k+1}}.$$

The degrees of freedom associated with u are determined by means of the Satterthwaite approximation. For the balanced model, this is an approximation because the degrees of freedom associated with u are approximate. For the unbalanced case, this is an approximation because of the approximate degrees of freedom and because u and MS_F are not necessarily independent. Examples are discussed in Chapter 23.

CONCLUDING REMARKS

This chapter used a mixed model to describe experiments where the factors in the treatment structure involved both fixed and random effects. The analysis of the mixed model involves estimating and making inferences toward the variance components. The analysis of the fixed part of the model was examined in order to make inferences about the fixed effects model. Methods were described for balanced models as well as unbalanced models.

23

Two Case Studies of a Mixed Model

CHAPTER OUTLINE

Chapter 22 discussed how to analyze the data for balanced and unbalanced mixed models. This chapter presents detailed analyses for both cases; the data for the unbalanced case is obtained by randomly deleting some observations from the data for the balanced situation.

23.1 BALANCED TWO-WAY MODEL

A company wanted to replace the machines used to make a certain component in one of its factories. Three different brands of machines were available, so the management designed an experiment to evaluate the productivity of the machines when operated by the company's own personnel. Six employees were randomly selected to participate in the experiment, each of whom was to operate each machine three different times. The data recorded were overall scores, which took into account the number and quality of components produced. The data are in Table 23.1.

Table 23.1 Productivity Scores for Machine—Person Example

		SCORE					
		BALANCED CASE (SECTION 23.1)			UNBALANCED CASE (SECTION 23.2)		
Machine	*Person*	*1*	*2*	*3*	*1*	*2*	*3*
1	1	52.0	52.8	53.1	52.0		
1	2	51.8	52.8	53.1	51.8	52.8	
1	3	60.0	60.2	58.4	60.0		
1	4	51.1	52.3	50.3	51.1	52.3	
1	5	50.9	51.8	51.4	50.9	51.8	51.4
1	6	46.4	44.8	49.2	46.4	44.8	49.2
2	1	62.1	62.6	64.0			64.0
2	2	59.7	60.0	59.0	59.7	60.0	59.0
2	3	68.6	65.8	69.7	68.6	65.8	
2	4	63.2	62.8	62.2	63.2	62.8	62.2
2	5	64.8	65.0	65.4	64.8	65.0	
2	6	43.7	44.2	43.0	43.7	44.2	43.0
3	1	67.5	67.2	66.9	67.5	67.2	66.9
3	2	61.5	61.7	62.3	61.5	61.7	62.3
3	3	70.8	70.6	71.0	70.8	70.6	71.0
3	4	64.1	66.2	64.0	64.1	66.2	64.0
3	5	72.1	72.0	71.1	72.1	72.0	71.1
3	6	62.0	61.4	60.5	62.0	61.4	60.5

The treatment structure for this experiment is two-way, machines being a fixed effect and people a random effect. The design structure is completely randomized. The model used to describe the data is

$$y_{ijk} = \mu + \tau_i + p_j + g_{ij} + \varepsilon_{ijk}$$

where μ denotes the overall mean, τ_i the effect of Machine type i, p_j the random variable for Person j, g_{ij} the random variable for the Machine by Person interaction, and ε_{ijk} denotes the random variable corresponding to the error associated with the jth person operating the ith machine type at the kth time. The random components and corresponding variances are $p_j \sim$ i.i.d. $N(0, \sigma_p^2)$, $g_{ij} \sim$ i.i.d. $N(0, \sigma_g^2)$, and $\varepsilon_{ijk} \sim$ i.i.d. $N(0, \sigma_\varepsilon^2)$.

The first step is to do the random part of the analysis. A computer code was used to obtain the method-of-moments, maximum likelihood, and MINQUE estimates of the variance components. The analysis of variance table for the method of moments, including the expected mean squares, is shown in Table 23.2. The method-of-moments and MINQUE estimators are

$$\hat{\sigma}_p^2 = 22.858, \quad \hat{\sigma}_g^2 = 13.909, \quad \text{and} \quad \hat{\sigma}_\varepsilon^2 = .925,$$

and the maximum likelihood estimates are

$$\hat{\sigma}_p^2 = 19.049, \quad \hat{\sigma}_g^2 = 11.540, \quad \text{and} \quad \hat{\sigma}_\varepsilon^2 = .925.$$

The first hypothesis to be tested is H_0: $\sigma_g^2 = 0$ versus H_a: $\sigma_g^2 > 0$. The statistic to test this hypothesis, constructed by using the expected mean squares in Table 23.2, is

$$F_c = \frac{\text{MSMACHINE} * \text{PERSON}}{\text{MSERROR}} = 46.13.$$

The second hypothesis to be tested is H_0: $\sigma_p^2 = 0$ versus H_a: $\sigma_p^2 > 0$. Again, the statistic to test this hypothesis is constructed by using the

Table 23.2 Analysis of Variance Table for a Balanced Mixed Model

Source of Variation	*df*	*SS*	*MS*	*EMS*
Machine	2	1,755.26	877.632	$\sigma_\varepsilon^2 + 3\sigma_g^2 + Q(\tau)$
Person	5	1,241.89	248.379	$\sigma_\varepsilon^2 + 3\sigma_g^2 + 9\sigma_p^2$
Machine * Person	10	462.53	42.653	$\sigma_\varepsilon^2 + 3\sigma_g^2$
Error	36	33.29	.925	σ_ε^2

expected mean squares in Table 23.2 and is given by

$$F_c = \frac{\text{MSPERSON}}{\text{MSMACHINE} * \text{PERSON}} = 5.82.$$

It is also of interest to construct approximate confidence intervals about the variance components. To demonstrate the technique described in Chapter 20, we construct an approximate 95% confidence interval about σ_p^2. The method-of-moments estimator of σ_p^2 is

$$\begin{aligned}\hat{\sigma}_p^2 &= \left(\frac{\text{SSPERSON}}{5} - \frac{\text{SSMACHINE} * \text{PERSON}}{10}\right)\frac{1}{9} \\ &= \frac{1}{45}(\text{SSPERSON}) - \frac{1}{90}(\text{SSMACHINE} * \text{PERSON}) \\ &= 22.858\end{aligned}$$

The degrees of freedom of the approximating chi-square distribution obtained through the Satterthwaite approximation is

$$r = \frac{\left(\hat{\sigma}_p^2\right)^2}{\dfrac{[(1/45)(\text{SSPERSON})]^2}{5} + \dfrac{[(1/90)(\text{SSMACHINE} * \text{PERSON})]^2}{10}}$$

$$= 3.43.$$

The interpolated χ^2 values are

$$\chi^2_{.025,3.43} = 10.10 \quad \text{and} \quad \chi^2_{.975,3.43} = .331.$$

Thus, an approximate 95% confidence interval for σ_p^2 is

$$\frac{(3.43)(22.8584)}{10.10} \leq \sigma_p^2 \leq \frac{(3.43)(22.8584)}{.331}$$

or

$$7.76 \leq \sigma_p^2 \leq 236.70,$$

and a 95% confidence interval for σ_p is

$$2.79 \leq \sigma_p \leq 15.38.$$

A confidence interval for σ_g^2 can be obtained in a similar manner.

The methods necessary to estimate and test hypotheses about the fixed effects in a mixed model, which depend on whether the data are balanced or unbalanced, are discussed in Section 22.3. The general mixed model can be expressed as

$$\mathbf{y} = \mathbf{X}\boldsymbol{\mu} + \mathbf{X}_2\mathbf{a}_2 + \mathbf{X}_3\mathbf{a}_3 + \cdots + \mathbf{X}_k\mathbf{a}_k + \boldsymbol{\varepsilon}$$

where $\mathbf{X} = [\mathbf{j}, \mathbf{X}_1]$ and $\boldsymbol{\mu} = (\mu, \boldsymbol{\beta}')'$.

If the elements of

$$\boldsymbol{\Sigma} = \sigma_2^2 \mathbf{X}_2 \mathbf{X}_2' + \sigma_3^2 \mathbf{X}_3 \mathbf{X}_3' + \cdots + \sigma_k^2 \mathbf{X}_k \mathbf{X}_k' + \sigma_\varepsilon^2 \mathbf{I}$$

are known, that is, if the variance components are known, then the best estimate of $\boldsymbol{\mu}$ is

$$\hat{\boldsymbol{\mu}} = (\mathbf{X}'\boldsymbol{\Sigma}^{-1}\mathbf{X})^{-1}\mathbf{X}'\boldsymbol{\Sigma}^{-1}\mathbf{y}.$$

For most balanced mixed models, the estimator of $\boldsymbol{\mu}$ simplifies to

$$\hat{\boldsymbol{\mu}} = (\mathbf{X}'\mathbf{X})^{-1}\mathbf{X}'\mathbf{y}.$$

The example below shows this simplification for the balanced two-way mixed model.

The covariance matrix for the balanced model has elements

$$\operatorname{Cov}(y_{ijk}, y_{i'j'k'}) = \begin{cases} \sigma_p^2 + \sigma_g^2 + \sigma_\varepsilon^2 & i = i', \quad j = j', \quad k = k' \\ \sigma_p^2 + \sigma_g^2 & i = i', \quad j = j', \quad k \neq k' \\ \sigma_p^2 & i = i', \quad j \neq j', \quad \text{any } k, k' \\ 0 & i \neq i', \quad \text{any } j, j', k, k'. \end{cases}$$

The model can be reparametrized as

$$y_{ijk} = \mu_i + p_j + g_{ij} + \varepsilon_{ijk}$$

where $\mu_i = \mu + \tau_i$ and $\mathbf{X}_1$, the $nbt \times b$ design matrix corresponding to the fixed effects part of the model $\boldsymbol{\mu}$, is

$$\mathbf{X}_1 = \begin{bmatrix} \mathbf{j}_{nt} & 0 & \cdots & \mathbf{0} \\ \mathbf{0} & \mathbf{j}_{nt} & \cdots & \mathbf{0} \\ \mathbf{0} & \mathbf{0} & \cdots & \mathbf{0} \\ \vdots & \vdots & & \vdots \\ \mathbf{0} & \mathbf{0} & \cdots & \mathbf{j}_{nt} \end{bmatrix}.$$

Then the matrices $\mathbf{X}_1'\boldsymbol{\Sigma}^{-1}\mathbf{X}_1$ and $\mathbf{X}_1'\boldsymbol{\Sigma}^{-1}\mathbf{y}$ are

$$\mathbf{X}_1'\boldsymbol{\Sigma}^{-1}\mathbf{X}_1 = \left(\sigma_\varepsilon^2 + n\sigma_g^2\right)^{-1}\mathbf{X}_1'\mathbf{X}_1'$$

and

$$\mathbf{X}_1'\boldsymbol{\Sigma}^{-1}\mathbf{y} = \left(\sigma_\varepsilon^2 + n\sigma_g^2\right)^{-1}\mathbf{X}_1'\mathbf{y}.$$

Thus the estimator of $\boldsymbol{\mu}$ is

$$\hat{\boldsymbol{\mu}} = \left(\mathbf{X}_1'\mathbf{X}_1\right)^{-1}\mathbf{X}_1'\mathbf{y},$$

which has elements $\hat{\mu}_i = \bar{y}_{i\cdot\cdot}$, $i = 1, 2, \ldots, b$. The variance of $\hat{\mu}$ is

$$\mathrm{Var}(\hat{\mu}) = (\sigma_\varepsilon^2 + n\sigma_g^2)(\mathbf{X}_1'\mathbf{X}_1)^{-1} = \left(\frac{\sigma_\varepsilon^2 + n\sigma_g^2}{nt}\right)\mathbf{I}_b.$$

Thus the variance of $\hat{\mu}_i$ is

$$\frac{\sigma_\varepsilon^2 + n\sigma_g^2}{nt}.$$

For the machine–person example, the variance of each sample mean is

$$\frac{\sigma_\varepsilon^2 + 3\sigma_g^2}{18}.$$

The estimate of the variance of $\hat{\mu}_i$ is MSMACHINE∗PERSON/18, since MSMACHINE∗PERSON has expected mean square $\sigma_\varepsilon^2 + 3\sigma_g^2$. The standard error of the difference between the two means is

$$\text{s.e.} = \sqrt{\frac{2(\text{MSMACHINE}*\text{PERSON})}{18}} = 2.177,$$

and an $\text{LSD}_{.05}$ for comparing the two means is

$$\text{LSD}_{.05} = (2.228)(2.177) = 4.85 \qquad \text{where} \quad t_{.025,10} = 2.228.$$

The machine means are

$$\hat{\mu}_1 = 52.36, \quad \hat{\mu}_2 = 60.32, \quad \text{and} \quad \hat{\mu}_3 = 66.27.$$

By using the LSD as a multiple comparison tool, the means of all three machine types are significantly different.

To test the hypothesis that the means are equal, use the expected mean squares from Table 23.2 to construct the test statistic

$$F_c = \frac{\text{MSMACHINE}}{\text{MSMACHINE}*\text{PERSON}} = 20.576.$$

This example shows that the analysis is quite straightforward for the balanced case. However, the unbalanced case is different.

23.2 UNBALANCED TWO-WAY MODEL

The data in Table 23.1 for this example are the same as those for the balanced example except that some observations have been randomly deleted. This has been done to demonstrate the problems that occur when analyzing unbalanced data sets and to compare the estimation procedures for the balanced and unbalanced cases.

The analysis of variance table for the unbalanced data is shown in Table 23.3. The sums of squares are those obtained by the method of

Table 23.3 Analysis of Variance Table for an Unbalanced Mixed Model

Source of Variation	*df*	*SS*	*MS*	*EMS*
Machine	2	1,648.663	824.332	$\sigma_\varepsilon^2 + 2.6115\sigma_g^2 + .1569\sigma_p^2 + Q(\tau)$
Person	5	1,008.764	201.753	$\sigma_\varepsilon^2 + 2.5866\sigma_g^2 + 7.2190\sigma_p^2$
Machine * Person	10	404.315	40.432	$\sigma_\varepsilon^2 + 2.3162\sigma_g^2$
Error	26	22.687	.873	σ_ε^2

Note: The sums of squares were derived by the method of fitting constants.

fitting constants, that is,

$$\text{SSMACHINE} = R(\tau|\mu), \quad \text{SSPERSON} = R(p|\mu,\tau),$$

$$\text{and} \quad \text{SSMACHINE*PERSON} = R(g|\mu,\tau,p).$$

The corresponding method-of-moments estimates are $\hat{\sigma}_p^2 = 21.707$, $\hat{\sigma}_g^2 = 17.079$, and $\hat{\sigma}_\varepsilon^2 = .873$.

The MINQUE and maximum likelihood estimates were obtained from SAS® VARCOMP. The one-iteration MINQUE estimators are $\hat{\sigma}_p^2 = 24.339$, $\hat{\sigma}_g^2 = 16.155$, and $\hat{\sigma}_\varepsilon^2 = -.631$. The maximum likelihood estimators are $\hat{\sigma}_p^2 = 125{,}497.421$, $\hat{\sigma}_g^2 = 172{,}688.369$, and $\hat{\sigma}_\varepsilon^2 = .516$. The maximum likelihood estimates for the unbalanced case are nonsensical; even though the procedure says it converges, it does not converge to acceptable estimates. In this case, the method-of-moments technique produces the only acceptable solution. The data in Table 23.1 are not too unbalanced, and yet the SAS®, MINQUE, and maximum likelihood methods are not acceptable. We question the applicability of these two algorithms when data are very unbalanced.

The techniques in Chapter 20 can be used to test hypotheses about σ_p^2 and σ_g^2 as well as to construct confidence intervals about the variance components.

The next step is to use the above estimates of variance components in the fixed effects analysis. The unbalanced case is more difficult than the balanced case, since the best estimator of $\boldsymbol{\mu}$ depends on the known values of the variance components. Because the values of the variance components are not generally known, in order to obtain a good estimate of $\boldsymbol{\mu}$, we must first estimate the variance components and then use those estimates to estimate $\boldsymbol{\Sigma}$ as

$$\hat{\boldsymbol{\Sigma}} = \hat{\sigma}_2^2\mathbf{X}_2\mathbf{X}_2' + \hat{\sigma}_3^2\mathbf{X}_3\mathbf{X}_3' + \cdots + \hat{\sigma}_k^2\mathbf{X}_k\mathbf{X}_k' + \hat{\sigma}_\varepsilon^2\mathbf{I}.$$

Care must be taken so that no negative estimates of variances are used (see Section 20.3 for a discussion on how to avoid negative estimates). Then a weighted least squares estimate of the fixed effects parameter vector is

$$\hat{\mu} = (\mathbf{X}'\hat{\boldsymbol{\Sigma}}^{-1}\mathbf{X})^{-1}\mathbf{X}'\hat{\boldsymbol{\Sigma}}^{-1}\mathbf{y}.$$

For almost all unbalanced models, the experimenter will have to obtain estimates of the fixed effects through the weighted least squares estimator. We shall continue with this example to demonstrate how to obtain the weighted least squares estimate.

Unfortunately, no popular computing package has a procedure that automatically provides a weighted means analysis. However, after obtaining the estimates of the variance components, it is possible to obtain a weighted means analysis by using a matrix manipulation code.

1. Vector of indices of PER:

```
DATA PER; SET SEPARAT; KEEP PER;
IF M=. THEN DELETE;
```

2. Vector of indices of MACH:

```
DATA MACH; SET SEPARAT ; KEEP MACH;
IF M=. THEN DELETE;
```

3. Vector of observations:

```
DATA M; SET SEPARAT; KEEP M;
IF M=. THEN DELETE;
```

4. Use of PROC MATRIX to compute $\hat{\mu}$:

```
PROC MATRIX; *USE PROC MATRIX TO EATIMATE MACH EFFECTS;
SE=0.8726;SP=21.7069;SPM=17.0791; * ENTER TYPE1 VAR COMP ESTIMA

FETCH  MACH DATA=MACH; *GET MACH CODES;
N=NROW(MACH); * N IS NUMBER OF OBSERVATIONS;
FETCH PER DATA=PER; *GET PER CODES;
FETCH M DATA=M; *GET DATA;
X1=DESIGN(MACH);* DESIGN MATRIX FOR MACH;
X2=DESIGN(PER);  * DESIGN MATRIX FOR PER;
INTER=(6#(MACH-1))+PER; *COMPUTE INTERACTION CODES;
X3=DESIGN(INTER); * DESIGN MATRIX FOR PER*MACH;
SIGMA=(SE#I(N))+(SP#(X2*(X2')))+SPM#(X3*(X3')); * COMPUTE SIGMA

SIGMA=INV(SIGMA); *COMPUTE SIGMA INVERSE;
SU=INV((X1')*SIGMA*X1); * VARIANCE OF MACH EFFECTS;
PRINT SU;
MACHHAT=SU*(X1')*SIGMA*M; *COMPUTE MACH HAT;
PRINT MACHHAT;
```

5. Computation of contrasts and the contrast variances:

```
L=1 -1 0/1 0 -1/0  1 -1;
V=L*SU*(L'); *COMPUTE  VARIANCE OF CONTRASTS;
E=L*MACHHAT;*COMPARE MACHHAT EFFECTS;
PRINT E V;
```

Figure 23.1 SAS® statements for obtaining the least squares estimates of fixed effects for the unbalanced mixed model.

Table 23.4 Results from Proc Matrix in Figure 23.1

SU	COL1	COL2	COL3
ROW1	6.55278	3.6179	3.61782
ROW2	3.6179	6.53676	3.61782
ROW3	3.61782	3.61782	6.51281

MACHHAT	COL1
ROW1	52.3538
ROW2	60.3193
ROW3	66.2722

E	COL1
ROW1	-7.96552
ROW2	-13.9184
ROW3	-5.95289

V	COL1	COL2	COL3
ROW1	5.85374	2.93488	-2.91885
ROW2	2.93488	5.82996	2.89508
ROW3	-2.91885	2.89508	5.81393

The statements required to use SAS® MATRIX to obtain the weighted least squares estimates of μ_i where $\mu_i = \mu + \beta_i$ are given in Figure 23.1. The necessary variables are in a data set named SEPARAT. Steps 1 and 2 generate two data sets, where PER contains the levels of the person codes and MAC contains the machine codes for each observation. These codes are used to generate the design matrices $\mathbf{X}_1$, $\mathbf{X}_2$, and $\mathbf{X}_3$ by using the DESIGN statement in MATRIX. Step 3 generates the data set M, which consists of the vector of observations, $\mathbf{y}$. Step 4 shows the MATRIX statements used to obtain $\hat{\mu}$. The statements are annotated by the comment statements, which are preceded by asterisks.

The variance of $\hat{\mu}$ is $(\mathbf{X}_1'\hat{\Sigma}^{-1}\mathbf{X}_1)^{-1}$, which is displayed in Table 23.4, where it is denoted by SU. The value of $\hat{\mu}$ is denoted by MACHHAT in Table 23.4. The diagonal elements of SU are the variances of the corresponding elements of $\hat{\boldsymbol{\mu}}$. For example, the variance of $\hat{\mu}_2 = 60.3193$ is 6.53676. To generate a multiple comparison method, we construct all pairwise differences, $\hat{\mu}_i - \hat{\mu}_{i'}$, and find their corresponding variances.

Step 5 of Figure 23.1 computes the comparisons $\hat{\mu}_1 - \hat{\mu}_2$, $\hat{\mu}_1 - \hat{\mu}_3$, and $\hat{\mu}_2 - \hat{\mu}_3$ and their respective variances. The estimates of the contrasts are denoted by E in Table 23.4. The variance of $\hat{\mu}_1 - \hat{\mu}_2 = -7.96552$ is 5.85374. To construct an LSD type multiple comparison, a conservative number of degrees of freedom associated with the variances is

$$u = \text{Min}[df(\text{MACHINE} * \text{PERSON}), df(\text{ERROR})].$$

Thus LSD values are constructed by using a t-value with $u = \text{Min}(10, 26)$ (degrees of freedom obtained from Table 23.3) or $u = 10$ degrees of freedom. The three LSD values are

1. $\mu_1 - \mu_2$, $\text{LSD}_{.05} = 2.228\sqrt{5.85374} = 5.390$
2. $\mu_1 - \mu_3$, $\text{LSD}_{.05} = 2.228\sqrt{5.82996} = 5.379$
3. $\mu_2 - \mu_3$, $\text{LSD}_{.05} = 2.228\sqrt{5.81393} = 5.372$

For the unbalanced case, there usually is no mean square from the fitting-of-constants sums of squares whose expectation is equal to the expectation of the $R(\tau|\mu)$ mean square when the null hypothesis is true. Hence, a divisor must be constructed so that the expectation of the divisor is equal to the expectation of the hypothesis mean square. A Satterthwaite approximation is used, as discussed in Section 20.1. For the unbalanced case, to test H_0: $\tau_1 = \tau_2 = \cdots = \tau_b$, compute $R(\tau|\mu)$ by the method of fitting constants. The expectation of this mean square, which is the hypothesis mean square, is of the form

$$E[\text{MSR}(\tau|\mu)] = \sigma_\varepsilon^2 + k_4\sigma_g^2 + k_5\sigma_p^2 + Q(\tau)$$

where $Q(\tau) = 0$ if and only if H_0 is true. The expectations of the other mean squares are

$$E[\text{MSR}(p|\mu, \tau)] = \sigma_\varepsilon^2 k_3\sigma_g^2 + k_2\sigma_p^2,$$

$$E[\text{MSR}(g|\mu, \tau, p)] = \sigma_\varepsilon^2 + k_1\sigma_g^2,$$

and

$$E(\text{MSERROR}) = \sigma_\varepsilon^2.$$

The appropriate mean square is

$$\begin{aligned} Q^* &= \left(1 - \frac{k_5}{k_2} - \frac{k_4}{k_1} + \frac{k_3 k_5}{k_1 k_2}\right)\text{MSERROR} \\ &\quad + \left(\frac{k_4/k_1 - k_3 k_5}{k_1 k_2}\right)\text{MSR}(g|\mu, \tau, p) + \left(\frac{k_5}{k_2}\right)\text{MSR}(p|\mu, \tau) \\ &= q_1(\text{MSERROR}) + q_2(\text{MSMACHINE} * \text{PERSON}) \\ &\quad + q_3(\text{MSPERSON}) \end{aligned}$$

where

$$\text{MSMACHINE}*\text{PERSON} = \text{MSR}(\mathbf{g}|\mu, \tau, p)$$

and

$$\text{MSPERSON} = \text{MSR}(p|\mu, \tau).$$

The test statistic is

$$F_c = \text{MSMACHINE}/Q^*,$$

which is approximately distributed as F with $b - 1$ and r degrees of freedom where

$$r = \frac{(Q^*)^2}{\dfrac{(q_1\text{MSERROR})^2}{df_{\text{MSERROR}}} + \dfrac{[q_2\text{MSR}(\mathbf{g}|\mu, \tau, p)]^2}{df_{\text{MSR}(\mathbf{g}|\mu, \tau, p)}} + \dfrac{[q_3\text{MSR}(p|\mu, \tau)]^2}{df_{\text{MS}(p|\mu, \tau)}}}.$$

Now the above procedure shall be applied to the unbalanced data case in Table 23.1. From Table 23.3, $k_1 = 2.316$, $k_2 = 7.219$, $k_3 = 2.587$, $k_4 = 2.611$, $k_5 = .157$, MSERROR $= .873$, MSMACHINE$*$PERSON $= 40.432$, MSPERSON $= 201.753$, and MSMACHINE $= 824.332$. Therefore,

$$\begin{aligned} q_1 &= 1 - \frac{.157}{7.219} - \frac{2.611}{2.316} + \frac{(2.587)(.157)}{(2.316)(7.219)} \\ &= -.125, \\ q_2 &= \frac{2.611}{2.316} - \frac{(2.587)(.157)}{(2.316)(7.219)} = 1.103, \quad \text{and} \\ q_3 &= \left(\frac{.157}{7.219}\right) = .022, \end{aligned}$$

and

$$\begin{aligned} Q^* &= (-.125)(.873) + 1.103(40.432) + .022(201.753) \\ &= 48.925 \end{aligned}$$

The approximate degrees of freedom are

$$\begin{aligned} r &= \frac{(48.925)^2}{\dfrac{[(-.125)(.873)]^2}{26} + \dfrac{[(1.103)(40.432)]^2}{10} + \dfrac{[(.022)(201.753)]^2}{5}} \\ &= \frac{2{,}397.5712}{203.186} = 11.8. \end{aligned}$$

The value of the test statistic is

$$F_c = \frac{824.332}{48.925} = 16.848,$$

which is approximately distributed as F with 2 and 11.8 degrees of freedom. The $F_{.01,2,11.8}$ value, obtained by interpolation, is 6.984; thus, the hypothesis of equal machine effects would be rejected at $\alpha < .01$.

CONCLUDING REMARKS

In this chapter, two data sets for a mixed model were used to demonstrate the analysis techniques described in Chapter 22. The first example involved a set of balanced data, showing that the analysis is quite straightforward. The second example involved a set of unbalanced data. Because the analysis of the unbalanced mixed model is computationally difficult, matrix methods were described that can be implemented for a computer analysis.

24

Methods for Analyzing Balanced Split-Plot Designs

CHAPTER OUTLINE

In this chapter, the analysis of split-plot designs is discussed by means of examples involving two sizes of experimental units. The examples demonstrate how to specify the model and how to construct proper F-tests. Since these designs involve different sizes of experimental units and different variances, the standard errors of various mean comparisons involve one or more of the variances. Methods of constructing standard errors for means, which are described in Section 24.3, can easily be applied to more complex models.

24.1 MODEL DEFINITION AND PARAMETER ESTIMATION

The key concepts in constructing models for split-plot designs are recognizing the different sizes of experimental units and then identifying the corresponding design structures and treatment structures. The overall model is constructed by incorporating models developed for each size of experimental unit.

Several examples of model construction were presented in Chapter 5, but the assumption underlying the models was not stated. The assumption is that the components denoting the error terms for the various experimental units are all distributed independently with mean zero and an associated variance (see Chapter 26 for assumptions that are more general). The objective of the analysis here is to use the model assumption to obtain estimates of the population parameters and to make inferences.

EXAMPLE 24.1: Varieties of Wheat on Fertility Regimes

The data in Figure 24.1 are the yields in pounds of two varieties of wheat (B) grown in four different fertility regimes (A). The field was divided into two blocks, each with four whole plots. Each of the four fertilizer levels was randomly assigned to one whole plot within each block. Thus, the whole plot experimental design consists of a one-way treatment

Block 1

Fertility Regime	Variety	B_1	B_2
	A_1	35.4	37.9
	A_2	36.7	38.2
	A_3	34.8	36.4
	A_4	39.5	40.0

Block 2

Fertility Regime	Variety	B_1	B_2
	A_1	41.6	40.3
	A_2	42.7	41.6
	A_3	43.6	42.8
	A_4	44.5	47.6

Subplot (circled: Block 1, A_4, B_2 = 40.0)

Whole Plot (circled: Block 2, A_4 = 44.5, 47.6)

Figure 24.1 Data for the split-plot example.

structure in a randomized complete block design structure with two blocks where each block contains four whole plot experimental units. Each whole plot was split into two parts (called subplots), and each variety of wheat was randomly assigned to one subplot within each whole plot. The subplot experimental design consists of a one-way treatment structure in a randomized complete block design with eight blocks, each block containing two subplot experimental units.

Let μ_{ik} denote the expected response (yield) and y_{ijk} denote the observed response of variety k grown with fertility regime i in block j. Express μ_{ik} as

$$\mu_{ik} = \mu + F_i + V_k + F * V_{ik}.$$

The model for this split-plot experiment is

$$\begin{aligned} y_{ijk} &= \mu + F_i + b_j + e_{ij} \} && \text{Whole plot part of model} \\ &\quad + V_k + F * V_{ik} + \varepsilon_{ijk} \} && \text{Subplot part of model} \end{aligned} \tag{24.1.1}$$

for $i = 1, 2, 3, 4$; $j = 1, 2$; and $k = 1, 2$ where b_j denotes the effect of the jth block, which consists of four whole plots; e_{ij} denotes the whole plot error, which is assumed to be distributed i.i.d. $N(0, \sigma_e^2)$; and ε_{ijk} denotes the subplot error which is assumed to be distributed i.i.d. $N(0, \sigma_\varepsilon^2)$; and e_{ij} and ε_{ijk} are assumed to be independently distributed.

An analysis of variance table for model (24.1.1) is shown in Table 24.1, where A denotes the whole plot treatment with a levels, B denotes

Table 24.1 Analysis of Variance Table for Model (24.1.1)

Source of Variation	*df*	*EMS*
Replication	$r - 1$	$\sigma_\varepsilon^2 + a\sigma_e^2 + ab\sigma_b^2$
A	$a - 1$	$\sigma_\varepsilon^2 + b\sigma_e^2 + \phi^2(\alpha)$
Error(Whole Plot)	$(a - 1)(r - 1)$	$\sigma_\varepsilon^2 + b\sigma_e^2$
B	$b - 1$	$\sigma_\varepsilon^2 + \phi^2(\beta)$
$A * B$	$(a - 1)(b - 1)$	$\sigma_\varepsilon^2 + \phi^2(\alpha\beta)$
Error(Subplot)	$a(b - 1)(r - 1)$	σ_ε^2

the subplot treatment with b levels, and r denotes the number of replications.

The whole-plot error mean square, denoted by Error(Whole Plot), estimates (that is, has expected mean square) $\sigma_\varepsilon^2 + 2\sigma_e^2$, where the 2 denotes the number of subplot treatments. The subplot error mean square, denoted by Error(Subplot), estimates σ_ε^2.

Using the method of moments (Section 19.1), estimators for σ_ε^2 and σ_e^2 are

$$\hat{\sigma}_\varepsilon^2 = \text{MSERROR(SUBPLOT)} \quad \text{and}$$

$$\hat{\sigma}_e^2 = \frac{\text{MSERROR(WHOLE PLOT)} - \hat{\sigma}_\varepsilon^2}{b}$$

The estimator of μ_{ik} is $\bar{y}_{i\cdot k}$, of $\bar{\mu}_{i\cdot}$ is $\bar{y}_{i\cdot\cdot}$, and of $\bar{\mu}_{\cdot k}$ is $\bar{y}_{\cdot\cdot k}$.

As can be seen from the expected mean squares, the comparisons between fertility regimes is a between-whole-plot experimental unit comparison; thus, the proper F-statistic for comparing equal fertility regime effects is MSA/MSERROR(WHOLE PLOT). The comparisons between varieties and Variety by Fertility interactions are between-subplot experimental unit comparisons; thus, the proper F-statistics for comparing varieties and for comparing the interactions are MSB/MSERROR(SUBPLOT) and MSA * B/MSERROR(SUBPLOT).

The analysis of variance table for the data in Figure 24.1 is shown in Table 24.2, which demonstrates how to use the expected mean squares in Table 24.1 to compute the proper F-tests.

Once the F-tests have been made to determine if there are significant differences between means, the next step is to carry out multiple comparisons to determine where the differences occur. The next section presents methods to compute standard errors for split-plot designs.

Table 24.2 Analysis of Variance Table for Wheat Yields in Figure 24.1

Source of Variation	*df*	*SS*	*F*	*p*
Block	1	131.1025		
A	3	40.1900	5.80	.0914
Block * A = Error(Whole Plot)	3	6.9275		
B	1	2.2500	1.07	.3599
$A * B$	3	1.5500	0.25	.8612
Error(Subplot)	4	8.4300		

24.2 STANDARD ERRORS FOR MULTIPLE COMPARISONS

When the analysis of variance of a split-plot experimental design shows significant effects, we want to determine which treatment combinations are significantly different. To solve this problem, the standard error of the difference between two means must be determined.

To demonstrate methods for producing standard errors, we shall use a split-plot experimental design where the whole plot experimental design is a one-way treatment structure in a randomized complete block design structure. The model for this situation is

$$y_{ijk} = \mu_{ik} + r_j + e_{ij} + \varepsilon_{ijk} \tag{24.2.1}$$

or

$$y_{ijk} = \mu + \alpha_i + r_j + e_{ij} + \beta_k + (\alpha\beta)_{ik} + \varepsilon_{ijk}$$

for $i = 1, 2, \ldots, a$; $j = 1, 2, \ldots, r$; and $k = 1, 2, \ldots, b$, where μ_{ik} denotes the mean of whole plot treatment i with subplot treatment k; r_j denotes the random effect of replication with $r_j \sim$ i.i.d. $N(0, \sigma_r^2)$; e_{ij} denotes the random effect of the ith whole plot in replication j with $e_{ij} \sim$ i.i.d. $N(0, \sigma_e^2)$; ε_{ijk} denotes the random effect of the kth subplot in the ith whole plot of the jth replication with $\varepsilon_{ijk} \sim$ i.i.d. $N(0, \sigma_\varepsilon^2)$; α_i denotes the effect of the ith whole plot treatment; β_k denotes the effect of the kth subplot treatment; and $(\alpha\beta)_{ik}$ denotes the interaction effect of the ith whole plot treatment with the kth subplot treatment. Table 24.1 contains the analysis of variance table for model (24.2.1) along with the expected mean squares for the two error terms.

Four types of comparisons may be of interest, depending on whether any interaction exists between the levels of α and the levels of β. If there is no interaction, then it is of interest to compare the levels of α and to compare the levels of β.

To compare the levels of β, one needs to compare the $\bar{\mu}_{\cdot k}$'s, which are estimated by the $\bar{y}_{\cdot\cdot k}$'s. Express $\bar{y}_{\cdot\cdot k}$ in terms of the quantities in model (24.2.1) by summing over i and j. By doing this, $\bar{y}_{\cdot\cdot k}$ can be expressed as

$$\bar{y}_{\cdot\cdot k} = \bar{\mu}_{\cdot k} + \bar{r}_{\cdot} + \bar{e}_{\cdot\cdot} + \bar{\varepsilon}_{\cdot\cdot k}. \tag{24.2.2}$$

To illustrate the method, consider the difference $\bar{\mu}_{\cdot 1} - \bar{\mu}_{\cdot 2}$. The estimate of $\bar{\mu}_{\cdot 1} - \bar{\mu}_{\cdot 2}$ is $\bar{y}_{\cdot\cdot 1} - \bar{y}_{\cdot\cdot 2}$, which can be expressed in terms of model (24.2.1) as

$$\bar{y}_{\cdot\cdot 1} - \bar{y}_{\cdot\cdot 2} = \bar{\mu}_{\cdot 1} - \bar{\mu}_{\cdot 2} + \bar{\varepsilon}_{\cdot\cdot 1} - \bar{\varepsilon}_{\cdot\cdot 2}.$$

Thus, the comparison $\bar{y}_{\cdot\cdot 1} - \bar{y}_{\cdot\cdot 2}$ does not depend on the whole plot error. The variance of $\bar{y}_{\cdot\cdot 1} - \bar{y}_{\cdot\cdot 2}$ is

$$\operatorname{Var}(\bar{y}_{\cdot\cdot 1} - \bar{y}_{\cdot\cdot 2}) = \operatorname{Var}(\bar{\varepsilon}_{\cdot\cdot 1} - \bar{\varepsilon}_{\cdot\cdot 2}) = \frac{2\sigma_\varepsilon^2}{ar},$$

where the variance of the mean $\bar{\varepsilon}_{\cdot\cdot 1}$ is $\operatorname{Var}(\varepsilon_{ijk})/n$ where $n = ar$ is the number of observations in the mean. Similarly, one obtains the same variance for any $\bar{y}_{\cdot\cdot k} - \bar{y}_{\cdot\cdot k'}$ for $k \neq k'$. The estimate of the standard error of $\bar{y}_{\cdot\cdot k} - \bar{y}_{\cdot\cdot k'}$ is

$$\widehat{\text{s.e.}}(\bar{y}_{\cdot\cdot k} - \bar{y}_{\cdot\cdot k'}) = \sqrt{\frac{2 \times \text{MSERROR(SUBPLOT)}}{ar}}\,.$$

The LSD value is

$$\text{LSD}_\alpha = \{t_{\alpha/2, a(b-1)(r-1)}\}\widehat{\text{s.e.}}(\bar{y}_{\cdot\cdot 1} - \bar{y}_{\cdot\cdot 2}).$$

The above LSD value can be used to construct a Fisher LSD multiple comparison procedure (see Chapter 3).

To compare the levels of α, one needs to compare $\bar{\mu}_{i\cdot}$'s, which are estimated by the $\bar{y}_{i\cdot\cdot}$'s. The quantity $\bar{y}_{i\cdot\cdot}$ can be expressed in terms of model (24.2.1) by summing over j and k as

$$\bar{y}_{i\cdot\cdot} = \bar{\mu}_{i\cdot} + \bar{r}_{\cdot} + \bar{e}_{i\cdot} + \bar{\varepsilon}_{i\cdot\cdot}. \tag{24.2.3}$$

The estimate of the contrast $\bar{\mu}_{1\cdot} - \bar{\mu}_{2\cdot}$ is $\bar{y}_{1\cdot\cdot} - \bar{y}_{2\cdot\cdot}$, which can be expressed in terms of model (24.2.3) as

$$\bar{y}_{1\cdot\cdot} - \bar{y}_{2\cdot\cdot} = \bar{\mu}_{1\cdot} - \bar{\mu}_{2\cdot} + \bar{e}_{1\cdot} - \bar{e}_{2\cdot} + \bar{\varepsilon}_{1\cdot\cdot} - \bar{\varepsilon}_{2\cdot\cdot}.$$

This comparison depends on both the whole plot and subplot error terms. The variance of $\bar{y}_{1\cdot\cdot} - \bar{y}_{2\cdot\cdot}$ is

$$\begin{aligned}\text{Var}(\bar{y}_{1\cdot\cdot} - \bar{y}_{2\cdot\cdot}) &= \text{Var}(\bar{e}_{1\cdot} - \bar{e}_{2\cdot} + \bar{\varepsilon}_{1\cdot\cdot} - \bar{\varepsilon}_{2\cdot\cdot}) \\ &= \frac{2\sigma_e^2}{r} + \frac{2\sigma_\varepsilon^2}{br} \\ &= \left(\frac{2}{br}\right)(\sigma_\varepsilon^2 + b\sigma_e^2).\end{aligned}$$

The estimate of the standard error of $\bar{y}_{i'\cdot\cdot} - \bar{y}_{i\cdot\cdot}$ is

$$\widehat{\text{s.e.}}(\bar{y}_{i'\cdot\cdot} - \bar{y}_{i\cdot\cdot}) = \sqrt{\frac{2 \times \text{MSERROR(WHOLE PLOT)}}{br}} \qquad \text{for} \quad i \neq i',$$

The LSD value for comparing $\bar{\mu}_{i'\cdot} - \bar{\mu}_{i\cdot}$ is

$$\text{LSD}_\alpha = \{t_{\alpha/2, (r-1)(a-1)}\}\widehat{\text{s.e.}}(\bar{y}_{i'\cdot\cdot} - \bar{y}_{i\cdot\cdot}).$$

When there is a significant $\alpha * \beta$ interaction, comparisons must be based on the set of two-way cell means. There are two different types of comparisons one must consider when studying the cell means. The first type arises when two subplot treatment (β) means are compared at the same level of a whole plot treatment (α), such as $\mu_{11} - \mu_{12}$. The best estimator of $\mu_{11} - \mu_{12}$ is $\bar{y}_{1\cdot 1} - \bar{y}_{1\cdot 2}$. The term $\bar{y}_{i\cdot k}$ can be expressed in terms of model (24.2.1) by summing over j as

$$\bar{y}_{i\cdot k} = \mu_{ik} + \bar{r}_{\cdot} + \bar{e}_{i\cdot} + \bar{\varepsilon}_{i\cdot k}. \tag{24.2.4}$$

Thus, the estimator $\bar{y}_{1\cdot 1} - \bar{y}_{1\cdot 2}$ can be expressed as

$$\bar{y}_{1\cdot 1} - \bar{y}_{1\cdot 2} = \mu_{11} - \mu_{12} + \bar{\varepsilon}_{1\cdot 1} - \bar{\varepsilon}_{1\cdot 2}$$

with variance

$$\text{Var}(\bar{y}_{1\cdot 1} - \bar{y}_{1\cdot 2}) = \text{Var}(\bar{\varepsilon}_{1\cdot 1} - \bar{\varepsilon}_{1\cdot 2}) = \frac{2\sigma_\varepsilon^2}{r}.$$

The variance of comparisons between subplot treatments at the same whole plot treatment depends only on the subplot error. The estimate of the standard error is

$$\widehat{\text{s.e.}}(\bar{y}_{1\cdot 1} - \bar{y}_{1\cdot 2}) = \sqrt{\frac{2 \times \text{MSERROR(SUBPLOT)}}{r}}$$

and the corresponding LSD value is

$$\text{LSD}_\alpha = \{t_{\alpha/2, a(r-1)(b-1)}\}\widehat{\text{s.e.}}(\bar{y}_{1\cdot 1} - \bar{y}_{1\cdot 2}).$$

This LSD value can be used to compare any subplot treatments at the same level of a whole plot treatment.

The second type of comparison occurs when two whole plot treatments are compared at the same level or different levels of the subplot treatments, such as $\mu_{11} - \mu_{21}$ or $\mu_{11} - \mu_{22}$. Comparison of these two types have the same standard errors. The best estimate of $\mu_{11} - \mu_{21}$ is $\bar{y}_{1\cdot 1} - \bar{y}_{2\cdot 1}$, which can be expressed in terms of model (24.2.1) as

$$\bar{y}_{1\cdot 1} - \bar{y}_{2\cdot 1} = \mu_{11} - \mu_{21} + \bar{e}_{1\cdot} - \bar{e}_{2\cdot} + \bar{\varepsilon}_{1\cdot 1} - \bar{\varepsilon}_{2\cdot 1},$$

and hence,

$$\begin{aligned}\text{Var}(\bar{y}_{1\cdot 1} - \bar{y}_{2\cdot 1}) &= \text{Var}(\bar{e}_{1\cdot} - \bar{e}_{2\cdot} + \bar{\varepsilon}_{1\cdot 1} - \bar{\varepsilon}_{2\cdot 1}) \\ &= \frac{2\sigma_e^2}{r} + \frac{2\sigma_\varepsilon^2}{r} \\ &= \left(\frac{2}{r}\right)(\sigma_\varepsilon^2 + \sigma_e^2). \end{aligned} \quad (24.2.5)$$

This comparison depends on both the whole plot error and the subplot error. An unbiased estimate of $\sigma_\varepsilon^2 + \sigma_e^2$ is

$$\widehat{\sigma_\varepsilon^2 + \sigma_e^2} = \frac{\text{MSERROR(WHOLE PLOT)} + (b-1)\text{MSERROR(SUBPLOT)}}{b}. \quad (24.2.6)$$

The sampling distribution associated with $\widehat{\sigma_\varepsilon^2 + \sigma_e^2}$ is not a chi-square distribution but a linear combination of chi-square distributions. There is not an exact LSD value for this comparison, but by using the method of moments, an approximate LSD value can be computed as

$$\text{LSD}_\alpha = t^*_{\alpha/2}\sqrt{\frac{2 \times \widehat{\left(\sigma_\varepsilon^2 + \sigma_e^2\right)}}{r}}$$

where

$$t^*_{\alpha/2} = \frac{\{t_{\alpha/2,(a-1)(r-1)}\}\text{MSERROR(WHOLE PLOT)} + \{t_{\alpha/2,a(r-1)(b-1)}\}(b-1)\text{MSERROR(SUBPLOT)}}{\text{MSERROR(WHOLE PLOT)} + (b-1)\text{MSERROR(SUBPLOT)}} \tag{24.2.7}$$

The value of $t^*_{\alpha/2}$ is a weighted average of

$$t_{\alpha/2,(a-1)(r-1)} \quad \text{and} \quad t_{\alpha/2,a(r-1)(b-1)}.$$

The standard error in (24.2.6) and the corresponding LSD value in (24.2.7) are also used to make comparisons of the form $\mu_{11} - \mu_{22}$.

From the data in Figure 24.1, the standard errors for the four types of comparisons are computed. Figure 24.2 contains the various means and necessary standard errors for making the comparisons.

1. Compare Variety means:

$$\widehat{\text{s.e.}}(\bar{y}_{\cdot\cdot 1} - \bar{y}_{\cdot\cdot 2}) = \sqrt{\frac{2 \times 2.1075}{8}} = .726$$

$$\text{LSD}_{.05} = 2.776 \times .726 = 2.015 \qquad (t_{.025}(4) = 2.776)$$

2. Compare Fertilizing means:

$$\widehat{\text{s.e.}}(\bar{y}_{1\cdot\cdot} - \bar{y}_{2\cdot\cdot}) = \sqrt{\frac{2 \times 6.9275/3}{4}} = 1.0745$$

$$\text{LSD}_{.05} = 3.182 \times 1.0745 = 3.419 \qquad (t_{.025}(3) = 3.182)$$

		Variety B_1	Variety B_2	
Fertilizing Regime	A_1	$\bar{y}_{11\cdot} = 38.5$	$\bar{y}_{12\cdot} = 39.1$	$\bar{y}_{1\cdot\cdot} = 38.80$
	A_2	$\bar{y}_{21\cdot} = 39.7$	$\bar{y}_{22\cdot} = 39.9$	$\bar{y}_{2\cdot\cdot} = 39.80$
	A_3	$\bar{y}_{31\cdot} = 39.2$	$\bar{y}_{32\cdot} = 39.6$	$\bar{y}_{3\cdot\cdot} = 39.40$
	A_4	$\bar{y}_{41\cdot} = 42.0$	$\bar{y}_{42\cdot} = 43.8$	$\bar{y}_{4\cdot\cdot} = 42.90$
		$\bar{y}_{\cdot 1\cdot} = 39.85$	$\bar{y}_{\cdot 2\cdot} = 40.60$	$\bar{y}_{\cdot\cdot\cdot} = 40.225$

Figure 24.2 Mean yields for the data in Figure 24.1.

3. Compare Variety means at same level of fertilizer:

$$\widehat{\text{s.e.}}(\bar{y}_{1.1} - \bar{y}_{1.2}) = \sqrt{\frac{2 \times 2.1075}{2}} = 1.452$$

$$\text{LSD}_{.05} = 2.776 \times 1.452 = 4.031$$

4. Compare Fertilizing means at same variety or different varieties:

$$\widehat{\text{s.e.}}(\bar{y}_{11.} - \bar{y}_{21.}) = \sqrt{\frac{2 \times [(6.9275/3) + 1 \times 2.1075]}{4}} = 1.486$$

$$t^* = \frac{3.182 \times (6.9275/3) + 2.776 \times 1 \times 2.1075}{(6.9275/3) + 1 \times 2.1075}$$

$$= \frac{13.198}{4.4167} = 2.988.$$

$$\text{LSD}_{.05} = 2.988 \times 1.486 = 4.440.$$

Below is listed a method for computing standard errors for complex split-plot type designs.

24.3 GENERAL METHOD FOR COMPUTING STANDARD ERRORS FOR MULTIPLE COMPARISONS

Applying the techniques presented here for computing standard errors for comparisons involving more than one size of experimental unit is not always straightforward. In this section, a general method for computing standard errors and approximate LSD values is described that can be applied to more complex situations. This section may be skipped by the casual reader, but the method is used to compute standard errors for the repeated measures designs in Chapter 26 and the crossover designs in Chapter 32. Since the method can be described by example, model (24.2.1) is used to demonstrate the technique.

Basically, the technique consists of expressing a given comparison of means as the sum of components where each component is a comparison of means involving only one size of experimental unit. Then the components of the comparison are independently distributed, and the variance of the comparison is obtained by summing the variances of the components.

Thus, the comparison $\mu_{11} - \mu_{21}$ can be expressed as the sum

$$\mu_{11} - \mu_{21} = (\bar{\mu}_{1.} - \bar{\mu}_{2.}) + [(\mu_{11} - \bar{\mu}_{1.}) - (\mu_{21} - \bar{\mu}_{2.})].$$

The component $\bar{\mu}_{1.} - \bar{\mu}_{2.}$ is a whole plot comparison, and the component

$$(\mu_{11} - \bar{\mu}_{1.}) - (\mu_{21} - \bar{\mu}_{2.})$$

is a subplot comparison. The estimates of these components are

$$\widehat{\bar{\mu}_{1.} - \bar{\mu}_{2.}} = \bar{y}_{1..} - \bar{y}_{2..}$$

and

$$\widehat{(\mu_{11}-\bar{\mu}_{1\cdot})-(\mu_{21}-\bar{\mu}_{2\cdot})} = (\bar{y}_{1\cdot 1}-\bar{y}_{1\cdot\cdot})-(\bar{y}_{2\cdot 1}-\bar{y}_{2\cdot\cdot});$$

thus, the estimate of $\mu_{11}-\mu_{21}$ is

$$\begin{aligned}\widehat{\mu_{11}-\mu_{21}} &= \widehat{\bar{\mu}_{1\cdot}-\bar{\mu}_{2\cdot}} + \left[\widehat{(\mu_{11}-\bar{\mu}_{1\cdot})-(\mu_{21}-\bar{\mu}_{21})}\right]\\ &= \bar{y}_{1\cdot 1}-\bar{y}_{2\cdot 1}.\end{aligned}$$

Since comparisons computed from the whole plot part of the model are independently distributed of comparisons computed from the subplot part, the variance of

$$\widehat{\mu_{11}-\mu_{21}}$$

is

$$\operatorname{Var}\left(\widehat{\mu_{11}-\mu_{21}}\right) = \operatorname{Var}(\bar{y}_{1\cdot\cdot}-\bar{y}_{2\cdot\cdot}) + \operatorname{Var}\left[(\bar{y}_{1\cdot 1}-\bar{y}_{1\cdot\cdot})-(\bar{y}_{2\cdot 1}-\bar{y}_{2\cdot\cdot})\right].$$

The quantity $\bar{y}_{1\cdot\cdot}-\bar{y}_{2\cdot\cdot}$ expressed in terms of model (24.2.1) is

$$\bar{y}_{1\cdot\cdot}-\bar{y}_{2\cdot\cdot} = \bar{\mu}_{1\cdot}-\bar{\mu}_{2\cdot}+\bar{e}_{1\cdot}-\bar{e}_{2\cdot}+\bar{\varepsilon}_{1\cdot\cdot}-\bar{\varepsilon}_{2\cdot\cdot},$$

and its variance is

$$\begin{aligned}\operatorname{Var}(\bar{y}_{1\cdot\cdot}-\bar{y}_{2\cdot\cdot}) &= \operatorname{Var}(\bar{e}_{1\cdot}-\bar{e}_{2\cdot}+\bar{\varepsilon}_{1\cdot\cdot}-\bar{\varepsilon}_{2\cdot\cdot})\\ &= \left(\frac{2}{br}\right)\left(\sigma_\varepsilon^2+b\sigma_e^2\right).\end{aligned}$$

An estimate of $\operatorname{Var}(\bar{y}_{1\cdot\cdot}-\bar{y}_{2\cdot\cdot})$ is $(2/br)$MSERROR(WHOLE PLOT). The quantity

$$(\bar{y}_{1\cdot 1}-\bar{y}_{1\cdot\cdot})-(\bar{y}_{2\cdot 1}-\bar{y}_{2\cdot\cdot})$$

expressed in terms of model (24.2.1) is

$$\begin{aligned}&(\bar{y}_{1\cdot 1}-\bar{y}_{1\cdot\cdot})-(\bar{y}_{2\cdot 1}-\bar{y}_{2\cdot\cdot})\\ &\quad= (\mu_{11}+\bar{r}_{\cdot}+\bar{e}_{1\cdot}+\bar{\varepsilon}_{1\cdot 1})-(\bar{\mu}_{1\cdot}+\bar{r}_{\cdot}+\bar{e}_{1\cdot}+\bar{\varepsilon}_{1\cdot\cdot})\\ &\qquad -(\mu_{21}+\bar{r}_{\cdot}+\bar{e}_{2\cdot}+\bar{\varepsilon}_{2\cdot 1})+(\bar{\mu}_{2\cdot}+\bar{r}_{\cdot}+\bar{e}_{2\cdot}+\bar{\varepsilon}_{2\cdot\cdot})\\ &\quad= (\mu_{11}-\bar{\mu}_{1\cdot})-(\mu_{21}-\bar{\mu}_{2\cdot})+(\bar{\varepsilon}_{1\cdot 1}-\bar{\varepsilon}_{1\cdot\cdot})-(\bar{\varepsilon}_{2\cdot 1}-\bar{\varepsilon}_{2\cdot\cdot})\end{aligned}$$

and its variance is

$$\begin{aligned}\operatorname{Var}\left[(\bar{y}_{1\cdot 1}-\bar{y}_{1\cdot\cdot})-(\bar{y}_{2\cdot 1}-\bar{y}_{2\cdot\cdot})\right] &= \operatorname{Var}\left[(\bar{\varepsilon}_{1\cdot 1}-\bar{\varepsilon}_{1\cdot\cdot})-(\bar{\varepsilon}_{2\cdot 1}-\bar{\varepsilon}_{2\cdot\cdot})\right]\\ &= \frac{2(b-1)}{b}\left(\sigma_\varepsilon^2\right).\end{aligned}$$

An estimate of the above variance is

$$\left[\frac{2(b-1)}{b}\right]\text{MSERROR(SUBPLOT)}.$$

Combining the whole plot component variance and the subplot component variance yields

$$\text{Var}(\bar{y}_{1\cdot 1} - \bar{y}_{2\cdot 1}) = \left(\frac{2}{br}\right)(\sigma_\varepsilon^2 + b\sigma_e^2) + \frac{2(b-1)}{b}(\sigma_\varepsilon^2),$$

and the estimate of the standard error of $\bar{y}_{1\cdot 1} - \bar{y}_{2\cdot 1}$ is

$$\widehat{\text{s.e.}}(\bar{y}_{1\cdot 1} - \bar{y}_{2\cdot 1}) = \sqrt{\left(\frac{2}{br}\right)\text{MSERROR(WHOLE PLOT)} + \left[\frac{2(b-1)}{b}\right]\text{MSERROR(SUBPLOT)}}$$

The approximate Fisher's α% LSD value is

$$\text{LSD} = t^*_{\alpha/2}\widehat{\text{s.e.}}(\bar{y}_{1\cdot 1} - \bar{y}_{2\cdot 1})$$

where

$$t^*_{\alpha/2} = \frac{\{t_{\alpha/2,v_1}(2/br)\text{MSERROR(WHOLE PLOT)} + t_{\alpha/2,v_2}[2(b-1)/b]\text{MSERROR(SUBPLOT)}\}}{(2/br)\text{MSERROR(WHOLE PLOT)} + [2(b-1)/b]\text{MSERROR(SUBPLOT)}}$$

and v_1 is the degrees of freedom associated with MSERROR(WHOLE PLOT) and v_2 is the degrees of freedom associated with MSERROR(SUBPLOT). With some simple algebra, one can show that the above $\widehat{\text{s.e.}}(\bar{y}_{1\cdot 1} - \bar{y}_{2\cdot 1})$ and $t^*_{\alpha/2}$ values are identical to those obtained in equations (24.2.6), and (24.2.7). Other applications of this method are demonstrated in Chapters 26 and 32.

24.4 COMPARISONS VIA CONTRASTS

Contrasts can be constructed for each of the comparisons discussed in Section 24.2. A contrast between the levels of β or $\bar{\mu}_{\cdot k}$'s is

$$\theta = C_1\bar{\mu}_{\cdot 1} + C_2\bar{\mu}_{\cdot 2} + \cdots + C_b\bar{\mu}_{\cdot b}$$

where $C_1 + C_2 + \cdots + C_b = 0$. An estimate of the contrast is

$$\hat{\theta} = C_1\bar{y}_{\cdot\cdot 1} + C_2\bar{y}_{\cdot\cdot 2} + \cdots + C_b\bar{y}_{\cdot\cdot b}$$

with variance $\sigma_\varepsilon^2(C_1^2 + C_2^2 + \cdots + C_b^2)/ar$. The estimate of the standard error of $\hat{\theta}$ is

$$\widehat{\text{s.e.}}(\hat{\theta}) = \sqrt{\frac{\text{MSERROR(SUBPLOT)}\left(C_1^2 + C_2^2 + \cdots + C_b^2\right)}{ar}}.$$

A contrast between the levels of α or $\bar{\mu}_{i\cdot}$'s is

$$\tau = C_1\bar{\mu}_{1\cdot} + C_2\bar{\mu}_{2\cdot} + \cdots + C_a\bar{u}_{a\cdot}$$

with estimate

$$\hat{\tau} = C_1\bar{y}_{1\cdot\cdot} + C_2\bar{y}_{2\cdot\cdot} + \cdots + C_a\bar{y}_{a\cdot\cdot},$$

variance $(\sigma_e^2 + b\sigma_\varepsilon^2)(C_1^2 + C_2^2 + \cdots + C_a^2)/br$, and the estimate of the standard error of $\hat{\tau}$ is

$$\widehat{\text{s.e.}}(\hat{\tau}) = \sqrt{\frac{\text{MSERROR(WHOLE PLOT)}\left(C_1^2 + C_2^2 + \cdots + C_a^2\right)}{br}}.$$

A contrast of subplot treatments at the same whole plot treatment is

$$\delta = C_1\bar{\mu}_{i1} + C_2\bar{\mu}_{i2} + \cdots + C_b\bar{\mu}_{ib}$$

and is estimated by

$$\hat{\delta} = C_1\bar{y}_{i\cdot 1} + C_2\bar{y}_{i\cdot 2} + \cdots + C_b\bar{y}_{i\cdot b}.$$

The variance of $\hat{\delta}$ is $\sigma_e^2(C_1^2 + C_2^2 + \cdots + C_b^2)/r$, and the estimated standard error of $\hat{\delta}$ is

$$\widehat{\text{s.e.}}(\hat{\delta}) = \sqrt{\frac{\text{MSERROR(SUBPLOT)}\left(C_1^2 + C_2^2 + \cdots + C_b^2\right)}{r}}.$$

A contrast of whole plot treatments at the same subplot treatment is

$$\gamma = C_1\bar{\mu}_{1k} + C_2\bar{\mu}_{2k} + \cdots + C_a\bar{\mu}_{ak}$$

and is estimated by

$$\hat{\gamma} = C_1\bar{y}_{1\cdot k} + C_2\bar{y}_{2\cdot k} + \cdots + C_a\bar{y}_{a\cdot k}.$$

The variance of $\hat{\gamma}$ is $(\sigma_\varepsilon^2 + \sigma_e^2)(C_1^2 + C_2^2 + \cdots + C_a^2)/br$, and the estimated standard error of $\hat{\gamma}$ is

$$\widehat{\text{s.e.}}(\hat{\gamma}) = \sqrt{\frac{\left(\widehat{\sigma_\varepsilon^2 + \sigma_e^2}\right)\left(C_1^2 + C_2^2 + \cdots + C_a^2\right)}{r}}$$

where

$$\widehat{\sigma_\varepsilon^2 + \sigma_e^2} = \frac{\text{MSERROR(WHOLE PLOT)} + (b-1)\text{MSERROR(SUBPLOT)}}{b}.$$

The above estimates and standard errors can be used to test hypotheses via t-tests or used to construct confidence intervals.

The estimate of $\sigma_\varepsilon^2 + \sigma_e^2$ is a function of two sums of squares with different degrees of freedom. In order to make inferences, the Satterthwaite approximation (see Chapter 20) is used to obtain approximate degrees of freedom to be associated with the estimate of $\sigma_\varepsilon^2 + \sigma_e^2$. The approximate degrees of freedom are

$$r = \frac{[\text{MSERROR(WHOLE PLOT)} + (b-1)\text{MSERROR(SUBPLOT)}]^2}{\dfrac{[\text{MSERROR(WHOLE PLOT)}]^2}{(a-1)(r-1)} + \dfrac{[(b-1)\text{MSERROR(SUBPLOT)}]^2}{a(b-1)(r-1)}}.$$

A t-value with r degrees of freedom can be used to test hypotheses and construct confidence intervals about contrasts of the form of γ.

There are many possible choices of the contrasts. For example, if the levels of a factor are quantitative, you can use the linear, quadratic, and similar contrasts to investigate the possible trends.

EXAMPLE 24.2.

The data in Figure 24.3 are from an experiment where the amount of dry matter was measured on wheat plants grown in different levels of moisture and different amounts of fertilizer. There were 48 different peat

			Level of Fertilizer			
		Tray	2	4	6	8
Level of Moisture	10	1	3.3458	4.3170	4.5572	5.8794
		2	4.0444	4.1413	6.5173	7.3776
		3	1.97584	3.8397	4.4730	5.1180
	20	4	5.0490	7.9419	10.7697	13.5168
		5	5.91310	8.5129	10.3934	13.9157
		6	6.95113	7.0265	10.9334	15.2750
	30	7	6.56933	10.7348	12.2626	15.7133
		8	8.29741	8.9081	13.4373	14.9575
		9	5.27853	8.6654	11.1372	15.6332
	40	10	6.8393	9.0842	10.3654	12.5144
		11	6.4997	6.0702	10.7486	12.5034
		12	4.0482	3.8376	9.4367	10.2811

Figure 24.3 Dry matter measurements per pot for Example 24.2.

pots and 12 plastic trays; Four pots could be put into each tray. The moisture treatments consisted of adding 10, 20, 30, or 40 ml of water per pot per day to the tray, where the water was absorbed by the peat pots. The levels of moisture were randomly assigned to the trays. The trays are the large size of experimental unit or whole plot, and the whole plot experimental design is a one-way treatment structure (the four levels of moisture) in a completely randomized design structure.

The levels of fertilizer were 2, 4, 6, or 8 mg per pot. The four levels of fertilizer were randomly assigned to the four pots in each tray so that each fertilizer occurred once in each tray. The pot is the smallest size of experimental unit or split plot, and the split-plot experimental design is a one-way treatment structure (the four levels of fertilizer) in a randomized complete block design structure where the 12 trays are the blocks.

The wheat seeds were planted in each pot and after 30 days the dry matter of each pot was measured. The model to describe the dry matter is

$$y_{ijk} = \mu_{ik} + t_{ij} + p_{ijk}$$

where μ_{ik} is the mean of level i of moisture with level k of fertilizer, t_{ij} is the tray error term distributed $N(0, \sigma_t^2)$, and p_{ijk} is the pot error term distributed $N(0, \sigma_p^2)$. The analysis of variance table is in Table 24.3, which shows a significant interaction between Moisture and Fertilizer.

Since there is an interaction between Moisture and Fertilizer, the treatment combination means, which are in Table 24.4, are used when making inferences. Since the levels of moisture and the levels of fertilizer are equally spaced quantitative levels, orthogonal polynomials are used to investigate the trends over the levels of fertilizer for each level of moisture and the trends over the levels of moisture for each level of fertilizer. The contrasts that measure the linear and quadratic trends of fertilizer at each moisture level are

$$\theta_{\text{LF}|\text{M}i} = -3\mu_{i1} - \mu_{i2} + \mu_{i3} + 3\mu_{i4}$$

Table 24.3 Analysis of Variance Table for Moisture—Fertility Example

Source of Variation	*df*	*SS*	*MS*	*F*	*EMS*
Moisture	3	269.19	89.73	26.32	$\sigma_p^2 + 4\sigma_t^2 + Q_3$
Error(Tray)	8	27.27	3.09		$\sigma_p^2 + 4\sigma_t^2$
Fertilizer	3	297.08	99.03	132.03	$\sigma_p^2 + Q_2$
Moisture * Fertilizer	9	38.05	4.22	5.64	$\sigma_p^2 + Q_1$
Error(Pot)	24	.75			σ_p^2

Note: Q_i is the noncentrality parameter.

Table 24.4 Moisture-by-Fertilizer Dry Matter Means

	FERTILIZER			
Moisture	*2*	*4*	*6*	*8*
10	3.12	4.10	5.18	6.12
20	5.97	7.83	10.70	14.24
30	6.72	9.44	12.28	15.43
40	5.79	6.33	10.18	11.77

and

$$\theta_{\text{QF}|\text{M}i} = \mu_{i1} - \mu_{i2} - \mu_{i3} + \mu_{i4}.$$

The estimates and corresponding variances are

$$\hat{\theta}_{\text{LF}|\text{M}i} = -3\bar{y}_{i\cdot1} - \bar{y}_{i\cdot2} + \bar{y}_{i\cdot3} + 3\bar{y}_{i\cdot4},$$

$$\hat{\theta}_{\text{QF}|\text{M}i} = \bar{y}_{i\cdot1} - \bar{y}_{i\cdot2} - \bar{y}_{i\cdot3} + \bar{y}_{i\cdot4},$$

$$\text{Var}(\hat{\theta}_{\text{LF}|\text{M}i}) = \frac{\sigma_p^2(3^2 + 1^2 + 1^2 + 3^2)}{3}$$

$$= \frac{20\sigma_p^2}{3},$$

and

$$\text{Var}(\hat{\theta}_{\text{QF}|\text{M}i}) = \frac{\sigma_p^2(1^2 + 1^2 + 1^2 + 1^2)}{3}$$

$$= \frac{4\sigma_p^2}{3}.$$

The estimated standard errors are obtained by replacing σ_p^2 with MSERROR(POT) in the square root of the respective variances. The estimates of the linear and quadratic trends of fertilizer for each level of moisture and the corresponding estimated standard errors and t-statistics (testing that the trends are zero) are in Table 24.5. The contrasts that measure the linear and quadratic trends of moisture at each fertilizer level are

$$\theta_{\text{LM}|\text{F}k} = -3\mu_{1k} - \mu_{2k} + \mu_{3k} + 3\mu_{4k}$$

and

$$\theta_{\text{QM}|\text{F}k} = \mu_{1k} - \mu_{2k} - \mu_{3k} + \mu_{4k}.$$

Table 24.5 Estimates of Linear and Quadratic Trends of Fertilizer for Each Level of Moisture

	LINEAR		QUADRATIC	
Moisture	*Estimate*	*t-value*	*Estimate*	*t-value*
10	10.08	4.51	−.04	−.04
20	27.68	12.38	1.68	1.68
30	27.97	12.96	.43	.43
40	21.79	9.74	1.05	1.05

Notes: The standard errors for the linear and quadratic trends are 2.236 and 1.000, respectively. Compare t-values to $t_{\alpha/2,24}$.

The estimates and corresponding variances are

$$\hat{\theta}_{\text{LM}|\text{F}k} = -3\bar{y}_{1\cdot k} - \bar{y}_{2\cdot k} + \bar{y}_{3\cdot k} + 3\bar{y}_{4\cdot k},$$

$$\hat{\theta}_{\text{QM}|\text{F}k} = \bar{y}_{1\cdot k} - \bar{y}_{2\cdot k} - \bar{y}_{3\cdot k} + \bar{y}_{4\cdot k},$$

$$\text{Var}(\hat{\theta}_{\text{LM}|\text{F}k}) = \frac{(\sigma_p^2 + \sigma_t^2)(3^2 + 1^2 + 1^2 + 3^2)}{3} = \frac{20(\sigma_p^2 + \sigma_t^2)}{3}$$

and

$$\text{Var}(\theta_{\text{QM}|\text{F}k}) = \frac{(\sigma_p^2 + \sigma_t^2)(1^2 + 1^2 + 1^2 + 1^2)}{3} = \frac{4(\sigma_p^2 + \sigma_t^2)}{3}.$$

The estimated standard errors are obtained by replacing $\sigma_p^2 + \sigma_t^2$ with

$$\frac{\text{MSERROR(TRAY)} + (4-1)\text{MSERROR(POT)}}{4}$$

in the root square of the respective variances. Since MSERROR(TRAY) and MSERROR(POT) are based on different degrees of freedom, the Satterthwaite approximation is used to obtain the approximate degrees of freedom to be associated with $\hat{\sigma}_p^2 + \hat{\sigma}_e^2$. For this example,

$$\hat{\sigma}_p^2 + \hat{\sigma}_e^2 = \frac{3.09 + (4-1)(.75)}{4} = 1.355$$

Table 24.6 Estimates of Linear and Quadratic Trends of Moisture for Each Level of Fertilizer

	LINEAR		QUADRATIC	
Fertilizer	*Estimate*	*t-value*	*Estimate*	*t-value*
2	8.76	2.94	−3.78	−2.83
4	8.30	2.78	−6.84	−5.13
6	16.58	5.56	−7.62	−5.71
8	18.14	6.08	−11.78	−8.83

Notes: The standard errors for the linear and quadratic trends are 2.983 and 1.334, respectively. Compare t-values to $t_{\alpha/2,20}$.

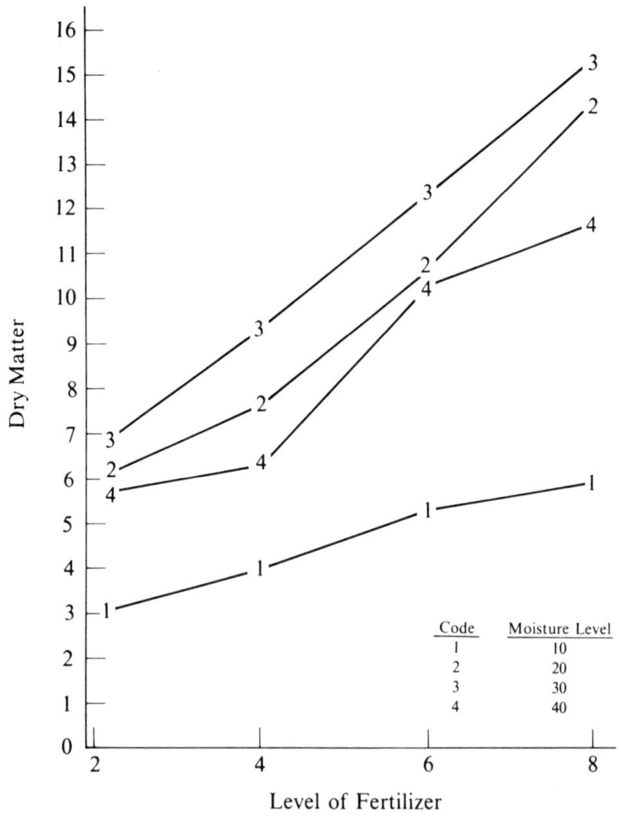

Figure 24.4 Graph of response to the levels of fertilizer for each level of moisture.

which is based on r degrees of freedom where

$$r = \frac{[3.09 + (4 - 1)(.75)]^2}{(3.09)^2/8 + [(4 - 1)(.75)]^2/24} = 20.3,$$

or use $r = 20$ degrees of freedom.

The estimates of linear and quadratic trends of moisture for each level of fertilizer and the corresponding estimated standard errors and t-values are in Table 24.6. Table 24.5 shows that there is a linear response to Fertilizer for each level of moisture and no significant quadratic trend; Table 24.6 shows that there are both linear and quadratic responses to Moisture for each level of fertilizer. The graphs in Figures 24.4 and 24.5 show the response to Fertilizer for each moisture level and the response to Moisture for each fertilizer level.

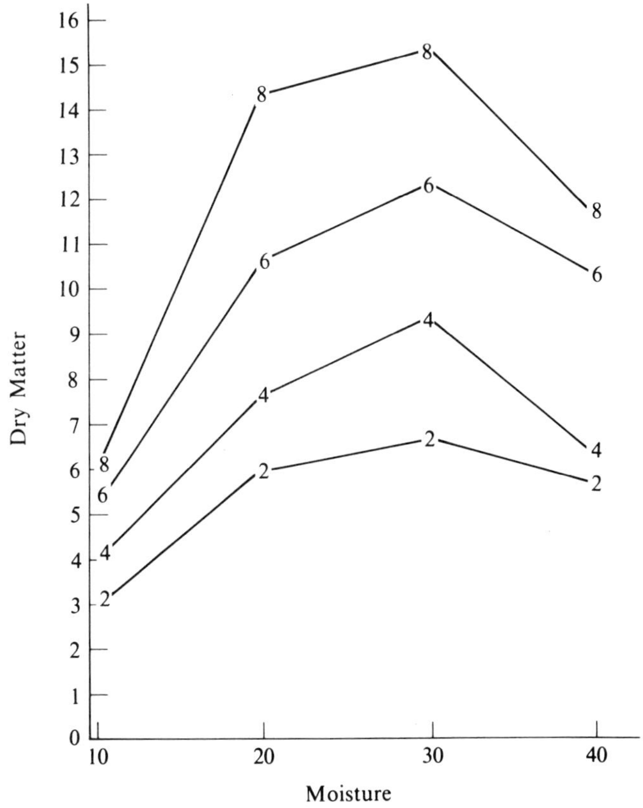

Figure 24.5 Graph of response to the levels of moisture for each level of fertilizer. Numbers on lines denote levels of fertilizer.

CONCLUDING REMARKS

In this chapter, split-plot designs were discussed in detail. The analysis of variance table for the designs discussed involves more than one error term, and standard errors of various estimates can involve more than one error term. Methods were presented for computing standard errors for the various types of comparisons that may be interesting to an experimenter. They can also be used for making multiple comparisons. Also described were methods for estimating linear, quadratic, and similar trends when the levels of a factor are quantitative. Two examples were discussed in detail.

25

Strip-Plot Experimental Designs

CHAPTER OUTLINE

Strip-plot experimental designs are used in situations where the experimental units are arranged in rectangles, such as in a field or a stack of cages. Then the levels of one factor are applied to the rows, and the levels of the other factor are applied to the columns. This design is useful when the method of application does not allow the experimental units to be treated individually. An example involving irrigation and applying levels of nitrogen shall be analyzed to demonstrate the strip-plot design.

25.1 DESCRIPTION OF A STRIP-PLOT DESIGN AND MODEL

A strip plot is like a split-plot experimental design but with differently constructed experimental units. The strip plot involves a two-way treatment structure where the basic experimental units are arranged in sets of rectangles. Each rectangle has s rows where s is the number of levels of the first factor and t columns where t is the number of levels of the second factor. As shown in Figure 25.1, the levels of factor A are randomly assigned to all of the experimental units within a row, while the levels of factor B are randomly assigned to all the experimental units within a column.

This assignment process generates a design with three sizes of experimental units. The rows are the experimental units for factor A, the columns are the experimental units for factor B; the experimental design for each factor consists of a one-way treatment structure in a randomized complete block design structure. Comparisons of the levels of factor A are between-row comparisons, those of factor B are between-column comparisons, and measures of interaction are within-row comparisons by within-column comparisons. Thus the size of the experimental units on which an interaction is measured are the units within a row and within a column.

A model to describe the strip trial is

$$y_{ijk} = \mu + s_k + \alpha_i + r_{ik} + \beta_j + c_{jk} + \gamma_{ij} + \varepsilon_{ijk} \qquad i = 1, 2, \ldots, s,$$

$$j = 1, 2, \ldots, t, \quad k = 1, 2, \ldots, r, \tag{25.1.1}$$

where α_i denotes the effect of the factor assigned to the rows; β_j denotes the effect of the factor assigned to the columns; γ_{ij} denotes the interaction between the two factors; r_{ik} is the error term associated with the rows, with $r_{ik} \sim$ i.i.d. $N(0, \sigma_r^2)$; c_{jk} is the error term associated with the columns, with $c_{jk} \sim$ i.i.d. $N(0, \sigma_c^2)$; s_k denotes the rectangle or the replication effect, with $s_k \sim$ i.i.d. $N(0, \sigma_s^2)$; and ε_{ijk} is the error term associated with a cell experimental unit, which is the intersection of a row and a column, with $\varepsilon_{ijk} \sim$ i.i.d. $N(0, \sigma_\varepsilon^2)$.

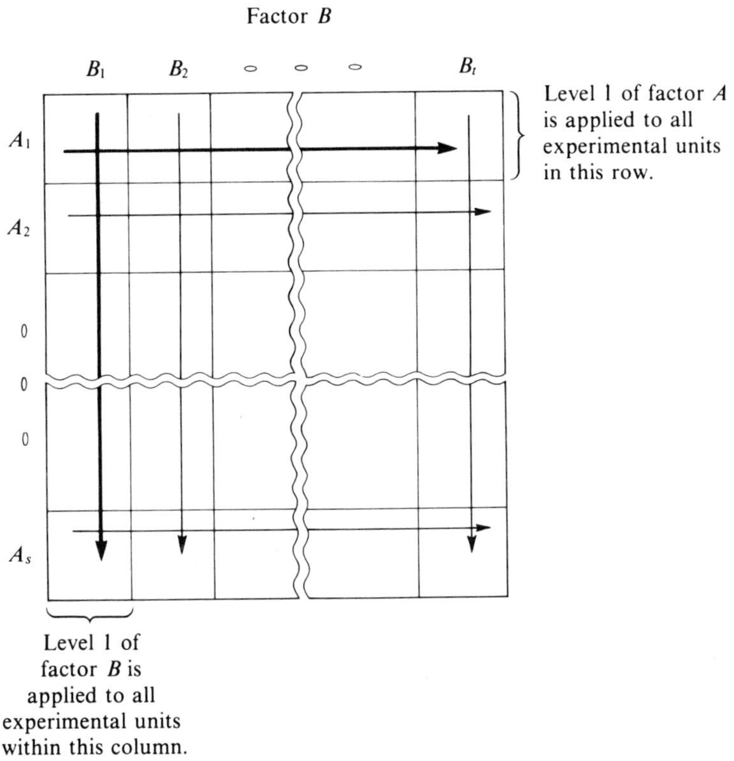

Factor B
B_1
B_2
B_t
A_1
A_2
A_s
Level 1 of factor A is applied to all experimental units in this row.
Level 1 of factor B is applied to all experimental units within this column.

Figure 25.1 Graphic showing how the levels of two factors are applied to strips (rows and columns) of experimental units arranged in a rectangle.

An analysis of variance table for the above model is shown in Table 25.1. Model (25.1.1) can also be expressed as

$$y_{ijk} = \mu_{ij} + s_k + r_{ik} + c_{jk} + \varepsilon_{ijk}, \qquad \text{where} \quad \mu_{ij} = \mu + \alpha_i + \beta_j + \gamma_{ij}.$$

Estimates of the population parameters are

$$\hat{\mu}_{ij} = \bar{y}_{ij\cdot},$$

$$\hat{\sigma}_\varepsilon^2 = \text{MSERROR(UNIT)},$$

$$\hat{\sigma}_c^2 = \frac{\text{MSERROR(COLUMN)} - \text{MSERROR(UNIT)}}{s},$$

$$\hat{\sigma}_r^2 = \frac{\text{MSERROR(ROW)} - \text{MSERROR(UNIT)}}{t}, \quad \text{and}$$

$$\hat{\sigma}_s^2 = \frac{\text{MSSETS} - \text{MSERROR(COLUMN)} - \text{MSERROR(ROW)} + \text{MSERROR(UNIT)}}{st}.$$

Table 25.1 Analysis of Variance Table Corresponding to Model (25.1.1)

Source of Variation	*df*	*EMS*
Set	$(r-1)$	$\sigma_\varepsilon^2 + t\sigma_r^2 + s\sigma_c^2 + st\sigma_s^2$
A	$(s-1)$	$\sigma_\varepsilon^2 + t\sigma_r^2 + Q_1^2(\alpha)$
Error(Row)	$(r-1)(s-1)$	$\sigma_\varepsilon^2 + t\sigma_r^2$
B	$(t-1)$	$\sigma_\varepsilon^2 + s\sigma_c^2 + Q_2^2(\beta)$
Error(Column)	$(r-1)(t-1)$	$\sigma_\varepsilon^2 + s\sigma_c^2$
$A*B$	$(s-1)(t-1)$	$\sigma_\varepsilon^2 + Q_3^2(\gamma)$
Error(Unit)	$(r-1)(s-1)(t-1)$	σ_ε^2

Various contrasts and sums (of means) involving μ_{ij} are estimated by taking the corresponding contrasts or sums of $\hat{\mu}_{ij}$.

Care must be taken when making inferences about the parameters for this problem. The necessary tests of hypotheses and standard errors for making comparisons among the means are discussed in the next section.

25.2 TECHNIQUES FOR MAKING INFERENCES

The column of expected mean squares in Table 25.1 provides the information necessary to make inferences about the factorial effects. To test the no-interaction hypothesis, divide the mean square $A*B$ by the MSERROR(UNIT). To test

$$\mu_{\cdot 1} = \mu_{\cdot 2} = \cdots = \mu_{\cdot t},$$

divide the mean square factor A by MSERROR(ROW). To test

$$\mu_{1\cdot} = \mu_{2\cdot} = \cdots = \mu_{s\cdot},$$

Table 25.2 Estimators and Their Variances for Strip-Plot Design for $i \neq m$ and $j \neq n$

Comparison	*Estimator*	*Variance*
$\bar{\mu}_{i\cdot} - \bar{\mu}_{m\cdot}$	$\bar{y}_{i\cdot\cdot} - \bar{y}_{m\cdot\cdot}$	$2(\sigma_\varepsilon^2 + t\sigma_r^2)/rt$
$\bar{\mu}_{\cdot j} - \bar{\mu}_{\cdot n}$	$\bar{y}_{\cdot j\cdot} - \bar{y}_{\cdot n\cdot}$	$2(\sigma_\varepsilon^2 + s\sigma_c^2)/rs$
$\mu_{ij} - \mu_{in} - \mu_{mj} + \mu_{mn}$	$\bar{y}_{ij\cdot} - \bar{y}_{in\cdot} - \bar{y}_{mj\cdot} + \bar{y}_{mn\cdot}$	$4\sigma_\varepsilon^2/r$
$\mu_{ij} - \mu_{in}$	$\bar{y}_{ij\cdot} - \bar{y}_{in\cdot}$	$2(\sigma_\varepsilon^2 + \sigma_c^2)/r$
$\mu_{ij} - \mu_{mj}$	$\bar{y}_{ij\cdot} - \bar{y}_{mj\cdot}$	$2(\sigma_\varepsilon^2 + \sigma_r^2)/r$
$\mu_{ij} - \mu_{mn}$	$\bar{y}_{ij\cdot} - \bar{y}_{mn\cdot}$	$2(\sigma_\varepsilon^2 + \sigma_r^2 + \sigma_c^2)/r$

Table 25.3 Estimates of Standard Errors for Comparisons in Table 25.2

Comparison	*Estimate of Standard Error*	*t- or t*-Value*
$\bar{\mu}_{i\cdot} - \bar{\mu}_{m\cdot}$	$\sqrt{\frac{2\text{MSERROR(ROW)}}{rt}}$	$t[(r-1)(s-1)] = t_R$
$\bar{\mu}_{\cdot j} - \bar{\mu}_{\cdot n}$	$\sqrt{\frac{2\text{MSERROR(COLUMN)}}{rs}}$	$t[(r-1)(t-1)] = t_C$
$\mu_{ij} - \mu_{in} - \mu_{mj} + \mu_{mn}$	$\sqrt{\frac{4\text{MSERROR(UNIT)}}{r}}$	$t[(r-1)(s-1)(t-1)] = t_I$
$\mu_{ij} - \mu_{in}$	$\sqrt{\frac{2[\text{MSERROR(COLUMN)} + (s-1)\text{MSERROR(UNIT)}]}{rs}}$	$\frac{t_C\text{MSERROR(COLUMN)} + (s-1)t_I\text{MSERROR(UNIT)}}{\text{MSERROR(COLUMN)} + (s-1)\text{MSERROR(UNIT)}}$
$\mu_{ij} - \mu_{mj}$	$\sqrt{\frac{2[\text{MSERROR(ROW)} + (t-1)\text{MSERROR(UNIT)}]}{rt}}$	$\frac{t_R\text{MSERROR(ROW)} + (t-1)t_I\text{MSERROR(UNIT)}}{\text{MSERROR(COLUMN)} + (t-1)\text{MSERROR(UNIT)}}$
$\mu_{ij} - \mu_{mn}$	$\sqrt{\frac{2[t\text{MSERROR(COLUMN)} + s\text{MSERROR(ROW)} + (ts-s-t)\text{MSERROR(UNIT)}]}{rts}}$	$\frac{\left[t_C xt \times \text{MSERROR(COLUMN)} + t_R \times s \times \text{MSERROR(ROW)} + t_I \times (ts-t-s)\text{MSERROR(UNIT)}\right]}{\left[t \times \text{MSERROR(COLUMN)} + s \times \text{MSERROR(ROW)} + (ts-t-s)\text{MSERROR(UNIT)}\right]}$

Nitrogen Level		Irrigation Method I_1	Irrigation Method I_2
	N_1	55	71
	N_3	69	78
	N_2	62	77

Nitrogen Level		Irrigation Method I_1	Irrigation Method I_2
	N_2	70	78
	N_3	79	80
	N_1	63	77

Nitrogen Level		Irrigation Method I_2	Irrigation Method I_1
	N_3	81	77
	N_1	77	63
	N_2	79	66

Nitrogen Level		Irrigation Method I_1	Irrigation Method I_2
	N_3	76	79
	N_2	66	76
	N_1	65	75

Figure 25.2 Field layout, treatment assignment, and data for the example in Section 25.3.

divide the mean square B by the MSERROR(COLUMN). When significant effects are obtained, various comparisons among the means must be made. Table 25.2 contains various comparisons of the μ_{ij}'s, the corresponding estimates, and the variances of the estimates.

The first three comparisons in Table 25.2 have variances that are estimated by entries in the analysis of variance table. For the other three comparisons, approximate confidence intervals are obtained as the variances are estimated by linear combinations of mean squares from the analysis of variance table. Thus, as in the split plot, approximate percentage points, denoted by t^*'s, are used to construct confidence intervals or multiple comparisons. Table 25.3 contains the estimates of the standard errors for the comparisons in Table 25.2 along with the corre-

Table 25.4 Analysis of Variance Table for Strip-Plot Data in Figure 25.2

Source of Variation	*df*	*SS*	*MS*	*F*
Replication or Set	3	123.5	41.4	
N	2	339.1	169.6	60.6
Error(Row)	6	16.9	2.8	
I	1	570.4	570.4	52.2
Error(Column)	3	32.8	10.9	
$I * N$	2	94.8	47.4	33.9
Error(Unit)	6	8.6	1.4	

Table 25.5 Estimates of Standard Errors and t- or t^*-Values for Yield Data in Figure 25.1

Comparison	*Estimate of Standard Error*	*t- or t^*-Value**
$\bar{\mu}_{i\cdot} - \bar{\mu}_{m\cdot}$	.8367	2.447
$\bar{\mu}_{\cdot j} - \bar{\mu}_{\cdot n}$	1.3478	3.182
$\mu_{ij} - \mu_{in} - \mu_{mj} + \mu_{mn}$	1.1832	2.447
$\mu_{ij} - \mu_{in}$	1.5111	3.098
$\mu_{ij} - \mu_{mj}$	1.0247	2.447
$\mu_{ij} - \mu_{mn}$	1.6228	2.954

*$\alpha = .05$.

sponding t- and t^*-values. As can be seen from Table 25.3, comparisons in the interaction part of the model are quite complex.

25.3 EXAMPLE OF A STRIP PLOT

An experiment was conducted to study the effects and relationships between two irrigation methods and three levels of nitrogen on the yield of wheat. The field layout for four replications, the randomization plan for treatment assignment, and the yields are shown in Figure 25.2. The analysis of variance table for these data are in Table 25.4.

The method-of-moments estimates of the variance components are $\hat{\sigma}_c^2 = 3.17$, $\hat{\sigma}_r^2 = .69$, and $\hat{\sigma}_\varepsilon^2 = 1.43$, and the estimates of the means are $\hat{\mu}_{11} = 61.50$, $\hat{\mu}_{12} = 75.00$, $\hat{\mu}_{21} = 66.00$, $\hat{\mu}_{22} = 77.50$, $\hat{\mu}_{31} = 75.25$, and $\hat{\mu}_{32} = 79.50$. The estimated standard errors and t- or t^*-values for the six types of comparisons are in Table 25.5.

The standard errors and t- or t^*-values can be used to compare various means in the treatment structure. An LSD procedure can easily be used with $\alpha = .05$, but a Bonferroni procedure could just as easily be used by selecting the proper α level (which depends on the number of comparisons).

CONCLUDING REMARKS

In this experiment, the strip-plot experimental design was used to show how several sizes of experimental units can occur by the method of assigning the levels of the factors to the experimental units. An example was analyzed that demonstrates the complexity in computing the standard errors for making various comparisons between the means.

26

Analysis of Repeated Measures Designs for Which the Usual Assumptions Hold

CHAPTER OUTLINE

Like experiments with split-plot designs, experiments utilizing repeated measures designs have structures that involve more than one size of experimental unit. For example, a subject may be measured over time where time is one of the factors in the treatment structure of the experiment. By measuring the subject at several different times, the subject is essentially being "split" into parts (time intervals), and the response is measured on each part. The larger experimental unit is the subject or the collection of time intervals. The smaller unit is the interval of time during which the subject is exposed to a treatment or an interval just between time measurements. The model and the analysis have two parts, one for the subjects and one for the time intervals.

Repeated measures designs differ from split-plot designs in that the levels of one or more factors cannot be randomly assigned to one or more of the sizes of experimental units in the experiment. In this case, the levels of time cannot be assigned at random to the time intervals, and thus the usual analysis of variance may not be valid. Because of this nonrandom assignment, the errors corresponding to the respective experimental units may have a covariance matrix that does not conform to those for which the usual split-plot analysis is valid.

In Section 26.1, the repeated measures models are described, and the assumptions necessary for the usual split-plot analysis of variance to be valid are given. Section 26.2 uses three examples to demonstrate the usual analysis of variance computations, including the computations of standard errors for making various comparisons between means. In the next chapter, methods are described for an analysis when the assumptions are not satisfied and ways are presented to determine when they are. A test is included that checks for an autocorrelated error structure for the repeated measures error terms.

26.1 MODEL SPECIFICATIONS AND ASSUMPTIONS

Repeated measures designs can be applied in numerous situations. Example 26.1, analyzed in the next section, investigates the effects of three drugs, where each drug was administered to eight people. Each person's heart rate was then measured every five minutes for four time intervals.

In the general model of this experiment, n_i subjects were randomly assigned to drug i, and each subject was measured t times. The larger size of experimental unit is the subject, and the smaller is the time interval. A model to describe the heart rate that reflects the two sizes of experimental units is

$$y_{ijk} = \mu + \delta_i + p_{ij} + \tau_k + (\delta\tau)_{ik} + \varepsilon_{ijk} \qquad (26.1.1)$$

or

$$y_{ijk} = \mu_{ik} + p_{ij} + \varepsilon_{ijk}$$

where

$$\mu + \delta_i + p_{ij}$$

is the subject part of the model and

$$\tau_k + (\delta\tau)_{ik} + \varepsilon_{ijk}$$

is the time interval part of the model. The mean heart rate of drug i at time k is

$$\mu_{ik} = \mu + \delta_i + \tau_k + (\delta\tau)_{ik}.$$

The p_{ij}'s represent the subject errors, and the ε_{ijk}'s represent the time interval errors. The assumptions often made for an analysis of model (26.1.1) are that $p_{ij} \sim$ i.i.d. $N(0, \sigma_p^2)$ and $\varepsilon_{ijk} \sim$ i.i.d. $N(0, \sigma_\varepsilon^2)$, which are the same assumptions made for the split-plot analysis.

Such assumptions may not always be appropriate for a repeated measures design. However, the split-plot analysis is also the correct analysis under more general assumptions. The more general assumptions require certain forms of the covariance matrix of the measurement errors of the time intervals and of the covariance matrix of the error terms of the subjects assigned to a given drug. Let

$$\boldsymbol{\varepsilon}_{ij} = (\varepsilon_{ij1}, \varepsilon_{ij2}, \ldots, \varepsilon_{ijt})'$$

be the vector of time errors for subject j assigned to drug i and

$$\mathbf{p}_i = (p_{i1}, p_{i2}, \ldots, p_{in_i})'$$

be the vector of subject errors for drug i. The covariance matrices of $\boldsymbol{\varepsilon}_{ij}$ and $\mathbf{p}_i$ are denoted by $\text{Cov}(\boldsymbol{\varepsilon}_{ij}) = \boldsymbol{\Sigma}$ and $\text{Cov}(\mathbf{p}_i) = \boldsymbol{\Lambda}$.

A sufficient condition for the F-tests of the usual split-plot analysis of variance to be valid is that the covariance matrices have a form called *compound symmetry*. A covariance matrix $\boldsymbol{\Sigma}$ is of compound symmetry form if it can be expressed as

$$\boldsymbol{\Sigma} = \sigma^2 \begin{bmatrix} 1 & \rho & \rho & \cdots & \rho \\ \rho & 1 & \rho & \cdots & \rho \\ \rho & \rho & 1 & \cdots & \rho \\ \vdots & \vdots & \vdots & & \vdots \\ \rho & \rho & \rho & \cdots & 1 \end{bmatrix}. \tag{26.1.2}$$

The compound symmetry condition implies that the random variables are equally correlated and have equal variances.

A more general condition on the form of $\boldsymbol{\Sigma}$ and $\boldsymbol{\Lambda}$ has been described by Huynh and Feldt (1970). The Huynh–Feldt (H–F) condition specifies that the elements of a covariance matrix $\boldsymbol{\Sigma}$ be expressed as

$$\sigma_{ii'} = \gamma_i + \gamma_{i'} + \lambda\delta_{ii'} \tag{26.1.3}$$

where

$$\delta_{ii'} = \begin{cases} 1 & \text{if } i = i' \\ 0 & \text{if } i \neq i', \end{cases}$$

or in terms of matrices,

$$\boldsymbol{\Sigma} = \lambda \mathbf{I}_t + \boldsymbol{\gamma}\mathbf{j}_t' + \mathbf{j}_t\boldsymbol{\gamma}'$$

where $\mathbf{I}_t$ is a $t \times t$ identity matrix, $\mathbf{j}_t$ is a $t \times 1$ vector of ones, and $\boldsymbol{\gamma}' = (\gamma_1, \gamma_2, \ldots, \gamma_t)$ are unknown parameters.

The H–F condition is both necessary and sufficient for the F-tests of the usual analysis of variance to be valid. The H–F condition is equivalent to specifying that the variances of the differences between pairs of errors, such as $\varepsilon_{ijk} - \varepsilon_{ijk'}$, are equal for all k and k', $k \neq k'$. The H–F condition allows for unequal variances and unequal correlations, but if the variances are all equal, then the condition is equivalent to compound symmetry. Compound symmetry and independent errors are special cases of the H–F condition.

When a model has more than one error term, the covariance matrices of all of the random vectors must satisfy the H–F condition before the F-tests of the usual split-plot analysis can be valid. In the heart rate example, the covariance matrices $\boldsymbol{\Sigma}$ and $\boldsymbol{\Lambda}$ must satisfy the H–F condition. A test of the hypothesis that a covariance matrix satisfies the H–F condition is given in Chapter 27.

In Example 26.3 of the next section, the attitudes among family members were studied in relation to whether the home was in a rural or an urban environment. The type of family required for the study consisted of three members, a son, a father, and a mother. Then n_1 such families were randomly selected from a rural environment and n_2 such families from an urban environment. The attitude of each family member toward a moral issue was determined. Also of interest was whether the attitudes changed over time; hence, the measurements were made on each person every six months for a total of t times.

There are three sizes of experimental units in this design. The experimental unit corresponding to the treatment effect environment is the family; the experimental unit corresponding to the type of family member effect is the person, and the experimental unit for time is the six-month time interval. A model to describe the attitudes has three error terms, one for each size of experimental unit in the experiment. One such model is

$$\begin{aligned} y_{ijkm} &= \mu + \alpha_i + f_{ij} && \text{Family experimental unit} \\ &\quad + \beta_k + (\alpha\beta)_{ik} + p_{ijk} && \text{Person experimental unit} \qquad (26.1.4) \\ &\quad + \tau_m + (\alpha\tau)_{im} + (\beta\tau)_{km} && \left\{\begin{array}{l}\text{Time interval} \\ \text{experimental unit}\end{array}\right. \\ &\quad + (\alpha\beta\tau)_{ikm} + \varepsilon_{ijkm} && \end{aligned}$$

or

$$y_{ijkm} = \mu_{ikm} + f_{ij} + p_{ijk} + \varepsilon_{ijkm}.$$

The error terms are f_{ij}, p_{ijk}, and ε_{ijkm}, which denote the family within environment error, the person within family error, and the time interval within person error, respectively. The covariance matrices for the error terms are $\text{Var}(\boldsymbol{\varepsilon}_{ijk}) = \Sigma$, $\text{Var}(\mathbf{p}_{ij}) = \Lambda$, and $\text{Var}(\mathbf{f}_i) = \Gamma$. For the F-tests of the usual analysis of variance to be valid, the matrices Σ, Λ, and Γ must all satisfy the H–F condition. The variances of the family members may not be equal in that a son's variance may be quite different from his parents', but if Λ satisfies the H–F condition, then the analysis is appropriate.

A method to test for the H–F condition is described in Chapter 27, and alternative analyses are described for when the conditions are not satisfied. For the experiment here, the multivariate procedure described in Chapter 30 can always be used when the number of families is greater than the number of time intervals.

26.2 THE USUAL ANALYSIS OF VARIANCE

The usual analysis of variance refers to the analyses discussed in Chapters 24 and 25 for split-plot and strip-plot designs, which are provided by most computer packages. If the covariance matrices of the various error vectors satisfy the H–F condition, then these analyses provide valid F-tests. However, when the errors are not independent, some problems occur when making multiple comparisons. These problems are described in detail in Example 26.1 below.

Three examples are used in this section to demonstrate the analysis of the repeated measures design and to show how to determine estimates, their estimated standard errors (which are necessary to make various multiple comparisons), and provide methods to study contrasts between the means. The examples used here are also used in Chapter 27 to demonstrate the techniques needed when the H–F condition is not satisfied.

EXAMPLE 26.1: Effect of Drugs on Heart Rate

An experiment involving d drugs was conducted to study each drug's effect on the heart rate of humans. After the drug was administered, the heart rate was measured every five minutes for a total of t times. At the start of the study, n female human subjects were randomly assigned to each drug. Model (26.1.1) was used to describe the data, a model with two error terms, p_{ij} and ε_{ijk}.

Let $\boldsymbol{\varepsilon}_{ij}$ denote the vector of errors for the jth subject assigned to the ith drug. The assumption is that the $\boldsymbol{\varepsilon}_{ij}$'s $\sim N(0, \Sigma)$ where Σ satisfies the H–F condition of equation (26.1.3). To simplify evaluating the expectations, express Σ as

$$\Sigma = \sigma_\varepsilon^2(\mathbf{I} + \boldsymbol{\lambda}\mathbf{j}' + \mathbf{j}'\boldsymbol{\lambda})$$

Table 26.1 Analysis of Variance Table for Example 26.1

Source of Variation	*df*	*SS*	*EMS*
Drug	$d-1$	$nt\sum_{i=1}^{d}(\bar{y}_{i\cdot\cdot}-\bar{y}_{\cdot\cdot\cdot})^2$	$\sigma_\varepsilon^2(1+2\lambda_\cdot)+t\sigma_p^2+Q_1$
Error(Subject)	$d(n-1)$	$t\sum_{i=1}^{d}\sum_{j=1}^{n}(\bar{y}_{ij\cdot}-\bar{y}_{i\cdot\cdot})^2$	$\sigma_\varepsilon^2(1+2\lambda_\cdot)+t\sigma_p^2$
Time	$t-1$	$dn\sum_{k=1}^{t}(\bar{y}_{\cdot\cdot k}-\bar{y}_{\cdot\cdot\cdot})^2$	$\sigma_\varepsilon^2+Q_2$
Drug * Time	$(d-1)(t-1)$	$n\sum_{i=1}^{d}\sum_{k=1}^{t}(\bar{y}_{i\cdot k}-\bar{y}_{i\cdot\cdot}-\bar{y}_{\cdot\cdot k}+\bar{y}_{\cdot\cdot\cdot})^2$	$\sigma_\varepsilon^2+Q_3$
Error(Time)	$d(n-1)(t-1)$	$\sum_{i=1}^{d}\sum_{j=1}^{n}\sum_{k=1}^{t}(y_{ijk}-\bar{y}_{ij\cdot}-\bar{y}_{i\cdot k}+\bar{y}_{i\cdot\cdot})^2$	σ_ε^2

which has parameters σ_ε^2 and $\boldsymbol{\lambda}$. The sums of squares corresponding to these assumptions are in Table 26.1. In the table, Q_1, Q_2, and Q_3 are noncentrality parameters measuring the Drug effect, Time effect, and the Drug * Time interaction, respectively, and $\lambda_\cdot$ is the sum of the elements of the vector $\boldsymbol{\lambda}$. The term $\lambda_\cdot$ occurs because the variance of $\bar{\varepsilon}_{ij\cdot}$, the mean of the errors of person j assigned to drug i, is

$$\operatorname{Var}(\bar{\varepsilon}_{ij\cdot})=\frac{1}{t}\sigma_\varepsilon^2(1+2\lambda_\cdot).$$

When we find effects in the data, we want to compare the various means. If there is no Drug * Time interaction, then we make comparisons between the Drug means and the Time means.

Since the repeated measures designs involve more than one size of experimental unit and more than one error term, the variance of each comparison could involve different functions of the variances of the respective error terms and thus need to be determined. One method to compute the variance is to express the comparison in terms of the model and evaluate the variance from the corresponding error terms (the same technique was used in Chapters 24 and 25). That technique is used to make comparisons in this example (and the others in this chapter) to demonstrate how it can be used for other experiments.

The model for comparing two Time means is

$$\bar{y}_{\cdot\cdot k}-\bar{y}_{\cdot\cdot k'}=\bar{\mu}_{\cdot k}-\bar{\mu}_{\cdot k'}+\bar{\varepsilon}_{\cdot\cdot k}-\bar{\varepsilon}_{\cdot\cdot k'}.$$

The variance of $\bar{y}_{\cdot\cdot k} - \bar{y}_{\cdot\cdot k'}$ is

$$\text{Var}(\bar{y}_{\cdot\cdot k} - \bar{y}_{\cdot\cdot k'}) = \text{Var}(\bar{\varepsilon}_{\cdot\cdot k} - \bar{\varepsilon}_{\cdot\cdot k'}) = \frac{2\sigma_\varepsilon^2}{dn}.$$

The estimate of the standard error of the difference between two Time means is

$$\widehat{\text{s.e.}}(\bar{y}_{\cdot\cdot k} - \bar{y}_{\cdot\cdot k'}) = \sqrt{\frac{2\text{MSERROR(TIME)}}{dn}},$$

and an LSD multiple comparison is constructed by

$$\text{LSD} = \{t_{\alpha/2,d(n-1)(t-1)}\}\widehat{\text{s.e.}}(\bar{y}_{\cdot\cdot k} - \bar{y}_{\cdot\cdot k'}).$$

To investigate a contrast between Time means as

$$\theta_{\text{T}} = c_1\bar{\mu}_{\cdot 1} + c_2\bar{\mu}_{\cdot 2} + \cdots + c_t\bar{\mu}_{\cdot t}$$

where $c_1 + c_2 + \cdots + c_t = 0$, the estimate of θ_{T} is

$$\hat{\theta}_{\text{T}} = c_1\bar{y}_{\cdot\cdot 1} + c_2\bar{y}_{\cdot\cdot 2} + \cdots + \bar{y}_{\cdot\cdot t}$$

with estimated standard error

$$\widehat{\text{s.e.}}(\hat{\theta}_{\text{T}}) = \sqrt{\frac{\text{MSERROR(TIME)}}{dn}(c_1^2 + c_2^2 + \cdots + c_t^2)}.$$

Such a contrast could be used to check for linear, quadratic, and similar trends over time.

Three inference procedures can be used to study the value of θ_{T}. A t-statistic to test H_0: $\theta_{\text{T}} = 0$ versus H_a: $\theta_{\text{T}} \neq 0$ is

$$t_c = \frac{\hat{\theta}_{\text{T}}}{\widehat{\text{s.e.}}(\theta_{\text{T}})}$$

which is compared to $t_{\alpha/2,d(n-1)(t-1)}$. A corresponding F-test is

$$F_c = \frac{\hat{\theta}_{\text{T}}^2}{[\widehat{\text{s.e.}}(\theta_{\text{T}})]^2} = t_c^2$$

which is compared to $F_{\alpha,1,d(n-1)(t-1)}$. A $(1 - \alpha)100\%$ confidence interval about θ_{T} is

$$\hat{\theta}_{\text{T}} - (t_{\alpha/2,v_t})\widehat{\text{s.e.}}(\hat{\theta}_{\text{T}}) \leq \theta_{\text{T}} \leq \hat{\theta}_{\text{T}} + (t_{\alpha/2,v_t})\widehat{\text{s.e.}}(\hat{\theta}_{\text{T}})$$

where $v_t = d(n - 1)(t - 1)$.

The model for comparing two Drug means is

$$\bar{y}_{i\cdot\cdot} - \bar{y}_{i'\cdot\cdot} = \bar{\mu}_{i\cdot} - \bar{\mu}_{i'\cdot} + \bar{p}_{i\cdot} - \bar{p}_{i'\cdot} + \bar{\varepsilon}_{i\cdot\cdot} - \bar{\varepsilon}_{i'\cdot\cdot},$$

and the variance of $\bar{y}_{i..} - \bar{y}_{i'..}$ is

$$\operatorname{Var}(\bar{y}_{i..} - \bar{y}_{i'..}) = \operatorname{Var}(\bar{p}_{i.} - \bar{p}_{i'.} + \bar{\varepsilon}_{i..} - \bar{\varepsilon}_{i'..}) = \frac{2\left[\sigma_\varepsilon^2(1 + 2\lambda_.) + t\sigma_p^2\right]}{nt}.$$

The estimate of the standard error of the difference of two Drug means is

$$\widehat{\text{s.e.}}(\bar{y}_{i..} - \bar{y}_{i'..}) = \sqrt{\frac{2\text{MSERROR(SUBJECT)}}{nt}},$$

and an LSD value is

$$\text{LSD} = \left[t_{\alpha/2,(d-1)(n-1)}\right]\widehat{\text{s.e.}}(\bar{y}_{i..} - \bar{y}_{i'..}).$$

A contrast between the drug means is

$$\theta_{\text{D}} = c_1\bar{\mu}_{1.} + c_2\bar{\mu}_{2.} + \cdots + c_d\bar{\mu}_{d.}$$

where $c_1 + c_2 + \cdots + c_d = 0$. The estimate of θ_{D} and the estimate of its standard error are

$$\hat{\theta}_{\text{D}} = c_1\bar{y}_{1..} + c_2\bar{y}_{2..} + \cdots + c_d\bar{y}_{d..}$$

and

$$\widehat{\text{s.e.}}(\hat{\theta}_{\text{D}}) = \sqrt{\frac{\text{MSERROR(SUBJECT)}}{tn}\left(c_1^2 + c_2^2 + \cdots + c_d^2\right)}.$$

Such a contrast could be used to compare the average of the new drugs to the control. The three inference procedures are

$$t_c = \frac{\hat{\theta}_{\text{D}}^2}{\widehat{\text{s.e.}}(\hat{\theta}_{\text{D}})},$$

which is compared with $t_{\alpha/2,(d-1)(n-1)}$,

$$F_c = \frac{\hat{\theta}_{\text{D}}^2}{\left[\widehat{\text{s.e.}}(\hat{\theta}_{\text{D}})\right]^2},$$

which is compared with $F_{\alpha,1,(d-1)(n-1)}$, and a $(1 - \alpha)100\%$ confidence interval about θ_{D} is

$$\hat{\theta}_{\text{D}} - (t_{\alpha/2,v_s})\widehat{\text{s.e.}}(\hat{\theta}_{\text{D}}) \le \theta_{\text{D}} \le \hat{\theta}_{\text{D}} + (t_{\alpha/2,v_s})\widehat{\text{s.e.}}(\hat{\theta}_{\text{D}})$$

where $v_s = (d - 1)(n - 1)$.

If there is a Drug∗Time interaction, then we need to compare the drugs at each time period and compare the time periods at the same level of Drug. The model for comparing two time means at the same level of

Drug is

$$\bar{y}_{i\cdot k} - \bar{y}_{i\cdot k'} = \mu_{ik} - \mu_{ik'} + \bar{\varepsilon}_{i\cdot k} - \bar{\varepsilon}_{i\cdot k'}.$$

The variance of the difference of the two means is

$$\text{Var}(\bar{y}_{i\cdot k} - \bar{y}_{i\cdot k'}) = \text{Var}(\bar{\varepsilon}_{i\cdot k} - \bar{\varepsilon}_{i\cdot k'}) = \frac{2\sigma_\varepsilon^2}{n},$$

and the estimate of the standard error is

$$\widehat{\text{s.e.}}(\bar{y}_{i\cdot k} - \bar{y}_{i\cdot k'}) = \sqrt{\frac{2\text{MSERROR(TIME)}}{n}}.$$

An LSD value for comparing times within each level of Drug is

$$\text{LSD} = \left[t_{\alpha/2,d(n-1)(t-1)}\right]\widehat{\text{s.e.}}(\bar{y}_{i\cdot k} - \bar{y}_{i\cdot k'}).$$

A contrast between time means at the same drug is

$$\theta_{\text{T}i} = c_1\mu_{i1} + c_2\mu_{i2} + \cdots + c_t\mu_{it}$$

where $c_1 + c_2 + \cdots + c_t = 0$. The estimate of $\theta_{\text{T}i}$ and the estimate of its standard error is

$$\hat{\theta}_{\text{T}i} = c_1\bar{y}_{i\cdot 1} + c_2\bar{y}_{i\cdot 2} + \cdots + c_t\bar{y}_{i\cdot t}$$

and

$$\widehat{\text{s.e.}}(\hat{\theta}_{\text{T}i}) = \sqrt{\frac{\text{MSERROR(TIME)}}{n}\left(c_1^2 + c_2^2 + \cdots + c_t^2\right)}.$$

Such a contrast can be used to study linear, quadratic, and similar trends over time at each level of Drug. The three inference procedures are

$$t_c = \frac{\hat{\theta}_{\text{T}i}}{\widehat{\text{s.e.}}(\theta_{\text{T}i})},$$

which is compared to $t_{\alpha/2,v_t}$,

$$F_c = \frac{\hat{\theta}_{\text{T}i}^2}{\left[\widehat{\text{s.e.}}(\hat{\theta}_{\text{T}i})\right]^2},$$

which is compared to $F_{\alpha,1,v_t}$, and a $(1-\alpha)100\%$ confidence interval about $\theta_{\text{T}i}$, which is

$$\hat{\theta}_{\text{T}i} - \left[t_{\alpha/2,v_t}\right]\widehat{\text{s.e.}}(\hat{\theta}_{\text{T}i}) \le \hat{\theta}_{\text{T}i} + t_{\alpha/2,v_t}\left[\widehat{\text{s.e.}}(\hat{\theta}_{\text{T}i})\right],$$

where $v_t = d(n-1)(t-1)$.

The model for comparing two drug means at the same time or at different times is

$$\bar{y}_{i\cdot k} - \bar{y}_{i'\cdot k} = \bar{\mu}_{ik} - \bar{\mu}_{i'k} + \bar{p}_{i\cdot} - \bar{p}_{i'\cdot} + \bar{\varepsilon}_{i\cdot k} - \bar{\varepsilon}_{i'\cdot k}.$$

The variance of the difference is

$$\begin{aligned}\operatorname{Var}(\bar{y}_{i\cdot k} - \bar{y}_{i'\cdot k}) &= \operatorname{Var}(\bar{p}_{i\cdot} - \bar{p}_{i'\cdot} + \bar{\varepsilon}_{i\cdot k} - \bar{\varepsilon}_{i'\cdot k}) \\ &= \frac{2(\sigma_p^2 + \sigma_\varepsilon^2)}{n}.\end{aligned}$$

There is no estimate of $\sigma_p^2 + \sigma_\varepsilon^2$ available from this analysis unless $\lambda_{.} = 0$. If $\lambda_{.} = 0$, then

$$m = \frac{\text{MSERROR(SUBJECT)} + (t-1)\text{MSERROR(TIME)}}{t}$$

is an unbiased estimate of $\sigma_p^2 + \sigma_\varepsilon^2$. If $\lambda_{.} > 0$ then $E(m) > \sigma_p^2 + \sigma_\varepsilon^2$; if $\lambda_{.} < 0$, then $E(m) < \sigma_p^2 + \sigma_\varepsilon^2$. Since the value of $\lambda_{.}$ is unknown, $\sqrt{2m/n}$ is an estimate of

$$\sqrt{\frac{2(\sigma_p^2 + \sigma_\varepsilon^2)}{n}}.$$

An approximate LSD value for comparing drugs at the same time or different times is

$$\text{LSD} = t^*\sqrt{\frac{2m}{n}},$$

where t^* is an approximate t-value computed by the technique in Section 24.3 as

$$t^* = \frac{[t_{a/2,(a-1)(n-1)}]\text{MSERROR(SUBJECT)} + (t-1)[t_{a/2,a(n-1)(t-1)}]\text{MSERROR(TIME)}}{\text{MSERROR(SUBJECT)} + (t-1)\text{MSERROR(TIME)}}.$$

A contrast between the drug means at the same or different times is

$$\theta_{\text{D}k} = c_1\mu_{1k} + c_2\mu_{2k} + \cdots + c_d\mu_{dk}$$

where $c_1 + c_2 + \cdots + c_d = 0$. The estimate of $\theta_{\text{D}k}$ and its estimated standard error is

$$\hat{\theta}_{\text{D}k} = c_1\bar{y}_{1\cdot k} + c_2\bar{y}_{2\cdot k} + \cdots + c_d\bar{y}_{d\cdot k}$$

and

$$\widehat{\text{s.e.}}(\hat{\theta}_{\text{D}k}) = \sqrt{m(c_1^2 + c_2^2 + \cdots + c_d^2)}.$$

An approximate t-statistic is

$$t_c = \frac{\hat{\theta}_{Dk}}{\widehat{\text{s.e.}}(\hat{\theta}_{Dk})},$$

which is compared with t^*. An approximate $(1-\alpha)100\%$ confidence interval about θ_{Dk} is

$$\hat{\theta}_{Dk} - t^*\widehat{\text{s.e.}}(\hat{\theta}_{Dk}) \le \theta_{Dk} \le \hat{\theta}_{Dk} + t^*\widehat{\text{s.e.}}(\hat{\theta}_{Dk}).$$

An approximate F-statistic is

$$F_c = \frac{\theta_{Dk}^2}{\left[\widehat{\text{s.e.}}(\theta_{Dk})\right]^2},$$

which is compared to $F_{\alpha,1,r}$ where r is determined via the Satterthwaite approximation as

$$r = \frac{[\text{MSERROR(SUBJECT)} + (t-1)\text{MSERROR(TIME)}]^2}{W}$$

where

$$W = \frac{[\text{MSERROR(SUBJECT)}]^2}{(a-1)(n-1)} + \frac{[(t-1)\text{MSERROR(TIME)}]^2}{d(n-1)(t-1)}.$$

Since an approximation is being used, t^{*2} does not necessarily equal $F_{\alpha,1,r}$.

Table 26.2 Heart Rate Data for Example 26.1

	DRUG											
Person within drug	AX23				BWW9				CONTROL			
	T_1	T_2	T_3	T_4	T_1	T_2	T_3	T_4	T_1	T_2	T_3	T_4
1	72	86	81	77	85	86	83	80	69	73	72	74
2	78	83	88	81	82	86	80	84	66	62	67	73
3	71	82	81	75	71	78	70	75	84	90	88	87
4	72	83	83	69	83	88	79	81	80	81	77	72
5	66	79	77	66	86	85	76	76	72	72	69	70
6	74	83	84	77	85	82	83	80	65	62	65	61
7	62	73	78	70	79	83	80	81	75	69	69	68
8	69	75	76	70	83	84	78	81	71	70	65	65

Note: T_i denotes the time period.

Table 26.3 Analysis of Variance Table for Data of Example 26.1

Source of Variation	*df*	*SS*	*MS*	*F*	*EMS*
Drug	2	1,333.00	666.5	5.99	$\sigma_\varepsilon^2 + 4\sigma_p^2 + Q_1$
Error(Person)	21	2,337.91	111.33		$\sigma_\varepsilon^2 + 4\sigma_p^2$
Time	3	289.61	96.54	12.96	$\sigma_\varepsilon^2 + Q_2$
Time * Drug	6	527.42	87.90	11.80	$\sigma_\varepsilon^2 + Q_3$
Error(Time)	63	469.22	7.45		σ_ε^2

Note: Q_i denotes the respective noncentrality parameter.

The data in Table 26.2 are used to demonstrate the analysis described above. In this experiment there were three drugs, eight people per drug, and four time periods. The analysis of variance table is in Table 26.3. There is a significant Time * Drug interaction; thus, we need to compare the times at each drug and the drugs at each time.

To compare times at each drug, the standard error of the difference of the two means is

$$\text{s.e.}(\bar{y}_{i\cdot k} - \bar{y}_{i\cdot k'}) = \sqrt{\frac{2\text{MSERROR(TIME)}}{8}}$$
$$= 1.365,$$

and a 5% LSD value is

$$\text{LSD}_{.05} = 2.00(1.365) = 2.729.$$

A comparison of the time means within a drug is in Table 26.4.

Table 26.4 Comparisons of Time Means at the Same Drug for Example 26.1

	DRUG		
TIME	AX23	BWW9	CONTROL
1	70.50 (a)	81.75 (a, b)	72.75 (a)
2	80.50 (b)	84.00 (a)	72.38 (a)
3	81.00 (b)	78.63 (c)	71.50 (a)
4	73.13 (a)	79.75 (b, c)	71.00 (a)

Note: Means within a column with the same letter are not significantly different; $\text{LSD}_{.05} = 2.726$.

Since the levels of time are quantitative and equally spaced, orthogonal polynomials can be used to check for linear and quadratic trends in the response to each drug. The measure of the linear trend for the first drug is

$$\theta_{\text{TL1}} = -3\mu_{11} - 1\mu_{12} + 1\mu_{13} + 3\mu_{14},$$

its estimate is

$$\hat{\theta}_{\text{TL1}} = -3(70.50) - 1(80.50) + 1(81.00) + 3(73.13) = 8.39,$$

and the estimated standard error is

$$\widehat{\text{s.e.}}(\hat{\theta}_{\text{TL1}}) = \sqrt{\frac{7.45}{8}(3^2 + 1^2 + 1^2 + 3^2)} = 4.315.$$

The corresponding t-statistic is $t_c = 8.39/4.315 = 1.94$.

The measure of the quadratic trend for the first drug is

$$\theta_{\text{TQ1}} = 1\mu_{11} - 1\mu_{12} - 1\mu_{13} + 1\mu_{14},$$

its estimate is

$$\theta_{\text{TQ1}} = 70.50 - 80.50 - 81.00 + 73.13 = -17.87,$$

and its estimated standard error is

$$\widehat{\text{s.e.}}(\hat{\theta}_{\text{TQ1}}) = \sqrt{\frac{7.45}{8}(1^2 + 1^2 + 1^2 + 1^2)} = 1.930.$$

The corresponding t-statistic is $t_c = -17.87/1.930 = 9.259$.

The linear and quadratic trends for all drugs are summarized in Table 26.5. Drug BWW9 shows a negative linear trend, and drug AX23 shows a strong quadratic trend. The graph in Figure 26.1 displays these relationships.

Table 26.5 Linear and Quadratic Trends for Example 26.1

	DRUG		
TREND	AX23	BWW9	CONTROL
Linear	8.39 (1.94)	−11.37 (−2.635)	−6.13 (−1.420)
Quadratic	−17.87 (9.259)	−1.13 (− .585)	−.13 (− .067)

Note: The t_c values are in parentheses.

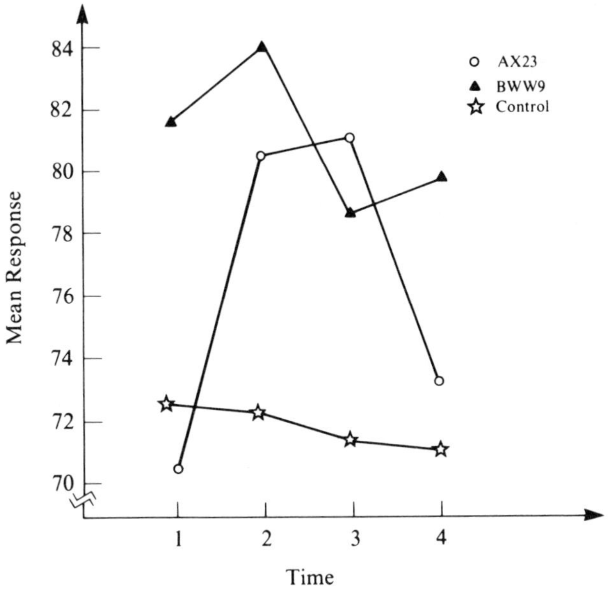

Figure 26.1 Graph of means over time for each drug for the heart rate data.

To compare drugs at each time (or at different times), the standard error is

$$\widehat{\text{s.e.}}(\bar{y}_{i\cdot k} - \bar{y}_{i'\cdot k}) = \sqrt{\frac{2[\text{MSERROR(PERSON)} + (4-1)\text{MSERROR(TIME)}]}{8(4)}} = 2.891.$$

The approximate t-value needed to construct a 5% LSD is

$$t^* = \frac{2.080(111.33) + (4-1)(2.00)(7.45)}{111.33 + (4-1)(7.45)} = 2.067.$$

An approximate 5% LSD value is

$$\text{LSD}_{.05} = (2.067)(2.891) = 5.974.$$

A comparison of the drugs at each time is in Table 26.6.

Suppose drugs AX23 and BWW9 are experimental and we want to compare them to the control at each time period. The contrast for this comparison at time 1 is

$$\theta_{\text{D1}} = \tfrac{1}{2}\mu_{11} + \tfrac{1}{2}\mu_{21} - 1\mu_{31},$$

its estimate is

$$\hat{\theta}_{\text{D1}} = \tfrac{1}{2}(70.50) + \tfrac{1}{2}(81.75) - 1(72.75) = 3.375,$$

Table 26.6 Comparisons of Drugs at the Same Time in Example 26.1

	TIME			
DRUG	1	2	3	4
AX23	70.50 (a)	80.50 (a)	81.00 (a)	73.12 (a)
BWW9	81.75 (b)	84.00 (a)	78.36 (a)	79.75 (b)
Control	72.75 (a)	72.38 (b)	71.50 (b)	71.00 (a)

Note: Means within a column with the same letter are not significantly different; $\text{LSD}_{.05} = 5.974$.

and the estimated standard error is

$$\text{s.e.}(\hat{\theta}_{\text{D1}}) = \sqrt{\left[\frac{111.33 + (3)(7.45)}{4}\right]\left[\frac{(1/2)^2 + (1/2)^2 + (-1)^2}{8}\right]}$$

$$= 2.503.$$

The corresponding t-statistic is $t_c = 3.375/2.503 = 1.348$. The t-statistic should be compared with $t^* = 2.067$.

The corresponding F-statistic is

$$F_c = \frac{(3.375)^2}{(2.503)^2} = 1.817,$$

which should be compared with $F_{\alpha,1,r}$ where

$$r = \frac{[111.33 + (3)(7.45)]^2}{(111.33)^2/21 + [3(7.45)]^2/63} = 29.8.$$

The value of $F_{.05,1,29.8}$ is approximately 4.17.

Table 26.7 Comparisons of Means of AX23 and BWW9 to the Control at Each Time in Example 26.1

	TIME			
STATISTIC	1	2	3	4
θ_{Dk}	3.375	9.870	8.315	5.440
t_c	1.348	3.943	3.321	2.173
F_c	1.817	15.547	11.029	4.723

The comparisons of the means of the two experimental drugs to the control at each time period is given in Table 26.7. The results show that the means of the drugs are significantly different at $\alpha = .05$ for the last three time intervals, but not for time 1.

EXAMPLE 26.2: A Complex Comfort Experiment

An engineer had three environments in which to test two types of clothing. Since responses to an environment also differ between males and females, sex of person was included as a factor. Four people (two males and two females) were put into an environmental chamber (which was assigned one of the three environments). One male and one female wore clothing type 1, and the other male and female wore clothing type 2. The comfort score of each person was recorded at the end of one hour, two hours, and three hours. The data for this experiment are shown in Table 26.8.

There are three sizes of experimental units. The largest experimental unit is a chamber or, equivalently, a group of four people. The chamber experimental unit experimental design is a one-way treatment structure (Environment) in a completely randomized design structure with three replications at each level of the environment. The middle-sized experimental unit is a person. The experimental design for a person is a two-way treatment structure (Sex * Clothing) in a randomized complete block design structure in nine blocks (each block contains four experimental units [people]). The smallest experimental unit is a one-hour time interval, which we will call hour. The experimental design for hour is a one-way treatment structure (Time) in a randomized complete block design structure in 36 blocks (each block contains three experimental units [one-hour time intervals]).

The model for this experiment (where the model is separated into parts corresponding to the three sizes of experimental units) is

$$y_{ijkmn} = \mu + E_i + r_{ij}\} \qquad \text{Chamber part}$$

$$\left.\begin{aligned} &+ S_k + C_m + (SC)_{km} + (ES)_{ik} \\ &+ (EC)_{im} + (ESC)_{ikm} + p_{ijkm} \end{aligned}\right\} \qquad \text{Person part}$$

$$\left.\begin{aligned} &+ T_n + (ET)_{in} + (ST)_{kn} + (CT)_{mn} \\ &+ (SCT)_{kmn} + (EST)_{ikn} + (ECT)_{imn} \\ &+ (ESCT)_{ikmn} + \varepsilon_{ijkmn} \end{aligned}\right\} \qquad \text{Hour part}$$

where E denotes environment, S denotes sex, C denotes clothing, T denotes time, r_{ij} denotes the chamber effect assumed to be distributed i.i.d. $N(0, \sigma_r^2)$, p_{ijkm} denotes the person effect assumed to be distributed

Table 26.8 Data for Comfort Experiment in Example 26.2

				SCORE OF ENVIRONMENT		
Replication	*Sex*	*Clothing Type*	*Time*	*1*	*2*	*3*
1	M	1	1	13.9001	10.2881	7.4205
1	M	1	2	7.5154	6.9090	7.1752
1	M	1	3	10.9742	8.4138	7.1218
1	M	2	1	18.3941	13.8631	12.5410
1	M	2	2	12.4151	10.1492	11.9157
1	M	2	3	15.2241	12.5372	12.2239
1	F	1	1	10.0149	6.1634	3.8293
1	F	1	2	3.7669	2.2837	3.5868
1	F	1	3	7.0326	4.0052	3.3004
1	F	2	1	16.4774	13.0291	11.0002
1	F	2	2	10.4104	9.7775	11.0282
1	F	2	3	13.1143	11.6576	10.5662
2	M	1	1	15.7185	11.9904	11.8158
2	M	1	2	9.7170	8.4793	11.9721
2	M	1	3	12.5080	9.8694	11.7187
2	M	2	1	19.7547	14.8587	16.4418
2	M	2	2	13.5293	11.0317	16.6355
2	M	2	3	16.5487	12.8317	16.6686
2	F	1	1	10.6902	6.7562	7.5707
2	F	1	2	4.8473	2.5634	7.3456
2	F	1	3	7.9829	4.7547	7.2404
2	F	2	1	17.1147	13.8977	13.5421
2	F	2	2	11.3858	9.6643	13.5672
2	F	2	3	14.1502	11.6034	14.0240
3	M	1	1	14.9015	9.7589	7.2364
3	M	1	2	9.1825	6.1772	7.8304
3	M	1	3	11.5819	8.0785	7.4147
3	M	2	1	18.0402	13.5513	12.0689
3	M	2	2	12.1004	9.3052	12.5003
3	M	2	3	15.4893	11.5259	12.1790
3	F	1	1	10.1944	4.5203	1.8330
3	F	1	2	4.1716	0.5913	1.6769
3	F	1	3	6.9688	2.8939	1.8065
3	F	2	1	16.0789	12.5057	9.4934
3	F	2	2	10.2357	7.7502	9.7000
3	F	2	3	12.4853	10.5226	10.0119

i.i.d. $N(0, \sigma_p^2)$, and ε_{ijkmn} denotes the measurement error for a given hour, which is assumed to be i.i.d. $N(0, \sigma_\varepsilon^2)$. In addition, all the error terms are assumed to be independently distributed.

The analysis of variance table is given in Table 26.9, where the expected mean squares are computed under the assumption of independent error terms. If the error terms have covariance matrices that satisfies the H–F condition, then the expectations will involve terms like $\lambda_{.}$, as occurred in Example 26.1. The analysis will be completed under the independent error assumption, deviations from which are discussed in Chapter 27.

The error sums of squares in Table 26.9 were computed as

$$\begin{aligned}\text{SSERROR(PERSON)} = {} & \text{Replication} * \text{Sex} * \text{Clothing(Environment)} \\ & + \text{Replication} * \text{Sex(Environment)} \\ & + \text{Replication} * \text{Clothing(Environment)}\end{aligned}$$

Table 26.9 Analysis of Variance Table for Example 26.2

Source of Variation	*df*	*SS*	*MS*	*F*	*Expected Mean Squares*
Environment	2	191.69	95.85	3.28	$\sigma_\varepsilon^2 + 3\sigma_p^2 + 12\sigma_r^2 + Q_1$
Error(Chamber)	6	175.26	29.21		$\sigma_\varepsilon^2 + 3\sigma_p^2 + 12\sigma_r^2$
Sex	1	289.46	289.46	501.89	$\sigma_\varepsilon^2 + 3\sigma_p^2 + Q_2$
Clothing	1	806.11	806.11	1,397.70	$\sigma_\varepsilon^2 + 3\sigma_p^2 + Q_3$
Sex * Clothing	1	55.96	55.96	97.04	$\sigma_\varepsilon^2 + 3\sigma_p^2 + Q_4$
Environment * Sex	2	.78	.39	.68	$\sigma_\varepsilon^2 + 3\sigma_p^2 + Q_5$
Environment * Clothing	2	4.31	2.16	3.73	$\sigma_\varepsilon^2 + 3\sigma_p^2 + Q_6$
Environment * Sex * Clothing	2	4.41	2.21	3.82	$\sigma_\varepsilon^2 + 3\sigma_p^2 + Q_7$
Error(Person)	18	10.38	.58		$\sigma_\varepsilon^2 + 3\sigma_p^2$
Time	2	194.60	97.3	1,672.24	$\sigma_\varepsilon^2 + Q_8$
Time * Environment	4	111.65	27.91	479.70	$\sigma_\varepsilon^2 + Q_9$
Time * Sex	2	.08	.04	.67	$\sigma_\varepsilon^2 + Q_{10}$
Time * Clothing	2	.08	.04	.71	$\sigma_\varepsilon^2 + Q_{11}$
Time * Sex * Clothing	2	.20	.10	1.68	$\sigma_\varepsilon^2 + Q_{12}$
Time * Environment * Sex	4	.18	.04	.78	$\sigma_\varepsilon^2 + Q_{13}$
Time * Environment * Clothing	4	.26	.06	1.14	$\sigma_\varepsilon^2 + Q_{14}$
Time * Environment * Sex * Clothing	4	.17	.04	.71	$\sigma_\varepsilon^2 + Q_{15}$
Error(Hour)	48	2.79	.06		σ_ε^2

Notes: All figures in the table are rounded to two significant figures, but the calculations were done indouble precision. Q_i denotes the respective noncentrality parameter.

and

$$\text{SSERROR(CHAMBER)} = \text{Replication(Environment)}.$$

The F-tests were conducted by using the expected mean squares as a guide.

The next step in the analysis is to make the needed comparisons. If we select $\alpha = .01$ as the probability of a Type I error, then there are two significant interactions, Environment ∗ Time and Sex ∗ Clothing. For the Sex ∗ Clothing interaction, we want to compare the four means. Since both treatments were applied to the same size of experimental unit (person), only one standard error needs to be computed. The difference $\bar{y}_{\cdot\cdot 11\cdot} - \bar{y}_{\cdot\cdot 22\cdot}$ can be expressed in terms of the model; thus, its variance is evaluated as

$$\begin{aligned} \text{Var}(\bar{y}_{\cdot\cdot 11\cdot} - \bar{y}_{\cdot\cdot 22\cdot}) &= \text{Var}[(\bar{r}_{\cdot\cdot} - \bar{r}_{\cdot\cdot}) + (\bar{p}_{\cdot\cdot 11} - \bar{p}_{\cdot\cdot 22}) + (\bar{\varepsilon}_{\cdot\cdot 11\cdot} - \bar{\varepsilon}_{\cdot\cdot 22\cdot})] \\ &= \frac{2\sigma_p^2}{9} + \frac{2\sigma_\varepsilon^2}{27} = \frac{2}{27}\left(\sigma_\varepsilon^2 + 3\sigma_p^2\right). \end{aligned}$$

The estimate of this variance is $\frac{2}{27}$MSERROR(PERSON). Fisher's LSD at the 1% level is

$$\begin{aligned} \text{LSD}_{.01} &= t_{.005,18}\sqrt{\frac{2}{27}\text{MSERROR(PERSON)}} \\ &= 2.878\sqrt{\frac{2}{27}\left(\frac{10.38128}{18}\right)} \\ &\doteq .595. \end{aligned}$$

Table 26.10 has the Sex ∗ Clothing means, all of which are significantly different.

The Environment ∗ Time interaction involves two different sizes of experimental units; thus, there are two types of comparisons. First, we

Table 26.10 Sex ∗ Clothing Means with LSD for Example 26.2

Sex	*Clothing*	*Score*
M	1	9.8396036
M	2	13.8638776
F	1	5.1255944
F	2	12.0294117

Note: $\text{LSD}_{.01} = .595$. All means are based on 27 observations.

can compare two Time means and to check for linear and quadratic trends between the Time means at the same level of Environment. The variance of a comparison of two Time means at the same Environment is

$$V(\bar{y}_{1\cdots 1} - \bar{y}_{1\cdots 2}) = V(\bar{r}_{1\cdot} - \bar{r}_{1\cdot} + \bar{p}_{1\cdots} - \bar{p}_{1\cdots} + \bar{\varepsilon}_{1\cdots 1} - \bar{\varepsilon}_{1\cdots 2})$$
$$= \frac{2\sigma_\varepsilon^2}{12}.$$

The estimate of this variance is $\frac{2}{12}$MSERROR(HOUR). The 1% Fisher's LSD for comparing Time means at the same level of Environment is

$$\text{LSD}_{.01} = t_{.005,48}\sqrt{\frac{2}{12}\text{MSERROR(HOUR)}}$$
$$= 2.683\sqrt{\frac{2}{12}(.05818512)} = .264.$$

Table 26.11 contains the Environment ∗ Time means, which allow us to make comparisons between different times within the same environment.

Since the three levels of Time are equally spaced, orthogonal polynomials can be used to measure linear and quadratic trends over time for each environment. The linear trend at environment 1 is measured by

$$\hat{\theta}_{\text{TL1}} = -\bar{y}_{1\cdots 1} + 0\bar{y}_{1\cdots 2} + \bar{y}_{1\cdots 3},$$
$$= -15.11 + 12.01 = -3.10,$$

and the estimate of the standard error is

$$\widehat{\text{s.e.}}(\hat{\theta}_{\text{TL1}}) = \sqrt{\frac{.058185}{12}\left[(-1)^2 + 0^2 + (1)^2\right]} = .264.$$

Table 26.11 Environment ∗ Time Means for Comparing Time Means within Each Level of Environment in Example 26.2

	ENVIRONMENT		
Time	*1*	*2*	*3*
1	15.11 (a)	10.93 (a)	9.57 (a)
2	9.11 (c)	7.07 (c)	9.57 (a)
3	12.01 (b)	9.06 (b)	9.52 (a)

Note: Within a given environment, Time means with the same letter are not significantly different; $\text{LSD}_{.01} = .264$.

Table 26.12 Measures of Linear and Quadratic Trends for Each Environment in Example 26.2

	ENVIRONMENT		
Trend	*1*	*2*	*3*
Linear	−3.10 (−11.74)	−.87 (−3.795)	−.05 (−.184)
Quadratic	8.90 (19.46)	5.85 (12.79)	−.05 (−.109)

Note: t-values are in parentheses.

The corresponding t-statistic is $t_c = -3.10/.264 = -11.742$. The quadratic trend at environment 1 is measured by

$$\hat{\theta}_{TQ1} = +\bar{y}_{1\cdots1} - 2\bar{y}_{1\cdots2} + \bar{y}_{1\cdots3}$$

$$= +15.11 - 2(9.11) + 12.01 = 8.90,$$

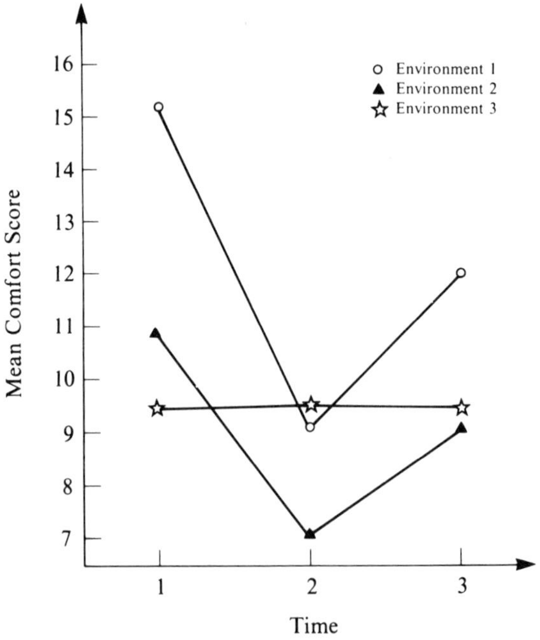

Figure 26.2 Graph showing response over time at each environment for the comfort data.

and the estimate of the standard error is

$$\widehat{\text{s.e.}}(\hat{\theta}_{\text{TQ1}}) = \sqrt{\frac{.058185}{12}(1^2 + 2^2 + 1^2)} = .457.$$

The corresponding t-statistic is $8.90/.457 = 19.46$. The linear and quadratic trends for each environment are in Table 26.12. There are significant linear and quadratic trends for environments 1 and 2, but none for environment 3. Figure 26.2 displays the relationships.

The second type of Environment ∗ Time comparison is to compare the different environments at the same time or at different times. The variance of such a comparison is

$$\begin{aligned} V(\bar{y}_{1\cdots1} - \bar{y}_{2\cdots2}) &= V(\bar{r}_{1\cdot} - \bar{r}_{2\cdot} + \bar{p}_{1\cdots} - \bar{p}_{2\cdots} + \bar{\varepsilon}_{1\cdots1} - \bar{\varepsilon}_{2\cdots2}) \\ &= \frac{2}{3}\sigma_r^2 + \frac{2}{12}\sigma_p^2 + \frac{2}{12}\sigma_\varepsilon^2 \\ &= \frac{2}{12}\left(\sigma_\varepsilon^2 + \sigma_p^2 + 4\sigma_r^2\right) \end{aligned}$$

An estimate of this variance is

$$\left(\frac{2}{12}\right)\frac{\text{MSERROR(CHAMBER)} + 2*\text{MSERROR(HOUR)}}{3}.$$

This variance is a linear combination of two error terms; thus, the approximate t-value for computing the 1% Fisher's LSD is

$$\begin{aligned} t^* &= \frac{t_{.005,6}\text{MSERROR(CHAMBER)} + t_{.005,48}(2)\text{MSERROR(HOUR)}}{\text{MSERROR(CHAMBER)} + 2\text{MSERROR(HOUR)}} \\ &= \frac{(2.969)(175.2559/6) + (2.683)(2)(.05818512)}{(175.2559/6) + (2)(.05818512)} = \frac{87.083}{29.326} = 2.968 \end{aligned}$$

The resulting approximate 1% LSD value is

$$\begin{aligned} &\text{LSD}_{.01} \\ &= t^*_{.005}\sqrt{\frac{(2/12)[\text{MSERROR(CHAMBER)} + 2\text{MSERROR(HOUR)}]}{3}} \\ &= 2.968\frac{(2/12)(29.326)}{3} \\ &= 3.788 \end{aligned}$$

Table 26.13 Environment * Time Means with Comparisons between Environment Means at the Same Times

	TIME		
Environment	*1*	*2*	*3*
1	15.11 (a)	9.11 (a)	12.01 (a)
2	10.93 (b)	7.07 (a)	9.06 (a)
3	9.57 (b)	9.57 (a)	9.52 (a)

Note: Within a given level of time, environment means with the same letter are not significantly different; $LSD_{.01} = 3.788$.

Table 26.13 contains the Environment * Time means with Fisher LSD comparisons made between different environments at the same times.

This example illustrates why it is very important to use correct error terms when comparing within a set of interaction means: The two LSD values may be extremely different. In this example, the two values are .264 and 3.788. Thus, the $LSD_{.01}$ value for comparing Environments at the same level of Time is 14 times larger than the $LSD_{.01}$ value for comparing Times at the same level of Environment.

To demonstrate how to use the information in Section 24.3 to compute standard errors and corresponding t^* values for cases where more than two sizes of experimental units are used, we construct an LSD for comparing environments and sexes at the same time interval. The comparison of interest is $\bar{y}_{1\cdot 1\cdot 1} - \bar{y}_{2\cdot 2\cdot 1}$, which can be expressed as

$$\begin{aligned}\bar{y}_{1\cdot 1\cdot 1} - \bar{y}_{2\cdot 2\cdot 1} &= [(\bar{y}_{1\cdot 1\cdot 1} - \bar{y}_{1\cdot 1\cdot\cdot}) - (\bar{y}_{2\cdot 2\cdot 1} - \bar{y}_{2\cdot 2\cdot\cdot})] \\ &\quad + [(\bar{y}_{1\cdot 1\cdot\cdot} - \bar{y}_{1\cdot\cdot\cdot\cdot}) - (\bar{y}_{2\cdot 2\cdot\cdot} - \bar{y}_{2\cdot\cdot\cdot\cdot})] + [\bar{y}_{1\cdot\cdot\cdot\cdot} - \bar{y}_{2\cdot\cdot\cdot\cdot}] \\ &= A + B + C\end{aligned}$$

where A, B, and C denote the respective terms in square brackets. Comparison A is an hour experimental unit comparison, comparison B is a person experimental unit comparison, and comparison C is a chamber experimental unit comparison. Thus, the variance of $\bar{y}_{1\cdot 1\cdot 1} - \bar{y}_{2\cdot 2\cdot 1}$ is

$$\operatorname{Var}(A) + \operatorname{Var}(B) + \operatorname{Var}(C).$$

The variance of A is

$$\operatorname{Var}(A) = \tfrac{2}{9}\sigma_\varepsilon^2,$$

which is estimated by $\frac{2}{9}$MSERROR(HOUR). The variance of B is

$$\operatorname{Var}(B) = \tfrac{1}{18}\left(\sigma_\varepsilon^2 + 3\sigma_p^2\right),$$

which is estimated by $\frac{1}{18}$MSERROR(PERSON). The variance of C is

$$\text{Var}(C) = \tfrac{1}{18}\left(\sigma_\varepsilon^2 + 3\sigma_p^2 + 12\sigma_r^2\right),$$

which is estimated by $\frac{1}{18}$MSERROR(CHAMBER). Therefore, the variance of $\bar{y}_{1\cdot 1\cdot 1} - \bar{y}_{2\cdot 2\cdot 1}$ is

$$\begin{aligned}\text{Var}(\bar{y}_{1\cdot 1\cdot 1} - \bar{y}_{2\cdot 2\cdot 1}) &= \tfrac{2}{9}\sigma_\varepsilon^2 + \tfrac{1}{18}\left(\sigma_\varepsilon^2 + 3\sigma_p^2\right) + \tfrac{1}{18}\left(\sigma_\varepsilon^2 + 3\sigma_p^2 + 12\sigma_r^2\right)\\ &= \tfrac{1}{3}\left(\sigma_\varepsilon^2 + \sigma_p^2 + 2\sigma_r^2\right).\end{aligned}$$

The estimated standard error of $\bar{y}_{1\cdot 1\cdot 1} - \bar{y}_{2\cdot 2\cdot 1}$ is

$$\begin{aligned}\widehat{\text{s.e.}}(\bar{y}_{1\cdot 1\cdot 1} - \bar{y}_{2\cdot 2\cdot 1}) = \big[&\tfrac{2}{9}\text{MSERROR(HOUR)}\\ &+ \tfrac{1}{18}\text{MSERROR(PERSON)}\\ &+ \tfrac{1}{18}\text{MSERROR(CHAMBER)}\big]^{1/2}.\end{aligned}$$

The approximate Fisher's $\alpha\%$ LSD value is

$$\text{LSD}_\alpha = t_{\alpha/2}^{*}\widehat{\text{s.e.}}(y_{1\cdot 1\cdot 1} - y_{2\cdot 2\cdot 1})$$

where

$$t_{\alpha/2}^{*} = \frac{\left[t_{\alpha/2,v_1}\right]\tfrac{2}{9}\text{MSERROR(HOUR)} + \left[t_{\alpha/2,v_2}\right]\tfrac{1}{18}\text{MSERROR(PERSON)} + \left[t_{\alpha/2,v_3}\right]\tfrac{1}{18}\text{MSERROR(CHAMBER)}}{\tfrac{2}{9}\text{MSERROR(HOUR)} + \tfrac{1}{18}\text{MSERROR(PERSON)} + \tfrac{1}{18}\text{MSERROR(CHAMBER)}},$$

v_1, v_2, and v_3 are the degrees of freedom associated with MSERROR(HOUR), MSERROR(PERSON), and MSERROR(CHAMBER), respectively. The methods used to compute the above two standard errors can be applied to other situations where a given comparison can be partitioned into the sum of components where each component is a comparison of only one size of experimental unit.

When we make comparisons across sizes of experimental units, the estimates of the variances are not necessarily unbiased when the errors are not independent. When the covariance matrices satisfy the H–F condition, the distributions associated with the F-tests are exact, but there can be bias in the standard errors.

EXAMPLE 26.3: Family Attitudes

The attitudes of families from rural and urban environments were measured every six months for three time periods. The data were obtained from seven rural families and ten urban families, each family consisting of a son, father, and mother. The model used to describe the data was previously given in equation (26.1.4). It is quite likely that the variances of the family members are unequal, and thus

$$\text{Var}(\mathbf{p}_{ij}) = \Lambda \neq \sigma_p^2 \mathbf{I}.$$

The analysis in this section assumes that the covariance matrices of the errors satisfy the H–F condition, and standard errors involving more than one size of experimental unit are computed assuming that $p_{ijk} \sim$

Table 26.14 Data for Family Attitude Study of Example 26.3

	PERSON								
	SON			FATHER			MOTHER		
Family	T_1	T_2	T_3	T_1	T_2	T_3	T_1	T_2	T_3
				Urban					
1	17	17	19	18	19	21	16	16	18
2	12	14	15	19	19	21	16	16	18
3	8	10	11	16	18	19	11	12	12
4	5	7	7	12	12	13	13	14	14
5	2	5	6	12	14	14	14	16	18
6	9	11	11	16	17	18	14	15	16
7	8	9	9	19	20	20	15	16	18
8	13	14	16	16	17	18	18	18	20
9	11	12	13	13	16	17	7	8	10
10	19	20	20	13	15	15	11	12	12
				Rural					
1	12	11	14	18	19	22	16	16	19
2	13	13	17	18	19	22	16	16	19
3	12	13	16	19	18	22	17	16	20
4	18	18	21	23	23	26	23	22	26
5	15	14	16	15	15	19	17	17	20
6	6	6	10	15	16	19	18	19	21
7	16	17	18	17	17	21	18	20	23

i.i.d. $N(0, \sigma_p^2)$ and $\varepsilon_{ijkm} \sim$ i.i.d. $N(0, \sigma_\varepsilon^2)$. The data for this example are in Table 26.14.

The analysis of variance table corresponding to the model in equation (26.1.4) is in Table 26.15. If we operate at $\alpha = .05$, there are four significant effects: Area, Person, Time, and Area ∗ Time. Thus, there are three comparisons we wish to make: between persons, between times at each area, and between areas at each time.

The variance of a comparison between people is

$$V(\bar{y}_{\cdot\cdot k\cdot} - \bar{y}_{\cdot\cdot k'\cdot}) = V(\bar{p}_{\cdot\cdot k} - \bar{p}_{\cdot\cdot k'} + \bar{\varepsilon}_{\cdot\cdot k\cdot} + \bar{\varepsilon}_{\cdot\cdot k'\cdot})$$

$$= \frac{2\sigma_p^2}{17} + \frac{2\sigma_\varepsilon^2}{3(17)} = \frac{2}{3(17)}\left(\sigma_\varepsilon^2 + 3\sigma_p^2\right).$$

The estimate of the standard error of the difference of two persons' means is

$$\widehat{\text{s.e.}}(\bar{y}_{\cdot\cdot k\cdot} - \bar{y}_{\cdot\cdot k'\cdot}) = \sqrt{\frac{2}{3(17)}\text{MSERROR(PERSON)}}$$

$$= .98.$$

A 5% LSD value is $\text{LSD}_{.05} = 2.042(.98) = 2.002$. Table 26.16 has a summary of the multiple comparison.

Table 26.15 Analysis of Variance Table for Family Attitude Data of Example 26.3

Source of Variation	*df*	*SS*	*MS*	*F*	*EMS*
Area	1	382.12	382.122	7.06	$\sigma_\varepsilon^2 + 3\sigma_p^2 + 9\sigma_f^2 + Q_1$
Error(Family)	15	812.32	54.155		$\sigma_\varepsilon^2 + 3\sigma_p^2 + 9\sigma_f^2$
Person	2	666.84	330.920	13.50	$\sigma_\varepsilon^2 + 3\sigma_p^2 + Q_2$
Person ∗ Area	2	60.27	30.135	1.23	$\sigma_\varepsilon^2 + 3\sigma_p^2 + Q_3$
Error(Person)	30	735.37	25.512		$\sigma_\varepsilon^2 + 3\sigma_p^2$
Time	2	204.28	102.14	268.94	$\sigma_\varepsilon^2 + Q_4$
Time ∗ Person	4	1.19	.298	.78	$\sigma_\varepsilon^2 + Q_5$
Time ∗ Area	2	30.95	15.475	40.74	$\sigma_\varepsilon^2 + Q_6$
Time ∗ Person ∗ Area	4	1.50	.375	.99	$\sigma_\varepsilon^2 + Q_7$
Error(Time)	90	34.18	.370		σ_ε^2

Note: Q_i denotes the noncentrality parameter of the respective sum of squares.

Table 26.16 Comparison of Person Means for Example 26.3

Person	*Mean*
Son	12.67 (a)
Father	17.47 (b)
Mother	16.53 (b)
$LSD_{.05}$	2.002

Note: Means within a column with the same letter are not significantly different.

The variance of the difference of two time means at the same area is

$$V(\bar{y}_{i\cdot\cdot m} - \bar{y}_{i\cdot\cdot m'}) = V(\bar{\varepsilon}_{i\cdot\cdot m} - \bar{\varepsilon}_{i\cdot\cdot m'})$$
$$= \frac{2\sigma_\varepsilon^2}{3n_i}$$

where n_i is the number of families in area i. The estimate of the standard error to compare two times in the urban area is

$$\widehat{\text{s.e.}}(\bar{y}_{1\cdot\cdot m} - \bar{y}_{1\cdot\cdot m'}) = \sqrt{\tfrac{2}{30}\text{MSERROR(TIME)}} = .159.$$

The standard error to compare two times in a rural area is

$$\widehat{\text{s.e.}}(\bar{y}_{2\cdot\cdot m} - \bar{y}_{2\cdot\cdot m'}) = \sqrt{\tfrac{2}{21}\text{MSERROR(TIME)}}$$
$$= .190.$$

Table 26.17 Comparison of Time Means at Each Area for Example 26.3

	AREA	
Time	*Rural*	*Urban*
1	16.33 (a)	13.10 (a)
2	16.38 (a)	14.30 (b)
3	19.61 (b)	15.30 (c)
$LSD_{.05}$	.378	.316

Note: Means within a column with the same letter are not significantly different.

The corresponding 5% LSDs are

$$\text{LSD}_{.05}(\text{Urban}) = 1.986(.159) = .316$$

and

$$\text{LSD}_{.05}(\text{Rural}) = 1.986(.190) = .378.$$

The multiple comparisons are summarized in Table 26.17.

The variance of the difference between urban and rural means at a given time is

$$\begin{aligned} V(\bar{y}_{1\cdot\cdot m} - \bar{y}_{2\cdot\cdot m}) &= \bar{f}_{1\cdot} - \bar{f}_{2\cdot} + \bar{p}_{1\cdot\cdot} - \bar{p}_{2\cdot\cdot} + \bar{\varepsilon}_{1\cdot\cdot m} - \bar{\varepsilon}_{2\cdot\cdot m} \\ &= \left(3\sigma_f^2 + \sigma_p^2 + \sigma_\varepsilon^2\right)\left(\tfrac{1}{30} + \tfrac{1}{21}\right). \end{aligned}$$

An estimate of the standard error of the difference between urban and rural means at time m is

$$\begin{aligned} &\widehat{\text{s.e.}}(\bar{y}_{1\cdot\cdot m} - \bar{y}_{2\cdot\cdot m}) \\ &= \sqrt{\frac{\text{MSERROR(FAMILY)} + 2\text{MSERROR(TIME)}}{3}} \cdot \sqrt{\frac{1}{30} + \frac{1}{21}} \\ &= \sqrt{\frac{54.155 + 2(.380)}{3}\left(\frac{1}{30} + \frac{1}{21}\right)} \\ &= 1.217. \end{aligned}$$

An approximate t-value is

$$t^* = \frac{2.131(54.155) + 1.986(2)(.380)}{54.155 + 2(.380)} = 2.129.$$

Table 26.18 Comparison of Area Means at Each Time for Example 26.3

	TIME		
Area	*1*	*2*	*3*
Rural	16.33 (a)	16.38 (a)	19.61 (a)
Urban	13.10 (b)	14.30 (a)	15.30 (b)
$\text{LSD}_{.05}$	2.590	2.590	2.590

Note: Means within a column with the same letter are not significantly different.

An approximate 5% LSD to compare area means at the same time is

$$\text{LSD}_{.05} = (2.129)(1.217) = 2.591.$$

The multiple comparison is summarized in Table 26.18.

The linear and quadratic trends of time of each area can be investigated using the method described in Example 26.1.

CONCLUDING REMARKS

The analysis of repeated measures designs was described for three examples where the usual assumptions hold. Included in the discussion were the models and assumptions for each example. Computational formulas for obtaining standard errors for multiple comparisons were given as well as methods for investigating various contrasts among the means.

27

Analysis of Repeated Measures Designs for Which the Usual Assumptions Do Not Hold

CHAPTER OUTLINE

Repeated measures designs involve one or more steps where the experimenter cannot randomly assign the levels of one or more treatments to a given size of experimental unit. The use of time as a factor is a common example where one does not use randomization. Nonrandomization influences the variances and covariances between the experimental units, and the usual assumptions described in Chapter 26 may not be valid.

This chapter presents strategies for analyzing data from experiments where those usual assumptions do not hold. Procedures are described for checking to see if the covariance matrix of the repeated measures satisfies the Huynh–Feldt conditions. Finally, a method is described in Section 27.2 for analyzing a repeated measures design when the repeated measures errors have an autocorrelated error structure.

27.1 WHEN THE ASSUMPTIONS DO NOT HOLD

When the H–F conditions are not satisfied, the analyses described in Section 26.2 are not valid. Several alternative analyses have been suggested. The multivariate analysis discussed in Chapter 31 is one appropriate alternative when the number of subjects exceed the number of repeated measures.

Box (1954) suggested a conservative test that amounts to using percentage points from F-distributions with fewer degrees of freedom rather than those associated with the respective sums of squares. For the simple two-factor repeated measures design, such as the heart rate experiment in Example 26.1, the statistic to test for Time effects is

$$F_{\mathrm{T}} = \frac{\mathrm{MSTIME}}{\mathrm{MSERROR(TIME)}}$$

which for the usual analysis is compared with a percentage point from an F-distribution with $t - 1$ and $d(n - 1)(t - 1)$ degrees of freedom. Box's correction is to divide each of the respective degrees of freedom by $t - 1$

Table 27.1 Box's Conservative Correction Applied to Results for Example 26.1

Source of Variation	*Actual df*	*Observed Significance Level*	*Box's df*	*New Significance Level*
Drug	2	.0088	2	.0088
Error(Person)	21		21	
Time	3	< .0001	1	.0015
Time * Drug	6	< .0001	2	.0025
Error(Time)	63		21	

Table 27.2 Box's Conservative Correction Applied to Results for Example 26.2

Source of Variation	*Actual df*	*Observed Significance Level*	*Box's df*	*New Significance Level*
Environment	2	.1090	2	.1090
Error(Chamber)	6		6	
Sex	1	< .0001	1*	< .0001
Clothing	1	< .0001	1*	< .0001
Sex * Clothing	1	< .0001	1*	< .0001
Environment * Sex	2	.5192	1*	.4411
Environment * Clothing	2	.0441	1*	.1017
Environment * Sex * Clothing	2	.0414	1*	.0984
Error(Person)	18		6	
Time	2	< .0001	1	< .0001
Time * Environment	4	< .0001	2	< .0001
Time * Sex	2	.5164	1	.4211
Time * Clothing	2	.4967	1	.4078
Time * Sex * Clothing	2	.1971	1	.2073
Time * Environment * Sex	4	.5437	2	.4697
Time * Environment * Clothing	4	.3491	2	.3365
Time * Environment * Sex * Clothing	4	.5891	2	.5017
Error(Hour)	48		24	

*Fractional degree of freedom rounded up to 1.

(the number of repeated measures minus one), that is, to compare F_T to a percentage point from an F-distribution with 1 and $d(n-1)$ degrees of freedom.

To apply the Box correction to designs with more than two sizes of experimental units, divide each of the respective degrees of freedom by $s-1$, where s is the number of repeated measures at that level. Box's correction is applied to Examples 26.1 through 26.3, and the results of this application are given in Tables 27.1, 27.2, and 27.3, respectively.

Box (1954) provided a measure of how far a covariance matrix deviates from compound symmetry (it also works for the H–F condition). Let Σ have elements σ_{ij}; then the measure θ is

$$\theta = \frac{t^2(\bar{\sigma}_{ii} - \bar{\sigma}_{..})^2}{(t-1)\left(\Sigma_{i=1}^t \Sigma_{j=1}^t \sigma_{ij}^2 - 2t\Sigma_{i=1}^t \bar{\sigma}_{i.}^2 + t^2\bar{\sigma}_{..}^2\right)}$$

Table 27.3 Box's Conservative Correction Applied to Results for Example 26.3

Source of Variation	*Actual df*	*Observed Significance Level*	*Box's df*	*New Significance Level*
Area	1	.02	1	.02
Error(Family)	15		15	
Person	2	< .0001	1	.0023
Person * Area	2	.3066	1	.2849
Error(Person)	30		15	
Time	2	< .0001	1	< .0001
Time * Person	4	.54	2	.3818
Time * Area	2	< .0001	1	< .0001
Time * Person * Area	4	.42	2	.3795
Error(Time)	90		45	

where

$$\bar{\sigma}_{..} = \frac{1}{t^2} \sum_{i=1}^{t} \sum_{j=1}^{t} \sigma_{ij}$$

is the mean of all of the elements of Σ,

$$\bar{\sigma}_{i.} = \frac{1}{t} \sum_{i=1}^{t} \sigma_{ij}$$

is the mean of the elements in the ith row of Σ, and

$$\bar{\sigma}_{ii} = \frac{1}{t} \sum_{i=1}^{t} \sigma_{ii}$$

is the mean of the diagonal elements of Σ. Box shows that for all $t \geq 2$, the range of θ is

$$\frac{1}{t-1} \leq \theta \leq 1$$

with $\theta = 1$ when Σ satisfies the H–F condition.

To use θ in the analysis, Box showed that F_T is approximately distributed as an F-distribution with $\theta(t-1)$ and $\theta(t-1)(d)(n-1)$ degrees of freedom. Thus, Box's conservative test uses the smallest value of θ, $1/(t-1)$, as the multiplier.

Collier, Baker, Mandeville, and Hayes (1967) carried out a simulation study of the Box correction where the value of θ was estimated from the data. Their study presents evidence that such an adjustment provides a test with the desired Type I error protection. The key in their adjustment is to estimate the elements of Σ and use the estimated values of the σ_{ij}'s to compute θ.

One problem is that Σ is generally not estimable from the data; however, we can obtain an estimate of

$$\boldsymbol{\pi} = \left[\mathbf{I}_t - \left(\frac{1}{t}\right)\mathbf{J}_t\right]\Sigma\left[\mathbf{I}_t - \left(\frac{1}{t}\right)\mathbf{J}_t\right].$$

Because of the form of $\boldsymbol{\pi}$, the value of θ is simplified as

$$\theta_\pi = \frac{t^2(\bar{\pi}_{ii})^2}{\left[(t-1)\Sigma_{i=1}^t \Sigma_{j=1}^t \pi_{ij}^2\right]}$$

where π_{ij} is the ijth element of the matrix $\boldsymbol{\pi}$ and $\bar{\pi}_{ii}$ is the mean of the diagonal elements of $\boldsymbol{\pi}$.

Thus, we can obtain an estimate of θ_π from $\hat{\boldsymbol{\pi}}$, but we cannot estimate Σ and hence cannot estimate θ. However, two results can be shown: (1) Σ satisfies the H–F condition if and only if $\boldsymbol{\pi}$ satisfies the H–F condition, and (2) $\theta = \theta_\pi$. Therefore, we can estimate θ by $\hat{\theta}_\pi$, which can be computed from the estimate of $\boldsymbol{\pi}$. We shall use the data from Examples 26.1 and 26.3 to demonstrate the computations.

For Example 26.1, an estimate of ε_{ijk} is

$$\hat{\varepsilon}_{ijk} = y_{ijk} - \bar{y}_{ij\cdot} - \bar{y}_{i\cdot k} + \bar{y}_{i\cdot\cdot}.$$

Let

$$\hat{\boldsymbol{\varepsilon}}_{ij} = (\hat{\varepsilon}_{ij1}, \hat{\varepsilon}_{ij2}, \ldots, \hat{\varepsilon}_{ijt})'.$$

The sample variance of the random vector $\hat{\boldsymbol{\varepsilon}}_{ij}$ provides an estimate of

$$\boldsymbol{\pi} = \left[\mathbf{I}_t - \left(\frac{1}{t}\right)\mathbf{J}_t\right]\Sigma\left[\mathbf{I}_t - \left(\frac{1}{t}\right)\mathbf{J}_t\right].$$

In particular,

$$\pi_{kk'} = \frac{\Sigma_{i=1}^3 \Sigma_{j=1}^8 \hat{\varepsilon}_{ijk}\hat{\varepsilon}_{ijk'}}{23}.$$

The estimate of $\boldsymbol{\pi}$ for Example 26.1 is

$$\hat{\boldsymbol{\pi}} = \begin{bmatrix} 5.4556 & -0.3556 & -0.9671 & -4.1328 \\ -0.3556 & 5.1974 & -1.5825 & -3.2592 \\ -0.9671 & -1.5825 & 2.4528 & 0.0968 \\ -4.1328 & -3.2592 & 0.0968 & 7.2952 \end{bmatrix},$$

and the estimate of θ is

$$\hat{\theta} = \frac{16(5.100)^2}{3(178.566)} = .777.$$

This value of $\hat{\theta}$ is applied to those sums of squares that are based on within-person comparisons. No adjustment is necessary for the between-person comparisons. The adjusted degrees of freedom are given in Table 27.4.

Example 26.3 presents another complexity, since there are three sizes of experimental units and there is a need to make a $\hat{\theta}$ adjustment for the within-family and the within-person comparisons. There is no adjustment for between-family comparisons.

For the within-person comparisons, the error vector is $\boldsymbol{\varepsilon}_{ijk}$, which has variance $\text{Var}(\boldsymbol{\varepsilon}_{ijk}) = \boldsymbol{\Sigma}$. An estimate of ε_{ijkm} is obtained from the residuals of the model in Equation (26.1.4). Again, we cannot estimate $\boldsymbol{\Sigma}$, but we can estimate

$$\boldsymbol{\pi} = \left[\mathbf{I}_t - \left(\frac{1}{t}\right)\mathbf{J}_t\right]\boldsymbol{\Sigma}\left[\mathbf{I}_t - \left(\frac{1}{t}\right)\mathbf{J}_t\right].$$

An estimate of $\pi_{mm'}$ is

$$\hat{\pi}_{mm'} = \frac{\sum_{i=1}^{2}\sum_{j=1}^{n_i}\sum_{k=1}^{3}\hat{\varepsilon}_{ijkm}\hat{\varepsilon}_{ijkm'}}{50}.$$

The estimate of $\hat{\boldsymbol{\pi}}$ is

$$\hat{\boldsymbol{\pi}} = \begin{bmatrix} 0.251746 & -0.106444 & -0.145302 \\ -0.0106444 & 0.196508 & -0.090063 \\ -0.145302 & -0.090063 & 0.235365 \end{bmatrix}$$

and the value of $\hat{\theta}$ is

$$\hat{\theta} = \frac{9(.0519)}{2(.2385)} = .980.$$

Table 27.4 Actual and Adjusted Degrees of Freedom for Data in Example 26.1

Source of Variation	*Actual df*	*$\hat{\theta}$-adjusted df*
Drug	2	2*
Error(Person)	21	21*
Time	3	2.33
Time * Drug	6	4.66
Error(Time)	63	48.95

*Unchanged (between-person comparison).

The within-person part of the analysis is hardly affected by the θ-adjusted degrees of freedom and thus is not presented. All degrees of freedom for terms involving Time and the Error(Time) term would be multiplied by .980.

For the within-family comparisons, the error vector is $\mathbf{s}_{ij}$ where

$$s_{ijk} = p_{ijk} + \bar{\varepsilon}_{ijk\cdot}.$$

The variance of $\mathbf{s}_{ij}$ is

$$\begin{aligned}\text{Var}(\mathbf{s}_{ij}) &= \text{Var}(\mathbf{p}_{ij}) + \text{Var}(\bar{\boldsymbol{\varepsilon}}_{ij})\\ &= \Lambda + \left(\frac{1}{n}\right)\mathbf{j}'\Sigma\mathbf{j}\mathbf{I}\\ &= \Delta.\end{aligned}$$

We cannot obtain an estimate of Δ, but we can obtain an estimate of

$$\Gamma = \left[\mathbf{I}_p - \left(\frac{1}{p}\right)\mathbf{J}_p\right]\Delta\left[\mathbf{I}_p - \left(\frac{1}{p}\right)\mathbf{J}_p\right]$$

where p is the number of people in a family (in this case, $p = 3$). The s_{ijk}'s are estimated from the residuals of the model

$$\bar{y}_{ijk\cdot} = \bar{\mu}_{ik\cdot} + f_{ij} + s_{ijk},$$

and the elements of Γ are estimated by

$$\gamma_{kk'} = \frac{\Sigma_i \Sigma_j \hat{s}_{ijk}\hat{s}_{ijk'}}{16}.$$

The estimate of Γ is

$$\hat{\Gamma} = \begin{array}{c} \begin{matrix}\text{Son} & \text{Father} & \text{Mother}\end{matrix} \\ \begin{bmatrix} 7.59086 & -2.65983 & -4.93103 \\ -2.65983 & 2.72906 & -0.069224 \\ -4.93103 & -0.069224 & 5.00025 \end{bmatrix}\end{array},$$

and the estimate of θ is

$$\hat{\theta} = \frac{9(5.107)^2}{2(152.861)} = .768.$$

The corresponding $\hat{\theta}$-adjusted degrees of freedom are given in Table 27.5.

This example shows how to apply the $\hat{\theta}$ correction to the various levels of an analysis. It may not be necessary to adjust the degrees of freedom, particularly if $\hat{\theta}$ is close to 1. Box (1954) provided a statistic to test H_0: $\theta = 1$ versus H_a: $\theta < 1$. Box showed that under the H–F conditions, the statistic

$$q = (1 - A)^* M$$

Table 27.5 Actual Adjusted Degrees of Freedom for Data in Example 26.3

Source of Variation	*Actual df*	$\hat{\theta}$ *adjusted df*
Person	2	1.536
Area * Person	2	1.536
Error(Person)	30	23.040

is approximately distributed as a chi-square random variable with f degrees of freedom where

$$M = -v \log_e(\hat{\theta}),$$

$$A = \frac{t(t+1)^2(2t-3)}{6v(t-1)(t^2+t-4)},$$

$$f = \frac{t^2+t-4}{2},$$

and v is the degrees of freedom associated with the estimated elements in Σ.

The chi-square approximation is less accurate for large t and small v, but it can be used to check for deviations from the H–F condition. If $q > \chi^2_{\alpha,f}$, then a correction should be used. If $g \leq \chi^2_{\alpha,f}$, then the uncorrected F-tests are most likely valid.

The values of the test statistics for Examples 26.1 and 26.3 have been computed and are listed below.

Heart Rate Example (Example 26.1):

$$v = 63, \quad \hat{\theta} = .777, \quad t = 4, \quad m = 15.896, \quad A = .0275,$$
$$q = 15.458, \quad f = 8, \quad \chi^2_{.05,8} = 15.51$$

Within-Family Analysis for Family Attitude Example (Example 26.3):

$$v = 30, \quad \hat{\theta} = .768, \quad t = 3, \quad m = 7.919, \quad A = .05,$$
$$q = 7.523, \quad f = 4, \quad \chi^2_{.05,4} = 9.49$$

Within-Person Analysis for Family Attitude Example (Example 26.3):

$$v = 90, \quad \hat{\theta} = .980, \quad t = 3, \quad m = 1.818, \quad a = .017,$$
$$q = 1.788, \quad f = 4, \quad \chi^2_{.05,4} = 9.49$$

Example 26.1 has a significance level close to .05, indicating that a $\hat{\theta}$ correction is appropriate; the tests for the two levels of analysis in the family attitude example, Example 26.3, indicate that a $\hat{\theta}$ correction may not be necessary.

When $\hat{\Sigma}$ has full rank, there is a test to see if Σ satisfies the H–F conditions. Compute $\mathbf{W} = \mathbf{C}\hat{\Sigma}_0\mathbf{C}'$ where $\mathbf{C}$ is a $(t-1) \times t$ matrix of orthogonal contrasts. If Σ satisfies the H–F condition, then

$$\mathbf{C}\Sigma\mathbf{C}' = \theta \mathbf{I}_{t-1}$$

for some θ. The test statistic is

$$u = \frac{|\mathbf{W}|}{Z}$$

where $Z = \text{Tr}(\mathbf{W})^{t-1}$ and $|\mathbf{W}|$ denotes the determinants of the matrix $\mathbf{W}$.

If the value of u exceeds a tabulated percentage point, then a multivariate analysis should be used on the data. If the value of u is less than the percentage point, then the usual analysis, described in Section 26.2, is probably appropriate.

The computations of the u-statistic are demonstrated below for the within-family comparisons of Example 26.3.

$$\hat{\pi} = (\mathbf{I}_3 - q\mathbf{J})\hat{\Delta}(\mathbf{I}_3 - q\mathbf{J}) = \begin{bmatrix} 7.6 & -2.7 & -4.9 \\ -2.7 & 2.7 & -.1 \\ -4.9 & -.1 & 5.0 \end{bmatrix}$$

$$\text{where} \quad q = \frac{1}{3},$$

$$\mathbf{C}\hat{\Delta}\mathbf{C}' = \begin{bmatrix} .707 & -.707 & 0 \\ .408 & .408 & -.816 \end{bmatrix} \hat{\pi} \begin{bmatrix} .707 & .408 \\ -.707 & .408 \\ 0 & -.816 \end{bmatrix}$$

$$= \begin{bmatrix} 7.85 & 4.19 \\ 4.19 & 7.85 \end{bmatrix} = \mathbf{W},$$

$|\mathbf{W}| = 41.16$, $\text{Tr}(\mathbf{W}) = 15.33$, and

$$u = \frac{41.16}{(15.33)^2} = .175.$$

Computing the value of θ in order to adjust the degrees of freedom is not that easy. Greenhouse and Geisser (1959) developed a three-step procedure that can prevent having to compute the value of θ. The three-step procedure is as follows:

1. Compare the F-ratio in question to the percentage point with the usual degrees of freedom. If it is not significant, stop—the adjusted degrees of freedom test will also be nonsignificant.
2. Compare the F-ratio in question to the percentage point with the conservative Box correction degrees of freedom. If it is significant, stop—the adjusted degrees of freedom test will also be significant.
3. If the F-ratio is significant with the usual degrees of freedom and nonsignificant with the conservative Box correction degrees of free-

dom, then the θ-adjusted degrees of freedom must be computed to make a decision.

The three-step procedure is an excellent way to reduce the number of computations, although with existing statistical computing packages, the residuals from fitting various models can be computed, output into a file, and then used to compute the estimates of the various covariance matrices. The three-step procedure can be used, and if either step 1 or 2 stops the process, you do not have to be concerned with the θ-adjusted degrees of freedom.

The three-step procedure can also be applied to multiple comparisons and the investigation of contrasts. If two means are not significantly different for the usual degrees of freedom, then they will not be significantly different for the Box conservative degrees of freedom; and if two means are significantly different for the Box conservative degrees of freedom then they will be significantly different for the θ-adjusted degrees of freedom. Contrasts such as those that measure linear, quadratic, and similar trends can also be investigated by using the same strategy.

When the H–F condition is not satisfied for the repeated measures of an experiment, an alternative analysis can be constructed. If there is certain contrast of the repeated measures of interest, then that contrast can be computed for each subject and an analysis of variance carried out on those contrasts. For Example 26.1, we can construct a measure of the linear trend over time where the contrast

$$Z_{ij} = -3y_{ij1} - y_{ij2} + y_{ij3} + 3y_{ij4}$$

is computed for each subject. The model for Z_{ij} is

$$Z_{ij} = \mu_i + s_{ij}$$

Table 27.6 Analysis of Variance Table for a Contrast of the Repeated Measures in Example 26.1

Source of Variation	*df*	*SS*	*EMS*
Drug	$D-1$	$n\sum_{i=1}^{d}(\bar{Z}_{i\cdot} - \bar{Z}_{\cdot\cdot})$	$\sigma_s^2 + Q$
Error(Subject)	$d(n-1)$	$\sum_{i=1}^{d}\sum_{j=1}^{n}(Z_{ij} - \bar{Z}_{i\cdot})^2$	σ_s^2

Note: $Z_{ij} = C_1 y_{ij1} + C_2 y_{ij2} + \cdots + C_t y_{ijt}$.

Table 27.7 Analysis of Variance Table for the Linear Trends in Example 26.1

Source of Variation	*df*	*SS*	*MS*	*F*
Drug(Linear)	2	872.58	436.29	40.47
Error(Subject, Linear)	21	226.38	10.78	

where μ_i is the mean response of the linear trend across Time for drug i and s_{ij} is the error term for subjects with variance σ_s^2. The analysis of variance table for d drugs and n subjects per drug that provides the analysis of a given contrast

$$Z_{ij} = C_1 y_{ij1} + C_2 y_{ij2} + \cdots + C_t y_{ijt}$$

where

$$C_1 + C_2 + \cdots + C_t = 0$$

is given in Table 27.6. The resulting model for the contrast computed for each subject corresponds to a model for a one-way treatment structure in a completely randomized design structure. The F-ratio tests the hypothesis that the contrast has the same value for all drugs.

The analysis of variance tables for the linear trend and the quadratic trend for the heart rate data in Example 26.1 are in Tables 27.7 and 27.8, respectively. The F-ratio in Table 27.7 tests that the linear components are equal for all drugs, and the F-ratio in Table 27.8 tests that the quadratic components are equal for all drugs. The drug means contrasts can be compared by means of a multiple comparison procedure. Those multiple comparisons are given in Tables 27.9 and 27.10.

The final topic in this section concerns the problem of autocorrelation. When the repeated measures are over time, the errors can be

Table 27.8 Analysis of Variance Table for Quadratic Trends in Example 26.1

Source of Variation	*df*	*SS*	*MS*	*F*
Drug(Quadratic)	2	360.75	180.38	4.77
Error(Subject, Quadratic)	21	793.87	37.80	

Table 27.9 Drug Means of the Linear Trend in Example 26.1

Drug	*Mean*
AX23	10.500 (a)
BWW9	−3.1250 (b)
Control	−1.250 (b)

Notes: $LSD_{.05} = 3.414$. Means with the same letter are not significantly different.

correlated through an autoregressive error structure. The covariance matrix corresponding to errors with a first-order autoregressive process is

$$\Sigma = \frac{\sigma^2}{1-\rho^2}\begin{bmatrix} 1 & \rho & \rho^2 & \rho^3 & \cdots & \rho^{t-1} \\ \rho & 1 & \rho & \rho^2 & \cdots & \rho^{t-2} \\ \rho^2 & \rho & 1 & \rho & \cdots & \rho^{t-3} \\ \rho^3 & \rho^2 & \rho & 1 & \cdots & \rho^{t-4} \\ \vdots & \vdots & \vdots & \vdots & \vdots & \vdots \\ \rho^{t-1} & \rho^{t-2} & \rho^{t-3} & \rho^{t-4} & \cdots & 1 \end{bmatrix}$$

The method discussed in Section 27.2 provides an analysis that incorporates the information about the autocorrelation where there is first-order autocorrelation. The θ-adjusted test statistics seem to be quite appropriate whatever the error structure, but a simulation study (Albohali, 1983) has shown that the method in Section 27.2 is more powerful than the θ-adjusted test.

Table 27.10 Drug Means of the Quadratic Trend in Example 26.1

Drug	*Mean*
AX23	−9.500 (a)
BWW9	−7.625 (a)
Control	−.500 (b)

Notes: $LSD_{.05} = 6.393$. Means with the same letter are not significantly different.

27.2 A TWO-FACTOR MODEL WITH AUTOCORRELATED ERRORS

In many experiments, a set of treatments is applied in sequence to the same subject over a specified period of time. The response of the subject to one treatment is observed before the next treatment is administered, and the process is continued until all treatments have been observed. Often the sequence of treatments are the times; that is, time is a factor in the treatment structure. When the responses of a single subject are measured sequentially, the errors are often correlated through an autocorrelated error structure rather than an error structure satisfying the H–F condition.

To develop this analysis, we shall consider a two-factor repeated measures experiment. The experiment consists of randomly assigning subjects to the levels of factor A and then applying the levels of factor B to each subject in some predetermined sequence.

A model to describe such data is

$$y_{ijkm} = \mu_{ik} + p_{ij} + \varepsilon_{ijm}$$

where $i = 1, 2, \ldots, a$; $j = 1, 2, \ldots, n$; $k = 1, 2, \ldots, b$; and $m = 1, 2, \ldots, b$. In this model, the number of time periods is equal to the number of levels of factor B, but that need not be the case, as shown by a three-period crossover design in Chapter 32. In the model, y_{ijkm} denotes the response from the jth subject assigned to the ith level of factor A and where the kth level of factor B was assigned to the mth time interval. The parameter μ_{ik} denotes the expected response of the $A_i \times B_k$ treatment combination. The p_{ij} terms denote the subjects' errors, which are assumed to be distributed i.i.d. $N(0, \sigma_p^2)$, and the ε_{ijm}'s denote the errors associated with the measurement at time m.

An example illustrating data from a levels of factor A, four levels of factor B, and four times is displayed in Table 27.11. The symbols B_k in the table denote the level of treatment B assigned to the corresponding subject at the given time.

Several different types of time series models might be appropriate for describing the within-subject error structure. We shall restrict our discussion here to a first-order autoregressive model, which is the only case considered in this book.

The assumptions about the error structure of the ε_{ijm}'s for the first-order autoregressive model are that the ε_{ijm}'s can be expressed as

$$\varepsilon_{ijm} = \rho\varepsilon_{ijm-1} + v_{ijm}$$

where $v_{ijm} \sim$ i.i.d. $N(0, \sigma_v^2)$ for all i, j, and m and

$$\varepsilon_{ij0} \sim \text{i.i.d. } N\left(0, \frac{\sigma_v^2}{1 - \rho^2}\right)$$

for all i and j. The corresponding covariance matrix for the errors

Table 27.11 Data Array for a Two-Way Repeated Measures Design

		TIME INTERVAL			
Level of A	*Subject*	*1*	*2*	*3*	*4*
	1	B_1	B_3	B_2	B_4
A_1	2	B_2	B_1	B_3	B_4
	⋮	⋮	⋮	⋮	⋮
	n	B_4	B_1	B_2	B_3
(Data arrays for levels A_2 through A_{a-1} appear here).					
	1	B_3	B_4	B_2	B_1
A_a	2	B_2	B_1	B_4	B_3
	⋮	⋮	⋮	⋮	⋮
	n	B_2	B_3	B_4	B_1

Note: This example shows an equal number of subjects assigned to each level of A, but the analysis works for unequal numbers of subjects.

associated with the data from the jth subject assigned to the ith level of A is

$$\operatorname{Var}(\boldsymbol{\varepsilon}_{ij}) = \frac{\sigma_v^2}{1-\rho^2}\begin{bmatrix} 1 & \rho & \rho^2 & \rho^3 & \cdots & \rho^{t-1} \\ \rho & 1 & \rho & \rho^2 & \cdots & \rho^{t-2} \\ \rho^2 & \rho & 1 & \rho & \cdots & \rho^{t-3} \\ \rho^3 & \rho^2 & \rho & 1 & \cdots & \rho^{t-4} \\ \vdots & \vdots & \vdots & \vdots & \vdots & \vdots \\ \rho^{t-1} & \rho^{t-2} & \rho^{t-3} & \rho^{t-4} & \cdots & 1 \end{bmatrix}$$

$$= \frac{\sigma_v^2}{1-\rho^2}\mathbf{R}.$$

The first-order autoregressive error structure is also referred to as being *serially correlated.*

The experiment has two sizes of experimental units, the larger size corresponding to the subjects assigned to the different levels of A and the smaller to the within-subject time intervals to which the levels of B are assigned. Thus the analysis of the two-factor model has two levels, a between-subject analysis and a within-subject or time interval analysis.

The between-subject analysis is carried out by using the subject means (observed) and their corresponding model. The subject means model is

$$\begin{aligned}\bar{y}_{ij\cdot} &= \bar{\mu}_{i\cdot} + p_{ij} + \bar{\varepsilon}_{ij\cdot} \\ &= \bar{\mu}_{i\cdot} + p^*_{ij}\end{aligned} \qquad (27.2.1)$$

where $p^*_{ij} = p_{ij} + \bar{\varepsilon}_{ij\cdot}$. The p^*_{ij}'s are distributed i.i.d. $N(0, \sigma^2_{p*})$ where

$$\sigma^2_{p*} = \sigma^2_e + \frac{1}{b} \cdot \frac{\sigma^2_v}{1 - \rho^2} q$$

where q is the sum of all of the elements in the matrix **R**. The model in equation (27.2.1) is that of a one-way treatment structure in a completely randomized design structure with independent observations with variance σ^2_{p*}. The corresponding between-subject analysis of variance table is displayed in Table 27.12.

The statistic to test for equal A effects is

$$F_A = \frac{\text{MSA}}{\text{MSERROR(SUBJECT)}},$$

which is distributed as an F with $a - 1$ and $a(n - 1)$ degrees of freedom. This analysis points out that irrespective of the form of the error structure for the ε_{ij}'s' there is an exact analysis for the between-subject analysis.

The estimate of the difference between two levels of A is

$$\widehat{\bar{\mu}_{i\cdot} - \bar{\mu}_{i'\cdot}} = \bar{y}_{i\cdot\cdot} - \bar{y}_{i'\cdot\cdot},$$

Table 27.12 Between-Subject Analysis of Variance Table for Two-Factor Model with Autocorrelated Errors

Source of Variation	*df*	*SS*	*EMS*
A	$a - 1$	$n \sum_{i=1}^{a} (\bar{y}_{i\cdot\cdot} - \bar{y}_{\cdot\cdot\cdot})^2$	$\sigma^2_{p*} + Q_A$
Error(Subject)	$a(n - 1)$	$\sum_{i=1}^{a} \sum_{j=1}^{n} (\bar{y}_{ij\cdot} - \bar{y}_{i\cdot\cdot})^2$	σ^2_{p*}

Note: Q_A is the noncentrality parameter associated with the sum of squares for factor A.

which has variance

$$\mathrm{Var}(\bar{y}_{i\cdot\cdot} - \bar{y}_{i'\cdot\cdot}) = \frac{2\sigma_{p*}^2}{n}.$$

The estimated standard error of the difference between means comparing two levels of A is

$$\widehat{\mathrm{s.e.}}(\bar{y}_{i\cdot\cdot} - \bar{y}_{i'\cdot\cdot}) = \sqrt{\frac{2\mathrm{MSERROR(SUBJECT)}}{n}}.$$

An LSD type of multiple comparison method for comparing the levels of A can be constructed by using

$$\mathrm{LSD} = t_{\alpha/2,a(n-1)}\widehat{\mathrm{s.e.}}(\bar{y}_{i\cdot\cdot} - \bar{y}_{i'\cdot\cdot}).$$

If appropriate, contrasts between the levels of A can also be investigated.

While there is an exact between-subject analysis, there is no exact within-subject analysis when the error structure follows a first-order autoregressive model. However, an approximate analysis (Albohali, 1983) is available that, with the aid of computer simulation experiments, has been shown to be appropriate for analyzing repeated measures data satisfying a first-order autoregressive structure. However, when there are many more subjects per level of A than there are time intervals, the multivariate approach described in Chapter 31 is more appropriate.

Albohali's method uses the maximum likelihood estimate of ρ and the least squares estimates of the ε_{ijm}'s to "filter" the data; in other words, it strives to remove the autocorrelation between the time intervals within subjects before carrying out the within-subject analysis. Then a usual repeated measures analysis, such as that described in Section 26.2, is used on the filtered data to complete the within-subject analysis.

To help motivate this approximate analysis, suppose that the ε_{ijm}'s and ρ are known; then

$$X_{ijkm} = y_{ijkm} - \rho\varepsilon_{ijm-1} = \mu_{ik} + p_{ij} + v_{ijm}$$

where the errors $p_{ij} \sim$ i.i.d. $N(0, \sigma_p^2)$ and $v_{ijm} \sim$ i.i.d. $N(0, \sigma_v^2)$. The model for X_{ijkm} has exactly the assumptions necessary for performing the usual analysis discussed in Section 26.2 for an error structure satisfying the H–F condition (in fact, the errors are independent for this case). Since ρ and the ε_{ijm}'s are unknown, an approximate analysis requires estimates of ρ and the ε_{ijm}'s, which can be used for estimating the X_{ijkm}'s as follows:

$$\hat{X}_{ijkm} = y_{ijkm} - \hat{\rho}\hat{\varepsilon}_{ijk}.$$

The first step in Albohali's method is to obtain least squares estimates of the ε_{ijm}'s by fitting the model

$$y_{ijkm} = \mu_{ik} + p_{ij} + \varepsilon_{ijm}$$

and computing

$$\hat{\varepsilon}_{ijk} = y_{ijkm} - \hat{\mu}_{ik} - \hat{p}_{ij}.$$

Next, an approximation to the maximum likelihood estimate of ρ is obtained (Hasza, 1980) as

$$\hat{\rho} = \frac{\frac{2}{3}\left(A_2^2 - 3A_1A_3\right)^{1/2}\cos(\theta) - \frac{1}{3}A_2}{A_3}$$

where

$$\theta = \frac{1}{3}\arccos\left[-\frac{1}{2}\left(2A_2^3 - 9A_1A_2A_3 + 27A_0A_3^2\right)\left(A_2^2 - 3A_1A_3\right)^{-3/2}\right] + \frac{4\pi}{3},$$

$$A_3 = \left(\frac{b-1}{b}\right)\sum_{i=1}^{a}\sum_{j=1}^{n}\sum_{m=2}^{b-1}\hat{\varepsilon}_{ijm}^2,$$

$$A_2 = \left(\frac{2-b}{b}\right)\sum_{i=1}^{a}\sum_{j=1}^{n}\sum_{m=2}^{b}\hat{\varepsilon}_{ijm}\hat{\varepsilon}_{ijm-1},$$

$$A_1 = \left(\frac{b}{b-1}\right)A_3 - \frac{1}{b}\sum_{i=1}^{a}\sum_{j=1}^{n}\sum_{m=1}^{b}\hat{\varepsilon}_{ijm}^2,$$

and

$$A_0 = -\frac{b}{b-2}A_2.$$

An approximation to the maximum likelihood estimate of σ_v^2 is

$$\sigma_v^2 = \frac{1}{anb}\left[\left(1-\hat{\rho}^2\right)\sum_{i=1}^{a}\sum_{j=1}^{n}\hat{\varepsilon}_{ij1}^2 + \sum_{i=1}^{a}\sum_{j=1}^{n}\sum_{m=2}^{b}\left(\hat{\varepsilon}_{ijm} - \hat{\rho}\hat{\varepsilon}_{ijm-1}\right)^2\right]$$

Next, the filtered data are determined as

$$\hat{X}_{ijkm} = y_{ijkm} - \hat{\rho}\hat{\varepsilon}_{ijm-1}$$

for $i = 1, 2, \ldots, a$; $j = 1, 2, \ldots, n$; $k = 1, 2, \ldots, b$; and $m = 1, 2, \ldots, b$ where $\hat{\varepsilon}_{ij0} = 0$. The approximating model for the filtered data is

$$\hat{X}_{ijkm} = \mu_{ij} + p_{ij} + v_{ijm}.$$

Finally, we use the $\hat{X}_{ijkm}$'s as data and do the usual within-subject analysis, as described in Section 26.2. The within-subject analysis of variance table is displayed in Table 27.13.

If there is no $A * B$ interaction, then the levels of treatment B can be compared by

$$\widehat{\mu_{\cdot k} - \mu_{\cdot k'}} = \hat{\bar{X}}_{\cdot\cdot k\cdot} - \hat{\bar{X}}_{\cdot\cdot k'\cdot},$$

Table 27.13 Within-Subject Analysis of Variance Table Based on the Filtered Data

Source of Variation	*df*	*SS*	*EMS*
B	$b-1$	$an\sum_{k=1}^{b}(\hat{\bar{X}}_{..k.}-\hat{\bar{X}}_{....})^2$	$\sigma_v^2+Q_B$
$A*B$	$(a-1)(b-1)$	$n\sum_{i=1}^{a}\sum_{k=1}^{b}(\hat{\bar{X}}_{i\cdot k\cdot}-\hat{\bar{X}}_{i...}-\hat{\bar{X}}_{..k.}+\hat{\bar{X}}_{....})^2$	$\sigma_v^2+Q_{AB}$
Error(Time)	$a(n-1)(b-1)$	$\sum_{i=1}^{a}\sum_{j=1}^{n}\sum_{k=1}^{b}(\hat{\bar{X}}_{ijk.}-\hat{\bar{X}}_{i\cdot k\cdot}-\hat{\bar{X}}_{ij..}+\hat{\bar{X}}_{i...})^2$	σ_v^2

Notes: Q_B and Q_{AB} are the noncentrality parameters associated with the respective sums of squares. The expected mean squares are based on the approximating model for the filtered data.

which has variance $2\sigma_v^2/an$. An LSD type multiple comparison can be constructed by using

$$\text{LSD} = \left[t_{\alpha/2,a(n-1)(b-1)}\right]\sqrt{\frac{2\text{MSERROR(TIME)}}{an}}.$$

If there is an $A*B$ interaction, then the levels of B can be compared at each level of A by

$$\widehat{\mu_{ik}-\mu_{ik'}}=\hat{\bar{X}}_{i\cdot k\cdot}-\hat{\bar{X}}_{i\cdot k'\cdot},$$

which has variance $2\sigma_v^2/n$. An LSD type multiple comparison can be constructed to compare the levels of B within each level of A by

$$\text{LSD} = \left[t_{\alpha/2,a(n-1)(b-1)}\right]\sqrt{\frac{2\text{MSERROR(TIME)}}{n}}.$$

To compare the levels of A at the same or different levels of B, a between-subject comparison must be combined with a within-subject comparison, as described in Section 24.3. The comparison of interest is $\mu_{ik}-\mu_{i'k}$, which can be expressed as

$$\begin{aligned}\mu_{ik}-\mu_{i'k} &= (\bar{\mu}_{i\cdot}-\bar{\mu}_{i'\cdot})+\left[(\mu_{ik}-\bar{\mu}_{i\cdot})-(\mu_{i'k}-\bar{\mu}_{i'\cdot})\right]\\ &= \theta+\phi\end{aligned}$$

where θ is a between-subject comparison and ϕ is a within-subject comparison. The estimate of θ is

$$\hat{\theta}=\bar{y}_{i...}-\bar{y}_{i'...}$$

(from the between-subject means model for unfiltered data), which has variance

$$\mathrm{Var}(\theta) = 2\sigma_{p*}^2/n.$$

The estimate of ϕ is

$$\hat{\phi} = \left(\bar{\hat{X}}_{i\cdot k\cdot} - \bar{\hat{X}}_{i\cdots}\right) - \left(\bar{\hat{X}}_{i'\cdot k\cdot} - \bar{\hat{X}}_{i'\cdots}\right)$$

which has approximate variance

$$\mathrm{Var}(\hat{\phi}) = \frac{2(b-1)\sigma_v^2}{b}.$$

Since θ is obtained from the between-subject part of the analysis and ϕ from the within-subject part, the variance of $\hat{\theta} + \hat{\phi}$ is approximated by

$$\mathrm{Var}(\hat{\theta} + \hat{\phi}) = \mathrm{Var}(\hat{\theta}) + \mathrm{Var}(\hat{\phi}) = \frac{2\sigma_{p*}^2}{n} + \frac{2(b-1)}{bn}\sigma_v^2.$$

An estimate of $\mathrm{Var}(\hat{\theta} + \hat{\phi})$ is

$$\mathrm{Var}(\widehat{\hat{\theta} + \hat{\phi}}) = \frac{2}{n}\mathrm{MSERROR(SUBJECT)} + \frac{2(b-1)}{bn}\mathrm{MSERROR(TIME)}.$$

Since the two mean squares are based on different degrees of freedom, an approximate critical point t^* can be constructed (by using the method of Section 24.3) as

$$t^* = \frac{[t_{\alpha/2,a(n-1)}]\frac{2}{n}\mathrm{MSERROR(SUBJECT)} + [t_{\alpha/2,a(n-1)(b-1)}]\frac{2(b-1)}{bn}\mathrm{MSERROR(TIME)}}{\frac{2}{n}\mathrm{MSERROR(SUBJECT)} + \frac{2(b-1)}{bn}\mathrm{MSERROR(TIME)}}.$$

An LSD value for comparing the levels of A at the same or different levels of B is

$$\mathrm{LSD} = t^*\sqrt{\mathrm{Var}(\widehat{\theta + \phi})}.$$

The description of the analysis of a two-way repeated measures design with a first-order autoregressive error structure can be applied to design structures with more complex between-subject treatment structures or within-subject design structures. In any event, there will be two levels of the analyses where the between-subject analysis provides an exact F-test and the within-subject analysis, which is based on the filtered data, provides approximate F-tests. If ρ is zero or $\hat{\rho}$ is close to zero, then the filtered values are not very different from the observed values, and the within-subject analysis will be very close to the usual analysis on the observed data. To conclude this section, we shall study an example that demonstrates the above ideas.

EXAMPLE 27.1

This experiment involved studying the effect of a dose of a drug on the growth of rats. The data in Table 27.14 consists of the growth of 50 rats, where 10 rats were randomly assigned to each of the five doses of the drug. The weights were obtained each week for 11 weeks.

These data conform to the pattern of an autocorrelated error structure. The model used for the analysis is

$$y_{ijk} = \mu_{ik} + p_{ij} + \varepsilon_{ijk}$$

where μ_{ik} is the expected mean response to dose i at time k, p_{ij} is the animal error term, and ε_{ijk} is the time interval error term.

The animal means model,

$$\bar{y}_{ij\cdot} = \bar{\mu}_{i\cdot} + p_{ij}^{*},$$

is used for the between-animal analysis, which is in Table 27.15. Thus, the residuals of the complete model are obtained by

$$\hat{\varepsilon}_{ijk} = y_{ijk} - \bar{y}_{ij\cdot} - \bar{y}_{i\cdot k} + \bar{y}_{i\cdot\cdot},$$

and the estimate of the autocorrelation can be obtained. The estimate of the autocorrelation is $\hat{\rho} = .604$. Next, the data are filtered by

$$\hat{x}_{ijk} = y_{ijk} - \hat{\rho}\hat{\varepsilon}_{ijk-1},$$

and the within-animal analysis is done. The results of the within-animal analysis on the filtered data are in Table 27.16. Since there is a significant interaction between Time and Dose, contrasts and multiple comparisons are made of the Dose∗Time means. The comparisons of times at each dose and of the doses at each time are shown in Tables 27.17 and 27.18.

The value of $\hat{\theta}$ for these data is $\hat{\theta} = .2634$. The $\hat{\theta}$-adjusted degrees of freedom for Time, Time∗Dose, and Error(Time) are 2.6, 10.5, and 118.5, respectively. If this adjustment was made for the unfiltered within-animal analysis, none of the decisions from the filtered analysis would change. However, the multiple comparisons would be computed based on the $\hat{\theta}$-adjusted degrees of freedom, meaning that the LSD values (or values from some other multiple comparison method) would be larger and fewer differences would be found. The analysis of variance table based on the unfiltered data is given in Table 27.19.

The between-animal sums of squares in Table 27.15 are 11 times larger than the corresponding sums of squares in Table 27.16 because those in Table 27.15 are based on the means of 11 observations (there are 11 time periods). It is important to note that in comparing the two tables, MSERROR(TIME) is smaller for the filtered data than for the unfiltered

Table 27.14 Body Weights of Rats in Example 27.1

		WEEK										
DOSE	RAT	*1*	*2*	*3*	*4*	*5*	*6*	*7*	*8*	*9*	*10*	*11*
0	1	54	60	63	74	77	89	93	100	108	114	124
	2	69	75	81	90	97	120	114	119	126	138	143
	3	77	81	87	94	101	110	117	124	134	141	151
	4	64	69	77	83	88	96	104	109	120	123	131
	5	51	58	62	71	74	81	88	93	99	103	113
	6	64	71	77	89	90	100	106	114	122	134	139
	7	80	91	97	101	111	119	129	131	137	147	154
	8	79	85	89	99	104	105	116	121	132	139	147
	9	77	82	88	92	101	109	119	127	135	144	158
	10	79	84	91	98	107	114	119	131	137	146	155
.5	1	62	71	75	79	87	91	100	105	111	121	124
	2	68	73	81	89	94	101	110	114	123	132	139
	3	94	102	109	110	128	133	147	151	153	171	184
	4	81	90	95	102	109	120	128	137	141	154	160
	5	64	69	72	76	84	89	97	103	108	114	124
	6	67	74	81	81	84	95	100	109	119	128	130
	7	73	80	86	89	97	101	110	116	117	135	141
	8	71	74	82	84	93	97	102	113	119	124	131
	9	69	74	79	89	94	100	107	113	124	134	139
	10	60	62	67	74	78	85	92	103	112	121	130
1	1	59	63	66	75	80	87	99	104	110	115	124
	2	56	66	70	81	77	88	96	100	113	120	130
	3	71	77	84	80	97	106	111	109	128	133	140
	4	59	64	69	76	85	88	96	104	110	119	126
	5	65	70	73	77	85	92	96	101	111	118	121
	6	61	69	77	81	89	92	107	111	118	127	132
	7	80	86	95	99	106	113	127	131	142	150	160
	8	74	80	84	90	99	101	108	117	126	133	140
	9	71	79	88	90	98	102	116	121	127	139	142
	10	69	75	80	86	96	97	104	113	122	129	138

Table 27.14 Continued

		WEEK										
DOSE	RAT	*1*	*2*	*3*	*4*	*5*	*6*	*7*	*8*	*9*	*10*	*11*
4	1	64	71	79	82	85	94	103	110	113	122	125
	2	53	57	61	72	74	76	81	91	99	100	105
	3	64	69	76	85	89	96	104	108	116	120	128
	4	68	69	78	82	91	97	104	108	115	122	132
	5	69	74	80	85	90	99	104	114	123	129	133
	6	85	91	98	100	105	104	118	121	130	141	141
	7	75	82	85	92	99	107	112	125	130	137	146
	8	57	61	65	68	77	81	87	91	95	101	107
	9	69	72	77	80	84	91	96	103	109	116	125
	10	66	68	76	81	88	95	103	106	112	119	130
8	1	57	64	70	76	80	90	93	99	105	113	118
	2	62	67	74	83	87	93	104	108	114	124	129
	3	60	68	73	80	83	94	101	106	112	122	131
	4	64	66	76	81	91	100	102	111	120	128	136
	5	57	60	67	73	67	64	75	85	89	98	105
	6	78	83	89	99	105	113	117	128	132	139	149
	7	81	81	92	100	108	119	120	133	138	149	157
	8	46	47	51	55	63	65	68	74	78	85	90
	9	69	72	74	76	77	82	82	90	95	101	103
	10	67	77	83	83	92	99	104	108	114	118	129

Table 27.15 Between-Animal Analysis of Variance Table for Original Data in Example 27.1

Source of Variation	*df*	*SS*	*MS*	*F*	*EMS*
Dose	4	935.97	233.97	1.53	$\sigma^2_{p*} + Q_A$
Error (Subject)	45	6,878.94	152.87		σ^2_{p*}

Table 27.16 Within-Animal Analysis of Variance Table for Filtered Data in Example 27.1

Source of Variation	*df*	*SS*	*MS*	*F*	*EMS*
Time	10	243,381.13	24,338.11	2,435	$\sigma_v^2 + Q_{\mathrm{T}}$
Time * Dose	40	1,517.88	37.95	3.80	$\sigma_v^2 + Q_{\mathrm{T*D}}$
Error(Time)	450	4,496.73	9.99		σ_v^2

data. Consequently, the standard errors for the filtered analysis are smaller than those for the unfiltered.

The response of body weight to the drugs over time is shown in Figure 27.1. Of interest are the linear and quadratic trends for each drug. The linear trend for drug i is measured by

$$\theta_{\mathrm{L}i} = -5\mu_{i1} - 4\mu_{i2} - 3\mu_{i3} - 2\mu_{i4} - 1\mu_{i5} - 0\mu_{i5}$$
$$+1\mu_{i7} + 2\mu_{i8} + 3\mu_{i9} + 4\mu_{i10} + 5\mu_{i11}.$$

Table 27.17 Comparison of Times at Each Dose in Example 27.1

	DOSE				
Time	*0*	*.5*	*1*	*4*	*8*
1	69.4	70.9	66.5	67.0	64.1
2	75.6	76.9	72.9	71.4	68.5
3	81.2	82.7	78.6	77.5	74.9
4	89.1	87.3	83.5	82.7	80.6
5	95.0	94.8	91.2	88.2	85.3
6	104.3	101.2	96.6	94.0	91.9
7	110.5	109.3	106.0	101.2	96.6
8	116.9	116.4	111.1	107.7	104.2
9	125.0	122.7	120.7	114.2	109.7
10	132.9	133.4	128.3	120.7	117.7
11	141.5	140.2	135.3	127.2	124.7

Notes: All means within a column are significantly different.

$\widehat{\text{s.e.}}(\widehat{\mu_{ik} - \mu_{ik'}}) = \sqrt{\dfrac{2(9.99)}{10}} = 1.1414.$

$\text{LSD}_{.05} = 1.96(1.414) = 2.77.$

Table 27.18 Comparison of Doses at Each Time in Example 27.1

Dose	TIME 1	2	3	4	5	6	7	8	9	10	11
0	69.4 (a)	75.6 (a)	81.2 (a)	89.1 (a)	95.0 (a)	104.3 (a)	110.5 (a)	116.9 (a)	125.0 (a)	132.9 (a)	141.5 (b)
.5	70.9 (a)	76.9 (a)	82.7 (a)	87.3 (a)	94.8 (a)	101.2 (a, b)	109.3 (a)	116.4 (a)	122.7 (a)	133.4 (a)	140.2 (a)
1	66.5 (a)	72.9 (a)	78.6 (a)	83.5 (a)	91.2 (a)	96.6 (a, b)	106.0 (a, b)	111.1 (a, b)	120.7 (a, b)	128.3 (a, b)	135.3 (a, b)
4	67.0 (a)	71.4 (a)	77.5 (a)	82.7 (a)	88.2 (a)	94.0 (a, b)	101.2 (a, b)	107.7 (a, b)	114.2 (a, b)	120.7 (b)	127.2 (b)
8	64.1 (a)	68.5 (a)	74.9 (a)	80.6 (a)	85.3 (a)	91.9 (b)	96.6 (b)	104.2 (b)	109.7 (b)	117.7 (b)	124.7 (b)

Notes: Means within a column with the same letter are not significantly different.

$$\widehat{\text{s.e.}}\left(\widehat{\mu_{ik} - \mu_{i'k}}\right) = \sqrt{\frac{2}{10}(152.87) + \frac{2(10)}{11(10)}(9.99)} = 5.691$$

$$t^* = \frac{2.014(2/10)152.87 + 1.96[2(10)/11(10)]9.99}{(2/10)152.87 + [2(10)/11(10)]9.99} = 2.011$$

$$\text{LSD}_{.05} = 2.011(5.691) = 11.44$$

Table 27.19 Analysis of Variance Table for the Unfiltered Data in Example 27.1

Source of Variation	*df*	*SS*	*MS*	*F*
Dose	4	10,295.73	2,573.93	1.53
Error(Animal)	45	75,668.30	1,681.51	
Time	10	243,381.13	24,338.81	1,783
Time * Dose	40	1,517.88	37.95	2.78
Error(Time)	450	6,140.80	13.84	

Note: The $\hat{\theta}$-adjusted degrees of freedom would be applied to this analysis of variance table (see the text).

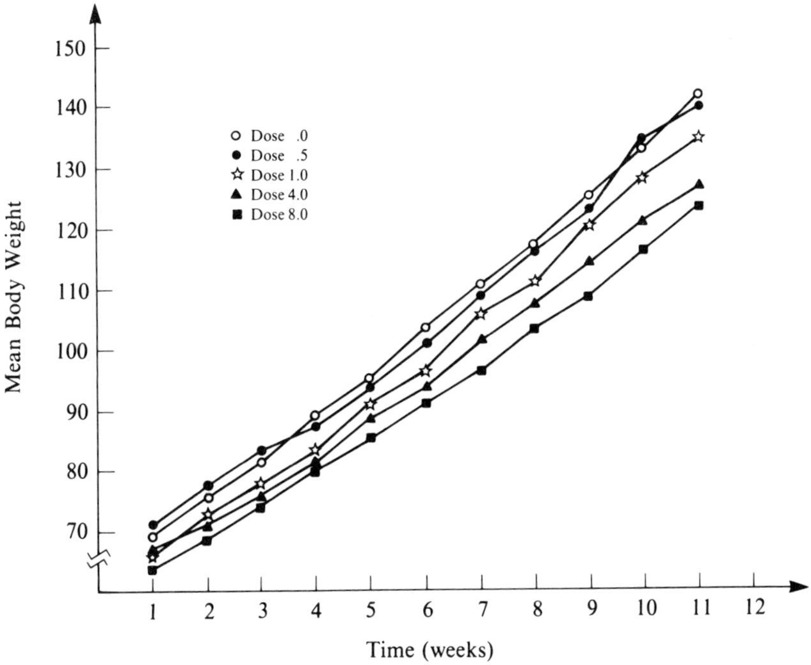

Figure 27.1 Graph of body weights over time for each level of drug.

Table 27.20 Estimates and t-Tests for Linear and Quadratic Trends for Each Drug Dose in Example 27.1

	LINEAR		QUADRATIC	
Dose	*Estimate*	*t-value*	*Estimate*	*t-value*
0	792.2	75.6	79.8	2.7
.5	765.2	73.0	151.8	5.2
1	761.9	72.7	126.5	4.3
4	671.3	64.0	86.9	3.0
8	662.7	63.2	99.7	3.4

The estimate of the standard error of θ_{Li} is

$$\widehat{\text{s.e.}}(\theta_{Li}) = \sqrt{\frac{9.99\left[(-5)^2 + (-4)^2 + \cdots + (5)^2\right]}{10}} = \sqrt{109.89} = 10.48.$$

The quadratic trend for drug i is measured by

$$\theta_{Qi} = 15\mu_{i1} + 6\mu_{i2} - 1\mu_{i3} - 6\mu_{i4} - 9\mu_{i5} - 10\mu_{i6} - 9\mu_{i7} - 6\mu_{i8} - 1\mu_{i9} + 6\mu_{i10} + 15\mu_{i11},$$

and the estimate of the standard error of θ_{Qi} is

$$\widehat{\text{s.e.}}(\theta_{Qi}) = \sqrt{\frac{9.99\left[15^2 + 6^2 + (-1)^2 + \cdots + (15)^2\right]}{10}} = \sqrt{857.14} = 29.27.$$

The estimates of the linear and quadratic trends and the corresponding t-statistics are in Table 27.20. Those values show a very strong linear trend with a slight quadratic trend for each of the doses of the drug.

CONCLUDING REMARKS

This chapter described the analysis of repeated measures designs when the usual assumptions do not hold. Methods for testing the assumptions were presented. The Box conservative adjusted degrees of freedom and the $\hat{\theta}$-adjusted degrees of freedom methods were explained, and the application to the three examples of Chapter 26 were presented in detail. A method was described for filtering the data to remove the correlation when the repeated measurements are from a first-order autoregressive process (serially correlated). Techniques were also presented for estimating contrasts and computing their variances. One example with autocorrelated errors was discussed in detail.

28

Analyzing Split-Plot and Certain Repeated Measures Experiments with Unbalanced and Missing Data

CHAPTER OUTLINE

In Chapters 24, 26, and 27, we discussed the analysis of the two-way treatment structure in split-plot and repeated measures designs with balanced data. In those chapters we assumed that there were equal numbers of replicates for each whole plot treatment and that the same set of subplot treatments was observed within each whole plot experimental unit.

Often, particularly when the whole plot design structure is completely randomized, different numbers of whole plot experimental units are assigned to each whole plot treatment. Likewise, in many repeated measures experiments where the larger experimental unit is a person, an animal, or a plant and measures are taken at several different times, different numbers of persons, animals, or plants may be assigned to each treatment. The analyses recommended in this chapter are appropriate only for repeated measures experiments that satisfy the Huynh–Feldt conditions discussed in Chapter 26.

Section 28.1 discusses the analysis of split-plot (and repeated measures) experiments where each whole plot treatment is applied to different numbers of whole plot experimental units. It is assumed that the set of subplot treatments is the same for each whole plot experimental unit.

Section 28.2 discusses the case when the set of subplot treatments differs for some of the whole plot experimental units. These differences may occur by design, but they usually occur because of lost data. For example, in experiments where subplot measurements are taken with respect to time on individuals, some individuals may quit before the experiment is completed, and hence missing subplots occur.

28.1 UNEQUAL NUMBERS OF WHOLE PLOT EXPERIMENTAL UNITS

We shall begin this section by studying an example in which the whole plot treatments are observed on unequal numbers of whole plot experimental units. In the experiment, an experimenter wanted to study the effects of three different herbicides and four fertilizers on the growth rate of corn. Fifteen plots of land were available for the experiment, and 5 plots were randomly assigned to each of the three herbicides. Each of the 15 plots was further divided into 4 subplots, and a different fertilizer treatment was randomly assigned to each. Unfortunately, before any measurements could be taken, 3 of the 15 whole plots were destroyed by excessive rainfall. Herbicide 1 had been assigned to two of these lost subplots, and herbicide 3 to the third. At the beginning of the third week, ten plants were selected at random from each subplot, and the height of each plant was measured. The average of the ten heights was recorded as the measurement from the subplot. The data are given in Table 28.1.

28.1.1 Model Definition and Assumptions

The model for this experiment is the same type as that given in (24.1.1); that is,

$$y_{ijk} = \mu + H_i + \delta_{ik} + F_j + (HF)_{ij} + \varepsilon_{ijk} \qquad (28.1.1)$$

where $i = 1, 2, 3$; $j = 1, 2, 3, 4$; $k = 1, 2, \ldots, n_i$; $n_1 = 3$; $n_2 = 5$; and

Table 28.1 Height Data (in cm) for Herbicide Experiment

		FERTILIZER			
Herbicide	*Plot*	*1*	*2*	*3*	*4*
	7	5	8	12	14
1	4	10	15	18	20
	6	6	10	14	17
	9	14	20	28	32
	8	10	14	18	20
2	2	13	15	18	24
	5	9	10	18	22
	13	11	13	20	21
	11	5	20	21	21
3	3	7	16	16	18
	1	10	22	21	23
	15	3	15	18	19

$n_3 = 4$. Accordingly, the analysis of this type of an experiment is quite similar to the analysis of balanced split-plot experiments discussed in Chapter 24. All hypotheses that were testable in the balanced case are also testable for this case, as are all linear contrasts of the parameters. The only differences in these two cases occur in the basic analysis of variance tables, the best estimates of the estimable functions, and the standard errors of the estimators.

A general model for the case under discussion is given by

$$y_{ijk} = \mu + T_i + \delta_{ik} + B_j + (TB)_{ij} + \varepsilon_{ijk} \qquad (28.1.2)$$

where $i = 1, 2, \ldots, t$; $j = 1, 2, \ldots, b$; and $k = 1, 2, \ldots, n_i$. We let

$$N = \sum_i n_i.$$

28.1.2 Analysis of Variance and Expected Mean Squares

One possible analysis of variance table for model (28.1.2) is given in Table 28.2. Forms of the expected mean squares of the rows in Table 28.2 are given in Table 28.3.

The analysis of variance given in Table 28.2, although inappropriate in most cases, is the one that is usually given in texts. It is equivalent to the one obtained by conducting a Type I or Type II analysis using the model

$$Y = T \quad PLOT(T) \quad B \quad T*B$$

Table 28.2 Analysis of Variance Table for Model (28.1.2)

Source of Variation	*df*	*SS*
Total	$Nb - 1$	$\sum_{i,j,k} y_{ijk}^2 - (y_{\cdot\cdot\cdot}^2/Nb)$
Whole Plot Analysis		
Whole Plot	$N - 1$	$\sum_{i,k} (y_{i\cdot k}^2/b) - (y_{\cdot\cdot\cdot}^2/Nb)$
T	$t - 1$	$\sum_{i} (y_{i\cdot\cdot}^2/n_i b) - (y_{\cdot\cdot\cdot}^2/Nb)$
Error	$N - t$	$\sum_{i,k} (y_{i\cdot k}^2/b) - \sum_{i} (y_{i\cdot\cdot}^2/bn_i)$
Subplot Analysis		
B	$b - 1$	$\sum_{j} (y_{\cdot j \cdot}^2/N) - (y_{\cdot\cdot\cdot}^2/Nb)$
$T * B$	$(t-1)(b-1)$	$\sum_{i,j} (y_{ij\cdot}^2/n_i) - \sum_{i} (y_{i\cdot\cdot}^2/bn_i) - \sum_{j} (y_{\cdot j\cdot}^2/N) + (y_{\cdot\cdot\cdot}^2/Nb)$
Error	$(N-t)(b-1)$	$\sum_{ijk} y_{ijk}^2 - \sum_{ik} (y_{i\cdot k}^2/b) - \sum_{ij} (y_{ij\cdot}^2/n_i) + \sum_{i} (y_{i\cdot\cdot}^2/bn_i)$

Table 28.3 Expected Mean Squares for Table 28.2

Source of Variation	*df*	*EMS*
Total	$Nb - 1$	
Whole Plot Analysis		
Whole Plots	$N - 1$	
T	$t - 1$	$\sigma_\varepsilon^2 + b\sigma_\delta^2 + Q_1(T, T * B)$
Error	$N - t$	$\sigma_\varepsilon^2 + b\sigma_\delta^2$
Subplot Analysis		
B	$b - 1$	$\sigma_\varepsilon^2 + Q_2(B, T * B)$
$T * B$	$(t-1)(b-1)$	$\sigma_\varepsilon^2 + Q_3(T * B)$
Error	$(N-t)(b-1)$	σ_ε^2

In this case, the hypotheses tested by the T and $T * B$ effects are the same as those tested in the balanced case; however, the hypothesis tested by the B effect is a weighted linear combination of the true cell means. To get the appropriate sum of squares for the usual hypothesis, we should construct a Type III analysis. Because the sums of squares for such an analysis are very tedious to calculate by hand, we shall not give any formulas; most statistical computing packages will calculate the sums of squares for you. The hypothesis on B tested by the Type I–Type II analysis is

$$\sum_i n_i \mu_{i1} = \sum_i n_i \mu_{i2} = \cdots = \sum_i n_i \mu_{ib},$$

whereas the hypothesis on B tested by a Type III analysis is

$$\bar{\mu}_{\cdot 1} = \bar{\mu}_{\cdot 2} = \cdots = \bar{\mu}_{\cdot b} \qquad \text{where} \quad \mu_{ij} = \mu + T_i + B_j + (T * B)_{ij}.$$

To compare the main effects of T, the whole plot treatment, we use $F = \text{TMS}/\text{WPEMS}$ where TMS is the T mean square and WPEMS is the whole plot error mean square. The hypothesis of equal T main effects is rejected if

$$F > F_{\alpha, t-1, N-t}.$$

To compare the B main effects and test for $T * B$ interaction, we divide their respective mean squares by the subplot error mean square (SPEMS). The hypothesis of no interaction is rejected if

$$F = \frac{(T * B)\text{MS}}{\text{SPEMS}} > F_{\alpha, (t-1)(b-1), (N-t)(b-1)},$$

and the hypothesis of equal B main effects is rejected if

$$F = \frac{\text{BMS}}{\text{SPEMS}} > F_{\alpha, b-1, (N-t)(b-1)}$$

where BMS is the Type III BMS.

28.1.3 Pairwise Comparisons

To compare means, let

$$\mu_{ij} = \mu + T_i + B_j + (TB)_{ij}.$$

The best estimate of μ_{ij} is $\hat{\mu}_{ij} = \bar{y}_{ij\cdot}$, and the estimated standard error of this estimate is

$$\widehat{\text{s.e.}}(\hat{\mu}_{ij}) = \sqrt{\frac{\hat{\sigma}_\varepsilon^2 + \hat{\sigma}_\delta^2}{n_i}}$$

where $\hat{\sigma}_\varepsilon^2 = \text{SPEMS}$ and $\hat{\sigma}_\delta^2 = (\text{WPEMS} - \text{SPEMS})/b$. The best estimate of $\bar{\mu}_{i\cdot}$ is

$$\hat{\bar{\mu}}_{i\cdot} = \bar{\hat{\mu}}_{i\cdot} = \frac{\hat{\mu}_{i1} + \hat{\mu}_{i2} + \cdots + \hat{\mu}_{ib}}{b},$$

and its estimated standard error is

$$\widehat{\text{s.e.}}(\hat{\bar{\mu}}_{i\cdot}) = \sqrt{\frac{\hat{\sigma}_{\varepsilon}^2 + b\hat{\sigma}_{\delta}^2}{bn_i}} = \sqrt{\frac{\text{WPEMS}}{bn_i}}.$$

The best estimate of $\bar{\mu}_{\cdot j}$ is

$$\hat{\bar{\mu}}_{\cdot j} = \bar{\hat{\mu}}_{\cdot j} = \frac{\hat{\mu}_{1j} + \hat{\mu}_{2j} + \cdots + \hat{\mu}_{tj}}{t},$$

and its estimated standard error is

$$\widehat{\text{s.e.}}(\hat{\bar{\mu}}_{\cdot j}) = \sqrt{\frac{\hat{\sigma}_{\varepsilon}^2 + \hat{\sigma}_{\delta}^2}{\tilde{n}t}} \qquad \text{where} \quad \tilde{n} = t\left(\sum_{i=1}^{t} n_i^{-1}\right)^{-1}.$$

The best estimate of $\bar{\mu}_{i'\cdot} - \bar{\mu}_{i'}$ is $\hat{\bar{\mu}}_{i\cdot} - \hat{\bar{\mu}}_{i'\cdot}$, and its estimated standard error is

$$\widehat{\text{s.e.}}(\hat{\bar{\mu}}_{i\cdot} - \hat{\bar{\mu}}_{i'\cdot}) = \sqrt{(\text{WPEMS})\left(\frac{1}{bn_i} + \frac{1}{bn_{i'}}\right)}.$$

An LSD_α for comparing $\bar{\mu}_{i\cdot}$ with $\bar{\mu}_{i'\cdot}$ is

$$\text{LSD}_\alpha(\bar{\mu}_{i\cdot}, \bar{\mu}_{i'\cdot}) = t_{\alpha/2, N-t}\sqrt{(\text{WPEMS})\left(\frac{1}{bn_i} + \frac{1}{bn_{i'}}\right)}. \qquad (28.1.3)$$

The best estimate of $\bar{\mu}_{\cdot j} - \bar{\mu}_{\cdot j'}$ is $\hat{\bar{\mu}}_{\cdot j} - \hat{\bar{\mu}}_{\cdot j'}$, and its estimated standard error is

$$\widehat{\text{s.e.}}(\hat{\bar{\mu}}_{\cdot j} - \hat{\bar{\mu}}_{\cdot j'}) = \sqrt{(\text{SPEMS})\left(\frac{2}{\tilde{n}t}\right)}.$$

An LSD_α for comparing $\bar{\mu}_{\cdot j}$ to $\bar{\mu}_{\cdot j'}$ is

$$\text{LSD}_\alpha\,(\bar{\mu}_{\cdot j}, \bar{\mu}_{\cdot j'}) = t_{\alpha/2,(N-t)(b-1)}\sqrt{(\text{SPEMS})\left(\frac{2}{\tilde{n}t}\right)}. \qquad (28.1.4)$$

If there is significant interaction, then one may want to compare the subplot treatments at each level of the whole plot treatment or vice versa (compare the whole plot treatments at each level of the subplot treatment). The best estimate of $\mu_{ij} - \mu_{ij'}$ is $\hat{\mu}_{ij} - \hat{\mu}_{ij'}$ and its standard error is

$$\widehat{\text{s.e.}}(\hat{\mu}_{ij} - \hat{\mu}_{ij'}) = \sqrt{\text{SPEMS}\left(\frac{2}{n_i}\right)}.$$

An LSD_α value for comparing two subplot treatments at a given level of a whole plot treatment is

$$LSD_\alpha(\mu_{ij}, \mu_{ij'}) = t_{\alpha/2,(N-t)(b-1)}\sqrt{(\text{SPEMS})\left(\frac{2}{n_i}\right)}. \quad (28.1.5)$$

The best estimate of $\mu_{ij} - \mu_{i'j}$ is $\hat{\mu}_{ij} - \hat{\mu}_{i'j}$, and its standard error is

$$\widehat{\text{s.e.}}(\hat{\mu}_{ij} - \hat{\mu}_{i'j}) = \sqrt{(\hat{\sigma}_\varepsilon^2 + \hat{\sigma}_\delta^2)\left(\frac{1}{n_i} + \frac{1}{n_{i'}}\right)}.$$

An approximate LSD_α for comparing two whole plot treatments at a given level of the subplot treatment is

$$LSD_\alpha(\mu_{ij}, \mu_{i'j}) = t_\alpha^*\sqrt{(\hat{\sigma}_\varepsilon^2 + \hat{\sigma}_\delta^2)\left(\frac{1}{n_i} + \frac{1}{n_{i'}}\right)} \quad (28.1.6)$$

where

$$t_\alpha^* = \frac{(N - t/N)(\text{SPEMS})t_{\alpha/2,(N-t)(b-1)} + (t/N)(\text{WPEMS})t_{\alpha/2,N-t}}{(N - t/N)(\text{SPEMS}) + (t/N)(\text{WPEMS})}$$

This LSD value is also the appropriate one for comparing any two-way means at different levels of both the whole plot treatments and subplot treatments.

In summary, we use the LSD_α in (28.1.5) for comparing two subplot treatments at a specified level of the whole plot treatments, and we use the LSD_α in (28.1.6) for comparing two whole plot treatments at the same or different levels of the subplot treatments.

28.1.4 Computer Analyses

Obtaining a complete analysis of the data in Table 28.1 by using SAS® requires the following statements:

```
PROC GLM; CLASSES HERB FERT PLOT;
MODEL HEIGHT = HERB PLOT(HERB) FERT HERB*FERT;
RANDOM PLOT(HERB);
LSMEANS HERB FERT HERB*FERT/STDERR PDIFF;
```

Using the RANDOM option instructs SAS® to print a table of expected mean squares for each type of sum of squares requested. This is very helpful in more complex experiments because it shows the user how to choose appropriate *F*-ratios with which to test hypotheses of interest.

Analysis of these data set with SAS® ANOVA gives the same analysis as that given by the SAS® GLM Type I analysis, although this analysis is generally not the one of interest. The correct analysis is given by the Type III analysis.

The means obtained by the MEANS option in either SAS® GLM or SAS® ANOVA do not give appropriate means for the FERT marginal means. The means obtained for the HERB levels by using the MEANS statement are the same as those that would be obtained by the LSMEANS option and are appropriate for those effects.

Unfortunately, at the present time, SAS® claims that the FERT marginal means are not estimable, which is incorrect. The problem occurs because of the coefficients that SAS® generates for the PLOT(HERB) parameters, which SAS® considers as fixed effects when doing computations, when in fact they are random effects. SAS® generates coefficients that are equal for each PLOT(HERB) parameter, when they should be weighted according to the number of whole plots assigned to each treatment. However, by using the ESTIMATE option, the user can specify appropriate coefficients and thus obtain correct FERT marginal means. For example, to get the FERT 1 marginal mean, we would use

```
ESTIMATE 'FERT 1 LSMEAN' INTERCEPT 1
HERB .33333 .33333 .33333 FERT 1 0 0 0
PLOT(HERB) .11111 .11111 .11111 .06667 .06667 .06667
           .06667 .06667 .08333 .08333 .08333 .08333
HERB*FERT .33333 0 0 0 .33333 0 0 0 .33333 0 0 0;
```

Most of the standard errors associated with the means that SAS® GLM prints are not correct for split-plot and repeated measures experiments. SAS® calculates these standard errors by assuming a completely randomized experimental design rather than a split-plot design and does not take the two error terms into account. Likewise, the significance levels printed for comparing two means are often incorrect. To obtain corrected standard errors, we can use the CONTRAST option and the ESTIMATE option along with the RANDOM option. These three options supply the user with enough information to calculate corrected standard errors and corrected significance levels. This is considered in detail in Section 28.2.5, where the missing data problem is considered. The results given there also apply to the case considered in this section.

28.2 UNEQUAL NUMBERS OF SUBPLOT EXPERIMENTAL UNITS

In this section we shall consider the case where some subplot measurements are missing but at least one subplot has been observed in each whole plot. Much of the material in this section does not presently appear in the literature. Thus, we shall present the information in this section in slightly more detail than we have previously.

Those readers that only want to use the results and are willing to take them on faith are advised to read the example in Section 28.2.1 and then turn to Section 28.2.5, where the results are summarized in usable form.

In Section 28.2.5 an SAS® analysis of the data in Table 28.4 is given and discussed. Sections 28.2.1 through 28.2.4 analyze the data by hand and are used to justify the analysis given in Section 28.2.5.

28.2.1 Drug Effects Example

This example is a small part of an experiment conducted to determine the effects of a drug on the scores obtained by depressed patients on a test designed to measure depression. Two patients were in the placebo group, and three in the drug group. Two of the patients did not return for examination after the first week. The data are provided in Table 28.4.

Let μ_{ij} represent the expected response to treatment i during week j. One model for μ_{ij} is

$$\mu_{ij} = \mu + T_i + W_j + (TW)_{ij}.$$

The corresponding model for an observed response is

$$y_{ijk} = \mu + T_i + \delta_{k(i)} + W_j + (TW)_{ij} + \varepsilon_{ijk} \qquad (28.2.1)$$

for $i = 1,2$; $k(1) = 1,2$; $k(2) = 3,4,5$; and $j = 1,2$. It is assumed that $\delta_{k(i)} \sim$ i.i.d. $N(0, \sigma_\delta^2)$, which represents the error contribution of the kth patient in the ith treatment group. Also as before, it is assumed that $\varepsilon_{ijk} \sim$ i.i.d. $N(0, \sigma_\varepsilon^2)$, which represents the error contribution of the jth week to the kth patient in the ith treatment group. It is also assumed that these two error components are distributed independently of each other. Note that the patients correspond to whole plot experimental units to which the treatments were applied, and weeks were the subplot treatments.

Table 28.4 Treatment Received and Depression Scores for Drug Effects Example

		SCORE	
Patient	*Treatment*	*Week 1*	*Week 2*
1	Placebo	24	18
2	Placebo	22	—
3	Drug	25	22
4	Drug	23	—
5	Drug	26	24

28.2.2 Estimating Functions of the Parameters

There are several possible effects that one might want to estimate for this experiment. For each effect, the best estimate and the variance of each estimate are both derived. The results are summarized later in the chapter in Table 28.5.

The first effect considered is μ_{11}. The best estimate of μ_{11} is

$$\hat{\mu}_{11} = \frac{y_{111} + y_{112}}{2} = \frac{24 + 22}{2} = 23.$$

The variance of $\hat{\mu}_{11}$ is

$$\begin{aligned} \mathrm{Var}(\hat{\mu}_{11}) &= \tfrac{1}{4}\mathrm{Var}(y_{111} + y_{112}) \\ &= \tfrac{1}{4}\mathrm{Var}(\delta_{1(1)} + \varepsilon_{111} + \delta_{2(1)} + \varepsilon_{112}) \\ &= \tfrac{1}{4}(\sigma_\delta^2 + \sigma_\varepsilon^2 + \sigma_\delta^2 + \sigma_\varepsilon^2) \\ &= \tfrac{1}{2}(\sigma_\varepsilon^2 + \sigma_\delta^2); \end{aligned}$$

hence, the standard error of $\hat{\mu}_{11}$ is

$$\widehat{\text{s.e.}}(\hat{\mu}_{11}) = \sqrt{\tfrac{1}{2}(\sigma_\varepsilon^2 + \sigma_\delta^2)}\,.$$

Before estimating μ_{12}, we must first estimate the missing value y_{122}. Since patients are the experimental units, there is no defined interaction between patients and treatments; this allows us to use the results observed for patient 1 to estimate the missing value at week 2 for patient 2. Since the week 2 value is 6 units less than the week 1 value for patient 1, we estimate the week 2 value for patient 2 to be 6 units less than the week 1 value for patient 2. Hence,

$$\hat{y}_{122} = 22 - 6 = 16.$$

In terms of the y_{ijk}'s,

$$\hat{y}_{122} = y_{112} - y_{111} + y_{121}.$$

The variance of $\hat{y}_{122}$ is

$$\begin{aligned} \mathrm{Var}(\hat{y}_{122}) &= \mathrm{Var}(y_{112} - y_{111} + y_{121}) \\ &= \mathrm{Var}(\delta_{1(1)} + \varepsilon_{112} - \delta_{1(1)} - \varepsilon_{111} + \delta_{2(1)} + \varepsilon_{121}) \\ &= \mathrm{Var}(\varepsilon_{112} - \varepsilon_{111} + \varepsilon_{121} + \delta_{2(1)}) \\ &= 3\sigma_\varepsilon^2 + \sigma_\delta^2 \\ &= 3(\sigma_\varepsilon^2 + \tfrac{1}{3}\sigma_\delta^2). \end{aligned}$$

The estimate of μ_{12} is

$$\hat{\mu}_{12} = \frac{y_{121} + \hat{y}_{122}}{2} = \frac{18 + 16}{2} = 17,$$

and

$$\begin{aligned}
\operatorname{Var}(\hat{\mu}_{12}) &= \tfrac{1}{4}\operatorname{Var}(y_{121} + \hat{y}_{122}) \\
&= \tfrac{1}{4}\operatorname{Var}(y_{121} + y_{112} - y_{111} + y_{121}) \\
&= \tfrac{1}{4}\operatorname{Var}(2y_{121} + y_{112} - y_{111}) \\
&= \tfrac{1}{4}\operatorname{Var}(2\delta_{1(1)} + 2\varepsilon_{121} + \delta_{2(1)} + \varepsilon_{112} - \delta_{1(1)} - \varepsilon_{111}) \\
&= \tfrac{1}{4}\operatorname{Var}(2\varepsilon_{121} + \varepsilon_{112} - \varepsilon_{111} + \delta_{1(1)} + \delta_{2(1)}) \\
&= \tfrac{1}{4}(6\sigma_\varepsilon^2 + 2\sigma_\delta^2) \\
&= \tfrac{3}{2}(\sigma_\varepsilon^2 + \tfrac{1}{3}\sigma_\delta^2).
\end{aligned}$$

Since there are no missing data for week 1, the best estimate of the effect of the drug at week 1, namely μ_{21}, is

$$\begin{aligned}
\hat{\mu}_{21} &= \frac{y_{213} + y_{214} + y_{215}}{3} \\
&= \frac{25 + 23 + 26}{3} = 24.667.
\end{aligned}$$

The variance of the estimate is

$$\begin{aligned}
\operatorname{Var}(\hat{\mu}_{21}) &= \tfrac{1}{9}\operatorname{Var}(\varepsilon_{213} + \delta_{3(2)} + \varepsilon_{214} + \delta_{4(2)} + \varepsilon_{215} + \delta_{5(2)}) \\
&= \tfrac{1}{9}\operatorname{Var}(3\sigma_\varepsilon^2 + 3\sigma_\delta^2) \\
&= \tfrac{1}{3}\operatorname{Var}(\sigma_\varepsilon^2 + \sigma_\delta^2).
\end{aligned}$$

Before μ_{22} can be estimated, we must first estimate the missing value y_{224}. The average of patients 3 and 4 at week 2 was 2.5 less than their average at week 1. Thus, the estimate for patient 4 at week 2 is 2.5 less than patient 4's response at week 1. Hence,

$$\hat{y}_{224} = 23 - 2.5 = 20.5.$$

In terms of y_{ijk}, we have

$$\begin{aligned}
\hat{y}_{224} &= y_{214} - \left(\frac{y_{213} + y_{215}}{2} - \frac{y_{223} + y_{225}}{2}\right) \\
&= y_{214} - \tfrac{1}{2}(y_{213} + y_{215} - y_{223} - y_{225}),
\end{aligned}$$

and the variance is

$$\begin{aligned}
\operatorname{Var}(\hat{y}_{224}) &= \operatorname{Var}\big[\varepsilon_{214} + \delta_{4(2)} - \tfrac{1}{2}(\varepsilon_{213} + \delta_{3(2)} + \varepsilon_{215} \\
&\qquad + \delta_{5(2)} - \varepsilon_{223} - \delta_{3(2)} - \varepsilon_{225} - \delta_{5(2)})\big] \\
&= \operatorname{Var}(\varepsilon_{214} - \tfrac{1}{2}\varepsilon_{213} - \tfrac{1}{2}\varepsilon_{215} + \tfrac{1}{2}\varepsilon_{223} + \tfrac{1}{2}\varepsilon_{225} + \delta_{4(2)}) \\
&= 2\sigma_\varepsilon^2 + \sigma_\delta^2 \\
&= 2(\sigma_\varepsilon^2 + \tfrac{1}{2}\sigma_\delta^2).
\end{aligned}$$

Now, using $\hat{y}_{224}$, the estimate of μ_{22} is

$$\begin{aligned}
\hat{\mu}_{22} &= \frac{y_{223} + \hat{y}_{224} + y_{225}}{3} \\
&= \frac{y_{223} + \left(y_{214} - \frac{1}{2}y_{213} - \frac{1}{2}y_{215} + \frac{1}{2}y_{223} + \frac{1}{2}y_{225}\right) + y_{225}}{3} \\
&= \frac{\frac{3}{2}y_{223} + y_{214} - \frac{1}{2}y_{213} - \frac{1}{2}y_{215} + \frac{3}{2}y_{225}}{3} \\
&= \frac{3y_{223} + 2y_{214} - y_{213} - y_{215} + 3y_{225}}{6} \\
&= \frac{3 \cdot 22 + 2 \cdot 23 - 25 - 26 + 3 \cdot 24}{6} = 22.167
\end{aligned}$$

and the variance is

$$\begin{aligned}
\mathrm{Var}(\hat{\mu}_{22}) &= \frac{\mathrm{Var}\left(3\varepsilon_{223} + 3\delta_{3(2)} + 2\varepsilon_{214} + 2\delta_{4(2)} - \varepsilon_{213} - \delta_{3(2)} - \varepsilon_{215} - \delta_{5(2)} + 3\varepsilon_{225} + 3\delta_{5(2)}\right)}{36} \\
&= \frac{\mathrm{Var}\left(3\varepsilon_{223} + 2\varepsilon_{214} - \varepsilon_{213} - \varepsilon_{215} + 3\varepsilon_{225} + 2\delta_{3(2)} + 2\delta_{4(2)} + 2\delta_{5(2)}\right)}{36} \\
&= \frac{24\sigma_\varepsilon^2 + 12\sigma_\delta^2}{36} \\
&= \tfrac{2}{3}\left(\sigma_\varepsilon^2 + \tfrac{1}{2}\sigma_\delta^2\right).
\end{aligned}$$

Next we examine the main effects of the drugs. The best estimate of $\bar{\mu}_{1\cdot}$, the placebo marginal mean, is

$$\hat{\bar{\mu}}_{1\cdot} = \frac{\hat{\mu}_{11} + \hat{\mu}_{12}}{2} = \frac{23 + 17}{2} = 20.$$

The variance of $\hat{\bar{\mu}}_{1\cdot}$ is

$$\begin{aligned}
\mathrm{Var}(\hat{\bar{\mu}}_{1\cdot}) &= \frac{\mathrm{Var}(\hat{\mu}_{11} + \hat{\mu}_{12})}{4} \\
&= \frac{\mathrm{Var}\left[(y_{111} + y_{112})/2 + (y_{121} + \hat{y}_{122})/2\right]}{4} \\
&= \frac{\mathrm{Var}(y_{112} + y_{121})}{4} \\
&= \frac{\mathrm{Var}(\varepsilon_{112} + \delta_{2(1)} + \varepsilon_{121} + \delta_{1(1)})}{4} \\
&= \frac{2\sigma_\varepsilon^2 + 2\sigma_\delta^2}{4} \\
&= \tfrac{1}{2}\left(\sigma_\varepsilon^2 + \sigma_\delta^2\right).
\end{aligned}$$

The best estimate of $\bar{\mu}_{2\cdot}$, the drug group marginal mean, is

$$\hat{\bar{\mu}}_{2\cdot} = \frac{\hat{\mu}_{21} + \hat{\mu}_{22}}{2} = \frac{24.667 + 22.167}{2} = 23.417,$$

and the variance is

$$\begin{aligned}\mathrm{Var}(\hat{\bar{\mu}}_{2\cdot}) &= \frac{\mathrm{Var}[(y_{213} + y_{214} + y_{215})/3 + (y_{223} + \hat{y}_{224} + y_{225})/3]}{4} \\ &= \frac{\mathrm{Var}(\frac{1}{6}y_{213} + \frac{2}{3}y_{214} + \frac{1}{6}y_{215} + \frac{1}{2}y_{223} + \frac{1}{2}y_{225})}{4} \\ &= \frac{\mathrm{Var}(\frac{1}{6}\varepsilon_{213} + \frac{2}{3}\varepsilon_{214} + \frac{1}{6}\varepsilon_{215} + \frac{1}{2}\varepsilon_{223} + \frac{1}{2}\varepsilon_{225} + \frac{2}{3}\delta_{3(2)} + \frac{2}{3}\delta_{4(2)} + \frac{2}{3}\delta_{5(2)})}{4} \\ &= \frac{(\frac{1}{36} + \frac{4}{9} + \frac{1}{36} + \frac{1}{4} + \frac{1}{4})\sigma_\varepsilon^2 + (\frac{4}{9} + \frac{4}{9} + \frac{4}{9})\sigma_\delta^2}{4} \\ &= \frac{1}{4}(\sigma_\varepsilon^2 + \frac{4}{3}\sigma_\delta^2).\end{aligned}$$

The drug main effect is measured by $\bar{\mu}_{1\cdot} - \bar{\mu}_{2\cdot}$, the difference between the placebo group and the drug group. This effect is estimated by

$$\widehat{\bar{\mu}_{1\cdot} - \bar{\mu}_{2\cdot}} = \hat{\bar{\mu}}_{1\cdot} - \hat{\bar{\mu}}_{2\cdot} = 20 - 23.417 = -3.417.$$

The variance of the estimate of the main drug effect is

$$\begin{aligned}\mathrm{Var}(\hat{\bar{\mu}}_{1\cdot} - \hat{\bar{\mu}}_{2\cdot}) &= \mathrm{Var}(\hat{\bar{\mu}}_{1\cdot}) + \mathrm{Var}(\hat{\bar{\mu}}_{2\cdot}) \\ &= \frac{1}{2}(\sigma_\varepsilon^2 + \sigma_\delta^2) + \frac{1}{4}(\sigma_\varepsilon^2 + \frac{4}{3}\sigma_\delta^2) \\ &= \frac{3}{4}(\sigma_\varepsilon^2 + \frac{10}{9}\sigma_\delta^2).\end{aligned}$$

In evaluating this variance, we used the fact that $\hat{\bar{\mu}}_{1\cdot}$ and $\hat{\bar{\mu}}_{2\cdot}$ depend on different sets of patients and hence are independent of each other.

Next, we examine the week main effect, which is estimated by the difference between the week 1 and week 2 marginal means. The best estimate of the week 1 marginal mean, $\bar{\mu}_{\cdot 1}$, is

$$\hat{\bar{\mu}}_{\cdot 1} = \frac{\hat{\mu}_{11} + \hat{\mu}_{21}}{2} = \frac{23 + 24.667}{2} = 23.833,$$

and its variance is

$$\begin{aligned}\mathrm{Var}(\hat{\bar{\mu}}_{\cdot 1}) &= \frac{\mathrm{Var}(\hat{\mu}_{11} + \hat{\mu}_{21})}{4} \\ &= \frac{\mathrm{Var}(\hat{\mu}_{11}) + \mathrm{Var}(\hat{\mu}_{21})}{4} \\ &= \frac{\frac{1}{2}(\sigma_\varepsilon^2 + \sigma_\delta^2) + \frac{1}{3}(\sigma_\varepsilon^2 + \sigma_\delta^2)}{4} \\ &= \frac{5}{24}(\sigma_\varepsilon^2 + \sigma_\delta^2).\end{aligned}$$

The best estimate of the week 2 marginal mean is

$$\hat{\bar{\mu}}_{\cdot 2} = \frac{\hat{\mu}_{12} + \hat{\mu}_{22}}{2} = \frac{17 + 22.167}{2} = 19.583,$$

and its variance is

$$\begin{aligned}
\text{Var}(\hat{\bar{\mu}}_{\cdot 2}) &= \frac{\text{Var}(\hat{\mu}_{12} + \hat{\mu}_{22})}{4} \\
&= \frac{\text{Var}(\hat{\mu}_{12}) + \text{Var}(\hat{\mu}_{22})}{4} \\
&= \frac{\frac{3}{2}\left(\sigma_\varepsilon^2 + \frac{1}{3}\sigma_\delta^2\right) + \frac{2}{3}\left(\sigma_\varepsilon^2 + \frac{1}{2}\sigma_\delta^2\right)}{4} \\
&= \tfrac{13}{24}\left(\sigma_\varepsilon^2 + \tfrac{5}{13}\sigma_\delta^2\right).
\end{aligned}$$

The best estimate for the Week main effect, $\bar{\mu}_{\cdot 1} - \bar{\mu}_{\cdot 2}$, is

$$\widehat{\bar{\mu}_{\cdot 1} - \bar{\mu}_{\cdot 2}} = \hat{\bar{\mu}}_{\cdot 1} - \hat{\bar{\mu}}_{\cdot 2} = 23.833 - 19.583 = 4.25,$$

and its variance is

$$\begin{aligned}
&\text{Var}(\hat{\bar{\mu}}_{\cdot 1} - \hat{\bar{\mu}}_{\cdot 2}) = \text{Var}\left[\tfrac{1}{2}(\hat{\mu}_{11} - \hat{\mu}_{12}) + \tfrac{1}{2}(\hat{\mu}_{21} - \hat{\mu}_{22})\right] \\
&= \tfrac{1}{4}\text{Var}(\hat{\mu}_{11} - \hat{\mu}_{12}) + \tfrac{1}{4}\text{Var}(\hat{\mu}_{21} - \hat{\mu}_{22}) \\
&= \tfrac{1}{4}\text{Var}(y_{111} - y_{121}) + \tfrac{1}{4}\text{Var}\left(\tfrac{1}{2}y_{213} + \tfrac{1}{2}y_{215} - \tfrac{1}{2}y_{223} - \tfrac{1}{2}y_{225}\right) \\
&= \tfrac{1}{4}\text{Var}\left(\varepsilon_{111} - \varepsilon_{121} + \delta_{1(1)} - \delta_{1(1)}\right) + \tfrac{1}{4}\text{Var}\left(\tfrac{1}{2}\varepsilon_{213} + \tfrac{1}{2}\varepsilon_{215} - \tfrac{1}{2}\varepsilon_{223} - \tfrac{1}{2}\varepsilon_{225}\right) \\
&= \tfrac{1}{4}\left(2\sigma_\varepsilon^2\right) + \tfrac{1}{4}\left(\sigma_\varepsilon^2\right) \\
&= \tfrac{3}{4}\sigma_\varepsilon^2.
\end{aligned}$$

The interaction between treatments and weeks is estimated by

$$\hat{\mu}_{11} - \hat{\mu}_{12} - \hat{\mu}_{21} + \hat{\mu}_{22} = 23 - 17 - 24.667 + 22.167 = 3.5,$$

and its variance is given by

$$\begin{aligned}
\text{Var}(\hat{\mu}_{11} - \hat{\mu}_{12} - \hat{\mu}_{21} + \hat{\mu}_{22}) &= \text{Var}\left[(\hat{\mu}_{11} - \hat{\mu}_{12}) - (\hat{\mu}_{21} - \hat{\mu}_{22})\right] \\
&= \text{Var}(\hat{\mu}_{11} - \hat{\mu}_{12}) + \text{Var}(\hat{\mu}_{21} - \hat{\mu}_{22}).
\end{aligned}$$

In evaluating $\text{Var}(\bar{\mu}_{\cdot 1} - \bar{\mu}_{\cdot 2})$, we showed that $\text{Var}(\hat{\mu}_{11} - \hat{\mu}_{12}) = 2\sigma_\varepsilon^2$ and $\text{Var}(\hat{\mu}_{21} - \hat{\mu}_{22}) = \sigma_\varepsilon^2$. Thus,

$$\begin{aligned}
\text{Var}(\hat{\mu}_{11} - \hat{\mu}_{12} - \hat{\mu}_{21} + \hat{\mu}_{22}) &= 2\sigma_\varepsilon^2 + \sigma_\varepsilon^2 \\
&= 3\sigma_\varepsilon^2.
\end{aligned}$$

If there is significant Treatment∗Week interaction, we might also want to estimate $\mu_{11} - \mu_{21}$ and $\mu_{12} - \mu_{22}$. The best estimates of these

Table 28.5 Results of Analysis for Missing Data in Drug Effects Example

Parameter Function	*Best Estimate*	*Variance*
μ_{11}	23	$\frac{1}{2}(\sigma_\varepsilon^2 + \sigma_\delta^2)$
μ_{12}	17	$\frac{3}{2}(\sigma_\varepsilon^2 + \frac{1}{3}\sigma_\delta^2)$
μ_{21}	24.667	$\frac{1}{3}(\sigma_\varepsilon^2 + \sigma_\delta^2)$
μ_{22}	22.167	$\frac{2}{3}(\sigma_\varepsilon^2 + \frac{1}{2}\sigma_\delta^2)$
$\bar{\mu}_{1\cdot}$	20	$\frac{1}{2}(\sigma_\varepsilon^2 + \sigma_\delta^2)$
$\bar{\mu}_{2\cdot}$	23.417	$\frac{1}{4}(\sigma_\varepsilon^2 + \frac{4}{3}\sigma_\delta^2)$
$\bar{\mu}_{1\cdot} - \bar{\mu}_{2\cdot}$	−3.417	$\frac{3}{4}(\sigma_\varepsilon^2 + \frac{10}{9}\sigma_\delta^2)$
$\bar{\mu}_{\cdot 1}$	23.833	$\frac{5}{24}(\sigma_\varepsilon^2 + \sigma_\delta^2)$
$\bar{\mu}_{\cdot 2}$	19.583	$\frac{13}{24}(\sigma_\varepsilon^2 + \frac{5}{13}\sigma_\delta^2)$
$\bar{\mu}_{\cdot 1} - \bar{\mu}_{\cdot 2}$	4.25	$\frac{3}{4}\sigma_\varepsilon^2$
$\mu_{11} - \mu_{12} - \mu_{21} + \mu_{22}$	3.5	$3\sigma_\varepsilon^2$
$\mu_{11} - \mu_{21}$	−1.667	$\frac{5}{6}(\sigma_\varepsilon^2 + \sigma_\delta^2)$
$\mu_{12} - \mu_{22}$	−5.167	$\frac{13}{6}(\sigma_\varepsilon^2 + \frac{5}{13}\sigma_\delta^2)$

functions of the parameters are

$$\hat{\mu}_{11} - \hat{\mu}_{21} = 23 - 24.667 = -1.667$$

and

$$\hat{\mu}_{12} - \hat{\mu}_{22} = 17 - 22.167 = -5.167,$$

respectively. Their respective variances are $\frac{5}{6}(\sigma_\varepsilon^2 + \sigma_\delta^2)$ and $\frac{13}{6}(\sigma_\varepsilon^2 + \frac{5}{13}\sigma_\delta^2)$.

Table 28.5 summarizes all of the above results.

28.2.3 Estimating Standard Errors

When the data are balanced, estimates of linear functions of the parameters have standard errors that depend on only a few functions of the two experimental error variances (see Section 24.3). However, when data are missing, almost every linear function of the parameters has an estimate whose standard error depends on a different function of the two experimental error variances.

To estimate these standard errors, we recommend that one first obtain estimates of σ_ε^2 and σ_δ^2. Estimating σ_ε^2 is relatively easy; its estimate is the error mean square after fitting the data with an appropriate model. However, estimating σ_δ^2 is much more complex. The problem is finding a sum of squares that estimates some linear function of σ_ε^2 and σ_δ^2 only, that is, a sum of squares that is free of the fixed treatment effects in the model. In order to construct confidence intervals and test hypotheses, it is also necessary to find a sum of squares that is statistically independent of the error mean square (see Chapters 18 and 19 for more discussion about estimating the two variances).

In order to satisfy these two objectives when estimating σ_ε^2 and σ_δ^2, we recommend that the model be fit in a stepwise fashion by first fitting all the fixed effects and then fitting the random effects. For the data in Table 28.4, we would use a model of the form

$$y_{ijk} = \mu + T_i + W_j + (TW)_{ij} + P_k(T_i) + \varepsilon_{ijk} \qquad (28.2.2)$$

rather than the model

$$y_{ijk} = \mu + T_i + P_k(T_i) + W_j + (TW)_{ij} + \varepsilon_{ijk}, \qquad (28.2.3)$$

which is recommended for balanced experiments.

Using the model in (28.2.2) gives a Patient(Treatment) Type I sum of squares that is free of Treatment and Week effects. Next we equate the observed values of the MSERROR and the MSPATIENT(TREATMENT) to their respective expected values. These expected values are obtained by using the techniques in Chapter 29 or by using the RANDOM option in SAS® GLM. The resulting equations are solved simultaneously for $\hat{\sigma}_\varepsilon^2$ and $\hat{\sigma}_\delta^2$.

An analysis of the data in Table 28.4 using the model in (28.2.2) gives (1) a MSERROR equal to .25, (2) a Type I MSPATIENT(TREATMENT) equal to $8.417/3 = 2.806$, and (3) an expected value of MSPATIENT(TREATMENT) equal to $\sigma_\varepsilon^2 + \frac{4}{3}\sigma_\delta^2$. Thus, to estimate σ_ε^2 and σ_δ^2, we solve

$$\hat{\sigma}_\varepsilon^2 = .25 \quad \text{and} \quad \hat{\sigma}_\varepsilon^2 + \tfrac{4}{3}\hat{\sigma}_\delta^2 = 2.806$$

for $\hat{\sigma}_\varepsilon^2$ and $\hat{\sigma}_\delta^2$, obtaining

$$\hat{\sigma}_\varepsilon^2 = .25 \quad \text{and} \quad \hat{\sigma}_\delta^2 = 1.917.$$

These estimates are then substituted for the parameters in Table 28.5 to obtain estimates of the standard errors of the estimates.

For example,

$$\text{Var}(\hat{\mu}_{11}) = \tfrac{1}{2}(\sigma_\varepsilon^2 + \sigma_\delta^2),$$

and hence the estimate of the standard error of $\hat{\mu}_{11}$ is

$$\widehat{\text{s.e.}}(\hat{\mu}_{11}) = \sqrt{\tfrac{1}{2}(\hat{\sigma}_\varepsilon^2 + \hat{\sigma}_\delta^2)} = \sqrt{\tfrac{1}{2}(.25 + 1.917)} = 1.041.$$

Table 28.6 Estimated Standard Errors for Data in Table 28.5

Parameter Function	*Best Estimate*	*Estimated Standard Error*
μ_{11}	23	1.041
μ_{12}	17	1.155
μ_{21}	24.667	.850
μ_{22}	22.167	1.009
$\bar{\mu}_{1\cdot}$	20	1.041
$\bar{\mu}_{2\cdot}$	23.417	.837
$\bar{\mu}_{1\cdot} - \bar{\mu}_{2\cdot}$	−3.417	1.336
$\bar{\mu}_{\cdot 1}$	23.833	.672
$\bar{\mu}_{\cdot 2}$	19.583	.731
$\bar{\mu}_{\cdot 1} - \bar{\mu}_{\cdot 2}$	4.25	.807
$\mu_{11} - \mu_{12} - \mu_{21} + \mu_{22}$	3.5	1.614
$\mu_{11} - \mu_{21}$	−1.667	1.344
$\mu_{12} - \mu_{22}$	−5.167	1.462

Table 28.6 gives the estimated standard errors for the estimators of each function in Table 28.5.

28.2.4 Tests of Hypotheses and Confidence Intervals

This section describes procedures for testing hypotheses and obtaining confidence intervals for unbalanced split-plot and repeated measures experiments, either with or without missing data. The procedures we recommend may not be best in a statistical sense, but they are the best we have discovered. We believe they are better than the results that are automatically provided by options available in computing packages.

The Type I and Type III analysis of variance tables for the data in Table 28.4 are given in Tables 28.7 and 28.8, respectively.

There is only one exact F-test that can be constructed from the Type I analysis that is meaningful. The hypothesis H_0: $\sigma_\delta^2 = 0$ versus H_a: $\sigma_\delta^2 > 0$ can be tested by

$$F = \frac{\text{Type I MSPATIENT(TREATMENT)}}{\text{MSERROR}} = \frac{2.806}{.25} = 11.22.$$

In this example, this F tests for variability among patients. In this case H_0 is accepted, but this is probably due to the fact that there is only 1

Table 28.7 Type I Analysis of Variance Table for Depression Scores

Source of Variation	*df*	*Type I MS*	*EMS*
Treatment	1	13.333	$\sigma_\varepsilon^2 + 1.717\sigma_\delta^2 + Q(T, W, T*W)$
Week	1	15.238	$\sigma_\varepsilon^2 + .248\sigma_\delta^2 + Q(W, T*W)$
Treatment * Week	1	4.762	$\sigma_\varepsilon^2 + .286\sigma_\delta^2 + Q(T*W)$
Patient(Treatment)	3	2.806	$\sigma_\varepsilon^2 + 1.333\sigma_\delta^2$
Error	1	.25	σ_ε^2

degree of freedom for estimating σ_ε^2 rather than all patients being nearly identical. This hypothesis is rejected when the full data set is analyzed.

All other so-called F-ratios from the Type I analysis test no meaningful hypotheses.

In the Type III analysis, the ratio

$$F = \frac{\text{MSWEEK}}{\text{MSERROR}} = 96.333$$

is a legitimate $F-$ test that tests H_0: $\bar{\mu}_{\cdot 1} = \bar{\mu}_{\cdot 2}$. The ratio

$$F = \frac{\text{MSTREATMENT} * \text{WEEK}}{\text{MSERROR}} = 19.05$$

is also legitimate, testing

$$H_0: \mu_{11} - \mu_{12} - \mu_{21} + \mu_{22} = 0;$$

that is, it tests for interaction.

The ratio

$$F = \frac{\text{MSTREATMENT}}{\text{MSERROR}}$$

Table 28.8 Type III Analysis of Variance Table for Depression Scores

Source of Variation	*df*	*Type III MS*	*EMS*
Treatment	1	15.565	$\sigma_\varepsilon^2 + 1.111\sigma_\delta^2 + Q(T, T*W)$
Week	1	24.083	$\sigma_\varepsilon^2 + Q(T, T*W)$
Treatment * Week	1	4.083	$\sigma_\varepsilon^2 + Q(T*W)$
Patient(Treatment)	3	2.806	$\sigma_\varepsilon^2 + 1.333\sigma_\delta^2$
Error	1	.25	σ_ε^2

is an F-test, but it could be large if σ_δ^2 is large or if there is a difference between treatments. Thus, a significant treatment F-ratio tells us that something is happening, but it does not say exactly what.

In balanced data sets, we used the ratio

$$\frac{\text{MSTREATMENT}}{\text{MSPATIENT(TREATMENT)}}$$

to test for significant treatment effects. This is not appropriate in unbalanced data sets, since the coefficients of σ_δ^2 in the expected values of the numerator and denominator mean squares usually differ. A statistic for testing for significant treatment effects is

$$\begin{aligned}\tilde{F} &= \frac{15.565}{\hat{\sigma}_\varepsilon^2 + 1.111\hat{\sigma}_\delta^2} \\ &= \frac{15.565}{.25 + 1.111(1.917)} \\ &= \frac{15.565}{2.380} = 6.54.\end{aligned}$$

We note that $\tilde{F}$ does not have an exact F-distribution since (1) the statistic in the denominator does not have a distribution that is proportional to an exact chi-square distribution, and (2) the numerator and denominator of $\tilde{F}$ may not be independently distributed. Still, $\tilde{F}$ should have an approximate F-distribution with the critical point obtained by a Satterthwaite approximation.

Recall from Chapter 2 that if

$$\frac{\nu_1\hat{\sigma}_1^2}{\sigma_1^2} \sim \chi^2(\nu_1) \quad \text{and} \quad \frac{\nu_2\hat{\sigma}_2^2}{\sigma_2^2} \sim \chi^2(\nu_2),$$

and if $\hat{\sigma}_1^2$ and $\hat{\sigma}_2^2$ are independent, then an unbiased estimate of $c_1\sigma_1^2 + c_2\sigma_2^2$ is

$$c_1\hat{\sigma}_1^2 + c_2\hat{\sigma}_2^2,$$

and

$$\frac{\hat{\nu}\left(c_1\hat{\sigma}_1^2 + c_2\hat{\sigma}_2^2\right)}{c_1\sigma_1^2 + c_2\sigma_2^2} \;\dot{\sim}\; \chi^2(\hat{\nu}) \tag{28.2.4}$$

where

$$\hat{\nu} = \frac{\left(c_1\hat{\sigma}_1^2 + c_2\hat{\sigma}_2^2\right)^2}{c_1^2\hat{\sigma}_1^4/\nu_1 + c_2^2\hat{\sigma}_2^4/\nu_2}. \tag{28.2.5}$$

Using this result, we can test H_0: $\bar{\mu}_{1\cdot} = \bar{\mu}_{2\cdot\cdot}$ If H_0 is true, then

$$\frac{\text{Type III SSTREATMENT}}{\sigma_\varepsilon^2 + 1.111\sigma_\delta^2} \sim \chi^2(1),$$

and from the Type I analysis we have that

$$\frac{3 \cdot \text{MSPATIENT(TREATMENT)}}{\sigma_\varepsilon^2 + 1.333\sigma_\delta^2} \sim \chi^2(3)$$

and

$$\frac{1 \cdot \text{EMS}}{\sigma_\varepsilon^2} \sim \chi^2(1),$$

and both mean squares are independently distributed. Let $\sigma_1^2 = \sigma_\varepsilon^2 + 1.333\sigma_\delta^2$ and $\sigma_2^2 = \sigma_\varepsilon^2$. Then in equations (28.2.4) and (28.2.5), let

$$\hat{\sigma}_1^2 = \hat{\sigma}_\varepsilon^2 + 1.333\hat{\sigma}_\delta^2, \quad \nu_1 = 3, \quad \hat{\sigma}_2^2 = \hat{\sigma}_\varepsilon^2, \quad \text{and} \quad \nu_2 = 1.$$

Note that

$$\sigma_\varepsilon^2 + 1.111\sigma_\delta^2 = \left(\frac{1.111}{1.333}\right)\sigma_1^2 + \left(1 - \frac{1.111}{1.333}\right)\sigma_2^2$$
$$= .833\sigma_1^2 + .167\sigma_2^2$$

so that c_1 and c_2 in equations (28.2.4) and (28.2.5) must have values $c_1 = .833$ and $c_2 = .167$. Upon applying Satterthwaite's result, we get

$$\frac{\hat{\nu}\left(.833\hat{\sigma}_1^2 + .167\hat{\sigma}_2^2\right)}{\sigma_\varepsilon^2 + 1.111\sigma_\delta^2} \doteq \chi^2(\hat{\nu})$$

where

$$\hat{\nu} = \frac{[(.833)(2.806) + (1.67)(.25)]^2}{(.833)^2(2.806)^2/3 + (1.67)^2(.25)^2/1} = 3.1.$$

Thus, since

$$.833\hat{\sigma}_1^2 + .167\hat{\sigma}_2^2 = \hat{\sigma}_\varepsilon^2 + 1.111\hat{\sigma}_\delta^2 = 2.380$$

we find that $\tilde{F}$ is approximately distributed as the F-distribution $F(1, 3.1)$ when H_0 is true. To test H_0: $\bar{\mu}_{1\cdot} = \bar{\mu}_{2\cdot}$, we compare the observed value of $\tilde{F}$, $\tilde{F} = 6.54$, to $F_{\alpha,1,3.1}$. From a table of F critical points, we determine that

$$F_{.10,1,3} = 5.54 \quad \text{and} \quad F_{.10,1,4} = 4.54.$$

By interpolation, we get

$$F_{.10,1,3.1} \doteq 5.44.$$

Since 6.54 > 5.44, we conclude that the drug had a significant effect at the 10% level.

In a similar manner, approximate t-statistics for testing any linear combinations of the μ_{ij}'s equal to zero can be obtained. The approximate

t-statistic for testing

$$H_0: \sum_{ij} c_{ij}\mu_{ij} = 0$$

is

$$\tilde{t} = \frac{\Sigma_{ij} c_{ij}\hat{\mu}_{ij}}{\widehat{\text{s.e.}}(\Sigma_{ij} c_{ij}\hat{\mu}_{ij})}.$$

The appropriate degrees of freedom for this t-statistic must now be estimated. If

$$\text{Var}\left(\sum_{ij} c_{ij}\hat{\mu}_{ij}\right) = k\left(\sigma_\varepsilon^2 + c\sigma_\delta^2\right),$$

$$\hat{\sigma}_1^2 = \hat{\sigma}_\varepsilon^2 + q\hat{\sigma}_\delta^2$$

with ν_1 degrees of freedom, and $\hat{\sigma}_2^2 = \hat{\sigma}_\varepsilon^2$ with ν_2 degrees of freedom, then

$$\hat{\nu} = \frac{\left[c\hat{\sigma}_1^2 + (q - c)\hat{\sigma}_2^2\right]}{c^2\hat{\sigma}_1^4/\nu_1 + (q - c)^2\hat{\sigma}_2^4/\nu_2}.$$

As an example, suppose we want to test H_0: $\mu_{12} = \mu_{22}$ or equivalently, that $\mu_{12} - \mu_{22} = 0$. From Table 28.6 we get

$$\tilde{t} = \frac{-5.167}{1.462} = -3.534.$$

From Table 28.5 we observe that $k = \frac{13}{6}$ and $c = \frac{5}{13}$, and from Table 28.7 that $q = 1.333 = \frac{4}{3}$. Hence, the degrees of freedom of $\tilde{t}$ can be approximated by

$$\hat{\nu} = \frac{\left[\frac{5}{13}(2.806) + \left(\frac{4}{3} - \frac{5}{13}\right)(.25)\right]^2}{\left(\frac{5}{13}\right)^2(2.806)^2/3 + \left(\frac{4}{3} - \frac{5}{13}\right)^2(.25)^2/1}$$
$$= \frac{1.733}{.445} = 3.9.$$

Using the table of percentage points of the t-distribution and interpolating, we get $t_{.025,3.9} \doteq 2.817$. Thus, H_0 would be rejected at the 5% level of significance.

The results presented in this section are summarized in the next section in "cookbook" fashion. Because they are required for tests of hypotheses and confidence intervals, we also show how to obtain the values for c, k, and q from a SAS® analysis.

28.2.5 Inference Procedures Using SAS®

This section describes an SAS® GLM analysis of the data in Table 28.4 and how to use the GLM procedure to get a complete analysis of all split-plot and certain repeated measures experiments with missing data. The important results from the last three sections will also be summarized in the course of the discussion.

Unfortunately, none of the well-known statistical packages correctly analyze split-plot and repeated measures experiments with missing data. However, SAS® GLM does provide enough information to enable an informed user to construct an appropriate analysis by doing additional calculations, which most likely will have to be done by hand. Non-SAS® users may skip this section.

In order to obtain the necessary information from SAS® GLM for a complete analysis of split-plot experiments with missing data, one should observe the following guidelines:

1. Always write the model in such a way so that sums of squares corresponding to the random components in the model are free of the fixed effects. This can be accomplished by listing the fixed effects in the model statement first and then listing the random effects. This allows one to obtain independent sums of squares from a Type I analysis, from which estimates of the different experimental error variances can be obtained.
2. Always include the RANDOM option. This allows one to determine whether the objectives in the first guideline have been met. It is also necessary in order to obtain expected mean squares, which are required to construct tests of hypotheses and confidence intervals.
3. Always include CONTRAST and ESTIMATE options for every linear combination of the treatment effects of interest. This includes such linear combinations as population marginal means, main effects, and all interaction contrasts of interest. It should be noted that the LSMEANS option, with or without PDIFF or STDERR, does not give enough information by itself to enable the experimenter to do a correct analysis.

The data in Table 28.4 was analyzed with SAS® GLM using the following statements:

```
PROC GLM; CLASSES TRT PATIENT WEEK;
MODEL Y = TRT|WEEK PATIENT(TRT)/SOLUTION E E3 SS1 SS3;
RANDOM P(T);
CONTRAST 'TW11' INTERCEPT 1 TRT 1 0 W 1 0 TRT*W 1 0 0 0;
ESTIMATE 'TW11' INTERCEPT 1 TRT 1 0 W 1 0 TRT*W 1 0 0 0;
CONTRAST 'TW12' INTERCEPT 1 TRT 1 0 W 0 1 T*W 0 1 0 0;
ESTIMATE 'TW12' INTERCEPT 1 TRT 1 0 W 0 1 T*W 0 1 0 0;
CONTRAST 'TW21' INTERCEPT 1 T 0 1 W 1 0 T*W 0 0 1 0;
```

```
ESTIMATE 'TW21' INTERCEPT 1 T 0 1 W 1 0 T*W 0 0 1 0;
CONTRAST 'TW22' INTERCEPT 1 T 0 1 W 0 1 T*W 0 0 0 1;
ESTIMATE 'TW22' INTERCEPT 1 T 0 1 W 0 1 T*W 0 0 0 1;
CONTRAST 'TRT1' INTERCEPT 1 TRT 1 0 W .5 .5 T*W .5 .5 0 0
  P(T) .5 .5 0 0 0;
ESTIMATE 'TRT1' INTERCEPT 1 TRT 1 0 W .5 .5 T*W .5 .5 0 0
  P(T) .5 .5 0 0 0;
CONTRAST 'TRT2' INTERCEPT 1 TRT 0 1 W .5 .5 T*W 0 0 .5 .5
  P(T) 0 0 .33333 .33333 .33333;
ESTIMATE 'TRT2' INTERCEPT 1 TRT 0 1 W .5 .5 T*W 0 0 .5 .5
  P(T) 0 0 .33333 .33333 .33333;
CONTRAST 'TRT DIFF' TRT 1 -1;
ESTIMATE 'TRT DIFF' TRT 1 -1;
CONTRAST 'WEEK1 M E' INTERCEPT 1 TRT .5 .5 W 1 0 TRT*WEEK
  .5 0 .5 0 P(T) .25 .25 .166667 .166667 .166667;
ESTIMATE 'WEEK1 M E' INTERCEPT 1 TRT .5 .5 W 1 0 TRT*WEEK
  .5 0 .5 0 P(T) .25 .25 .166667 .166667 .166667;
CONTRAST 'WEEK2 M E' INTERCEPT 1 TRT .5 .5 W 0 1 T*W 0
  .5 0 .5 P(T) .25 .25 .166667 .166667 .166667;
ESTIMATE 'WEEK2 M E' INTERCEPT 1 TRT .5 .5 W 0 1 T*W 0
  .5 0 .5 P(T) .25 .25 .166667 .166667 .166667;
CONTRAST 'WEEK DIFF' WEEK 1 -1 T*W .5 -.5 .5 -.5;
ESTIMATE 'WEEK DIFF' WEEK 1 -1 T*W .5 -.5 .5 -.5;
CONTRAST 'INTERACTION' TRT*WEEK 1 -1 -1 1;
ESTIMATE 'INTERACTION' TRT*WEEK 1 -1 -1 1;
CONTRAST 'WEEK1 DIFF' TRT 1 -1 TRT*WEEK 1 0 -1 0;
ESTIMATE 'WEEK1 DIFF' TRT 1 -1 TRT*WEEK 1 0 -1 0;
CONTRAST 'WEEK2 DIFF' TRT 1 -1 TRT*WEEK 0 1 0 -1;
ESTIMATE 'WEEK2 DIFF' TRT 1 -1 TRT*WEEK 0 1 0 -1;
```

Portions of the resulting analysis are given in Tables 28.9 through 28.13.

Table 28.9 Expected Mean Squares Resulting from RANDOM Option of SAS® GLM

SOURCE	TYPE I EXPECTED MEAN SQUARE
TRT	VAR(ERROR) + 1.71666667 VAR(PATIENT(TRT)) + Q(TRT,WEEK,TRT*WEEK)
WEEK	VAR(ERROR) + 0.247619048 VAR(PATIENT(TRT)) + Q(WEEK,TRT*WEEK)
TRT*WEEK	VAR(ERROR) + 0.285714286 VAR(PATIENT(TRT)) + Q(TRT*WEEK)
PATIENT(TRT)	VAR(ERROR) + 1.33333333 VAR(PATIENT(TRT))

SOURCE	TYPE III EXPECTED MEAN SQUARE
TRT	VAR(ERROR) + 1.11111111 VAR(PATIENT(TRT)) + Q(TRT,TRT*WEEK)
WEEK	VAR(ERROR) + Q(WEEK,TRT*WEEK)
TRT*WEEK	VAR(ERROR) + Q(TRT*WEEK)
PATIENT(TRT)	VAR(ERROR) + 1.33333333 VAR(PATIENT(TRT))

CONTRAST	EXPECTED MEAN SQUARE
TW11	VAR(ERROR) + VAR(PATIENT(TRT)) + Q(INTERCEPT,TRT,WEEK,TRT*WEEK)
TW12	VAR(ERROR) + 0.333333333 VAR(PATIENT(TRT)) + Q(INTERCEPT,TRT,WEEK,TRT*WEEK)
TW21	VAR(ERROR) + VAR(PATIENT(TRT)) + Q(INTERCEPT,TRT,WEEK,TRT*WEEK)
TW22	VAR(ERROR) + 0.5 VAR(PATIENT(TRT)) + Q(INTERCEPT,TRT,WEEK,TRT*WEEK)
TRT1	VAR(ERROR) + VAR(PATIENT(TRT)) + Q(INTERCEPT,TRT,WEEK,TRT*WEEK)
TRT2	VAR(ERROR) + 1.33331556 VAR(PATIENT(TRT)) + Q(INTERCEPT,TRT,WEEK,TRT*WEEK)
TRT DIFF	VAR(ERROR) + 1.11111111 VAR(PATIENT(TRT)) + Q(TRT,TRT*WEEK)
WEEK1 M E	VAR(ERROR) + 1.0000016 VAR(PATIENT(TRT)) + Q(INTERCEPT,TRT,WEEK,TRT*WEEK)
WEEK2 M E	VAR(ERROR) + 0.384615882 VAR(PATIENT(TRT)) + Q(INTERCEPT,TRT,WEEK,TRT*WEEK)
WEEK DIFF	VAR(ERROR) + Q(WEEK,TRT*WEEK)
INTERACTION	VAR(ERROR) + Q(TRT*WEEK)
WEEK1 DIFF	VAR(ERROR) + VAR(PATIENT(TRT)) + Q(TRT,TRT*WE
WEEK2 DIFF	VAR(ERROR) + 0.384615385 VAR(PATIENT(TRT)) + Q(TRT,TRT*WEEK)

The general form of estimable functions for the data in Table 28.4 provided by SAS® GLM is

$$L_1 \cdot \mu + L_2 \cdot T_1 + (L_1 - L_2) \cdot T_2 + L_4 \cdot W_1 + (L_1 - L_4) \cdot W_2$$

$$+ L_6(T * W)_{11} + (L_2 - L_6) \cdot (T * W)_{12} + (L_4 - L_6) \cdot (T * W)_{21}$$

$$+ (L_1 - L_2 - L_4 + L_6)(T * W)_{22} + L_{10} \cdot P_1(T_1)$$

$$+ (L_2 - L_{10}) \cdot P_2(T_1) + L_{12} \cdot P_3(T_2) + L_{13} P_4(T_2)$$

$$+ P_5(T_2).$$

Recall that $P_k(T_i)$ is a random effect in the split-plot model rather than a fixed effect and thus should not be involved in any judgments about whether a function of the fixed effects parameters (μ, T_1, T_2, W_1, W_2, $(T * W)_{11}$, $(T * W)_{12}$, $(T * W)_{21}$, and $(T * W)_{22}$) is estimable. However, when SAS® checks for estimability, it considers all effects to be fixed.

Table 28.10 Type I and Type III Analysis of Variance Tables Generated by SAS® GLM

SOURCE	DF	SUM OF SQUARES	MEAN SQUARE
MODEL	6	41.75000000	6.95833333
ERROR	1	0.25000000	0.25000000
CORRECTED TOTAL	7	42.00000000	

MODEL F =	27.83	PR > F = 0.1441

R-SQUARE	C.V.	ROOT MSE	Y MEAN
0.994048	2.1739	0.50000000	23.00000000

SOURCE	DF	TYPE I SS	F VALUE	PR > F
TRT	1	13.33333333	53.33	0.0866
WEEK	1	15.23809524	60.95	0.0811
TRT*WEEK	1	4.76190476	19.05	0.1434
PATIENT(TRT)	3	8.41666667	11.22	0.2152

SOURCE	DF	TYPE III SS	F VALUE	PR > F
TRT	1	15.56481481	62.26	0.0803
WEEK	1	24.08333333	96.33	0.0646
TRT*WEEK	1	4.08333333	16.33	0.1544
PATIENT(TRT)	3	8.41666667	11.22	0.2152

Table 28.11 Results from CONTRAST Options of SAS® GLM

CONTRAST	DF	SS	F VALUE	PR > F
TW11	1	1058.00000000	4232.00	0.0098
TW12	1	192.66666667	770.67	0.0229
TW21	1	1825.33333333	7301.33	0.0075
TW22	1	737.04166667	2948.17	0.0117
TRT1	1	800.00000000	3200.00	0.0113
TRT2	1	2193.37869938	8773.51	0.0068
TRT DIFF	1	15.56481481	62.26	0.0803
WEEK1 M E	1	2726.53297106	10906.13	0.0061
WEEK2 M E	1	708.01248818	2832.05	0.0120
WEEK DIFF	1	24.08333333	96.33	0.0646
INTERACTION	1	4.08333333	16.33	0.1544
WEEK1 DIFF	1	3.33333333	13.33	0.1702
WEEK2 DIFF	1	12.32051282	49.28	0.0901

This is one reason why SAS® claimed that the FERT marginal means in the herbicidal example were not estimable, as pointed out in Section 28.1.4.

The Type III estimable functions given in Table 28.13 can be used to determine the hypotheses being tested by using Type III sums of squares. When considering these hypotheses, one can ignore the coefficients of the random effects in the model. As an aside, note that estimable functions having nonzero coefficients for the Patient(Treatment) terms are comparisons involving the whole plot experimental units; hence, their standard errors involve both of the error variances, σ_ε^2 and σ_δ^2. Those estimable functions having zero coefficients for these terms are within-whole-plot or subplot experimental unit comparisons and involve the subplot errors only. Hence, their standard errors involve only the subplot error variance, σ_ε^2.

The first thing one should do when analyzing split-plot experiments with missing data is to estimate the two error variances, σ_ε^2 and σ_δ^2. This

Table 28.12 Results from ESTIMATE Options of SAS® GLM

PARAMETER	ESTIMATE	T FOR H0: PARAMETER=0	PR > |T|	STD ERROR OF ESTIMATE
TW11	23.00000000	65.05	0.0098	0.35355339
TW12	17.00000000	27.76	0.0229	0.61237244
TW21	24.66666667	85.45	0.0075	0.28867513
TW22	22.16666667	54.30	0.0117	0.40824829
TRT1	20.00000000	56.57	0.0113	0.35355339
TRT2	23.41668250	93.67	0.0068	0.24999917
TRT DIFF	-3.41666667	-7.89	0.0803	0.43301270
WEEK1 M E	23.83333175	104.43	0.0061	0.22821773
WEEK2 M E	19.58333175	53.22	0.0120	0.36799009
WEEK DIFF	4.25000000	9.81	0.0646	0.43301270
INTERACTION	3.50000000	4.04	0.1544	0.86602540
WEEK1 DIFF	-1.66666667	-3.65	0.1702	0.45643546
WEEK2 DIFF	-5.16666667	-7.02	0.0901	0.73598007

Table 28.13 Type III Estimable Functions Generated by SAS® GLM

EFFECT		COEFFICIENTS	COEFFICIENTS	COEFFICIENTS
INTERCEPT		0	0	0
TRT	1	L2	0	0
	2	-L2	0	0
WEEK	1	0	L4	0
	2	0	-L4	0
TRT*WEEK	1 1	0.5*L2	0.5*L4	L6
	1 2	0.5*L2	-0.5*L4	-L6
	2 1	-0.5*L2	0.5*L4	-L6
	2 2	-0.5*L2	-0.5*L4	L6
PATIENT(TRT)	1 1	0.5*L2	0	0
	2 1	0.5*L2	0	0
	3 2	-0.3333*L2	0	0
	4 2	-0.3333*L2	0	0
	5 2	-0.3333*L2	0	0

can be done by equating the observed MSERROR and the Type I MSPATIENT(TREATMENT) to their respective expected mean squares and then solving for the individual estimates.

Generally, the expected mean squares will be of the form σ_ε^2 and $\sigma_\varepsilon^2 + q\sigma_\delta^2$ for some constant q. From Table 28.9, $q = 1.333$, and the resulting equations are $\hat{\sigma}_\varepsilon^2 = .25$ (from the error mean square) and

$$\hat{\sigma}_\varepsilon^2 + 1.3333\sigma_\delta^2 = \frac{8.4167}{3}$$

(from MSPATIENT(TREATMENT). (Note that the MSPATIENT(TREATMENT) is the ratio of the Patient(Treatment) sum of squares and its corresponding degrees of freedom.) Solving the above two equations for $\hat{\sigma}_\varepsilon^2$ and $\hat{\sigma}_\delta^2$ yields $\hat{\sigma}_\varepsilon^2 = .25$ and

$$\hat{\sigma}_\delta^2 = \frac{\text{MSPATIENT(TREATMENT)} - \hat{\sigma}_\varepsilon^2}{q} = 1.917.$$

Next, we consider the hypotheses that are automatically tested by SAS®. All F-values calculated by SAS® are obtained by dividing each "effect" mean square by $\hat{\sigma}_\varepsilon^2$. If the Type III expected mean square of an effect involves σ_ε^2 only and not σ_δ^2, then the resulting F-value is a proper one to use.

However, if the expected mean square of an effect involves σ_δ^2 as well as σ_ε^2, the F-value given by SAS® is not legitimate. In balanced split-plot experiments, one can use the TEST option to obtain appropriate F's for whole plot comparisons. SAS® GLM also allows this option and, unless the user specifies otherwise, divides by the error mean square corresponding to the highest type that was computed in the model. More often than not for messy data cases, this is not a legitimate F-test either, since the expected mean squares of the numerator and denominator are not identical when H_0 is true; in addition, the Type III sums of squares are generally not distributed independently of one another.

Hence, we recommend that one calculate an approximate F-statistic and estimate its degrees of freedom. The procedure, which was discussed in detail in Section 28.2.4, can be summarized as follows:

1. From the table of expected mean squares (Table 28.9), examine the one corresponding to the effect being considered. It will be of the form $\sigma_\varepsilon^2 + c\sigma_\delta^2 + Q(\cdot)$ for some constant c. Thus, an appropriate divisor for obtaining an approximate F-statistic will be $\hat{\sigma}_\varepsilon^2 + c\hat{\sigma}_\delta^2$ where $\hat{\sigma}_\varepsilon^2 = .25$ and $\hat{\sigma}_\delta^2 = 1.917$, the estimates of the variance components. An approximate F-statistic is
$$\tilde{F} = \frac{\text{MSEFFECT}}{\hat{\sigma}_\varepsilon^2 + c\hat{\sigma}_\delta^2}.$$
2. Compare $\tilde{F}$ to F_{α, f_1, f_2} where f_1 is the degrees of freedom of the MSEFFECT and
$$f_2 = \frac{\left[c\hat{\sigma}_1^2 + (q-c)\hat{\sigma}_2^2\right]^2}{c^2\hat{\sigma}_1^4/\nu_1 + (q-c)^2\hat{\sigma}_2^4/\nu_2}, \qquad (28.5.1)$$
$\hat{\sigma}_1^2 = \hat{\sigma}_\varepsilon^2 + q\hat{\sigma}_\delta^2$ and $\hat{\sigma}_2^2 = \hat{\sigma}_\varepsilon^2$. Note that $\hat{\sigma}_1^2$ is the Type I MSPATIENT(TREATMENT), which is based on ν_1 degrees of freedom, and $\hat{\sigma}_2^2$ is the observed MSERROR, which is based on ν_2 degrees of freedom.

From Table 28.9 the Type III MSTREATMENT has an expectation given by $\sigma_\varepsilon^2 + 1.111\sigma_\delta^2$. An approximate F-statistic for comparing treatment main effects is
$$\tilde{F} = \frac{\text{Type III MSTREATMENT}}{\hat{\sigma}_\varepsilon^2 + 1.111\hat{\sigma}_\delta^2} = \frac{15.5648/1}{.25 + (1.1111)(1.917)} = 6.54.$$
Its degrees of freedom are $f_1 = 1$ and $f_2 = 3.1$. To obtain f_2, note that
$$\hat{\sigma}_1^2 = \frac{8.4167}{3} = 2.806, \quad \hat{\sigma}_2^2 = .25, \quad \nu_1 = 3, \quad \text{and} \quad \nu_2 = 1.$$
Thus,
$$f_2 = \frac{[(1.111)(2.806) + (1.333 - 1.111)(.25)]^2}{(1.111)^2(2.806)^2/3 + (1.333 - 1.111)^2(.25)^2/1} = 3.1.$$
Interpolating in the F-table provides
$$F_{.10,1,3.1} \doteq 5.44.$$

Since $\tilde{F} > 5.44$, we conclude that the treatment has a significant effect at the 10% significance level.

In missing-data situations, the hypotheses generated by SAS® may not be meaningful. Hence, the experimenter will often need to specify his or her own hypotheses. Furthermore, if one uses the SAS® GLM LSMEANS option along with its options, PDIFF and STDERR, the information obtained is not always correct. In particular, the standard errors are often wrong, and hence, the significance probabilities are also wrong.

Information obtained with SAS® GLM can be used to get corrected standard errors and approximate significance probabilities. Suppose the experimenter is interested in estimating an estimable linear function of the parameters denoted by L. The standard error of the least squares estimator, $\hat{L}$, is always of the form

$$\left[k\left(\sigma_\varepsilon^2 + c\sigma_\delta^2\right)\right]^{1/2}.$$

Thus, in order to estimate the standard error of $\hat{L}$, the constants k and c must be determined. One of these constants can be found by using the CONTRAST 'L' option, while the other can be found by using the ESTIMATE 'L' option.

When one uses the CONTRAST 'L' option, the expected mean square of the observed mean square for testing H_0: $L = 0$ is printed provided that the RANDOM option has also been specified. This expected mean square is always of the form $\sigma_\varepsilon^2 + c\sigma_\delta^2 + Q(\cdot)$ and hence provides the value of c.

When one uses the ESTIMATE 'L' option, $\hat{L}$ and a standard error (which is incorrect) is printed. The printed standard error is always of the form $(k\hat{\sigma}_\varepsilon^2)^{1/2}$. Since $\hat{\sigma}_\varepsilon^2$ is known, this printed standard error can be used to determine k as

$$k = \frac{[\text{Printed standard error}]^2}{\hat{\sigma}_\varepsilon^2}.$$

Combining the results obtained from these two options, a corrected standard error of $\hat{L}$ is

$$\widehat{\text{s.e.}}(\hat{L}) = \sqrt{k\left(\hat{\sigma}_\varepsilon^2 + c\hat{\sigma}_\delta^2\right)}\,.$$

To test hypotheses about L or to construct a confidence interval for L, we note that

$$\frac{\hat{L} - L}{\widehat{\text{s.e.}}(\hat{L})} \;\dot{\sim}\; t(f_2)$$

where f_2 is given by equation (28.5.1). Thus, to test H_0: $L = 0$, H_0 is rejected at the $\alpha 100\%$ significance level if

$$|t_c| = \left|\frac{\hat{L}}{\widehat{\text{s.e.}}(\hat{L})}\right| > t_{\alpha/2, f_2}.$$

An approximate $(1-\alpha)100\%$ confidence interval for L is given by

$$\hat{L} \mp t_{\alpha/2, f_2} \cdot \widehat{\text{s.e.}}(\hat{L}).$$

Alternatively, one could also test H_0: $L = 0$ by using the F-test procedure described in Section 1.4. This does not require the use of the ESTIMATE option, since the CONTRAST option alone gives all the necessary information. The CONTRAST option is sufficient since only a value for c is required, not a value for k. One does need the ESTIMATE option to construct confidence intervals.

As an example, consider the data in Table 28.4 and let

$$\mu_{ij} = \mu + T_i + W_j + (TW)_{ij}$$

for all i and j. For each of the parameter functions in Table 28.5, we give the least squares estimate, its standard error, a t-statistic for testing the function equal to zero, and the approximate degrees of freedom of the t-statistic. The results are summarized in Table 28.14.

In Table 28.14, the values under the column labeled "Best Estimate" are obtained from the ESTIMATE option of the SAS® analysis and were given in Table 28.12. The values under the column labeled "c" are

Table 28.14 Required Calculations for Specified Contrasts from Table 28.5

Parameter Function	*Best Estimate*	*c*	*Printed Standard Error*	*k*	*Corrected Standard Error*	*Estimated Degrees of Freedom*	$\tilde{t}$
μ_{11}	23	1	.3536	.500	1.041	3.2	22.09
μ_{12}	17	.3333	.6124	1.500	1.155	4.0	14.72
μ_{21}	24.667	1	.2887	.333	.849	3.2	29.05
μ_{22}	22.167	.5	.4082	.667	.898	3.7	24.68
$\bar{\mu}_{1\cdot}$	20	1	.3536	.500	1.041	3.2	18.21
$\bar{\mu}_{2\cdot}$	23.417	1.3333	.2500	.250	.837	3.0	47.19
$\bar{\mu}_{1\cdot} - \bar{\mu}_{2\cdot}$	−3.417	1.1111	.4330	.750	1.336	3.1	−2.56
$\bar{\mu}_{\cdot 1}$	23.833	1	.2282	.208	.671	3.2	35.52
$\bar{\mu}_{\cdot 2}$	19.583	.3846	.3680	.542	.731	3.9	26.79
$\bar{\mu}_{\cdot 1} - \bar{\mu}_{\cdot 2}$	4.25	0	.4330	.750	.433	1	9.81
$\bar{\mu}_{11} - \bar{\mu}_{12} - \bar{\mu}_{21} + \bar{\mu}_{22}$	3.5	0	.8660	3.000	.866	1.0	4.04
$\bar{\mu}_{11} - \bar{\mu}_{21}$	−1.667	1	.4564	.833	1.343	3.2	−1.24
$\bar{\mu}_{12} - \bar{\mu}_{22}$	−5.167	.3846	.7360	2.167	1.463	3.9	−3.53

obtained from EXPECTED MEAN SQUARES of the SAS® analysis, where they are the coefficients of the VAR(PATIENT(TRT)) term. The values of c are given in Table 28.9. The values under the column labeled "Printed Standard Error" are obtained from the "STD ERROR OF ESTIMATE" column of Table 28.12 and correspond to the estimates just obtained. The values below the column labeled "k" are obtained from

$$k = \frac{[\text{Printed standard error}]^2}{\hat{\sigma}_\varepsilon^2}.$$

The column labeled "Corrected Standard Error" is obtained by

$$\text{Corrected standard error} = \sqrt{k\left(\hat{\sigma}_\varepsilon^2 + c\hat{\sigma}_\delta^2\right)}.$$

The column labeled "Estimated Degrees of Freedom" is obtained by formula (28.5.1); finally, the column labeled "$\tilde{t}$" is obtained from

$$\tilde{t} = \frac{[\text{Best estimate}]}{[\text{Corrected standard error}]}.$$

To obtain an approximate confidence interval for the parametric function given in column 1, one should use

$$[\text{Best estimate}] \pm \left(t_{\alpha/2, f_2}\right)[\text{Corrected standard error}].$$

Admittedly, such analyses are tedious, but they should be done to obtain correct information.

CONCLUDING REMARKS

In this chapter, we considered the analysis of messy split-plot experiments. The results given can be applied to repeated measures experiments provided that they satisfy the Huyhn–Feldt conditions. Since existing computing packages do not do correct analyses on these kinds of data, we introduced procedures that enable experimenters to do a better job of analyzing these kinds of messy experiments. Admittedly, the analyses recommended are not simple or easy to obtain, but if an experiment is worth doing, it should be worth analyzing by the best method available.

At the present time, SAS® GLM provides all of the information necessary for constructing an approximate analysis. We showed how to use this information to obtain the analyses recommended. The procedures discussed have been extended to more complicated design structures such as a split-split-plot design, but these extensions are not discussed in this book.

29

Computing the Variances of Contrasts for Repeated Measures and Split-Plot Designs by Using Hartley's Method of Synthesis

CHAPTER OUTLINE

It is very difficult to compute the coefficients of the variance components in the variance of a contrast (or any linear combination) of the parameter vector of an unbalanced model. The method of synthesis developed by Hartley (1967) for determining the coefficients of variance components in the expected sums of squares, discussed in Chapter 18, can also be used to obtain the coefficients of variance components for the variance of any estimator provided by a computer program.

29.1 USING HARTLEY'S METHOD OF SYNTHESIS

A general model involving more than one error term can be expressed in matrix notation as

$$\mathbf{y} = \mathbf{X}\boldsymbol{\beta} + \mathbf{X}_1\mathbf{a}_1 + \mathbf{X}_2\mathbf{a}_2 + \cdots + \mathbf{X}_k\mathbf{a}_k + \boldsymbol{\varepsilon} \tag{29.1.1}$$

where $\mathbf{a}_i \sim N(\mathbf{0}, \sigma_i^2\mathbf{I}_{n_i})$, $i = 1, 2, \ldots, k$; $\boldsymbol{\varepsilon} \sim N(\mathbf{0}, \sigma_\varepsilon^2\mathbf{I}_n)$; and $\mathbf{a}_i$ and $\boldsymbol{\varepsilon}$ are independent random variables. The variance of $\mathbf{y}$ is

$$\text{Var}(\mathbf{y}) = \sigma_\varepsilon^2\mathbf{I} + \sigma_1^2\mathbf{X}_1\mathbf{X}_1' + \sigma_2^2\mathbf{X}_2\mathbf{X}_2' + \cdots + \sigma_k^2\mathbf{X}_k\mathbf{X}_k'. \tag{29.1.2}$$

When a computer program is used to obtain estimates of the parameter $\boldsymbol{\beta}$ or contrasts or linear combinations of $\boldsymbol{\beta}$ as $\mathbf{h}'\boldsymbol{\beta}$, the resulting estimate can be expressed as

$$\widehat{\mathbf{h}'\boldsymbol{\beta}} = \mathbf{b}'\mathbf{y} \tag{29.1.3}$$

for some choice of $\mathbf{b}$. Determining $\mathbf{b}$ explicitly is generally very difficult, but it is generated by the computer program (although the value of $\mathbf{b}$ is not necessarily provided as output). The important point is that the computer will compute the value of $\mathbf{b}'\mathbf{y} = \widehat{\mathbf{h}'\boldsymbol{\beta}}$.

The variance of the linear combination $\mathbf{b}'\mathbf{y}$ (or $\widehat{\mathbf{h}'\boldsymbol{\beta}}$) is

$$\begin{aligned}\text{Var}(\mathbf{b}'\mathbf{y}) &= \text{Var}\left(\widehat{\mathbf{h}'\boldsymbol{\beta}}\right)\\ &= \sigma_\varepsilon^2\mathbf{b}'\mathbf{b} + \sigma_1^2\mathbf{b}'\mathbf{X}_1\mathbf{X}_1'\mathbf{b} + \sigma_2^2\mathbf{b}'\mathbf{X}_2\mathbf{X}_2'\mathbf{b} + \cdots + \sigma_k^2\mathbf{b}'\mathbf{X}_k\mathbf{X}_k'\mathbf{b}.\end{aligned}$$

Many computer programs compute $\mathbf{b}'\mathbf{y}$, giving its variance (or standard error) as $\sigma_\varepsilon^2\mathbf{b}'\mathbf{b}$. Thus the value of $\mathbf{b}'\mathbf{b}$ is generally available as output in the form of

$$\mathbf{b}'\mathbf{b} = \frac{\widehat{\text{Var}(\mathbf{b}'\mathbf{y})}}{\hat{\sigma}_\varepsilon^2}.$$

The coefficient of σ_1^2 (or any σ_i^2) can be expressed as

$$\begin{aligned}\mathbf{b}'\mathbf{X}_1\mathbf{X}_1'\mathbf{b} &= \mathrm{Tr}\left[\mathbf{b}'\mathbf{X}_1\mathbf{X}_1'\mathbf{b}\right] \\ &= \mathrm{Tr}\left[\mathbf{X}_1'\mathbf{b}\mathbf{b}'\mathbf{X}_1\right] \\ &= \sum_{j=1}^{n_1} \mathbf{x}_{1j}'\mathbf{b}\mathbf{b}'\mathbf{x}_{1j} \\ &= \sum_{j=1}^{n_1} \left(\mathbf{b}'\mathbf{x}_{1j}\right)^2\end{aligned}$$

where $\mathbf{x}_{1j}$ denotes the jth column of $\mathbf{X}_1$; that is,

$$\mathbf{X}_1 = [\mathbf{x}_{11}, \mathbf{x}_{12}, \ldots, \mathbf{x}_{1n_1}].$$

Thus, the coefficient of σ_1^2 is the sum of the values $(\mathbf{b}'\mathbf{x}_{1j})^2$, $j = 1, 2, \ldots, n_1$, where $\mathbf{b}'\mathbf{x}_{1j}$ is the value of the linear combination $\widehat{\mathbf{h}'\boldsymbol{\beta}}$ when $\mathbf{x}_{ij}$ is used as data in place of the actual data, $\mathbf{y}$. Similarly, the coefficients of the other variance components can be determined by computing the value of $\widehat{\mathbf{h}'\boldsymbol{\beta}}$ where each column of $\mathbf{X}_1, \mathbf{X}_2, \ldots, \mathbf{X}_k$ is used as data in the computations. Then the respective values of $(\widehat{\mathbf{h}'\boldsymbol{\beta}})^2$ are summed to provide the coefficients of the variance components.

29.2 A SIMPLE EXAMPLE

Consider Example 28.1 in which the model is

$$y_{ijk} = \mu_{ijk} + p_{ik} + \varepsilon_{ijk}$$

where $p_{ik} \sim$ i.i.d. $N(0, \sigma_p^2)$ and $\varepsilon_{ijk} \sim$ i.i.d. $N(0, \sigma_\varepsilon^2)$. The covariance matrix of the observation vector is

$$V(\mathbf{y}) = \sigma_\varepsilon^2 \mathbf{I}_8 + \sigma_p^2 \mathbf{X}_1\mathbf{X}_1'$$

where

$$\mathbf{X}_1 = \begin{bmatrix} 1 & 0 & 0 & 0 & 0 \\ 1 & 0 & 0 & 0 & 0 \\ 0 & 1 & 0 & 0 & 0 \\ 0 & 0 & 1 & 0 & 0 \\ 0 & 0 & 1 & 0 & 0 \\ 0 & 0 & 0 & 1 & 0 \\ 0 & 0 & 0 & 0 & 1 \\ 0 & 0 & 0 & 0 & 1 \end{bmatrix} \quad \text{and} \quad \mathbf{y} = \begin{bmatrix} y_{111} \\ y_{121} \\ y_{112} \\ y_{213} \\ y_{223} \\ y_{214} \\ y_{215} \\ y_{225} \end{bmatrix}.$$

Let the columns of $\mathbf{X}_1$ be denoted by $\mathbf{X}_1 = [\mathbf{x}_1, \mathbf{x}_2, \mathbf{x}_3, \mathbf{x}_4, \mathbf{x}_5]$. To compute the variance of a linear combination of $\mathbf{y}$ as $\mathbf{b}'\mathbf{y}$, compute

$\mathbf{b}'\mathbf{x}_1, \mathbf{b}'\mathbf{x}_2, \ldots, \mathbf{b}'\mathbf{x}_5$ and then combine them into the variance as

$$\mathrm{Var}(\mathbf{b}'\mathbf{y}) = \sigma_\varepsilon^2 \sum_{i=1}^{8} b_i^2 + \sum_{k=1}^{5} \left(\mathbf{b}'\mathbf{x}_k\right)^2 \sigma_p^2$$

From the example, $\hat{\mu}_{11} = \frac{1}{2}(y_{111} + y_{112}) = \mathbf{b}'\mathbf{y}$. The next step is to compute $\hat{\mu}_{11}$ by considering each column of $\mathbf{X}_1$ as data. Thus,

$$\hat{\mu}_{11}(\mathbf{x}_1) = \tfrac{1}{2}(1 + 0) = \tfrac{1}{2}, \quad \hat{\mu}_{11}(\mathbf{x}_2) = \tfrac{1}{2}(0 + 1) = \tfrac{1}{2},$$
$$\hat{\mu}_{11}(\mathbf{x}_3) = \tfrac{1}{2}(0 + 0) = 0,$$
$$\hat{\mu}_{11}(\mathbf{x}_4) = \tfrac{1}{2}(0 + 0) = 0, \quad \hat{\mu}_{11}(\mathbf{x}_5) = \tfrac{1}{2}(0 + 0) = 0, \quad \text{and}$$
$$\sum b_i^2 = \left(\tfrac{1}{2}\right)^2 + \left(\tfrac{1}{2}\right)^2 = \tfrac{1}{2}.$$

Then

$$\begin{aligned}\mathrm{Var}(\hat{\mu}_{11}) &= \tfrac{1}{2}\sigma_\varepsilon^2 + \left[\left(\tfrac{1}{2}\right)^2 + \left(\tfrac{1}{2}\right)^2 + (0)^2 + (0)^2 + (0)^2\right]\sigma_p^2 \\ &= \tfrac{1}{2}\sigma_\varepsilon^2 + \tfrac{1}{2}\sigma_p^2.\end{aligned}$$

Now, compute the variance of

$$\begin{aligned}\hat{\mu}_{12} - \hat{\mu}_{22} = {} & \tfrac{1}{2}(2y_{121} + y_{112} - y_{111}) \\ & - \tfrac{1}{6}(3y_{223} + 2y_{214} - y_{213} - y_{215} + 3y_{225}).\end{aligned}$$

Compute $\hat{\mu}_{12} - \hat{\mu}_{22}$ for each column of $\mathbf{X}_1$ as follows:

$$\mathbf{x}_1\colon \tfrac{1}{2}(2 + 0 - 1) - \tfrac{1}{6}[3(0) + 2(0) - 0 - 0 + 3(0)] = \tfrac{1}{2},$$
$$\mathbf{x}_2\colon \tfrac{1}{2}(0 + 1 - 0) - \tfrac{1}{6}[3(0) + 2(0) - 0 - 0 + 3(0)] = \tfrac{1}{2},$$
$$\mathbf{x}_3\colon \tfrac{1}{2}(0 + 0 - 0) - \tfrac{1}{6}[3(1) + 2(0) - 1 - 0 + 3(0)] = -\tfrac{1}{3},$$
$$\mathbf{x}_4\colon \tfrac{1}{2}(0 + 0 - 0) - \tfrac{1}{6}[3(0) + 2(1) - 0 - 0 + 3(0)] = -\tfrac{1}{3}, \quad \text{and}$$
$$\mathbf{x}_5\colon \tfrac{1}{2}(0 + 0 - 0) - \tfrac{1}{6}[3(0) + 2(0) - 0 - 1 + 3(1)] = -\tfrac{1}{3}.$$

Then

$$\begin{aligned}\mathrm{Var}(\hat{\mu}_{12} - \hat{\mu}_{22}) = {} & \left[1^2 + \left(\tfrac{1}{2}\right)^2 + \left(-\tfrac{1}{2}\right)^2 + \left(-\tfrac{1}{2}\right)^2 + \left(-\tfrac{1}{3}\right)^2\right. \\ & \left. + \left(\tfrac{1}{6}\right)^2 + \left(\tfrac{1}{6}\right)^2 + \left(-\tfrac{1}{2}\right)^2\right]\sigma_\varepsilon^2 \\ & + \left[\left(\tfrac{1}{2}\right)^2 + \left(\tfrac{1}{2}\right)^2 + \left(-\tfrac{1}{3}\right)^2 + \left(-\tfrac{1}{3}\right)^2 + \left(-\tfrac{1}{3}\right)^2\right]\sigma_p^2 \\ = {} & \tfrac{13}{6}\sigma_\varepsilon^2 + \tfrac{5}{6}\sigma_p^2.\end{aligned}$$

Thus, if you have a computer program that does the analysis on $\mathbf{y}$, also do the analysis on each column of the matrices corresponding to the random effects. Such analyses will provide enough information to compute the variances of various estimates or linear combinations of parameters.

CONCLUDING REMARKS

A computer algorithm was described for computing the coefficients of variance components in the variance of a linear combination of the parameters of an unbalanced model. The method uses the idea of synthesis, which was described for computing expected mean squares in Chapter 18.

30

Analysis of Nested Designs

CHAPTER OUTLINE

Nested effects can occur in either the design structure or the treatment structure of an experimental design. For nesting to occur in the design structure, there must be more than one size of experimental unit where a small experimental unit is nested within a larger one. For nesting to occur in the treatment structure, there must be two or more factors. These factors may be all fixed effects, all random effects, or both and thus can be studied as fixed, random, or mixed models. The concept of nested factors (or experimental units) was introduced in Chapter 5. This chapter presents some examples that demonstrate model construction, parameter estimation, and hypothesis testing. Nested designs are often referred to as hierarchical designs.

30.1 DEFINITIONS, ASSUMPTIONS, AND MODELS

In the treatment structure, the levels of factor A are nested within the levels of factor B if each level of A occurs with only one level of factor B. The following example demonstrates nesting in the treatment structure.

EXAMPLE 30.1: Companies and Insecticides

Four chemical companies produce certain insecticides. Company A produces three such products, companies B and C produce two such products each, and company D produces four such products. No company produces a product exactly like that of another. The treatment structure is a two-way with Company as one factor and Product as the other. Such a treatment structure is shown in Table 30.1, where each level of Product occurs only once within each level of Company. Thus the levels of Product are nested within the levels of Company.

Table 30.1 Treatment Structure for Insecticide Experiment

	PRODUCT										
Company	*1*	*2*	*3*	*4*	*5*	*6*	*7*	*8*	*9*	*10*	*11*
A	X	X	X								
B				X	X						
C						X	X				
D								X	X	X	X

Note: An "X" denotes that the particular product is from the corresponding company.

The levels of both factors in the treatment structure are fixed. To conduct the experiment, a box of soil with bluegrass and 400 mosquitoes were put into each of 33 glass containers. Three glass containers were then randomly assigned to each product. The glass containers were treated with the product, and after 4 hours the number of live mosquitoes was counted.

The design structure for this experiment is completely randomized. The model is

$$y_{ijk} = \mu + \gamma_i + \rho_{j(i)} + \varepsilon_{ijk} \qquad i = 1,2,3,4; \quad j = 1,2,\ldots,m_i;$$
$$k = 1,2,\ldots,n_{j(i)} \qquad (30.1.1)$$

where y_{ijk} is the observed number of mosquitoes from the kth replication of the jth product of Company i, μ is the overall mean, γ_i is the effect of the ith company, $\rho_{j(i)}$ is the effect of the jth product in Company i, and $\varepsilon_{ijk} \sim$ i.i.d. $N(0, \sigma_\varepsilon^2)$ denotes the error associated with measuring y_{ijk}.

The model has only one size of experimental unit and thus only one error term. The experiment has a two-way nested treatment structure (both factors fixed) in a completely randomized design structure. The parameters of the model to be estimated are σ_ε^2, estimable functions of μ,

Table 30.2 Data for Example 30.1 Insecticide Experiment

	PRODUCT										
Company	*1*	*2*	*3*	*4*	*5*	*6*	*7*	*8*	*9*	*10*	*11*
	151	118	131								
A	135	132	137								
	137	135	121								
				140	151						
B				152	132						
				133	139						
						96	84				
C						108	87				
						94	82				
								79	67	90	83
D								74	78	81	89
								73	63	96	94

γ_i, and $\rho_{j(i)}$. The data for this example are in Table 30.2, and the analysis is discussed in Sections 30.2 and 30.3.

EXAMPLE 30.2: Comfort Experiment Revisited

The comfort experiment in Example 5.10 is an example of nesting in the design structure. The large experimental unit is an environmental chamber, and the small experimental unit is a person. The model is

$$y_{ijkm} = \mu_{ij} + c_{j(i)} + p_{m(ijk)} \tag{30.1.2}$$

where the terms in the model are as described in Example 5.10. The terms $c_{j(i)}$ and $p_{m(ijk)}$ denote the errors of observing temperature i on chamber j (chambers are nested within temperature) and person m with the ijk treatment combination (persons are nested within chambers). Since there are two sizes of experimental units, the analysis has two levels and two error terms. The parameters that need to be estimated are μ_{ij}, σ_c^2, and σ_p^2. This example will also be analyzed in Sections 30.2 and 30.3.

EXAMPLE 30.3: Coffee Price Example Revisited

The coffee price example in Chapter 18 is an experimental design with a three-way treatment structure, where the levels of Store are nested within the levels of City, which are nested within the levels of State (all three factors are random effects). The parameters of interest are the variance components σ_s^2, σ_c^2, and σ_a^2. The coffee price example is an example of a multistage sampling experiment, a very common application of nested designs. Again, the analysis of this example will occur in the next two sections.

The above examples demonstrate the nesting that can occur in the treatment structure, design structure, or both. The treatment structure can involve random and/or fixed effects, and the design structure can involve several sizes of experimental units. Thus, the analysis of such designs involves using the techniques of fixed effects models, random models, mixed models, and split-plot–repeated measures models, as discussed in the next two sections.

30.2 PARAMETER ESTIMATION

A model involving nesting falls into one of the classes of models already discussed. That is, if there are fixed effects, the means need to be estimated; if there are random effects, variance components need to be estimated; and if there are several sizes of experimental units, the analysis involves more than one error term. Thus the design may be quite simple or quite complex. The examples in this chapter are used to demonstrate the application of the previously discussed techniques to the analysis of designs involving nesting.

EXAMPLE 30.1: Continuation

The estimable functions of model (30.1.1) are contrasts of $\rho_{j(i)}$ for each i, that is, contrasts of the products within a company. Estimates of these contrasts are

$$\widehat{\rho_{j(i)} - \rho_{j'(i)}} = \bar{y}_{ij\cdot} - \bar{y}_{ij'\cdot}.$$

Model (30.1.1) can also be expressed as

$$y_{ijk} = \mu_{j(i)} + \varepsilon_{ijk}$$

where $\mu_{j(i)}$ is the mean of product j of company i. The estimators of the $\mu_{j(i)}$'s are the $\bar{y}_{ij\cdot}$'s. Any contrast between the $\mu_{j(i)}$'s is estimable, but one has to be careful in interpreting selected contrasts. The estimator of σ_ε^2 comes from pooling the variances of the observations within the $j(i)$ combinations as

$$\hat{\sigma}_\varepsilon^2 = \frac{1}{N-q} \sum_{i=1}^{m} \sum_{j=1}^{p_i} \sum_{k=1}^{n_{j(i)}} (y_{ijk} - \bar{y}_{ij\cdot})^2$$

where there are $n_{j(i)}$ observations in the $j(i)$ cell,

$$N = \sum_{j(i)} n_{j(i)},$$

$$\hat{\sigma}_\varepsilon^2 = \frac{1}{33-11} \sum_{i=1}^{4} \sum_{j=1}^{m_i} \sum_{k=1}^{n_{j(i)}} (y_{ijk} - \bar{y}_{ij\cdot})^2$$

$$= \frac{\sum_{i=1}^{4}\sum_{j=1}^{m_i}\sum_{k=1}^{n_{j(i)}} y_{ijk}^2 - \sum_{i=1}^{4}\sum_{j=1}^{m_i} n_{j(i)}\bar{y}_{ij\cdot}^2}{22} = 60.818$$

SAMPLE MEAN $\hat{\mu}_{j(i)}$

1(1)	*2(1)*	*3(1)*	*1(2)*	*2(2)*	*1(3)*	*2(3)*	*1(4)*	*2(4)*	*3(4)*	*4(4)*
141.0	128.3	129.7	141.7	140.7	99.3	84.3	75.3	69.3	89.0	89.7

Figure 30.1 Estimates of the parameters in Example 30.1.

and q is the number of $j(i)$ combinations. A summary of the estimates of various parameters for the data in Table 30.2 is in Figure 30.1.

Multiple comparisons can be made between the $\mu_{j(i)}$'s. Other hypotheses can be tested about the $\mu_{j(i)}$'s such as comparing companies or products within a company. For example, one could compare company B to company D via

$$\bar{\mu}_{\cdot(2)} = \bar{\mu}_{\cdot(4)} \qquad \text{where} \quad \bar{\mu}_{\cdot(2)} = \frac{\mu_{1(2)} + \mu_{2(2)}}{2}$$

and

$$\bar{\mu}_{\cdot(4)} = \frac{\mu_{1(4)} + \mu_{2(4)} + \mu_{3(4)} + \mu_{4(4)}}{4};$$

that is, the comparison would be between the mean of the two products from company B and the mean of the four products from company D. Hypothesis testing for this nested treatment structure is discussed in Section 30.3.

If there are unequal numbers of observations in the $j(i)$ cells, then the techniques for analyzing unbalanced models can be used to obtain estimates of $\mu_{j(i)}$. The estimates of the population marginal means provide $\hat{\mu}_{j(i)}$, and $\hat{\sigma}_\varepsilon^2$ is obtained from pooling the variances, which can be obtained from an analysis of variance.

EXAMPLE 30.2: Continuation

Data for the comfort study are in Figure 30.2. The nesting occurs in the design structure with Person nested within Chamber and Chamber nested

Temperature	Chamber	Male	Female
65	Chamber 1	5 4	1 2
65	Chamber 2	5 4	5 5
65	Chamber 3	4 2	1 3
70	Chamber 4	8 8	10 7
70	Chamber 5	6 3	8 8
70	Chamber 6	5 7	8 8
75	Chamber 7	12 8	11 13
75	Chamber 8	8 7	8 8
75	Chamber 9	6 6	6 7

Figure 30.2 Data for Example 30.2. The values are comfort scores; 1 = cold, 8 = comfortable, and 15-hot.

Table 30.3 Analysis of Variance Table and Estimates of Variance Components for Data in Figure 30.2

Source of Variation	*df*	*MS*	*EMS*
Temperature	2	79.19	$\sigma_p^2 + 4\sigma_c^2 + Q$ (Temperature)
Sex	1	3.36	$\sigma_p^2 + Q$ (Sex)
Temperature * Sex	2	7.86	$\sigma_p^2 + Q$ (Temperature * Sex)
Chamber (Temperature)	6	11.08	$\sigma_p^2 + 4\sigma_c^2$
Person (Chamber)	24	1.65	σ_p^2

Notes: The method-of-moments estimates are $\hat{\sigma}_p^2 = 1.65$ and $\hat{\sigma}_c^2 = 2.36$. The maximum likelihood estimates are $\hat{\sigma}_p^2 = 1.47$ and $\hat{\sigma}_c^2 = 1.48$.

within Temperature. The analysis of variance table with expected mean squares is shown in Table 30.3. The method-of-moments estimates (which are also MINQUE estimates) and the maximum likelihood estimates of the two components of variance are also in Table 30.3. The methods used for split-plot experiments can be used to compare the μ_{ij}'s. If the design is unbalanced, then a mixed-model maximum likelihood analysis would be appropriate. Also, the methods described for unbalanced split-plot designs could be used to construct an appropriate analysis.

EXAMPLE 30.3: Continuation

The coffee price study involves a random model whose parameters of interest are μ, σ_s^2, σ_c^2, and σ_ε^2. A method of maximum likelihood would provide a very appropriate analysis and suitable estimators of the parameters, as would the MINQUE technique. However, a method-of-moments analysis employing Type I sums of squares has typically been used for this type of multistage sampling design. Table 30.4 contains the Type I sums of squares, their expected mean squares, and the method-of-moments estimators.

As described in Chapter 20, preliminary tests of the hypotheses about each variance component can be used to prevent the method-of-moments estimators from being negative. For models with nesting, all factors random, and possibly several sizes of experimental units, the methods described for random models are appropriate for the analysis.

Table 30.4 Analysis of Variance Table with Type I Sums of Squares and Method-of-Moments Estimators for Example 30.3

Source of Variation	*df*	*SS*	*EMS*
State	$r-1$	$\sum_{i=1}^{r} n_{i\cdot}\bar{y}_{i\cdot\cdot}^2 - n_{\cdot\cdot}\bar{y}_{\cdot\cdot\cdot}^2$	$\sigma_\varepsilon^2 + \frac{a_1 - a_3}{r-1}\sigma_c^2 + \frac{n_{\cdot\cdot} - a_2}{r-1}\sigma_s^2$
City (State)	$\sum_{i=1}^{r}(t_i - 1)$	$\sum_{i=1}^{r}\sum_{j=1}^{t_i} n_{ij}\bar{y}_{ij\cdot}^2 - \sum_{i=1}^{r} n_{i\cdot}\bar{y}_{i\cdot\cdot}^2$	$\sigma_\varepsilon^2 + \frac{n_{\cdot\cdot} - a_1}{t_\cdot - r}\sigma_c^2$
Store (City)	$\sum_{i=1}^{r}\sum_{j=1}^{t_i}(n_{ij} - 1)$	$\sum_{i=1}^{r}\sum_{j=1}^{t_i}\sum_{k=1}^{n_{ij}} y_{ijk}^2 - \sum_{i=1}^{r}\sum_{j=1}^{t_i} n_{ij}\bar{y}_{ij\cdot}^2$	σ_ε^2

Source: Searle, 1971.

Note:

$$a_1 = \sum_{i=1}^{r}\left(\sum_{j=1}^{t_i}\frac{n_{ij}^2}{n_{i\cdot}}\right)^2, \quad a_2 = \sum_{i=1}^{r}\frac{n_{i\cdot}^2}{n_{\cdot\cdot}}, \quad a_3 = \sum_{i=1}^{r}\sum_{j=1}^{t_i}\frac{n_{ij}^2}{n_{\cdot\cdot}},$$

$$\hat{\sigma}_\varepsilon^2 = \frac{\text{MSSTORE(CITY)}}{(n_{\cdot\cdot} - t_\cdot)}, \quad \hat{\sigma}_c^2 = \frac{\text{MSCITY(STATE)} - \hat{\sigma}_\varepsilon^2}{(n_{\cdot\cdot} - a_1/t_\cdot - r)} \quad \hat{\sigma}_s^2 = \frac{\text{MSSTATE} - \hat{\sigma}_\varepsilon^2 - (a_1 - a_3/r - 1)\hat{\sigma}_c^2}{(n_{\cdot\cdot} - a_2/r - 1)}$$

30.3 TESTING HYPOTHESES AND CONFIDENCE INTERVALS

For balanced experimental designs with nesting in one or both of the structures, the methods for balanced random models, balanced mixed models, and balanced split-plot–repeated measures models can be used to construct confidence intervals and make comparisons between parameters. The unbalanced methods are needed for unbalanced nested designs, which involve using the Satterthwaite approximation for constructing confidence intervals about the variance components and weighted least squares for estimating and constructing confidence intervals about the estimable functions of the fixed effects. In general, the nested design is analyzed as either a fixed, random, or mixed model, as required.

When the nesting involves random factors or when the nesting is in the design structure, the methods for testing hypotheses are those used for random models, mixed models (for testing about fixed effects), and the split-plot–repeated measures models.

When the nesting is in the treatment structure and involves fixed factors, the method of analyzing fixed effects models is used. The major change is in the types of hypotheses that can be (or are) tested with the analysis of variance table, as demonstrated with the data of Example 30.1.

EXAMPLE 30.1: Continuation

There are two ways to write the model for the insecticide data

$$y_{ijk} = \mu_{j(i)} + \varepsilon_{ijk} \tag{30.3.1}$$

and

$$y_{ijk} = \mu + \gamma_i + \rho_{j(i)} + \varepsilon_{ijk}. \tag{30.3.2}$$

Table 30.5 contains the analysis of variance table for model (30.3.1). The

Table 30.5 Analysis of Variance Table for Example 30.1 Based on Model 30.3.1

Source of Variation	*df*	*SS*
Product	10	24,201.0
Error	22	1,338.0

Note: $\text{SSPRODUCT} = \sum_{i=1}^{4} \sum_{j=1}^{m_i} n_{j(i)} \bar{y}_{ij\cdot}^2 - n_{\cdot\cdot} \bar{y}_{\cdot\cdot\cdot}^2$ and

$\text{SSERROR} = 22\hat{\sigma}_\varepsilon^2$ from Figure 30.1.

hypothesis being tested by the sum of squares due to products is

$$\mu_{1(1)} = \mu_{1(2)} = \mu_{1(3)} = \mu_{2(1)} = \mu_{2(2)} = \mu_{1(3)} = \mu_{2(3)}$$
$$= \mu_{1(4)} = \mu_{2(4)} = \mu_{3(4)} = \mu_{4(4)}.$$

Table 30.6 contains the analysis of variance table for model (30.3.2). The sum of squares due to Product in Table 30.5 has been partitioned into the sum of squares due to Company and the sum of squares due to products within companies. Since the two-way treatment structure is nested (products nested within companies), there is no measure for interaction between Product and Company. The sum of squares due to Company tests

$$\bar{\mu}_{\cdot(1)} = \bar{\mu}_{\cdot(2)} = \bar{\mu}_{\cdot(3)} = \bar{\mu}_{\cdot(4)}$$

where

$$\bar{\mu}_{\cdot(i)} = \sum_{j=1}^{t_i} \frac{\mu_{j(i)}}{t_i} = \mu + \gamma_i + \bar{\rho}_{\cdot(i)}.$$

The sum of squares due to products within companies tests

$$\mu_{1(1)} = \mu_{2(1)} = \mu_{3(1)}, \quad \mu_{1(2)} = \mu_{2(2)}, \quad \mu_{1(3)} = \mu_{2(3)}, \quad \text{and}$$
$$\mu_{1(4)} = \mu_{2(4)} = \mu_{3(4)} = \mu_{4(4)},$$

or

$$\rho_{1(1)} = \rho_{2(1)} = \rho_{3(1)}, \quad \rho_{1(2)} = \rho_{2(2)}, \quad \rho_{1(3)} = \rho_{2(3)}, \quad \text{and}$$
$$\rho_{1(4)} = \rho_{2(4)} = \rho_{3(4)} = \rho_{4(4)}.$$

Table 30.6 Analysis of Variance Table for Example 30.1 Based on Model 30.3.2

Source of Variation	*df*	*SS*
Company	3	22,649.6
Product (Company)	7	1,551.4
Error	22	1,338.0

Note:

$$\text{SSCOMPANY} = \sum_{i=1}^{4} n_{\cdot(i)}\bar{y}_{i\cdot\cdot}^2 - n_{\cdot\cdot}\bar{y}_{\cdot\cdot\cdot}^2,$$

$$\text{SSPRODUCT(COMPANY)} = \sum_{i=1}^{4}\sum_{j=1}^{m_i} n_{j(i)}\bar{y}_{ij\cdot}^2 - \sum_{i=1}^{4} n_{\cdot(i)}\bar{y}_{i\cdot\cdot}^2, \quad \text{and}$$

$$\text{SSERROR} = 22\hat{\sigma}_\varepsilon^2 \text{ from Figure 30.1.}$$

Any of the appropriate multiple comparison procedures can be used where one may want to (1) compare the $\bar{\mu}_{\cdot(i)}$'s, (2) compare the $\mu_{j(i)}$'s, (3) compare the $\mu_{j(i)}$'s for each i, or (4) compare any combination of the above. For unbalanced models, the Type IV sums of squares are appropriate in testing the various hypotheses, and the estimates of the population marginal means provide estimates of the $\bar{\mu}_{\cdot(j)}$'s and $\mu_{j(i)}$'s.

For other nested models, the expected mean squares in the analysis of variance tables can be used as guides for constructing proper F-tests. If the design involves unbalanced data, then mean squares may need to be constructed to provide appropriate divisors for the F-test statistics. This case is like that discussed in Chapters 20 and 23.

CONCLUDING REMARKS

In this chapter, the concept of nesting was defined, and examples were given where there was nesting in the design structure, nesting in the treatment structure, and nesting between the two structures. Depending on the situation, nested models can be either fixed, random, or mixed. The examples demonstrated the methods necessary to carry out the analysis of such models.

31

Analysis of Repeated Measures Experiments by Using Multivariate Methods

CHAPTER OUTLINE

In this chapter, we shall discuss an analysis of repeated measures and other nonrandomized experiments by using multivariate methods. The multivariate methods can only be applied to experiments where either all repeated measures variables on a given experimental unit are present or all are missing. The numbers of experimental units per treatment need not be balanced, however. To understand this chapter, one will need to understand the matrix form of the model, discussed in Chapter 6.

31.1 GENERAL MULTIVARIATE MODEL

A general multivariate model for a repeated measures experiment is a generalization of the matrix form of the model defined in (6.1.1). The multivariate model can be expressed as

$$\mathbf{Y} = \mathbf{XB} + \mathbf{E} \tag{31.1.1}$$

where $\mathbf{Y}$ denotes all of the data measured in the experiment. Each row of the data matrix $\mathbf{Y}$ corresponds to a particular experimental unit, and each column corresponds to one of the repeated measures. We assume that $\mathbf{Y}$ has N rows and p columns. The matrix $\mathbf{X}$ is an $N \times r$ design matrix and is assumed to be of rank t. Each column of $\mathbf{B}$ is an $r \times 1$ vector of unknown parameters with each column corresponding to a particular repeated measure. The matrix $\mathbf{E}$ is an $N \times p$ matrix of unobservable random errors. It is assumed that the rows of $\mathbf{E}$ are independently and identically distributed as multivariate normal distributions with zero means and a common covariance matrix. Thus, while the rows in $\mathbf{E}$ are independent, the elements in the columns of $\mathbf{E}$ may be correlated.

31.2 HYPOTHESIS TESTING

For the multivariate model (31.1.1), we can test general hypotheses of the form

$$\mathrm{H}_0\colon \mathbf{CBM} = \mathbf{0} \quad \text{versus} \quad \mathrm{H}_a\colon \mathbf{CBM} \neq \mathbf{0} \tag{31.2.1}$$

where $\mathbf{C}$ is a $g \times r$ matrix of rank g and $\mathbf{M}$ is a $p \times q$ matrix of rank q.

To test the hypothesis in (31.2.1), we first need to determine least squares estimates of the parameters in $\mathbf{B}$ and an observed residual-sum-of-squares-and-cross-products matrix. These are denoted by $\hat{\mathbf{B}}$ and $\hat{\mathbf{E}}$, respectively. We get

$$\hat{\mathbf{B}} = (\mathbf{X}'\mathbf{X})^{-}\mathbf{X}'\mathbf{Y} \tag{31.2.2}$$

and

$$\hat{\mathbf{E}} = \mathbf{Y}'[\mathbf{I} - \mathbf{X}(\mathbf{X}'\mathbf{X})^{-}\mathbf{X}']\mathbf{Y}.$$

A likelihood ratio test statistic for testing (31.2.1) is given by

$$\Lambda = \frac{|\mathbf{R}|}{|\mathbf{H} + \mathbf{R}|} \tag{31.2.3}$$

where

$$\mathbf{R} = \mathbf{M}'\hat{\mathbf{E}}\mathbf{M}, \quad \mathbf{H} = \mathbf{M}'\left\{\hat{\mathbf{B}}'\mathbf{C}'\left[\mathbf{C}(\mathbf{X}'\mathbf{X})^{-}\mathbf{C}'\right]^{-1}\mathbf{C}\hat{\mathbf{B}}\right\}\mathbf{M},$$

and $|\mathbf{W}|$ denotes the determinant of the matrix $\mathbf{W}$.

The statistic Λ is called Wilk's likelihood ratio criterion (Morrison, 1976). The sampling distribution of Λ is quite complicated, but for most practical purposes, an approximate α-level test can be obtained by rejecting H_0 when

$$-\left(N - t - \frac{|q - g| + 1}{2}\right)\ln \Lambda > \chi^2_{\alpha, qg}.$$

A better approximation, to be used only when both q and g are greater than 2, is to reject H_0 when

$$F > F_{\alpha, qg, ab-c}$$

where

$$\left.\begin{aligned} F &= \frac{(1 - \Lambda^{1/b})(ab - c)}{qg\Lambda^{1/b}}, \\ a &= N - t - \frac{|q - s| + 1}{2}, \\ b &= \left(\frac{q^2s^2 - 4}{(q^2 + s^2 - 5)}\right)^{1/2} \\ c &= \frac{qs - 2}{2}, \quad \text{and} \\ s &= \text{Min}(q, g). \end{aligned}\right\} \tag{31.2.4}$$

Exact F-tests for (31.2.1) exist whenever $q = 1$ or 2 or whenever $g = 1$ or 2. These tests are as follows:

1. For $g = 1$ and any q, reject H_0 if

$$F = \left(\frac{1 - \Lambda}{\Lambda}\right)\left(\frac{N - t - q + 1}{q}\right) > F_{\alpha, q, N-t-q+1}. \tag{31.2.5}$$

2. For $q = 1$ and any g, reject H_0 if

$$F = \left(\frac{1 - \Lambda}{\Lambda}\right)\left(\frac{N - t}{g}\right) > F_{\alpha, g, N-t}. \tag{31.2.6}$$

3. For $g = 2$ and any $q \geq 2$, reject H_0 if

$$F = \left(\frac{1 - \sqrt{\Lambda}}{\sqrt{\Lambda}}\right)\left(\frac{N - t - q + 1}{q}\right) > F_{\alpha, 2q, 2(N-t-q+1)}. \qquad (31.2.7)$$

4. For $q = 2$ and any $g \geq 2$, reject H_0 if

$$F = \left(\frac{1 - \sqrt{\Lambda}}{\sqrt{\Lambda}}\right)\left(\frac{N - t - 1}{g}\right) > F_{\alpha, 2g, 2(N-t-1)}. \qquad (31.2.8)$$

One drawback to these multivariate methods is that one must have $p < N - t$. When $p \geq N - t$, it is often possible to combine adjacent repeated measures into p^* new variables or to analyze only a size p^* subset of the repeated measures with $p^* < N - t$.

31.3 A REPEATED MEASURES EXAMPLE

To illustrate the analysis described in Section 31.2, consider an experiment conducted to study the differences among varieties of sorghum on a leaf area index. Suppose there are four varieties of sorghum, denoted by V_1, V_2, V_3, and V_4. Also suppose that these four varieties were planted in a randomized block design with five blocks. Finally, assume that leaf area index measurements are made on each Variety∗Block plot at weekly intervals for five weeks beginning two weeks after emergence. The data obtained are given in Table 31.1.

For the data in Table 31.1, the data matrix is

$$\mathbf{Y} = \begin{bmatrix} 5.00 & 4.84 & 4.02 & 3.75 & 3.13 \\ 4.42 & 4.30 & 3.67 & 3.23 & 2.83 \\ 4.42 & 4.10 & 3.46 & 3.09 & 2.82 \\ 4.01 & 3.89 & 3.21 & 2.89 & 2.56 \\ 3.36 & 3.10 & 2.67 & 2.47 & 2.16 \\ 5.82 & 5.60 & 5.05 & 4.72 & 4.46 \\ 5.73 & 5.59 & 5.00 & 4.65 & 4.42 \\ 5.31 & 5.19 & 4.86 & 4.44 & 4.22 \\ 4.92 & 4.66 & 4.56 & 4.16 & 3.99 \\ 3.96 & 3.86 & 3.50 & 3.13 & 2.95 \\ 5.65 & 5.97 & 5.27 & 5.07 & 4.52 \\ 5.39 & 5.49 & 5.08 & 4.87 & 4.32 \\ 5.15 & 5.28 & 4.93 & 4.67 & 4.15 \\ 4.50 & 4.89 & 4.74 & 4.49 & 4.10 \\ 3.75 & 3.74 & 3.55 & 3.28 & 3.00 \\ 5.86 & 5.60 & 5.37 & 5.00 & 4.37 \\ 5.82 & 5.55 & 5.29 & 4.95 & 4.07 \\ 5.26 & 5.06 & 4.76 & 4.48 & 3.94 \\ 4.87 & 4.75 & 4.55 & 4.33 & 3.83 \\ 3.96 & 3.76 & 3.56 & 3.18 & 2.96 \end{bmatrix},$$

the matrix of parameters is

$$\mathbf{B} = \begin{bmatrix} \mu^{(1)} & \mu^{(2)} & \mu^{(3)} & \mu^{(4)} & \mu^{(5)} \\ \tau_1^{(1)} & \tau_1^{(2)} & \tau_1^{(3)} & \tau_1^{(4)} & \tau_1^{(5)} \\ \tau_2^{(1)} & \tau_2^{(2)} & \tau_2^{(3)} & \tau_2^{(4)} & \tau_2^{(5)} \\ \tau_3^{(1)} & \tau_3^{(2)} & \tau_3^{(3)} & \tau_3^{(4)} & \tau_3^{(5)} \\ \tau_4^{(1)} & \tau_4^{(2)} & \tau_4^{(3)} & \tau_4^{(4)} & \tau_4^{(5)} \\ \beta_1^{(1)} & \beta_1^{(2)} & \beta_1^{(3)} & \beta_1^{(4)} & \beta_1^{(5)} \\ \beta_2^{(1)} & \beta_2^{(2)} & \beta_2^{(3)} & \beta_2^{(4)} & \beta_2^{(5)} \\ \beta_3^{(1)} & \beta_3^{(2)} & \beta_3^{(3)} & \beta_3^{(4)} & \beta_3^{(5)} \\ \beta_4^{(1)} & \beta_4^{(2)} & \beta_4^{(3)} & \beta_4^{(4)} & \beta_4^{(5)} \\ \beta_5^{(1)} & \beta_5^{(2)} & \beta_5^{(3)} & \beta_5^{(4)} & \beta_5^{(5)} \end{bmatrix},$$

and the design matrix is

$$\mathbf{X} = \begin{bmatrix} 1 & 1 & 0 & 0 & 0 & 1 & 0 & 0 & 0 & 0 \\ 1 & 1 & 0 & 0 & 0 & 0 & 1 & 0 & 0 & 0 \\ 1 & 1 & 0 & 0 & 0 & 0 & 0 & 1 & 0 & 0 \\ 1 & 1 & 0 & 0 & 0 & 0 & 0 & 0 & 1 & 0 \\ 1 & 1 & 0 & 0 & 0 & 0 & 0 & 0 & 0 & 1 \\ 1 & 0 & 1 & 0 & 0 & 1 & 0 & 0 & 0 & 0 \\ 1 & 0 & 1 & 0 & 0 & 0 & 1 & 0 & 0 & 0 \\ 1 & 0 & 1 & 0 & 0 & 0 & 0 & 1 & 0 & 0 \\ 1 & 0 & 1 & 0 & 0 & 0 & 0 & 0 & 1 & 0 \\ 1 & 0 & 1 & 0 & 0 & 0 & 0 & 0 & 0 & 1 \\ 1 & 0 & 0 & 1 & 0 & 1 & 0 & 0 & 0 & 0 \\ 1 & 0 & 0 & 1 & 0 & 0 & 1 & 0 & 0 & 0 \\ 1 & 0 & 0 & 1 & 0 & 0 & 0 & 1 & 0 & 0 \\ 1 & 0 & 0 & 1 & 0 & 0 & 0 & 0 & 1 & 0 \\ 1 & 0 & 0 & 1 & 0 & 0 & 0 & 0 & 0 & 1 \\ 1 & 0 & 0 & 0 & 1 & 1 & 0 & 0 & 0 & 0 \\ 1 & 0 & 0 & 0 & 1 & 0 & 1 & 0 & 0 & 0 \\ 1 & 0 & 0 & 0 & 1 & 0 & 0 & 1 & 0 & 0 \\ 1 & 0 & 0 & 0 & 1 & 0 & 0 & 0 & 1 & 0 \\ 1 & 0 & 0 & 0 & 1 & 0 & 0 & 0 & 0 & 1. \end{bmatrix}$$

Note that each column of **B** represents the parameters needed for a randomized complete block model for the jth response column of the data matrix **Y**, $j = 1, 2, 3, 4, 5$. The design matrix **X** is 20×10 and has

Table 31.1 Leaf Area Index on Four Sorghum Varieties

		TIME				
Variety	*Block*	*1*	*2*	*3*	*4*	*5*
	1	5.00	4.84	4.02	3.75	3.13
	2	4.42	4.30	3.67	3.23	2.83
V_1	3	4.42	4.10	3.46	3.09	2.82
	4	4.01	3.89	3.21	2.89	2.56
	5	3.36	3.10	2.67	2.47	2.16
	1	5.82	5.60	5.05	4.72	4.46
	2	5.73	5.59	5.00	4.65	4.42
V_2	3	5.31	5.19	4.86	4.44	4.22
	4	4.92	4.66	4.56	4.16	3.99
	5	3.96	3.86	3.50	3.13	2.95
	1	5.65	5.97	5.27	5.07	4.52
	2	5.39	5.49	5.08	4.87	4.32
V_3	3	5.15	5.28	4.93	4.67	4.15
	4	4.50	4.89	4.74	4.49	4.10
	5	3.75	3.74	3.55	3.28	3.00
	1	5.86	5.60	5.37	5.00	4.37
	2	5.82	5.55	5.29	4.95	4.07
V_4	3	5.26	5.06	4.76	4.48	3.94
	4	4.87	4.75	4.55	4.33	3.83
	5	3.96	3.76	3.56	3.18	2.96

rank 8; thus $N = 20$, $p = 10$, and $t = 8$. The value of $\hat{\mathbf{B}}$ is

$$\hat{\mathbf{B}} = \begin{bmatrix} 3.350 & 3.283 & 3.003 & 2.788 & 2.150 \\ .222 & .106 & -.198 & -.260 & -.312 \\ 1.128 & 1.040 & .990 & .874 & .996 \\ .868 & 1.134 & 1.110 & 1.130 & 1.006 \\ 1.134 & 1.004 & 1.102 & 1.042 & .822 \\ 1.395 & 1.398 & 1.173 & 1.150 & .982 \\ 1.152 & 1.128 & 1.006 & .940 & .772 \\ .847 & .803 & .748 & .685 & .645 \\ .387 & .443 & .511 & .483 & .482 \\ -.430 & -.489 & -.434 & -.470 & -.370 \end{bmatrix},$$

and the value of $\hat{\mathbf{E}}$ is

$$\hat{\mathbf{E}} = \begin{bmatrix} .237 & .171 & .162 & .228 & .129 \\ .171 & .247 & .163 & .231 & .135 \\ .162 & .163 & .268 & .303 & .184 \\ .228 & .231 & .303 & .392 & .241 \\ .129 & .135 & .184 & .241 & .247 \end{bmatrix}.$$

The test for equal Variety marginal means is obtained from (31.2.3) by taking

$$\mathbf{C} = \begin{bmatrix} 0 & 1 & -1 & 0 & 0 & 0 & 0 & 0 & 0 & 0 \\ 0 & 1 & 0 & -1 & 0 & 0 & 0 & 0 & 0 & 0 \\ 0 & 1 & 0 & 0 & -1 & 0 & 0 & 0 & 0 & 0 \end{bmatrix} \quad \text{and} \quad \mathbf{M} = \begin{bmatrix} 1 \\ 1 \\ 1 \\ 1 \\ 1 \end{bmatrix}.$$

We get $\Lambda = .04345$ with $g = 3$ and $q = 1$. Since $q = 1$, we can use (31.2.6) and get

$$F = \frac{1 - .04345}{.04345} \cdot \frac{12}{3} = 88.06$$

with 3 and 12 degrees of freedom. The significance probability is $\hat{\alpha} < .0001$.

The test for equal Time marginal means is obtained from (31.2.3) by taking

$$\mathbf{C} = \begin{bmatrix} 1 & \frac{1}{4} & \frac{1}{4} & \frac{1}{4} & \frac{1}{4} & \frac{1}{5} & \frac{1}{5} & \frac{1}{5} & \frac{1}{5} & \frac{1}{5} \end{bmatrix}$$

and

$$\mathbf{M} = \begin{bmatrix} 1 & 1 & 1 & 1 \\ -1 & 0 & 0 & 0 \\ 0 & -1 & 0 & 0 \\ 0 & 0 & -1 & 0 \\ 0 & 0 & 0 & -1 \end{bmatrix}.$$

We get $\Lambda = .00502$ with $g = 1$ and $q = 4$. Since $g = 1$, we can use (31.2.5) and get

$$F = \frac{1 - .00502}{.00502} \cdot \frac{9}{4} = 445.96$$

with 4 and 9 degrees of freedom. The observed significance level is $\hat{\alpha} < .0001$.

The test for the Variety * Time interaction is obtained from (31.2.3) by taking

$$\mathbf{C} = \begin{bmatrix} 0 & 1 & -1 & 0 & 0 & 0 & 0 & 0 & 0 & 0 \\ 0 & 1 & 0 & -1 & 0 & 0 & 0 & 0 & 0 & 0 \\ 0 & 1 & 0 & 0 & -1 & 0 & 0 & 0 & 0 & 0 \end{bmatrix}$$

and

$$\mathbf{M} = \begin{bmatrix} 1 & 1 & 1 & 1 \\ -1 & 0 & 0 & 0 \\ 0 & -1 & 0 & 0 \\ 0 & 0 & -1 & 0 \\ 0 & 0 & 0 & -1 \end{bmatrix}.$$

We obtain $\Lambda = .01426$ with $g = 3$ and $q = 4$. Using (31.2.4), we get $s = 3$, $a = 11$, $b = 2.646$, $c = 5$, and

$$F = \frac{(1 - .2006)(11 \cdot 2.646 - 5)}{4 \cdot 3(.2006)} = 8.00$$

with 12 and 24.1 degrees of freedom. The significance probability is $\hat{\alpha} < .0001$.

The test for equal Variety means is the same as that obtained by doing a split-plot analysis. It is the only test of the three that is the same as the corresponding test from a split-plot analysis.

31.4 MARGINAL MEANS AND SPECIAL LINEAR COMBINATIONS

Many special functions of the parameters in $\mathbf{B}$ are estimable, and inferences can be made on them. Most interesting linear functions of the parameters in $\mathbf{B}$ can be written in the form $\mathbf{c}'\mathbf{Bm}$ where $\mathbf{c}$ is an $r \times 1$ vector and $\mathbf{m}$ is a $p \times 1$ vector. For $\mathbf{c}'\mathbf{Bm}$ to be estimable, $\mathbf{c}$ must satisfy $\mathbf{X}'\mathbf{Xu} = \mathbf{c}$ for some vector $\mathbf{u}$, as discussed in Chapter 6. There are no restrictions on $\mathbf{m}$.

The best estimate of an estimable function of the form $\mathbf{c}'\mathbf{Bm}$ is $\mathbf{c}'\hat{\mathbf{B}}\mathbf{m}$ where $\hat{\mathbf{B}}$ is given in (31.2.2). The estimated standard error of $\mathbf{c}'\hat{\mathbf{B}}\mathbf{m}$ is given by

$$\widehat{\text{s.e.}}(\mathbf{c}'\hat{\mathbf{B}}\mathbf{m}) = \{[\mathbf{c}'(\mathbf{X}'\mathbf{X})^{-}\mathbf{c}]\,(\mathbf{m}'\hat{\mathbf{E}}\mathbf{m})/(\mathbf{N}-\mathbf{t})\}^{1/2}, \qquad (31.4.1)$$

and its corresponding degrees of freedom are $N - t$. Thus, a $(1 - \alpha)100\%$ confidence interval for $\mathbf{c}'\mathbf{Bm}$ is obtained from

$$\mathbf{c}'\hat{\mathbf{B}}\mathbf{m} \pm t_{\alpha/2,N-t} \cdot \widehat{\text{s.e.}}(\mathbf{c}'\hat{\mathbf{B}}\mathbf{m}), \qquad (31.4.2)$$

and a t-statistic for testing H_0: $\mathbf{c}'\mathbf{Bm} = a_0$ is given by

$$t = \frac{\mathbf{c}'\hat{\mathbf{B}}\mathbf{m} - a_0}{\widehat{\text{s.e.}}(\mathbf{c}'\hat{\mathbf{B}}\mathbf{m})}.$$

If

$$t > t_{\alpha/2,N-t},$$

then H_0 is rejected.

As an example, consider estimating the V_1 marginal mean in the example in Section 31.3. For this marginal mean, we take

$$\mathbf{c}' = [1 \quad 1 \quad 0 \quad 0 \quad 0 \quad .2 \quad .2 \quad .2 \quad .2 \quad .2]$$

and

$$\mathbf{m}' = [.2 \quad .2 \quad .2 \quad .2 \quad .2].$$

The value of $\mathbf{c}'\hat{\mathbf{B}}\mathbf{m}$ is 3.496, and its estimated standard error is .059. A 95% confidence interval for the V_1 marginal mean is found from (31.4.2):

$$3.496 \pm (2.179)\,(.059).$$

As a second example, consider estimating the Time 1 marginal mean for the same example. For this marginal mean, we take

$$\mathbf{c}' = [1 \quad .25 \quad .25 \quad .25 \quad .25 \quad .2 \quad .2 \quad .2 \quad .2 \quad .2]$$

and

$$\mathbf{m}' = [1 \quad 0 \quad 0 \quad 0 \quad 0].$$

The value of $\mathbf{c}'\hat{\mathbf{B}}\mathbf{m}$ is 4.858, and its estimated standard error is .031.

As a third example, consider estimating the difference between the V_1 and V_2 marginal means. For this difference, we take

$$\mathbf{c}' = [0 \quad 1 \quad -1 \quad 0 \quad 0 \quad 0 \quad 0 \quad 0 \quad 0 \quad 0]$$

and

$$\mathbf{m}' = [.2 \quad .2 \quad .2 \quad .2 \quad .2].$$

The value of $\mathbf{c}'\hat{\mathbf{B}}\mathbf{m}$ is -1.094, and its estimated standard error is .084. A t-statistic for comparing these two marginal means is

$$t = \frac{-1.094}{.291} = -13.02,$$

and its observed significance level is $\hat{\alpha} = .0001$

CONCLUDING REMARKS

In this chapter, we discussed using a multivariate analysis of variance to analyze repeated measures experiments. The only drawback to a multivariate approach is that it may have low power. However, all other analyses of repeated measures experiments require that the errors satisfy some fairly strong restrictions. While a multivariate approach may have low power, there is some risk that other procedures will indicate differences that do not really exist. Such risks can be minimized by first testing the assumptions. If the assumptions do not appear to be satisfied, then the multivariate approach should be considered.

32

Analysis of Crossover Designs

CHAPTER OUTLINE

Crossover designs are used to compare treatments that are administered to an experimental unit (such as an animal) in a sequence. That is, each experimental unit is subjected to each treatment in a predetermined sequence. The objective of crossover designs is to eliminate variation in comparing treatments by observing all treatments on the same experimental unit.

Although crossover designs eliminate between-experimental-unit variation from treatment comparisons, other problems arise in the form of carry-over or residual effects. Carry-over effects occur, say, when treatment A is given first and its effect has not worn off by the time treatment B is applied. If this lingering effect of A interferes with the response of the subject to treatment B (either positively or negatively), then there is a residual effect of treatment A on the response of treatment B.

The crossover design model must contain a sequence effect, a time effect, a treatment effect, carry-over effects, an experimental unit error term, and a time interval error term. The first section discusses a general model and its assumptions, and the last two sections discuss crossover designs for two periods and more than two periods, respectively.

32.1 DEFINITIONS, ASSUMPTIONS, AND MODELS

In the general crossover design, t treatments are compared where each treatment is observed on each of the experimental units; that is, the treatments are applied in sequence to the experimental units. The experimenter constructs s sequences of the t treatments and randomly assigns n_i experimental units to the ith sequence. Table 32.1 contains a set of sequences of three treatments that could be applied to experimental units. The assignment of sequences to animals (a possible experimental unit) means that the animal is the experimental unit for sequences; the

Table 32.1 Possible Set of Sequences for Applying Three Treatments (A, B, and C) to Experimental Units

	TIME		
Sequence	*1*	*2*	*3*
1	A	B	C
2	A	C	B
3	B	A	C
4	B	C	A
5	C	A	B
6	C	B	A

assignment of a treatment to a time interval means that the time interval is the experimental unit for treatments.

A model to describe the response of an observation from the mth animal receiving the ith sequence and kth treatment in time period j is

$$y_{ijkm} = \mu_{ijk} + \varepsilon_{im} + e_{ijm} \qquad i = 1, 2, \ldots, s \quad j = 1, 2, \ldots, t$$
$$k = 1, 2, \ldots, t \quad m = 1, 2, \ldots, n_i$$

where μ_{ijk} is the mean response of treatment k applied in sequence i at time period j, ε_{im} is the random error associated with the mth experimental unit in sequence i, and e_{ijm} is the random error associated with time interval j of experimental unit m in sequence i.

The model can be reparameterized as

$$y_{ijkm} = \mu + \tau_k + \Pi_j + \sum_{r=0}^{j-1} \lambda_{jk_r} + \varepsilon_{im} + e_{ijm}$$

where τ_k is the treatment effect, Π_j is the time interval effect, and λ_{jk_r} denotes the carry-over effect of the k_rth treatment (which is administered in time period $r = 0, 1, \ldots, j - 1$) on time period j where $\lambda_{jk_0} = 0$. Most applications of the crossover design assume $\lambda_{jk_r} = 0$ for all r where $r < j - 1$, $j = 2, 3, \ldots, t$. Let

$$\lambda_{jk_{j-1}} = \lambda_{k_{j-1}};$$

then the model can be written as

$$y_{ijkm} = \mu + \tau_k + \Pi_j + \lambda_{k_{j-1}} + \varepsilon_{im} + e_{ijk}.$$

The assumptions about the distributions of ε_{im} and e_{ijm} are very important in developing an appropriate analysis. Since the experimental units are randomly assigned to the sequences, an appropriate assumption is that $\varepsilon_{im} \sim$ i.i.d. $N(0, \sigma_\varepsilon^2)$ (where normality is assumed for convenience). Since e_{ijm} is the error of the time interval, it is most likely that these errors are not independently distributed. One can assume that the covariance satisfies the compound symmetry condition (see Section 26.2). In that case, the usual analysis of variance methods for independent e_{ijm}'s can be used to carry out the analysis. The compound symmetry assumption is necessarily met for two-period designs and may be met for some three-period designs. A time series error structure may be more appropriate for crossover designs with three or more periods, and the methods discussed in Chapter 27 should be utilized.

32.2 TWO-PERIOD CROSSOVER DESIGNS

The two-period crossover experimental design is described in Section 5.3; it is classified as a repeated measures design since the same experimental unit is measured more than once. Model (5.3.2) is used to describe the data in Table 5.15 and is reparameterized in the form used in Grizzle

(1965) to aid in the interpretation of the μ_{ij}'s. The reparameterized model is

$$\left.\begin{aligned} y_{1jA} &= \mu + \pi_1 + \phi_A + \varepsilon_{1j} + e_{1jA} \\ y_{1jB} &= \mu + \pi_2 + \phi_B + \lambda_A + \varepsilon_{1j} + e_{1jB} \end{aligned}\right\} \quad j = 1, 2, \ldots, n_1$$

$$\left.\begin{aligned} y_{2jB} &= \mu + \pi_1 + \phi_B + \varepsilon_{2j} + e_{2jB} \\ y_{2jA} &= \mu + \pi_2 + \phi_A + \lambda_B + \varepsilon_{2j} + e_{2jA} \end{aligned}\right\} \quad j = 1, 2, \ldots, n_2$$

where π_m denotes the effect of period m, ϕ_k denotes the direct effect of treatment k, and λ_k denotes the carry-over effect of treatment k (given in the first period) into the second period.

Since there are two sizes of experimental units, there are two possible types of comparisons, between-animal comparisons and within-animal (time) comparisons.

Several contrasts of the μ_{ik}'s are of interest, and their interpretation is enhanced by expressing the contrasts in terms of Grizzle's (1965) parameterization. The sequence effect,

$$\begin{aligned} \bar{\mu}_{1\cdot} - \bar{\mu}_{2\cdot} &= \frac{1}{2}(\mu_{1A} + \mu_{1B} - \mu_{2B} - \mu_{2A}) \\ &= \frac{1}{2}(\lambda_A - \lambda_B), \end{aligned}$$

compares the residual or carry-over effects of the two treatments. The best estimate of the sequence effect is $\bar{y}_{1\cdot\cdot} - \bar{y}_{2\cdot\cdot}$, which is a between-animal comparison.

The Treatment effect is

$$\begin{aligned} \bar{\mu}_{\cdot A} - \bar{\mu}_{\cdot B} &= \frac{1}{2}(\mu_{1A} - \mu_{1B} - \mu_{2B} + \mu_{2A}) \\ &= \phi_A - \phi_B + \frac{1}{2}(\lambda_B - \lambda_A). \end{aligned}$$

The estimate of the Treatment effect is $\bar{y}_{\cdot A\cdot} - \bar{y}_{\cdot B\cdot}$, which is a within-animal comparison. The Treatment * Sequence interaction effect is

$$\mu_{1A} - \mu_{2A} - \mu_{1B} + \mu_{2B} = 2(\pi_1 - \pi_2) - (\lambda_A + \lambda_B).$$

The best estimate of this interaction effect is

$$\bar{y}_{1\cdot A} - \bar{y}_{2\cdot A} - \bar{y}_{1\cdot B} + \bar{y}_{2\cdot B},$$

which is a within-animal comparison. The variance associated with between-animal comparisons is a scalar multiple of $\sigma_e^2 + 2\sigma_\varepsilon^2$, and the variance of within-animal comparisons is a multiple of σ_ε^2. These results are summarized in Table 32.2.

If there is no Sequence effect, that is $\lambda_B - \lambda_A = 0$, the effects of the treatments can be compared through $\bar{\mu}_{\cdot A} - \bar{\mu}_{\cdot B}$. The estimate of the

Table 32.2 Analysis of Variance Table for a Two-Period Crossover Design

Source of Variation	*df*	*EMS*
Sequence	1	$\sigma_e^2 + 2\sigma_\varepsilon^2 + Q(\text{Sequence})$
Error(Animal)	$n_1 + n_2 - 2$	$\sigma_e^2 + 2\sigma_\varepsilon^2$
Treatment	1	$\sigma_e^2 + Q(\text{Treatment})$
Treatment * Sequence	1	$\sigma_e^2 + Q(\text{Treatment} * \text{Sequence})$
Error(Time)	$n_1 + n_2 - 2$	σ_e^2

Treatment effect is $\bar{y}_{\cdot\cdot A} - \bar{y}_{\cdot\cdot B}$ and has variance $2\sigma_e^2/(n_1 + n_2)$. An LSD is constructed by

$$\text{LSD}_\alpha = t_{\alpha/2, n_1+n_2-2}\sqrt{\frac{2 \times \text{MSERROR(TIME)}}{n_1 + n_2}}\,.$$

If there is a Treatment * Sequence interaction or a Sequence effect, then the treatments should be compared through their direct effects, which is a comparison of the two treatments at the first period only. The contrast of interest is $\mu_{1A} - \mu_{2B}$, which can be expressed as the sum of a contrast estimated by a within-animal comparison,

$$(\mu_{1A} - \bar{\mu}_{1\cdot}) - (\mu_{2B} - \bar{\mu}_{2\cdot}),$$

with a contrast estimated by a between-animal comparison $(\bar{\mu}_{1\cdot} - \bar{\mu}_{2\cdot})$. The estimator of $\mu_{1A} - \mu_{2B}$ is

$$\begin{aligned}\widehat{\mu_{1A} - \mu_{2B}} &= [(\bar{y}_{1\cdot A} - \bar{y}_{1\cdot\cdot}) - (\bar{y}_{2\cdot B} - \bar{y}_{2\cdot\cdot})] + (\bar{y}_{1\cdot\cdot} - \bar{y}_{2\cdot\cdot})\\ &= \bar{y}_{1\cdot A} - \bar{y}_{2\cdot B}.\end{aligned}$$

The advantage of expressing $\mu_{1A} - \mu_{2B}$ as the sum of two contrasts is that the variance of the estimator can be evaluated by the method described in Section 24.3. Since the two contrasts are independent, then

$$\begin{aligned}\text{Var}\left(\widehat{\mu_{1A} - \mu_{2B}}\right) &= \text{Var}(\bar{y}_{1\cdot A} - \bar{y}_{1\cdot\cdot} - \bar{y}_{2\cdot B} + \bar{y}_{2\cdot\cdot}) + \text{Var}(\bar{y}_{1\cdot\cdot} - \bar{y}_{2\cdot\cdot})\\ &= \frac{\frac{1}{2}(n_1 + n_2)(\sigma_e^2 + 2\sigma_\varepsilon^2)}{n_1 n_2} + \frac{\frac{1}{2}(n_1 + n_2)\sigma_e^2}{n_1 n_2}.\end{aligned}$$

This variance reduces to

$$\frac{(n_1 + n_2)(\sigma_e^2 + \sigma_\varepsilon^2)}{n_1 n_2},$$

but the above form is used to estimate the variance. The estimate of the variance is

$$\operatorname{Var}\left(\widehat{\bar{y}_{1\cdot A} - \bar{y}_{2\cdot B}}\right) = \frac{1}{2}(n_1 + n_2)\frac{\text{MSERROR(ANIMAL)} + \text{MSERROR(TIME)}}{n_1 n_2}.$$

An approximate LSD can be constructed by

$$\text{LSD}_\alpha = t_\alpha^* \sqrt{\operatorname{Var}\left(\widehat{\bar{y}_{1\cdot A} - \bar{y}_{2\cdot B}}\right)}$$

where

$$t_{\alpha/2}^* = \frac{\left[\frac{1}{2}(n_1 + n_2)/n_1 n_2\right]\left[t_{\alpha/2,v}\text{MSERROR(ANIMAL)} + t_{\alpha/2,u}\text{MSERROR(TIME)}\right]}{[\text{MSERROR(ANIMAL)} + \text{MSERROR(TIME)}]\left[\frac{1}{2}(n_1 + n_2)/n_1 n_2\right]}, \quad (32.2.1)$$

where v and u are the degrees of freedom associated with MSERROR(ANIMAL) and MSERROR(TIME), respectively.

For this design structure, $u = v$; thus,

$$t_{\alpha/2}^* = t_{\alpha/2, n_1 + n_2 - 2}.$$

This approach to constructing the LSD is consistent with that used to obtain approximate LSD values for split-plot and repeated measures designs (Section 24.3). In contrast, Grizzle obtained an estimate of $\sigma_e^2 + \sigma_\varepsilon^2$ by using only the data from the first period of each sequence. The following example illustrates the above ideas.

EXAMPLE 32.1: Grizzle's Data

Grizzle used the data in Table 32.3 to demonstrate his method of analysis. Table 32.4 gives an analysis of variance table for these data in the manner of Table 32.2. Since there is a moderate Sequence effect, the comparison of the two treatments are made through

$$\widehat{\mu_{1A} - \mu_{2B}} = 1.54.$$

The variance of the estimate is

$$\operatorname{Var}\left(\widehat{\mu_{1A} - \mu_{2B}}\right) = \frac{\frac{1}{2}(6 + 8)(1.00 + 1.245)}{(6 \cdot 8)} = .33.$$

Table 32.3 Data from Grizzle's (1965) Paper

	PERSON							
Treatment	*1*	*2*	*3*	*4*	*5*	*6*	*7*	*8*
Sequence 1								
A	0.2	0.0	−0.8	0.6	0.3	1.5	—	—
B	1.0	−0.7	0.2	1.1	0.4	1.2	—	—
Sequence 2								
B	1.3	−2.3	0.0	−0.8	−0.4	−2.9	−1.9	−2.9
A	0.9	1.0	0.6	−0.3	−1.0	1.7	−0.3	0.9

Table 32.4 Analysis of Variance Table for Example 32.1

Source of Variation	*df*	*SS*	*MS*	*F*
Sequence	1	4.57	4.57	4.57 (.10)
Error(Person)	12	12.00	1.00	
Treatment	1	5.14	5.14	4.13 (.10)
Treatment * Sequence	1	6.24	6.24	5.01 (.05)
Error(Time)	12	14.94	1.245	

Note: $\bar{y}_{1 \cdot A} = .30$, $\bar{y}_{1 \cdot B} = .53$, $\bar{y}_{2 \cdot A} = .44$, and $\bar{y}_{2 \cdot B} = -1.24$.

The value of $t^*_{.025}$ is

$$t^*_{.025} = \frac{\left[\frac{1}{2}(6+8)/(6 \cdot 8)\right]\left[t_{.025,12}(1.00) + t_{.025,12}(1.245)\right]}{(1.00 + 1.245)\left[\frac{1}{2}(6+8)/(6)(8)\right]}$$

$$= t_{.025,12} = 2.179.$$

The $\text{LSD}_{.05}$ is

$$\text{LSD}_{.05} = (2.179)\sqrt{.33} = 1.25.$$

Thus at $\alpha = .05$, the approximate LSD detects a significant difference between the treatments.

Table 32.5 Analysis of Variance Table for a Crossover Design with a Randomized Complete Block, Whole Plot Design Structure

Source of Variation	*df*	*EMS*
Block	$p - 1$	
Sequence	1	
Error(Animal)	$n_1 + n_2 - p - 1$	$\sigma_e^2 + 2\sigma_\varepsilon^2$
Treatment	1	
Sequence * Treatment	1	
Error(Time)	$n_1 + n_2 - 2$	σ_e^2

To generalize upon the above crossover design, suppose that the persons are grouped into blocks. Such a blocking could occur because facilities are not large enough to handle all the experimental units in one phase, thus requiring the experiment to be conducted in several phases. The persons assigned to phase 1 form one block, those assigned to phase 2 form a second block, and so on, until there are p phases.

In this revised experiment, the person experimental unit design structure consists of blocks and could be classified as a randomized complete block design structure with a one-way treatment structure. Only the person part of the analysis will be affected by the blocking, since the persons are the experimental units being blocked. The sources of variation and their degrees of freedom for this experiment are given in the analysis of variance table shown in Table 32.5.

The treatment effects are compared as in the nonblocked design structure except that when making the direct treatment comparison, the t^* value in equation (32.2.1) must be computed by using

$$u = n_1 + n_2 - p - 1 \quad \text{and} \quad v = n_1 + n_2 - 2.$$

32.3 CROSSOVER DESIGNS WITH MORE THAN TWO PERIODS

Crossover designs with more than two time periods are like two-period crossover designs in that there are two sizes of experimental units, two error terms, and two levels of analysis. In this discussion, the large experimental unit is called the "experimental unit," and the small experimental unit is called the "time interval." As the structure of the crossover design becomes more involved, obtaining estimates of the parameters of interest requires more effort.

The level of complexity depends on the design chosen. In some crossover designs, the estimates of the treatment comparisons are obtained by combining estimates from the within-experimental-unit part of the analysis with estimates from the between-experimental-unit part (see

Example 32.1). For other designs, estimates of the treatment effects are available from only one part of the analysis.

To demonstrate these ideas, two examples are presented below. Each has three periods, but the first has two treatments while the second has three.

EXAMPLE 32.2: Two Treatments in a Three-Period Crossover Design

Several possible sequences can be constructed by assigning two treatments to three time periods. The four sequences selected for this example are in Table 32.6 where n_1, n_2, n_3, and n_4 experimental units have been assigned to the respective sequences.

Two levels of analysis need to be carried out for this experiment. The first model is

$$y_{ijkm} = \mu + p_{im} + \Pi_k + \tau_j + \lambda_{i(k-1)} + \varepsilon_{ijkm}$$

where p_{im} denotes the random error associated with the mth experimental unit assigned to sequence i, which is assumed to be distributed i.i.d. $N(0, \sigma_p^2)$; Π_k denotes the Period effect; τ_j denotes the Treatment effect; $\lambda_{i(k-1)}$ denotes the carry-over effect of the treatment occurring in time period $k - 1$ of sequence i; and ε_{ijkm} denotes the random error associated with the time period within the experimental unit assumed to be distributed i.i.d. $N(0, \sigma_\varepsilon^2)$.

A convenient method for analyzing this model is to define

$$x_{ik} = \begin{Bmatrix} 0 & \text{if } k = 1 \\ 1 & \text{if the treatment in period } k - 1 \text{ is } A \\ -1 & \text{if the treatment in period } k - 1 \text{ is } B \end{Bmatrix}$$

and fit the model

$$y_{ijkm} = \mu + p_{im} + \Pi_k + \tau_j + \lambda x_{ik} + \varepsilon_{ijkm}$$

Table 32.6 Sequences of Two Treatments in Example 32.2

	SEQUENCE			
Time	*1*	*2*	*3*	*4*
1	*A*	*A*	*B*	*B*
2	*B*	*B*	*A*	*A*
3	*A*	*B*	*B*	*A*

Table 32.7 Means of Model for Example 32.2

	SEQUENCE			
Time	*1*	*2*	*3*	*4*
1	$\Pi_1 + \tau_A$	$\Pi_1 + \tau_A$	$\Pi_1 + \tau_B$	$\Pi_1 + \tau_B$
2	$\Pi_2 + \tau_B + \lambda_A$	$\Pi_2 + \tau_B + \lambda_A$	$\Pi_2 + \tau_A + \lambda_B$	$\Pi_2 + \tau_A + \lambda_B$
3	$\Pi_3 + \tau_A + \lambda_B$	$\Pi_3 + \tau_B + \lambda_B$	$\Pi_3 + \tau_B + \lambda_A$	$\Pi_3 + \tau_A + \lambda_A$

Table 32.8 Observed Means of Model for Example 32.2

	SEQUENCE			
Time	*1*	*2*	*3*	*4*
1	$\bar{Y}_{1A1\cdot}$	$\bar{Y}_{2A1\cdot}$	$\bar{Y}_{3B1\cdot}$	$\bar{Y}_{4B1\cdot}$
2	$\bar{Y}_{1B2\cdot}$	$\bar{Y}_{2B2\cdot}$	$\bar{Y}_{3A2\cdot}$	$\bar{Y}_{4A2\cdot}$
3	$\bar{Y}_{1A3\cdot}$	$\bar{Y}_{2B3\cdot}$	$\bar{Y}_{3B3\cdot}$	$\bar{Y}_{4A3\cdot}$

where x_{ik} is treated as a continuous variable or a covariate. The model provides within-experimental-unit estimates of $\tau_A - \tau_B$, $\lambda = \lambda_A - \lambda_B$, and σ_ε^2 where $\hat{\sigma}_\varepsilon^2$ denotes the within-experimental-unit mean square error. The means for each cell of the above model are in Table 32.7, and the observed means are in Table 32.8.

The second model is constructed by taking the means of the three responses (one from each time period) of each experimental unit. Each sequence has a different model, as shown below:

$$\begin{aligned}
\bar{y}_{1\cdot\cdot m} &= \mu + p_{1m} + \bar{\Pi}_{.} + \frac{2}{3}\tau_A + \frac{1}{3}\tau_B + \frac{1}{3}\lambda_A + \frac{1}{3}\lambda_B + \bar{\varepsilon}_{1\cdot\cdot m} \\
&= \mu^* + \frac{2}{3}\tau_A + \frac{1}{3}\tau_B + e_{1m}^*, \\
\bar{y}_{2\cdot\cdot m} &= \mu + p_{2m} + \bar{\Pi}_{.} + \frac{1}{3}\tau_A + \frac{2}{3}\tau_B + \frac{1}{3}\lambda_A + \frac{1}{3}\lambda_B + \bar{\varepsilon}_{2\cdot\cdot m} \\
&= \mu^* + \frac{1}{3}\tau_A + \frac{2}{3}\tau_B + e_{2m}^*, \\
\bar{y}_{3\cdot\cdot m} &= \mu + p_{3m} + \bar{\Pi}_{.} + \frac{1}{3}\tau_A + \frac{2}{3}\tau_B + \frac{1}{3}\lambda_A + \frac{1}{3}\lambda_B + \bar{\varepsilon}_{3\cdot\cdot m} \\
&= \mu^* + \frac{1}{3}\tau_A + \frac{2}{3}\tau_B + e_{3m}^*,
\end{aligned}$$

and

$$\bar{y}_{4\cdot\cdot m} = \mu + p_{4m} + \bar{\Pi}_{\cdot} + \frac{2}{3}\tau_A + \frac{1}{3}\tau_B + \frac{1}{3}\lambda_A + \frac{1}{3}\lambda_B + \bar{\varepsilon}_{4\cdot\cdot m}$$

$$= \mu^* + \frac{2}{3}\tau_A + \frac{1}{3}\tau_B + e^*_{4m}$$

where

$$\mu^* = \mu + \bar{\Pi}_{\cdot} + \frac{1}{3}\lambda_A + \frac{1}{3}\lambda_B \quad \text{and} \quad e^*_{im} = p_{im} + \bar{\varepsilon}_{i\cdot\cdot m},$$

where the e^*_{im} are distributed i.i.d. $N[0, \frac{1}{3}(\sigma_\varepsilon^2 + 3\sigma_p^2)]$.

Since the only difference between the means of the experimental unit mean models is in the coefficients of the τ's, there is between-experimental-unit information about the τ's but not about the λ's. An analysis of variance of the experimental unit means model provides a between-experimental-unit estimate of $\tau_A - \tau_B$ and an estimate of $\frac{1}{3}(\sigma_\varepsilon^2 + 3\sigma_p^2)$ from the between mean square error. The coefficient $\frac{1}{3}$ occurs because the experimental unit means are based on the mean of the data from three time intervals.

Next, the above ideas are applied to the data in Table 32.9, for which the computations were done by using SAS® GLM. There are two estimates of $\tau_A - \tau_B$ available, one a within-experimental-unit compari-

Table 32.9 Data from a Crossover Design with Two Treatments in Three Periods

			PERSON WITHIN SEQUENCE				
	Period	*Treatment*	*1*	*2*	*3*	*4*	*5*
Sequence 1	1	*A*	25.1	22.0	25.3		
	2	*B*	27.6	24.3	27.7		
	3	*A*	24.5	21.6	25.7		
Sequence 2	1	*A*	26.9	20.3	25.9	25.2	
	2	*B*	28.7	24.0	28.7	26.6	
	3	*B*	28.1	25.0	28.0	28.5	
Sequence 3	1	*B*	25.5	27.4	26.2		
	2	*A*	23.7	27.9	27.1		
	3	*B*	24.9	24.6	25.0		
Sequence 4	1	*B*	20.3	25.1	22.2	25.8	22.5
	2	*A*	22.2	26.2	25.0	26.5	23.6
	3	*A*	20.6	25.7	22.9	24.5	20.9

son and the other a between-experimental-unit comparison. The objective of this analysis is to combine the two estimates into a single estimate.

The within-experimental-unit analysis of variance table is shown in Table 32.10, which provides the within-experimental-unit estimates of $\lambda_A - \lambda_B$ and $\tau_A - \tau_B$. The between-experimental-unit analysis of variance table is shown in Table 32.11, which provides the between-experimental-unit estimate of $\tau_A - \tau_B$.

Using the method for combining two estimates of the same parameter with different variances, we get for the combined within–between-experimental-unit estimate of $\tau_A - \tau_B$

$$\left(\widehat{\tau_A - \tau_B}\right)_C = \frac{\left(1/\hat{\sigma}_B^2\right)\left(\widehat{\tau_A - \tau_B}\right)_b + \left(1/\hat{\sigma}_W^2\right)\left(\widehat{\tau_A - \tau_B}\right)_w}{\left(1/\hat{\sigma}_b^2\right) + \left(1/\hat{\sigma}_w^2\right)}$$

$$= \frac{\left[1/(2.8602)^2\right](-6.089) + \left[1/(.3277)^2\right](-1.189)}{(1/2.8602)^2 + (1/.3277)^2}$$

$$= -1.252.$$

The large sample variance of the combined estimate is

$$\operatorname{Var}\left[\left(\widehat{\tau_A - \tau_B}\right)_C\right] = \frac{\sigma_b^2\sigma_w^2}{\sigma_b^2 + \sigma_w^2}$$

and can be estimated by

$$\frac{\hat{\sigma}_b^2\hat{\sigma}_w^2}{\hat{\sigma}_b^2 + \hat{\sigma}_w^2} = \hat{\sigma}_c^2.$$

Table 32.10 Within-Experimental-Unit Analysis of Variance Table for Data in Table 32.9

Source of Variation	*df*	*SS*	*MS*	*F*
Experimental Unit	14	178.49		
Treatment	1	12.37	12.37	13.07
Carry-over	1	4.78	4.78	5.09
Period	2	22.14	11.07	11.78
Error(Within)	26	24.43	.94	

Note: $\left(\widehat{\tau_A - \tau_B}\right)_w = -1.189$ $\quad \widehat{\text{s.e.}}\left(\widehat{\tau_A - \tau_B}\right)_w = .3277 = \hat{\sigma}_w$

$\widehat{\lambda_A - \lambda_B} = .427$ $\quad \widehat{\text{s.e.}}\left(\widehat{\lambda_A - \lambda_B}\right) = .1892$

$\hat{\sigma}_\varepsilon^2 = .94$

Table 32.11 Between-Experimental-Unit Analysis of Variance Table for Data in Table 32.9

Source of Variation	*df*	*SS*	*MS*	*F*
Treatment	1	15.38	15.38	4.53
Error(Between)	13	44.12	3.39	

Note: $\widehat{(\tau_A - \tau_B)_{\text{b}}} = -6.089$ $\widehat{\text{s.e.}}\widehat{(\tau_A - \tau_B)_{\text{b}}} = 2.8602 = \hat{\sigma}_{\text{b}}$

$\hat{\sigma}_\varepsilon^2 + 3\hat{\sigma}_p^2 = 3(3.39) = 10.17$

To construct a confidence interval about $\tau_A - \tau_B$, compute an approximate t-value as

$$t_{\alpha/2}^* = \frac{(1/\hat{\sigma}_{\text{B}}^2)t_{\alpha/2,u} + (1/\hat{\sigma}_{\text{W}}^2)t_{\alpha/2,v}}{(1/\hat{\sigma}_{\text{B}}^2) + (1/\hat{\sigma}_{\text{W}}^2)}$$

where u and v are the degrees of freedom associated with the between error mean square and the within error mean square, respectively. For this example,

$$t_{.025}^* = \frac{\left[1/(2.8602)^2\right](2.160) + \left[1/(.3277)^2\right](2.056)}{\left[1/(2.8602)^2\right] + \left[1/(.3277)^2\right]}$$

$$= 2.057$$

The estimate of σ_{C}^2 is

$$\hat{\sigma}_{\text{C}}^2 = \frac{(2.8602)^2(.3277)^2}{(.3277)^2 + (2.8602)^2}$$

$$= .1060 \; (\hat{\sigma}_{\text{C}} = .3256).$$

Thus, a 95% confidence interval about $\tau_A - \tau_B$ is

$$-1.252 \pm 2.057\sqrt{.1060} \quad \text{or} \quad -1.922 \text{ to } -.582.$$

EXAMPLE 32.3: Three Treatments in a Three-Period Crossover Design

Many different sequences of three treatments can be constructed for a three-period crossover design. In the sequences chosen for this example, each treatment occurs in each sequence. There are six possible sequences,

Table 32.12 Sequences of Three Treatments in Example 32.3

	SEQUENCE					
Time	*1*	*2*	*3*	*4*	*5*	*6*
1	*A*	*A*	*B*	*B*	*C*	*C*
2	*B*	*C*	*A*	*C*	*A*	*B*
3	*C*	*B*	*C*	*A*	*B*	*A*

Table 32.13 Expected Means of Response (Without μ's) for Example 32.3

	SEQUENCE					
Time	*1*	*2*	*3*	*4*	*5*	*6*
1	$\Pi_1 + \tau_A$	$\Pi_1 + \tau_A$	$\Pi_1 + \tau_B$	$\Pi_1 + \tau_B$	$\Pi_1 + \tau_C$	$\Pi_1 + \tau_C$
2	$\Pi_2 + \tau_B + \lambda_A$	$\Pi_2 + \tau_C + \lambda_A$	$\Pi_2 + \tau_A + \lambda_B$	$\Pi_2 + \tau_C + \lambda_B$	$\Pi_2 + \tau_A + \lambda_C$	$\Pi_2 + \tau_B + \lambda_C$
3	$\Pi_3 + \tau_C + \lambda_B$	$\Pi_3 + \tau_B + \lambda_C$	$\Pi_3 + \tau_C + \lambda_A$	$\Pi_3 + \tau_A + \lambda_C$	$\Pi_3 + \tau_B + \lambda_A$	$\Pi_3 + \tau_A + \lambda_B$

which are listed in Table 32.12. Table 32.13 contains the expected mean responses for each cell, and Table 32.14 contains the observed cell means. In this example, the between-experimental-unit comparisons contain information about the carry-over effects (in contrast to Example 32.2, where they contained information about the treatment effects).

A model to describe the data from this design is

$$y_{ijkm} = \mu + p_{im} + \Pi_k + \tau_j + \lambda_{i(k-1)} + \varepsilon_{ijkm}$$

Table 32.14 Observed Means of Response for Example 32.3

	SEQUENCE					
Time	*1*	*2*	*3*	*4*	*5*	*6*
1	$\bar{Y}_{1A1}$	$\bar{Y}_{2A1}$	$\bar{Y}_{3B1}$	$\bar{Y}_{4B1}$	$\bar{Y}_{5C1}$	$\bar{Y}_{6C1}$
2	$\bar{Y}_{1B2}$	$\bar{Y}_{2C2}$	$\bar{Y}_{3A2}$	$\bar{Y}_{4C2}$	$\bar{Y}_{5A2}$	$\bar{Y}_{6B2}$
3	$\bar{Y}_{1C3}$	$\bar{Y}_{2B3}$	$\bar{Y}_{3C3}$	$\bar{Y}_{4A3}$	$\bar{Y}_{5B3}$	$\bar{Y}_{6A3}$

Table 32.15 Data from a Three-Period Crossover Design with Three Treatments

		EXPERIMENTAL UNIT					
Period	*Treatment*	*1*	*2*	*3*	*4*	*5*	*6*
Sequence 1							
1	*A*	20.1	23.3	23.4	19.7	19.2	22.2
2	*B*	20.3	24.8	24.8	21.3	20.9	22.0
3	*C*	25.6	28.7	28.3	25.7	25.9	26.2
Sequence 2							
1	*A*	24.7	23.8	23.6	20.2	19.8	21.5
2	*C*	29.4	28.7	26.4	26.2	23.7	25.5
3	*B*	27.5	24.1	25.0	21.4	23.3	20.8
Sequence 3							
1	*B*	24.3	26.4	19.9	23.9	20.5	21.8
2	*A*	23.2	26.4	23.7	26.8	23.2	23.6
3	*C*	30.1	32.3	25.5	30.8	26.3	29.1
Sequence 4							
1	*B*	20.9	21.9	22.0	23.3	18.8	24.6
2	*C*	27.5	28.6	27.4	30.7	27.9	29.8
3	*A*	24.3	23.1	24.5	26.6	24.6	26.6
Sequence 5							
1	*C*	24.0	25.9	25.5	27.9	25.3	25.7
2	*A*	21.8	23.7	22.0	25.4	26.4	24.7
3	*B*	21.6	23.9	23.4	24.4	25.8	24.9
Sequence 6							
1	*C*	23.2	23.9	28.0	24.6	27.7	21.5
2	*B*	18.9	21.5	25.3	22.7	23.5	18.1
3	*A*	23.8	25.4	28.1	23.8	25.6	22.8

where the parameters are defined as in Example 32.2. To fit this model, define a new carry-over parameter as

$$\lambda_{i(k-1)} = \lambda_A x_{Ai(k-1)} + \lambda_B x_{Bi(k-1)} + \lambda_C x_{Ci(k-1)}$$

where

$$x_{A_i(k-1)} = \begin{cases} 1 & \text{if treatment } A \text{ occurred in time period } k-1 \text{ of sequence } i \\ 0 & \text{if otherwise} \end{cases}$$

and $x_{Bi(k-1)}$ and $x_{Ci(k-1)}$ are defined similarly. Then fit the model

$$y_{ijkm} = \mu + p_{im} + \Pi_k + \tau_j + \lambda_A x_{Ai(k-1)} + \lambda_B x_{Bi(k-1)} + \lambda_C x_{Ci(k-1)} + \varepsilon_{ijkm}$$

to obtain the within-experimental-unit estimates of $\tau_A - \tau_B$, $\tau_A - \tau_C$,

Table 32.16 Within-Experimental-Unit Analysis of Variance Table for Data in Table 32.15

Source of Variation	*df*	*SS*	*MS*	*F*
Experimental Unit	35	360.97		
Treatment	2	249.72	124.86	124.86
Carry-over	2	4.45	2.23	2.23
Period	2	106.64	53.32	53.32
Error(Within)	66	66.19	1.00	

Note: $\hat{\sigma}_\varepsilon^3 = 1.00$.

Table 32.17 Analysis of Treatment Differences for Data in Table 32.16

Parameter	*Estimate*	*Standard Error*
$\tau_A - \tau_B$	.83	.264
$\tau_A - \tau_C$	−3.12	.264
$\tau_B - \tau_C$	−3.95	.264
$\lambda_A - \lambda_B$	−.27	.354
$\lambda_A - \lambda_C$	.46	.354
$\lambda_B - \lambda_C$	.73	.354

$\tau_B - \tau_C$, $\lambda_A - \lambda_B$, $\lambda_A - \lambda_C$, $\lambda_B - \lambda_C$, and σ_ε^2 where $\hat{\sigma}_\varepsilon^2$ denotes the within mean square error. The experimental unit means generate the between-experimental-unit model, where each sequence has its own model as follows:

$$\bar{Y}_{1\cdot\cdot m} = \mu + \bar{\Pi}_{.} + \bar{\tau}_{.} + \frac{\lambda_A + \lambda_B}{3} + p_{1m} + \bar{\varepsilon}_{1\cdot\cdot m}$$

$$= \mu^* + \frac{1}{3}\lambda_A + \frac{1}{3}\lambda_B + e_{1m}^*$$

$$\bar{Y}_{2\cdot\cdot m} = \mu + \bar{\Pi}_{.} + \bar{\tau}_{.} + \frac{\lambda_A + \lambda_C}{3} + p_{2m} + \bar{\varepsilon}_{2\cdot\cdot m}$$

$$= \mu^* + \frac{1}{3}\lambda_A + \frac{1}{3}\lambda_C + e_{2m}^*$$

$$\bar{Y}_{3\cdot\cdot m} = \mu + \bar{\Pi}_{.} + \bar{\tau}_{.} + \frac{\lambda_A + \lambda_B}{3} + p_{3m} + \bar{\varepsilon}_{3\cdot\cdot m}$$

$$= \mu^* + \frac{1}{3}\lambda_A + \frac{1}{3}\lambda_B + e_{3m}^*$$

$$\bar{Y}_{4\cdot\cdot m} = \mu + \bar{\Pi}_{.} + \bar{\tau}_{.} + \frac{\lambda_B + \lambda_C}{3} + p_{4m} + \bar{\varepsilon}_{4\cdot\cdot m}$$

$$= \mu^* + \frac{1}{3}\lambda_B + \frac{1}{3}\lambda_C + e_{4m}^*$$

$$\bar{Y}_{5\cdot\cdot m} = \mu + \bar{\Pi}_{.} + \bar{\tau}_{.} + \frac{\lambda_A + \lambda_C}{3} + p_{5m} + \bar{\varepsilon}_{5\cdot\cdot m}$$

$$= \mu^* + \frac{1}{3}\lambda_A + \frac{1}{3}\lambda_C + e_{5m}^*$$

$$\bar{Y}_{6\cdot\cdot m} = \mu + \bar{\Pi}_{.} + \bar{\tau}_{.} + \frac{\lambda_B + \lambda_C}{3} + p_{6m} + \bar{\varepsilon}_{6\cdot\cdot m}$$

$$= \mu^* + \frac{1}{3}\lambda_B + \frac{1}{3}\lambda_C + e_{6m}^*$$

Table 32.18 Between-Experimental-Unit Analysis of Variance Table for Data in Table 32.15

Source of Variation	*df*	*SS*	*MS*	*F*
Carry-over	2	.0615	.0307	.01
Error(Between)	33	120.2609	3.6443	

Note: $\widehat{\frac{1}{3}(\sigma_\varepsilon^2 + 3\sigma_p^2)} = 3.6443$.

Table 32.19 Analysis of Treatment Differences for Data in Table 32.18

Parameter	*Estimate*	*Standard Error*
$\lambda_A - \lambda_B$	−.30	2.34
$\lambda_A - \lambda_C$	−.11	2.34
$\lambda_B - \lambda_C$	.19	2.34

where

$$\mu^* = \mu + \overline{\Pi}_{.} + \bar{\tau}_{.} \quad \text{and} \quad e^*_{im} = p_{im} + \bar{\varepsilon}_{i \cdot \cdot m},$$

which are distributed i.i.d. $N[0, \frac{1}{3}(\sigma_\varepsilon^2 + 3\sigma_p^2)]$. The means of the models differ only in the coefficients of the λ's; thus, there is information only about the λ's from the between-experimental-unit comparisons. Fitting the experimental unit mean model provides between-experimental-unit estimates of $\lambda_A - \lambda_B$, $\lambda_A - \lambda_C$, $\lambda_B - \lambda_C$, and $\frac{1}{3}(\sigma_\varepsilon^2 + 3\sigma_p^2)$, which is estimated by the between mean square error.

Table 32.15 contains data from a three-period crossover design with three treatments. The within-experimental-unit analysis is in Table 32.16. The only information from the design about the treatments is contained in this analysis; thus, the estimates of the treatment differences and the corresponding standard errors, shown in Table 32.17, are all derived from the information in Table 32.16. The between-experimental-unit analysis is in Table 32.18, and the between-experimental-unit estimates of the carry-over effects are in Table 32.19. The within and between experimental unit estimates of differences between the carry-over effects could be combined into single estimates. The technique used in Example 32.2 could be used to combine the $\lambda_i - \lambda_j$ estimates in this case.

Before a given crossover design is used, one needs to determine (1) the kinds of comparisons that can be made, (2) at which level of the analysis the estimators will be obtained, and (3) the types of combinations of estimators necessary to obtain estimators of the parameters of interest.

CONCLUDING REMARKS

Cross-over designs are used in many experimental settings to compare several treatments that are administered to experimental units in a given sequence. There are many models involving treatment effects and carry-over effects. The choice of a model depends on the number of periods and the number of treatments used to construct the sequences.

The three examples discussed in this chapter demonstrate the methods needed to analyze crossover designs. Which comparisons will estimate the parameters of interest vary with each crossover design. Both of the models, the model for within-experimental-unit comparisons and the model for between-experimental-unit comparisons, must be investigated to obtain an appropriate analysis.

Appendices

Table A.1 Percentage Points of the Maximum *F*-Ratio

UPPER 5% POINTS

ν \ k	2	3	4	5	6	7	8	9	10	11	12
2	39.0	87.5	142	202	266	333	403	475	550	626	704
3	15.4	27.8	39.2	50.7	62.0	72.9	83.5	93.9	104	114	124
4	9.60	15.5	20.6	25.2	29.5	33.6	37.5	41.1	44.6	48.0	51.4
5	7.15	10.8	13.7	16.3	18.7	20.8	22.9	24.7	26.5	28.2	29.9
6	5.82	8.38	10.4	12.1	13.7	15.0	16.3	17.5	18.6	19.7	20.7
7	4.99	6.94	8.44	9.70	10.8	11.8	12.7	13.5	14.3	15.1	15.8
8	4.43	6.00	7.18	8.12	9.03	9.78	10.5	11.1	11.7	12.2	12.7
9	4.03	5.34	6.31	7.11	7.80	8.41	8.95	9.45	9.91	10.3	10.7
10	3.72	4.85	5.67	6.34	6.92	7.42	7.87	8.28	8.66	9.01	9.34
12	3.28	4.16	4.79	5.30	5.72	6.09	6.42	6.72	7.00	7.25	7.48
15	2.86	3.54	4.01	4.37	4.68	4.95	5.19	5.40	5.59	5.77	5.93
20	2.46	2.95	3.29	3.54	3.76	3.94	4.10	4.24	4.37	4.49	4.59
30	2.07	2.40	2.61	2.78	2.91	3.02	3.12	3.21	3.29	3.36	3.39
60	1.67	1.85	1.96	2.04	2.11	2.17	2.22	2.26	2.30	2.33	2.36
∞	1.00	1.00	1.00	1.00	1.00	1.00	1.00	1.00	1.00	1.00	1.00

UPPER 1% POINTS

ν \ k	2	3	4	5	6	7	8	9	10	11	12
2	199	448	729	1036	1362	1705	2063	2432	2813	3204	3605
3	47.5	85	120	151	184	21(6)	24(9)	28(1)	31(0)	33(7)	36(1)
4	23.2	37	49	59	69	79	89	97	106	113	120
5	14.9	22	28	33	38	42	46	50	54	57	60
6	11.1	15.5	19.1	22	25	27	30	32	34	36	37
7	8.89	12.1	14.5	16.5	18.4	20	22	23	24	26	27
8	7.50	9.9	11.7	13.2	14.5	15.8	16.9	17.9	18.9	19.8	21
9	6.54	8.5	9.9	11.1	12.1	13.1	13.9	14.7	15.3	16.0	16.6
10	5.85	7.4	8.6	9.6	10.4	11.1	11.8	12.4	12.9	13.4	13.9
12	4.91	6.1	6.9	7.6	8.2	8.7	9.1	9.5	9.9	10.2	10.6
15	4.07	4.9	5.5	6.0	6.4	6.7	7.1	7.3	7.5	7.8	8.0
20	3.32	3.8	4.3	4.6	4.9	5.1	5.3	5.5	5.6	5.8	5.9
30	2.63	3.0	3.3	3.4	3.6	3.7	3.8	3.9	4.0	4.1	4.2
60	1.96	2.2	2.3	2.4	2.4	2.5	2.5	2.6	2.6	2.7	2.7
∞	1.00	1.0	1.0	1.0	1.0	1.0	1.0	1.0	1.0	1.0	1.0

Table A.2 Table of Bonferroni Critical Points

VALUES OF c FOR $1 - \alpha = .95$

$$\int_{-\infty}^{c} f^{(v)}(t)\,dt = 1 - \frac{.05}{2m}$$

m \ v	5	7	10	12	15	20	24	30	40	60	120	∞
2	3.17	2.84	2.64	2.56	2.49	2.42	2.39	2.36	2.33	2.30	2.27	2.24
3	3.54	3.13	2.87	2.78	2.69	2.61	2.58	2.54	2.50	2.47	2.43	2.39
4	3.81	3.34	3.04	2.94	2.84	2.75	2.70	2.66	2.62	2.58	2.54	2.50
5	4.04	3.50	3.17	3.06	2.95	2.85	2.80	2.75	2.71	2.66	2.62	2.58
6	4.22	3.64	3.28	3.15	3.04	2.93	2.88	2.83	2.78	2.73	2.68	2.64
7	4.38	3.76	3.37	3.24	3.11	3.00	2.94	2.89	2.84	2.79	2.74	2.69
8	4.53	3.86	3.45	3.31	3.18	3.06	3.00	2.94	2.89	2.84	2.79	2.74
9	4.66	3.95	3.52	3.37	3.24	3.11	3.05	2.99	2.93	2.88	2.83	2.77
10	4.78	4.03	3.58	3.43	3.29	3.16	3.09	3.03	2.97	2.92	2.86	2.81
15	5.25	4.36	3.83	3.65	3.48	3.33	3.26	3.19	3.12	3.06	2.99	2.94
20	5.60	4.59	4.01	3.80	3.62	3.46	3.38	3.30	3.23	3.16	3.09	3.02
25	5.89	4.78	4.15	3.93	3.74	3.55	3.47	3.39	3.31	3.24	3.16	3.09
30	6.15	4.95	4.27	4.04	3.82	3.63	3.54	3.46	3.38	3.30	3.22	3.15
35	6.36	5.09	4.37	4.13	3.90	3.70	3.61	3.52	3.43	3.34	3.27	3.19
40	6.56	5.21	4.45	4.20	3.97	3.76	3.66	3.57	3.48	3.39	3.31	3.23
45	6.70	5.31	4.53	4.26	4.02	3.80	3.70	3.61	3.51	3.42	3.34	3.26
50	6.86	5.40	4.59	4.32	4.07	3.85	3.74	3.65	3.55	3.46	3.37	3.29
100	8.00	6.08	5.06	4.73	4.42	4.15	4.04	3.90	3.79	3.69	3.58	3.48
250	9.68	7.06	5.70	5.27	4.90	4.56	4.4*	4.2*	4.1*	3.97	3.83	3.72

Table A.2 (continued)

VALUES OF c FOR $1 - \alpha = .99$

$$\int_{-\infty}^{c} f^{(v)}(t)\,dt = -\frac{.01}{2m}$$

m \ v	5	7	10	12	15	20	24	30	40	60	120	∞
2	4.78	4.03	3.58	3.43	3.29	3.16	3.09	3.03	2.97	2.92	2.86	2.81
3	5.25	4.36	3.83	3.65	3.48	3.33	3.26	3.19	3.12	3.06	2.99	2.94
4	5.60	4.59	4.01	3.80	3.62	3.46	3.38	3.30	3.23	3.16	3.09	3.02
5	5.89	4.78	4.15	3.93	3.74	3.55	3.47	3.39	3.31	3.24	3.16	3.09
6	6.15	4.95	4.27	4.04	3.82	3.63	3.54	3.46	3.38	3.30	3.22	3.15
7	6.36	5.09	4.37	4.13	3.90	3.70	3.61	3.52	3.43	3.34	3.27	3.19
8	6.56	5.21	4.45	4.20	3.97	3.76	3.66	3.57	3.48	3.39	3.31	3.23
9	6.70	5.31	4.53	4.26	4.02	3.80	3.70	3.61	3.51	3.42	3.34	3.26
10	6.86	5.40	4.59	4.32	4.07	3.85	3.74	3.65	3.55	3.46	3.37	3.29
15	7.51	5.79	4.86	4.56	4.29	4.03	3.91	3.80	3.70	3.59	3.50	3.40
20	8.00	6.08	5.06	4.73	4.42	4.15	4.04	3.90	3.79	3.69	3.58	3.48
25	8.37	6.30	5.20	4.86	4.53	4.25	4.1*	3.98	3.88	3.76	3.64	3.54
30	8.68	6.49	5.33	4.95	4.61	4.33	4.2*	4.13	3.93	3.81	3.69	3.59
35	8.95	6.67	5.44	5.04	4.71	4.39	4.3*	4.26	3.97	3.84	3.73	3.63
40	9.19	6.83	5.52	5.12	4.78	4.46	4.3*	4.1*	4.01	3.89	3.77	3.66
45	9.41	6.93	5.60	5.20	4.84	4.52	4.3*	4.2*	4.1*	3.93	3.80	3.69
50	9.68	7.06	5.70	5.27	4.90	4.56	4.4*	4.2*	4.1*	3.97	3.83	3.72
100	11.04	7.80	6.20	5.70	5.20	4.80	4.7*	4.4*	4.5*		4.00	3.89
250	13.26	8.83	6.9*	6.3*	5.8*	5.2*	5.0*	4.9*	4.8*			4.11

*Obtained by graphical interpolation.

Reproduced from: O. J. Dunn, Multiple Comparisons among Means. *Journal of the American Statistical Association*, *56*: 52–64, 1961. With permission from the American Statistical Association and the author.

Table A.3 Probability Points for Multivariate *t*-Distribution

Entries are $t_{\alpha/2:q,m}$, where $P[\max|T_i| \leq t_{\alpha/2:q,m}$ for $i = 1, 2, \ldots, q] = 1 - \alpha$ and $\mathbf{t} = [T_1, T_2, \ldots, T_q]'$ is distributed $\mathbf{S}(\mathbf{t}: q, m; \mathbf{I})$.

m \ q	1	2	3	4	5	6	8	10	12	15	20
					$1 - \alpha = .90$						
3	2.353	2.989	3.369	3.637	3.844	4.011	4.272	4.471	4.631	4.823	5.066
4	2.132	2.662	2.976	3.197	3.368	3.506	3.722	3.887	4.020	4.180	4.383
5	2.015	2.491	2.769	2.965	3.116	3.239	3.430	3.576	3.694	3.837	4.018
6	1.943	2.385	2.642	2.822	2.961	3.074	3.249	3.384	3.493	3.624	3.790
7	1.895	2.314	2.556	2.725	2.856	2.962	3.127	3.253	3.355	3.478	3.635
8	1.860	2.262	2.494	2.656	2.780	2.881	3.038	3.158	3.255	3.373	3.522
9	1.833	2.224	2.447	2.603	2.723	2.819	2.970	3.086	3.179	3.292	3.436
10	1.813	2.193	2.410	2.562	2.678	2.771	2.918	3.029	3.120	3.229	3.360
11	1.796	2.169	2.381	2.529	2.642	2.733	2.875	2.984	3.072	3.178	3.313
12	1.782	2.149	2.357	2.501	2.612	2.701	2.840	2.946	3.032	3.136	3.268
15	1.753	2.107	2.305	2.443	2.548	2.633	2.765	2.865	2.947	3.045	3.170
20	1.725	2.065	2.255	2.386	2.486	2.567	2.691	2.786	2.863	2.956	3.073
25	1.708	2.041	2.226	2.353	2.450	2.528	2.648	2.740	2.814	2.903	3.016
30	1.697	2.025	2.207	2.331	2.426	2.502	2.620	2.709	2.781	2.868	2.978
40	1.684	2.006	2.183	2.305	2.397	2.470	2.585	2.671	2.741	2.825	2.931
60	1.671	1.986	2.160	2.278	2.368	2.439	2.550	2.634	2.701	2.782	2.884
					$1 - \alpha = .95$						
3	3.183	3.960	4.430	4.764	5.023	5.233	5.562	5.812	6.015	6.259	6.567
4	2.777	3.382	3.745	4.003	4.203	4.366	4.621	4.817	4.975	5.166	5.409
5	2.571	3.091	3.399	3.619	3.789	3.928	4.145	4.312	4.447	4.611	4.819
6	2.447	2.916	3.193	3.389	3.541	3.664	3.858	4.008	4.129	4.275	4.462
7	2.365	2.800	3.056	3.236	3.376	3.489	3.668	3.805	3.916	4.051	4.223
8	2.306	2.718	2.958	3.128	3.258	3.365	3.532	3.660	3.764	3.891	4.052
9	2.262	2.657	2.885	3.046	3.171	3.272	3.430	3.552	3.651	3.770	3.923
10	2.228	2.609	2.829	2.984	3.103	3.199	3.351	3.468	3.562	3.677	3.823

Table A.3 (continued)

Entries are $t_{\alpha/2:q,m}$, where $P[\max|T_i| \leq t_{\alpha/2:q,m}$, for $i = 1, 2, \ldots, q] = 1 - \alpha$ and $\mathbf{t} = [T_1, T_2, \ldots, T_q]'$ is distributed $\mathbf{S}(\mathbf{t} : q, m; \mathbf{I})$.

m \ q	1	2	3	4	5	6	8	10	12	15	20
					$1 - \alpha = .95$						
11	2.201	2.571	2.784	2.933	3.048	3.142	3.288	3.400	3.491	3.602	3.743
12	2.179	2.540	2.747	2.892	3.004	3.095	3.236	3.345	3.433	3.541	3.677
15	2.132	2.474	2.669	2.805	2.910	2.994	3.126	3.227	3.309	3.409	3.536
20	2.086	2.411	2.594	2.722	2.819	2.898	3.020	3.114	3.190	3.282	3.399
25	2.060	2.374	2.551	2.673	2.766	2.842	2.959	3.048	3.121	3.208	3.320
30	2.042	2.350	2.522	2.641	2.732	2.805	2.918	3.005	3.075	3.160	3.267
40	2.021	2.321	2.488	2.603	2.690	2.760	2.869	2.952	3.019	3.100	3.203
60	2.000	2.292	2.454	2.564	2.649	2.716	2.821	2.900	2.964	3.041	3.139
					$1 - \alpha = .99$						
3	5.841	7.127	7.914	8.479	8.919	9.277	9.838	10.269	10.616	11.034	11.559
4	4.604	5.462	5.985	6.362	6.656	6.897	7.274	7.565	7.801	8.087	8.451
5	4.032	4.700	5.106	5.398	5.625	5.812	6.106	6.333	6.519	6.744	7.050
6	3.707	4.271	4.611	4.855	5.046	5.202	5.449	5.640	5.796	5.985	6.250
7	3.500	3.998	4.296	4.510	4.677	4.814	5.031	5.198	5.335	5.502	5.716
8	3.355	3.809	4.080	4.273	4.424	4.547	4.742	4.894	5.017	5.168	5.361
9	3.250	3.672	3.922	4.100	4.239	4.353	4.532	4.672	4.785	4.924	5.103
10	3.169	3.567	3.801	3.969	4.098	4.205	4.373	4.503	4.609	4.739	4.905
11	3.106	3.485	3.707	3.865	3.988	4.087	4.247	4.370	4.470	4.593	4.750
12	3.055	3.418	3.631	3.782	3.899	3.995	4.146	4.263	4.359	4.475	4.625
15	2.947	3.279	3.472	3.608	3.714	3.800	3.935	4.040	4.125	4.229	4.363
20	2.845	3.149	3.323	3.446	3.541	3.617	3.738	3.831	3.907	3.999	4.117
25	2.788	3.075	3.239	3.354	3.442	3.514	3.626	3.713	3.783	3.869	3.978
30	2.750	3.027	3.185	3.295	3.379	3.448	3.555	3.637	3.704	3.785	3.889
40	2.705	2.969	3.119	3.223	3.303	3.367	3.468	3.545	3.607	3.683	3.780
60	2.660	2.913	3.055	3.154	3.229	3.290	3.384	3.456	3.515	3.586	3.676

Reprinted from "A Table of Percentage Points of the Distribution of the Largest Absolute Value of q Student t Variates and Its Applications," *Biometrika*, Vol. 58. It is reprinted here with the kind permission of the authors, G. J. Hahn and R. W. Hendrickson, and the Trustees of *Biometrika*.

Table A.4 Percentage Points of the Studentized Range

The entries are $q_{.05:m,n}$, where $P[Q < q_{.05:m,n}] = .95$.

n \ m	2	3	4	5	6	7	8	9	10
1	17.97	26.98	32.82	37.08	40.41	43.12	45.40	47.36	49.07
2	6.08	8.33	9.80	10.88	11.74	12.44	13.03	13.54	13.99
3	4.50	5.91	6.82	7.50	8.04	8.48	8.85	9.18	9.46
4	3.93	5.04	5.76	6.29	6.71	7.05	7.35	7.60	7.83
5	3.64	4.60	5.22	5.67	6.03	6.33	6.58	6.80	6.99
6	3.46	4.34	4.90	5.30	5.63	5.90	6.12	6.32	6.49
7	3.34	4.16	4.68	5.06	5.36	5.61	5.82	6.00	6.16
8	3.26	4.04	4.53	4.89	5.17	5.40	5.60	5.77	5.92
9	3.20	3.95	4.41	4.76	5.02	5.24	5.43	5.59	5.74
10	3.15	3.88	4.33	4.65	4.91	5.12	5.30	5.46	5.60
11	3.11	3.82	4.26	4.57	4.82	5.03	5.20	5.35	5.49
12	3.08	3.77	4.20	4.51	4.75	4.95	5.12	5.27	5.39
13	3.06	3.73	4.15	4.45	4.69	4.88	5.05	5.19	5.32
14	3.03	3.70	4.11	4.41	4.64	4.83	4.99	5.13	5.25
15	3.01	3.67	4.08	4.37	4.59	4.78	4.94	5.08	5.20
16	3.00	3.65	4.05	4.33	4.56	4.74	4.90	5.03	5.15
17	2.98	3.63	4.02	4.30	4.52	4.70	4.86	4.99	5.11
18	2.97	3.61	4.00	4.28	4.49	4.67	4.82	4.96	5.07
19	2.96	3.59	3.98	4.25	4.47	4.65	4.79	4.92	5.04
20	2.95	3.58	3.96	4.23	4.45	4.62	4.77	4.90	5.01
24	2.92	3.53	3.90	4.17	4.37	4.54	4.68	4.81	4.92
30	2.89	3.49	3.85	4.10	4.30	4.46	4.60	4.72	4.82
40	2.86	3.44	3.79	4.04	4.23	4.39	4.52	4.63	4.73
60	2.83	3.40	3.74	3.98	4.16	4.31	4.44	4.55	4.65
120	2.80	3.36	3.68	3.92	4.10	4.24	4.36	4.47	4.56
∞	2.77	3.31	3.63	3.89	4.03	4.17	4.29	4.39	4.47

Table A.4 (continued)

The entries are $q_{.05:m,n}$, where $P[Q < q_{.05:m,n}] = .95$.

11	*12*	*13*	*14*	*15*	*16*	*17*	*18*	*19*	*20*
50.59	51.96	53.20	54.33	55.36	56.32	57.22	58.04	58.83	59.56
14.39	14.75	15.08	15.38	15.65	15.91	16.14	16.37	16.57	16.77
9.72	9.95	10.15	10.35	10.52	10.69	10.84	10.98	11.11	11.24
8.03	8.21	8.37	8.52	8.66	8.79	8.91	9.03	9.13	9.23
7.17	7.32	7.47	7.60	7.72	7.83	7.93	8.03	8.12	8.21
6.65	6.79	6.92	7.03	7.14	7.24	7.34	7.43	7.51	7.59
6.30	6.43	6.55	6.66	6.76	6.85	6.94	7.02	7.10	7.17
6.05	6.18	6.29	6.39	6.48	6.57	6.65	6.73	6.80	6.87
5.87	5.98	6.09	6.19	6.28	6.36	6.44	6.51	6.58	6.64
5.72	5.83	5.93	6.03	6.11	6.19	6.27	6.34	6.40	6.47
5.61	5.71	5.81	5.90	5.98	6.06	6.13	6.20	6.27	6.33
5.51	5.61	5.71	5.80	5.88	5.95	6.02	6.09	6.15	6.21
5.43	5.53	5.63	5.71	5.79	5.86	5.93	5.99	6.05	6.11
5.36	5.46	5.55	5.64	5.71	5.79	5.85	5.91	5.97	6.03
5.31	5.40	5.49	5.57	5.65	5.72	5.78	5.85	5.90	5.96
5.26	5.35	5.44	5.52	5.59	5.66	5.73	5.79	5.84	5.90
5.21	5.31	5.39	5.47	5.54	5.61	5.67	5.73	5.79	5.84
5.17	5.27	5.35	5.43	5.50	5.57	5.63	5.69	5.74	5.79
5.14	5.23	5.31	5.39	5.46	5.53	5.59	5.65	5.70	5.75
5.11	5.20	5.28	5.36	5.43	5.49	5.55	5.61	5.66	5.71
5.01	5.10	5.18	5.25	5.32	5.38	5.44	5.49	5.55	5.59
4.92	5.00	5.08	5.15	5.21	5.27	5.33	5.38	5.43	5.47
4.82	4.90	4.98	5.04	5.11	5.16	5.22	5.27	5.31	5.36
4.73	4.81	4.88	4.94	5.00	5.06	5.11	5.15	5.20	5.24
4.64	4.71	4.78	4.84	4.90	4.95	5.00	5.04	5.09	5.13
4.55	4.62	4.68	4.74	4.80	4.85	4.89	4.93	4.97	5.01

Table A.4 (continued)

The entries are $q_{.01:m,n}$, where $P[Q < q_{.01:m,n}] = .99$.

n \ m	2	3	4	5	6	7	8	9	10
1	90.03	135.0	164.3	185.6	202.2	215.8	227.	237.0	245.6
2	14.04	19.02	22.29	24.72	26.63	28.20	29.53	30.68	31.69
3	8.26	10.62	12.17	13.33	14.24	15.00	15.64	16.20	16.69
4	6.51	8.12	9.17	9.96	10.58	11.10	11.55	11.93	12.27
5	5.70	6.98	7.80	8.42	8.91	9.32	9.67	9.97	10.24
6	5.24	6.33	7.03	7.56	7.97	8.32	8.61	8.87	9.10
7	4.95	5.92	6.54	7.01	7.37	7.68	7.94	8.17	8.36
8	4.75	5.64	6.20	6.62	6.96	7.24	7.47	7.68	7.86
9	4.60	5.43	5.96	6.35	6.66	6.91	7.13	7.33	7.49
10	4.48	5.27	5.77	6.14	6.43	6.67	6.87	7.05	7.21
11	4.39	5.15	5.62	5.97	6.25	6.48	6.67	6.84	6.99
12	4.32	5.05	5.50	5.84	6.10	6.32	6.51	6.67	6.81
13	4.26	4.96	5.40	5.73	5.98	6.19	6.37	6.53	6.67
14	4.21	4.89	5.32	5.63	5.88	6.08	6.26	6.41	6.54
15	4.17	4.84	5.25	5.56	5.80	5.99	6.16	6.31	6.44
16	4.13	4.79	5.19	5.49	5.72	5.92	6.08	6.22	6.35
17	4.10	4.74	5.14	5.43	5.66	5.85	6.01	6.15	6.27
18	4.07	4.70	5.09	5.38	5.60	5.79	5.94	6.08	6.20
19	4.05	4.67	5.05	5.33	5.55	5.73	5.89	6.02	6.14
20	4.02	4.64	5.02	5.29	5.51	5.69	5.84	5.97	6.09
24	3.96	4.55	4.91	5.17	5.37	5.54	5.69	5.81	5.92
30	3.89	4.45	4.80	5.05	5.24	5.40	5.54	5.65	5.76
40	3.82	4.37	4.70	4.93	5.11	5.26	5.39	5.50	5.60
60	3.76	4.28	4.59	4.82	4.99	5.13	5.25	5.36	5.45
120	3.70	4.20	4.50	4.71	4.87	5.01	5.12	5.21	5.30
∞	3.64	4.12	4.40	4.60	4.76	4.88	4.99	5.08	5.16

From E. S. Pearson and H. O. Hartley, *Biometrika Tables for Statisticians*, Vol. 1, Third Edition, Table 29, published by the Biometrika Trustees, Cambridge University Press, London, 1966. Reproduced with the permission of the authors and the publishers.

Table A.4 (continued)

The entries are $q_{.05:m,n}$, where $P[Q < q_{.05:m,n}] = .95$.

11	*12*	*13*	*14*	*15*	*16*	*17*	*18*	*19*	*20*
253.2	260.0	266.2	271.8	277.0	281.8	286.3	290.4	294.3	298.0
32.59	33.40	34.13	34.81	35.43	36.00	36.53	37.03	37.50	37.95
17.13	17.53	17.89	18.22	18.52	18.81	19.07	19.32	19.55	19.77
12.57	12.84	13.09	13.32	13.53	13.73	13.91	14.08	14.24	14.40
10.48	10.70	10.89	11.08	11.24	11.40	11.55	11.68	11.81	11.93
9.30	9.48	9.65	9.81	9.95	10.08	10.21	10.32	10.43	10.54
8.55	8.71	8.86	9.00	9.12	9.24	9.35	9.46	9.55	9.65
8.03	8.18	8.31	8.44	8.55	8.66	8.76	8.85	8.94	9.03
7.65	7.78	7.91	8.03	8.13	8.23	8.33	8.41	8.49	8.57
7.36	7.49	7.60	7.71	7.81	7.91	7.99	8.08	8.15	8.23
7.13	7.25	7.36	7.46	7.56	7.65	7.73	7.81	7.88	7.95
6.94	7.06	7.17	7.26	7.36	7.44	7.52	7.59	7.66	7.73
6.79	6.90	7.01	7.10	7.19	7.27	7.35	7.42	7.48	7.55
6.66	6.77	6.87	6.96	7.05	7.13	7.20	7.27	7.33	7.39
6.55	6.66	6.76	6.84	6.93	7.00	7.07	7.14	7.20	7.26
6.46	6.56	6.66	6.74	6.82	6.90	6.97	7.03	7.09	7.15
6.38	6.48	6.57	6.66	6.73	6.81	6.87	6.94	7.00	7.05
6.31	6.41	6.50	6.58	6.65	6.73	6.79	6.85	6.91	6.97
6.25	6.34	6.43	6.51	6.58	6.65	6.72	6.78	6.84	6.89
6.19	6.28	6.37	6.45	6.52	6.59	6.65	6.71	6.77	6.82
6.02	6.11	6.19	6.26	6.33	6.39	6.45	6.51	6.56	6.61
5.85	5.93	6.01	6.08	6.14	6.20	6.26	6.31	6.36	6.41
5.69	5.76	5.83	5.90	5.96	6.02	6.07	6.12	6.16	6.21
5.53	5.60	5.67	5.73	5.78	5.84	5.89	5.93	5.97	6.01
5.37	5.44	5.50	5.56	5.61	5.66	5.71	5.75	5.79	5.83
5.23	5.29	5.35	5.40	5.45	5.49	5.54	5.57	5.61	5.65

Table A.5 Percentage Points of the Duncan New Multiple Range Test

Error		r = number of ordered steps between means													
df	α	2	3	4	5	6	7	8	9	10	12	14	16	18	20
1	.05	18.0	18.0	18.0	18.0	18.0	18.0	18.0	18.0	18.0	18.0	18.0	18.0	18.0	18.0
	.01	90.0	90.0	90.0	90.0	90.0	90.0	90.0	90.0	90.0	90.0	90.0	90.0	90.0	90.0
2	.05	6.09	6.09	6.09	6.09	6.09	6.09	6.09	6.09	6.09	6.09	6.09	6.09	6.09	6.09
	.01	14.0	14.0	14.0	14.0	14.0	14.0	14.0	14.0	14.0	14.0	14.0	14.0	14.0	14.0
3	.05	4.50	4.50	4.50	4.50	4.50	4.50	4.50	4.50	4.50	4.50	4.50	4.50	4.50	4.50
	.01	8.26	8.5	8.6	8.7	8.8	8.9	8.9	9.0	9.0	9.0	9.1	9.2	9.3	9.3
4	.05	3.93	4.01	4.02	4.02	4.02	4.02	4.02	4.02	4.02	4.02	4.02	4.02	4.02	4.02
	.01	6.51	6.8	6.9	7.0	7.1	7.1	7.2	7.2	7.3	7.3	7.4	7.4	7.5	7.5
5	.05	3.64	3.74	3.79	3.83	3.83	3.83	3.83	3.83	3.83	3.83	3.83	3.83	3.83	3.83
	.01	5.70	5.96	6.11	6.18	6.26	6.33	6.40	6.44	6.5	6.6	6.6	6.7	6.7	6.8
6	.05	3.46	3.58	3.64	3.68	3.68	3.68	3.68	3.68	3.68	3.68	3.68	3.68	3.68	3.68
	.01	5.24	5.51	5.65	5.73	5.81	5.88	5.95	6.00	6.0	6.1	6.2	6.2	6.3	6.3
7	.05	3.35	3.47	3.54	3.58	3.60	3.61	3.61	3.61	3.61	3.61	3.61	3.61	3.61	3.61
	.01	4.95	5.22	5.37	5.45	5.53	5.61	5.69	5.73	5.8	5.8	5.9	5.9	6.0	6.0
8	.05	3.26	3.39	3.47	3.52	3.55	3.56	3.56	3.56	3.56	3.56	3.56	3.56	3.56	3.56
	.01	4.74	5.00	5.14	5.23	5.32	5.40	5.47	5.51	5.5	5.6	5.7	5.7	5.8	5.8
9	.05	3.20	3.34	3.41	3.47	3.50	3.52	3.52	3.52	3.52	3.52	3.52	3.52	3.52	3.52
	.01	4.60	4.86	4.99	5.08	5.17	5.25	5.32	5.36	5.4	5.5	5.5	5.6	5.7	5.7

Table A.5 (continued)

10	.05	3.15	3.30	3.37	3.43	3.46	3.47	3.47	3.47	3.47	3.47	3.47	3.47	3.47	3.48
	.01	4.48	4.73	4.88	4.96	5.06	5.13	5.20	5.24	5.28	5.36	5.42	5.48	5.54	5.55
11	.05	3.11	3.27	3.35	3.39	3.43	3.44	3.45	3.46	3.46	3.46	3.46	3.46	3.47	3.48
	.01	4.39	4.63	4.77	4.86	4.94	5.01	5.06	5.12	5.15	5.24	5.28	5.34	5.38	5.39
12	.05	3.08	3.23	3.33	3.36	3.40	3.42	3.44	3.44	3.46	3.46	3.46	3.46	3.47	3.48
	.01	4.32	4.55	4.68	4.76	4.84	4.92	4.96	5.02	5.07	5.13	5.17	5.22	5.23	5.26
13	.05	3.06	3.21	3.30	3.35	3.38	3.41	3.42	3.44	3.45	3.45	3.46	3.46	3.47	3.47
	.01	4.26	4.48	4.62	4.69	4.74	4.84	4.88	4.94	4.98	5.04	5.08	5.13	5.14	5.15
14	.05	3.03	3.18	3.27	3.33	3.37	3.39	3.41	3.42	3.44	3.45	3.46	3.46	3.47	3.47
	.01	4.21	4.42	4.55	4.63	4.70	4.78	4.83	4.87	4.91	4.96	5.00	5.04	5.06	5.07
15	.05	3.01	3.16	3.25	3.31	3.36	3.38	3.40	3.42	3.43	3.44	3.45	3.46	3.47	3.47
	.01	4.17	4.37	4.50	4.58	4.64	4.72	4.77	4.81	4.84	4.90	4.94	4.97	4.99	5.00
16	.05	3.00	3.15	3.23	3.30	3.34	3.37	3.39	3.41	3.43	3.44	3.45	3.46	3.47	3.47
	.01	4.13	4.34	4.45	4.54	4.60	4.67	4.72	4.76	4.79	4.84	4.88	4.91	4.93	4.94
17	.05	2.98	3.13	3.22	3.28	3.33	3.36	3.38	3.40	3.42	3.44	3.45	3.46	3.47	3.47
	.01	4.10	4.30	4.41	4.50	4.56	4.63	4.68	4.72	4.75	4.80	4.83	4.86	4.88	4.89
18	.05	2.97	3.12	3.21	3.27	3.32	3.35	3.37	3.39	3.41	3.43	3.45	3.46	3.47	3.47
	.01	4.07	4.27	4.38	4.46	4.53	4.59	4.64	4.68	4.71	4.76	4.79	4.82	4.84	4.85
19	.05	2.96	3.11	3.19	3.26	3.31	3.35	3.37	3.39	3.41	3.43	3.44	3.46	3.47	3.47
	.01	4.05	4.24	4.35	4.43	4.50	4.56	4.61	4.64	4.67	4.72	4.76	4.79	4.81	4.82
20	.05	2.95	3.10	3.18	3.25	3.30	3.34	3.36	3.38	3.40	3.43	3.44	3.46	3.46	3.47
	.01	4.02	4.22	4.33	4.40	4.47	4.53	4.58	4.61	4.65	4.69	4.73	4.76	4.78	4.79
22	.05	2.93	3.08	3.17	3.24	3.29	3.32	3.35	3.37	3.39	3.42	3.44	3.45	3.46	3.47
	.01	3.99	4.17	4.28	4.36	4.42	4.48	4.53	4.57	4.60	4.65	4.68	4.71	4.74	4.75

Table A.5 (continued)

Error		r = number of ordered steps between means													
df	α	*2*	*3*	*4*	*5*	*6*	*7*	*8*	*9*	*10*	*12*	*14*	*16*	*18*	*20*
24	.05	2.92	3.07	3.15	3.22	3.28	3.31	3.34	3.37	3.38	3.41	3.44	3.45	3.46	3.47
	.01	3.96	4.14	4.24	4.33	4.39	4.44	4.49	4.53	4.57	4.62	4.64	4.67	4.70	4.72
26	.05	2.91	3.06	3.14	3.21	3.27	3.30	3.34	3.36	3.38	3.41	3.43	3.45	3.46	3.47
	.01	3.93	4.11	4.21	4.30	4.36	4.41	4.46	4.50	4.53	4.58	4.62	4.65	4.67	4.69
28	.05	2.90	3.04	3.13	3.20	3.26	3.30	3.33	3.35	3.37	3.40	3.43	3.45	3.46	3.47
	.01	3.91	3.08	4.18	4.28	4.34	4.39	4.43	4.47	4.51	4.56	4.60	4.62	4.65	4.67
30	.05	2.89	3.04	3.12	3.20	3.25	3.29	3.32	3.35	3.37	3.40	3.43	3.44	3.46	3.47
	.01	3.89	4.06	4.16	4.22	4.32	4.36	4.41	4.45	4.48	4.54	4.58	4.61	4.63	4.65
40	.05	2.86	3.01	3.10	3.17	3.22	3.27	3.30	3.33	3.35	3.39	3.42	3.44	3.46	3.47
	.01	3.82	3.99	4.10	4.17	4.24	4.30	4.34	4.37	4.41	4.46	4.51	4.54	4.57	4.59
60	.05	2.83	2.98	3.08	3.14	3.20	3.24	3.28	3.31	3.33	3.37	3.40	3.43	3.45	3.47
	.01	3.76	3.92	4.03	4.12	4.17	4.23	4.27	4.31	4.34	4.39	4.44	4.47	4.50	4.53
100	.05	2.80	2.95	3.05	3.12	3.18	3.22	3.26	3.29	3.32	3.36	3.40	3.42	3.45	3.47
	.01	3.71	3.86	3.93	4.06	4.11	4.17	4.21	4.25	4.29	4.35	4.38	4.42	4.45	4.48
∞	.05	2.77	2.92	3.02	3.09	3.15	3.19	3.23	3.26	3.29	3.34	3.38	3.41	3.44	3.47
	.01	3.64	3.80	3.90	3.98	4.04	4.09	4.14	4.17	4.20	4.26	4.31	4.34	4.38	4.41

Reproduced from: D. B. Duncan, Multiple Range and Multiple F Tests. *Biometrics*, 11: 1–42, 1955. With permission from the Biometric Society and the author.

References

Albohali, M. N. 1983. *A time series approach to the analysis of repeated measures designs*. Ph.D. dissertation, Kansas State University.

Beyer, W. H. 1968. *CRC handbook of tables for probability and statistics*. 2nd ed. Boca Raton, FL, CRC Press, Inc.

Boardman, T. J. 1974. Confidence intervals for variance components—A comparative Monte Carlo study. *Biometrics* 30: 251–62.

Box, G. E. P. 1954. Some theorems on quadratic forms applied in the study of analysis of variance problems. *Annals of Mathematical Statistics* 25: 290–302.

Carmer, S. G., and Swanson, M. R. 1973. An evaluation of ten pairwise multiple comparison procedures by Monte Carlo methods. *Journal of the American Statistical Association* 68: 66–74.

Cochran, W. G., and Cox, G. M. 1957. *Experimental Design*. 2nd Ed. New York: Wiley.

Collier, R. O., Jr.; Baker, F. B.; Mandeville, G. K.; and Hayes, T. F. 1967. Estimates of test size for several test procedures based on variance ratios in repeated measures experiments. *Psychometrika* 32: 339–53.

Conover, W. J.; Johnson, M. E.; and Johnson, M. M. 1981. A comparative study of tests for homogeneity of variances, with applications to the Outer Continental Shelf bidding data. *Technometrics* 23: 351–61.

Corbeil, R. R., and Searle, S. R. 1976. A comparison of variance component estimators. *Biometrics* 32: 779–91.

Davies, O. L. 1954. *Design and analysis of industrial experiments*. London: Oliver and Boyd.

Dixon, W. J.; Brown, M. D.; Engelman, L.; Frane, J.; Hill, M.; Jennrick, R.; and Toporek, J.; eds. 1981. *BMDP biomedical computer programs*. Berkeley, Calif.: University of California Press.

Eisenhart, C. 1947. Assumptions underlying analysis of variance. *Biometrics* 3: 1–22.

Federer, W. T. 1955. *Experimental design*. New York: Macmillan.

Graybill, F. A. 1976. *Theory and application of the linear model*. North Scituate, Mass.: Duxbury.

Greenhouse, S. W., and Geisser, S. 1959. On methods in the analysis of profile data. *Psychometrika* 24: 95–112.

Grizzle, J. E. 1965. The two-period change-over design and its use in clinical trials. *Biometrics* 21: 467–80.

Hartley, H. O. 1967. Expectations, variances and covariances of ANOVA mean squares by "synthesis." *Biometrics* 23: 105–14; Corrigenda, 853.

Hartley, H. O., and Rao, J. N. K. 1967. Maximum likelihood estimation for the mixed analysis of variance model. *Biometrika* 54: 93–108.

Hasza, D. P. 1980. A note on maximum likelihood estimation for the first order autoregressive process. *Communications in Statistics, Series A*, 1411–15.

Hemmerle, W. J., and Hartley, H. O. 1973. Computing maximum likelihood estimates for the mixed A.O.V. model using the *W* transformation. *Technometrics* 15: 819–31.

Henderson, C. R. 1953. Estimation of variance and covariance components. *Biometrics* 9: 226–52.

Hicks, C. R. 1973. *Fundamental concepts in the design of experiments*. 2nd ed. New York: Holt, Rinehart and Winston.

Hull, C. H., and Nie, N. H. 1981. *SPSS Update 7–9: New procedures and facilities for Releases 7–9*. New York: McGraw-Hill.

Huynh, H., and Feldt, L. S. 1970. Conditions under which mean square ratios in repeated measures designs have exact *F*-distributions. *Journal of the American Statistical Association* 65: 1582–89.

John, P. W. M. 1971. *Statistical design and analysis of experiments*. New York: Wiley.

Johnson, D. E. 1973. A derivation of Scheffé's *S*-method by maximizing a quadratic form. *The American Statistician* 27: 27–29.

Johnson, D. E. 1976. Some new multiple comparison procedures for the two-way AoV model with interaction. *Biometrics* 328: 929–34.

Johnson, D. E., and Graybill, F. A. 1972. An analysis of a two-way model with interaction and no replication. *Journal of the American Statistical Association* 67: 862–68.

Kirk, R. E. 1968. *Experimental design: Procedures for the behavioral sciences*. Belmont, Calif.: Brooks/Cole.

Milliken, G. A., and Graybill, F. A. 1970. Extensions of the general linear hypothesis model. *Journal of the American Statistical Association* 65: 797–807.

Morrison, D. F. 1976. *Multivariate statistical methods*. New York: McGraw-Hill.

Nie, N. H.; Hull, C. H.; Jenkins, J.; Steinbrenner, K.; and Best, D. H. 1975. *Statistical package for the social sciences*. 2nd ed. New York: McGraw-Hill.

Ott, L. 1977. *An introduction to statistical methods and data analysis*. North Scituate, Mass.: Duxbury.

Rao, C. R. 1971. Minimum variance quadratic unbiased estimation of variance components. *Journal of Multivariate Analysis* 1: 445–56.

Rao, J. N. K. 1968. On expectations, variances and covariances of ANOVA mean squares by "synthesis." *Biometrics* 24: 963–78.

SAS Institute, Inc. 1982. *SAS user's guide: Statistics, 1982 edition*. Cary, N.C.: SAS Institute, Inc.

Satterthwaite, F. E. 1946. An approximate distribution of estimates of variance components. *Biometrics Bulletin* 2: 110–14.

Searle, S. R. 1971. *Linear Models*. New York: Wiley.

Searle, S. R.; Speed, F. M.; and Milliken, G. A. 1980. Population marginal means in the linear model: An alternative to least squares means. *The American Statistician* 34: 216–21.

Spjøtvoll, E., and Stoline, M. R. 1973. An extension of the *T*-method of multiple comparison to include the cases with unequal sample sizes. *Journal of the American Statistical Association* 68: 975–78.

Swallow, W. H., and Searle, S. R. 1978. Minimum variance quadratic unbiased estimation (MINVQUE) of variance components. *Technometrics* 20(3): 265–72.

Welch, B. L. 1951. On the comparison of several mean values: An alternative approach. *Biometrika* 38: 330–36.

Williams, J. S. 1962. A confidence interval for variance components. *Biometrika* 49: 278–81.

Winer, B. J. 1971. *Statistical principles in experimental design*. 2nd ed. New York: McGraw-Hill.

Index